Springer Proceedings in Mathematics & Statistics

Volume 253

T0255955

Springer Proceedings in Mathematics & Statistics

This book series features volumes composed of selected contributions from workshops and conferences in all areas of current research in mathematics and statistics, including operation research and optimization. In addition to an overall evaluation of the interest, scientific quality, and timeliness of each proposal at the hands of the publisher, individual contributions are all refereed to the high quality standards of leading journals in the field. Thus, this series provides the research community with well-edited, authoritative reports on developments in the most exciting areas of mathematical and statistical research today.

More information about this series at http://www.springer.com/series/10533

Debdas Ghosh · Debasis Giri
Ram N. Mohapatra · Kouichi Sakurai
Ekrem Savas · Tanmoy Som
Editors

Mathematics and Computing

ICMC 2018, Varanasi, India, January 9–11, Selected Contributions

 Springer

Editors
Debdas Ghosh
Department of Mathematical Sciences
Indian Institute of Technology (BHU)
Varanasi, Uttar Pradesh, India

Debasis Giri
Department of Computer Science
  and Engineering
Haldia Institute of Technology
Haldia, West Bengal, India

Ram N. Mohapatra
Department of Mathematics
University of Central Florida
Orlando, FL, USA

Kouichi Sakurai
Faculty of Information Science
  and Electrical Engineering
Kyushu University
Fukuoka, Japan

Ekrem Savas
Uşak University
Uşak, Turkey

Tanmoy Som
Department of Mathematical Sciences
Indian Institute of Technology (BHU)
Varanasi, Uttar Pradesh, India

ISSN 2194-1009   ISSN 2194-1017   (electronic)
Springer Proceedings in Mathematics & Statistics
ISBN 978-981-13-4731-3   ISBN 978-981-13-2095-8   (eBook)
https://doi.org/10.1007/978-981-13-2095-8

Mathematics Subject Classification (2010): 35-xx, 65-xx, 76-xx, 90-xx, 94-xx

This Springer imprint is published by the registered company Springer Nature Singapore Pte Ltd.
The registered company address is: 152 Beach Road, #21-01/04 Gateway East, Singapore 189721, Singapore

Dedicated to
Pandit Madan Mohan Malaviya—The
Founder of Banaras Hindu University

Preface

The Fourth International Conference on Mathematics and Computing (ICMC—2018) was organized in the Department of Mathematical Sciences, Indian Institute of Technology (Banaras Hindu University), Varanasi, India, during January 9–11, 2018, under the dynamic leadership of Dr. Debdas Ghosh along with the support of Prof. R. N. Mohapatra, Prof. D. Giri, Prof. T. Som, Prof. S. Mukhopadhyay, Prof. S. Das, Dr. A. Banerjee, and the faculty members of the Department of Mathematical Sciences, IIT (BHU), India. There was an overwhelming response to the program, and one hundred and twenty papers all over the country and abroad were submitted for the consideration of presentation and later publication in the proceedings. Taking into account the norms of the proceedings, the papers were gone through strict blind reviewing process by at least two referees in the respective areas and only forty-seven papers were selected for the presentation and twenty-nine for inclusion in the Proceeding of Mathematics and Statistics, Springer. The areas covered by the papers are the latest works in the field of cryptography, security, abstract algebra, functional analysis, fluid dynamics, fuzzy modeling and optimization, etc. The ICMC—2018 was attended by several experts of international repute from the nation as well as from USA, UK, Japan, China, Finland, etc., as invited speakers with their high-quality research presentations. Experts were from IIT Madras, ISI Chennai, University of Central Florida, Orlando, USA, Kettering University, USA, University of Surrey, UK, Auburn University, Alabama, USA, Kyushu University, Japan, Tianjin University of Science and Technology, China, Oracle's System of Technology, USA, University of Turku, Finland, Haldia Institute of Technology (HIT), India, Banaras Hindu University (BHU), India, and IIT (BHU), India. Most of the experts have submitted their contributions for the proceeding. The Organizing Committee of ICMC—2018 is truly thankful to all experts and paper presenters for their academic support.

Distinguished Prof. Anthony T. S. Ho of the Tianjin University, of the University of Surrey, also of the Wuhan University of Technology has nicely elaborated and explained the applications of Benford's law for multimedia security and forensics. Professor R. N. Mohapatra, University of Central Florida, has beautifully explored the various aspects of epidemiological models with mutating pathogens with basic SIR model, diffusion equation, the Fisher–Kolmogoroff equation, spatial epidemic models, and his proposed model supported with some nice examples. Professor S. R. Chakravarty of the Kettering University elaborately presented the different aspects of non-preemptive stochastic priority queuing model for two different types of customers and with a new threshold. Professor K. Sakurai of the Kyushu University has discussed non-commutative approach using ring for enhancing the security of cryptosystems. Professor Matti Vuorinen of the University of Turku gave insightful elaboration on computation of condenser capacity. Dr. Srinivas Pyda of Oracle's System of Technology has discussed well the mathematics in machine learning. Professor Dr. Parisa Hariri of the University of Turku has explored the hyperbolic metric of plane domain to a subdomain of R^n ($n \geq 2$), discussed the geometry and topology of metric balls, and compared different hyperbolic type metrics and gave an application to solve Ptolemy–Alhazen problem. Professor S. Ponnusamy of IIT Madras has described the classical Bohr's theorem for bounded functions, bounded n-symmetric functions, half-plane mappings, half-plane n-symmetric mappings, and added some nice examples supporting the theory; Prof. Debasis Giri of HIT has elaborated on authenticated encryption of long messages; Prof. Chris Rodger of the Auburn University has explored the various aspects of graph embedding and construction of Hamilton's decomposition of graphs and elaborated with nice examples having several applications. Professor S. K. Mishra of BHU talked about the properties and relations of strong pseudomonotone and strong quasimonotone operators. Professor T. Som of IIT (BHU) has contributed to convergence of generalized Mann type of iterates to common fixed point though he has dealt with soft relation and fuzzy soft relation with application to decision-making problems in the conference program. The submitted contributions of the experts are included in the proceeding. The organizing committee is truly thankful to all the experts for their valuable contribution to the conference.

I, on behalf of the organizing committee, gratefully acknowledge the financial support to the conference by

- Science and Engineering Research Board, India
- Defence Research and Development Organization, India
- Indian Institute of Technology (BHU), India
- Council of Research and Industrial Research, India
- SCUBE India.

Varanasi, India   Prof. Tanmoy Som
                                                                                   Organizing Secretary

Contents

Contributors

Rachid Ait Maalem Lahcen Department of Mathematics, University of Central Florida, Orlando, FL, USA

Anushri A. Aserkar Department of Mathematics, Rajiv Gandhi College of Engineering and Research, Nagpur, India

A. Banerjee Department of Mathematical Sciences, Indian Institute of Technology (BHU), Varanasi, Uttar Pradesh, India

Subrata Bera Department of Mathematics, National Institute of Technology Silchar, Silchar, India

S. Bhattacharyya Department of Mathematics, Indian Institute of Technology Kharagpur, Kharagpur, West Bengal, India

Sushil Kumar Bhuiya Department of Mathematics, Indian Institute of Technology Kharagpur, Kharagpur, West Bengal, India

Debjani Chakraborty Department of Mathematics, Indian Institute of Technology Kharagpur, Kharagpur, West Bengal, India

Srinivas R. Chakravarthy Departments of Industrial and Manufacturing Engineering and Mathematics, Kettering University, Flint, MI, USA

Subrato Chakravorty Department of Mechanical Engineering, Indian Institute of Technology (BHU), Varanasi, Uttar Pradesh, India

Amalendu Choudhury Department of Mathematics and Statistics, Haflong Government College, Haflong, Dima Hasao, Assam, India

B. C. Das Department of Applied Mathematics, Calcutta University, Kolkata, India

Barun Das Department of Mathematics, Sidho Kanho Birsha University, Purulia, West Bengal, India

S. De Department of Applied Mathematics, Calcutta University, Kolkata, India

Manjusha P. Gandhi Department of Mathematics, Yeshwantrao Chavan College of Engineering, Nagpur, India

Debdas Ghosh Department of Mathematical Sciences, Indian Institute of Technology (BHU), Varanasi, Uttar Pradesh, India

Debdulal Ghosh Department of Mathematics, Indian Institute of Technology Kharagpur, Kharagpur, West Bengal, India

Debasis Giri Department of Computer Science and Engineering, Haldia Institute of Technology, Haldia, East Midnapore, India

G. K. Gupta Department of Mathematical Sciences, Indian Institute of Technology (BHU), Varanasi, Uttar Pradesh, India

Nitin Gupta Indian Institute of Technology Kharagpur, Kharagpur, West Bengal, India

Sharmistha Halder (Jana) Department of Mathematics, Midnapore College [Autonomous], Midnapore, India

Rajib Haloi Department of Mathematical Sciences, Tezpur University, Tezpur, Sonitpur, Assam, India

Margareta Heilmann School of Mathematics and Natural Sciences, University of Wuppertal, Wuppertal, Germany

P. Helen Chandra Jayaraj Annapackiam College for Women (Autonomous), Theni, Tamil Nadu, India

Jordan Hristov Department of Chemical Engineering, University of Chemical Technology and Metallurgy (UCTM), Sofia, Bulgaria

Salisu Ibrahim Department of Mathematics Northwest University, Kano, Nigeria

Biswapati Jana Department of Computer Science, Vidyasagar University, Midnapore, West Bengal, India

Jagan Mohan Jonnalagadda Department of Mathematics, Birla Institute of Technology and Science, Pilani, Hyderabad, Telangana, India

Srinivas Kareenhalli Oracle India, Bengaluru, India

Rupinderjit Kaur Department of Mathematics, Birla Institute of Technology and Science, Pilani, Hyderabad, India

Takeshi Koshiba Faculty of Education and Integrated Arts and Sciences, Waseda University, Shinjuku-ku, Tokyo, Japan

Ajeet Kumar Department of Mathematics, Banaras Hindu University, Varanasi, India

Jitendra Kumar Department of Mathematics, Indian Institute of Technology Kharagpur, Kharagpur, West Bengal, India

Manish Kumar Department of Mathematics, Birla Institute of Technology and Science-Pilani, Hyderabad, Telangana, India

Somesh Kumar Indian Institute of Technology Kharagpur, Kharagpur, West Bengal, India

Nyurgun Lazarev North-Eastern Federal University, Yakutsk, Russia; Lavrentyev Institute of Hydrodynamics SB RAS, Novosibirsk, Russia

Manoranjan Maiti Department of Applied Mathematics with Oceanology and Computer Programming, Vidyasagar University, Midnapore, West Bengal, India

B. N. Mandal Physics and Applied Mathematics Unit, Indian Statistical Institute, Kolkata, India

Atanu Manna Faculty of Mathematics, Indian Institute of Carpet Technology, Bhadohi, Uttar Pradesh, India

A. Mary Imelda Jayaseeli Jayaraj Annapackiam College for Women (Autonomous), Theni, Tamil Nadu, India

S. K. Mishra Department of Mathematics, Institute of Science, Banaras Hindu University, Varanasi, India

Vinaytosh Mishra Indian Institute of Technology (BHU), Varanasi, India

Ram Mohapatra Department of Mathematics, University of Central Florida, Orlando, FL, USA

Fadel Nasaireh Department of Mathematics, Technical University, Cluj-Napoca, Romania

Srinivasan Natesan Department of Mathematics, Indian Institute of Technology Guwahati, Guwahati, India

Natalia Neustroeva North-Eastern Federal University, Yakutsk, Russia

H. Ohshima Faculty of Pharmaceutical Sciences, Tokyo University of Science, Noda, Chiba, Japan

S. K. Pal Department of Mathematics, Indian Institute of Technology Kharagpur, Kharagpur, West Bengal, India

Bijaya Laxmi Panigrahi Department of Mathematics, Sambalpur University, Sambalpur, Odisha, India

Goutam Panigrahi Department of Mathematics, National Institute of Technology, Durgapur, West Bengal, India

Tapas Ranjan Panigrahi Department of Mathematics, Birla Institute of Technology and Science, Pilani, Hyderabad, India

Lakshmi Kanta Patra Indian Institute of Information Technology Ranchi, Ranchi, India

Swaraj Paul Department of Mathematics, Visva Bharati, Santiniketan, West Bengal, India

Maharage Nisansala Sevwandi Perera Graduate School of Science and Engineering, Saitama University, Saitama, Japan

J. Philomenal Karoline Jayaraj Annapackiam College for Women (Autonomous), Theni, Tamil Nadu, India

Srinivas Pyda Oracle America, Redwood Shores, CA, USA

Abedallah Rababah Department of Mathematical Sciences, United Arab Emirates University, Al Ain, UAE

Ioan Raşa Department of Mathematics, Technical University, Cluj-Napoca, Romania

C. A. Rodger Department of Mathematics and Statistics, Auburn University, Baltimore, AL, USA

D. R. Sahu Department of Mathematics, Banaras Hindu University, Varanasi, India

Cherian Samuel Indian Institute of Technology (BHU), Varanasi, India

S. M. Saroja Theerdus Kalavathy Jayaraj Annapackiam College for Women (Autonomous), Theni, Tamil Nadu, India

Ekrem Savas Department of Mathematics, Usak University, Usak, Turkey

Rabia Savas Department of Mathematics, Sakarya University, Sakarya, Turkey

Avanish Shahi Department of Mathematics, Institute of Science, Banaras Hindu University, Varanasi, India

S. K. Sharma Indian Institute of Technology (BHU), Varanasi, India

Shivam Shreevastava Department of Mathematical Sciences, Indian Institute of Technology (BHU), Varanasi, India

Gautam Singh Department of Mathematics, Indian Institute of Technology Guwahati, Guwahati, India

Sanjeev Kumar Singh Department of Mathematics, Institute of Science, Banaras Hindu University, Varanasi, India

T. Som Department of Mathematical Sciences, Indian Institute of Technology (BHU), Varanasi, India

Tanmoy Som Department of Mathematical Sciences, Indian Institute of Technology (BHU), Varanasi, India

Karthikeyan Subbiah Department of Computer Science, Institute of Science (BHU), Varanasi, India

Anoop Kumar Tiwari Department of Computer Science, Institute of Science (BHU), Varanasi, India

Sumit Kumar Vishwakarma Department of Mathematics, Birla Institute of Technology and Science, Pilani, Hyderabad, India

Chapter 1
Constructions and Embeddings of Hamilton Decompositions of Families of Graphs

C. A. Rodger

Abstract In this paper, a discussion of the use of amalgamations in constructing Hamilton decompositions of graphs is presented. Edge-colorings that are fair in various senses are critical to this endeavor, so some discussion of them is also included. Finally, the power of amalgamations is demonstrated in the overview of results in the literature that take a given edge-coloring of a graph and extend it to one of a family of graphs (e.g., a complete graph or a complete multipartite graph) in which each color class is a Hamilton cycle.

Keywords Hamilton cycles · Amalgamations · Fair edge-colorings · Embeddings

1 Introduction

Colorings of graphs are very useful in a variety of settings, especially scheduling problems. In such problems, sharing objects (vertices or edges) out evenly in various ways usually has beneficial effects in the application being considered. For example, the most basic of these fairness notions is to ensure that the coloring is proper (no two adjacent objects receive the same color). But other notions also play a vital role. One could ask for the coloring to be equalized; that is, the number of objects of each color is within one of the number of objects of each other color. Two examples illustrate this.

The first example is the scheduling problem where various companies send representatives to a central location, such as Chicago Airport, where they are to meet other companies for one-on-one discussions. All representatives are in the same industry so, while not every pair of companies' representatives need to meet, there is a lot of congestion to manage. The aim is to schedule the meetings (each is to last 30 min) to minimize the number of time slots needed to satisfy all needs to meet. The number of rooms is also an issue, partly due to availability and partly due to

C. A. Rodger (✉)
Department of Mathematics and Statistics, Auburn University,
221 Parker Hall, Baltimore, AL 36849-5310, USA
e-mail: rodgec1@auburn.edu

© Springer Nature Singapore Pte Ltd. 2018
D. Ghosh et al. (eds.), *Mathematics and Computing*, Springer Proceedings
in Mathematics & Statistics 253, https://doi.org/10.1007/978-981-13-2095-8_1

expense. This problem can be modeled by a graph G formed by letting each company (representative) be represented by a vertex, two vertices being joined if and only if the corresponding companies need to meet. A proper edge-coloring with k colors provides a schedule using k time slots: Representatives i and j meet at time slot k if the edge $\{i, j\}$ is colored k. Clearly the fact that the edge-coloring is proper ensures that each representative is scheduled to meet at most one other representative at each time. The number of rooms needed is decided by the size of the biggest color class, and this is minimized if the edge-coloring is also equalized. Results in the literature come close to immediately answering this problem: k can be any value at least $\chi'(G)$, which by Vizing's Theorem is either the maximum degree $\Delta = \Delta(G)$ of G, or is $\Delta + 1$, and a result by McDiarmid [1] guarantees that if there exists a proper k-edge-coloring, then there exists an equalized proper k-edge-coloring. Deciding if a schedule with Δ timeslots is possible may be difficult to determine, as this falls in the class of NP-complete problems; but rather than working hard to save just one time period, simply using $\Delta + 1$ timeslots often may not be a problem.

The second example contrasts with the first quite nicely. Various university clubs are to meet one evening to plan their efforts to help Auburn collect enough food to win the Auburn-Alabama Food Fight, designed to help the hungry in Alabama. Ideally, each club would only meet if all its representatives attending that evening are able to be present at the meeting. Again the plan is to schedule the meetings (each is to last 30 min) in a way that minimizes the number of time slots needed for each club to meet, having all members present; as before, the number of rooms is also an issue. In this case, the model is a graph in which each club is represented by a vertex and two vertices are joined by an edge if the corresponding clubs have a member in common. So a proper vertex-coloring with k colors provides a schedule using k time slots: Club i meets at time slot k if vertex i is colored k. The fact that the vertex-coloring is proper ensures that clubs with members in common are scheduled at different times. Minimizing the number of rooms needed again calls for an equalized vertex-coloring (often called an equitable vertex-coloring in the literature). Unfortunately, results in the literature have more trouble solving this problem; both answering the question of how many colors are needed ($\chi(G)$ is not easily determined) and of whether or not an equalized vertex-coloring exists. The number of time slots can be any value at least $\chi(G)$; if it is chosen to be more than $\Delta(G)$, then it is known that the vertex-coloring can be equalized ([2]). Other efforts over the past 40 years to find conditions guaranteeing the existence of equalized vertex-colorings have been found, but much work remains to understand this property.

Many interesting problems associated with fair colorings of various sorts remain open and are of practical use. Several more will be introduced later in the paper as they are needed.

A third practical problem addressed by graph theory is the famous traveling salesman problem. A salesman has to visit a predetermined set of cities, one by one, then return home, following a route that minimizes the distance travelled. It is modeled by a graph, G, in which the vertices represent the cities, edges represent various routes to get from city to city, and all edges are weighted by the distance of the corresponding route. In the unworldy case where all the edges have weight 1, this problem asks whether or not G contains a Hamilton cycle (a cycle in G which includes each vertex in $V(G)$). This too falls into the family of NP-complete problems, so is difficult to solve, even in this seemingly far simpler situation. Related to the Hamiltonicity of a graph is a stronger property. A Hamilton decomposition of G is a partition of $E(G)$, each element of which induces a Hamilton cycle. Since each Hamilton cycle includes exactly two edges incident with each vertex, clearly G needs to be regular in order to have a Hamilton decomposition. Around 125 years ago, Walecki proved that the complete graph K_n has a Hamiltonian decomposition if and only if n is odd [3]. This too has an interpretation in an applied setting, related to the traveling salesman problem. In this case, the salesman wants to visit certain important cities on every trip, but other towns along the way can be visited less often. The Hamilton cycles in a Hamilton decomposition ensure that each time out the salesman visits the important cities (the vertices of the graph), and then since each edge is in exactly one Hamilton cycle, towns along the roads corresponding to the edges will be visited as the road is traversed.

For each of these three problems, interest eventually turned from complete graphs to another natural family of graphs, namely the complete multipartite graphs: The vertices in each such graph are partitioned into p parts, with two vertices being joined by an edge if and only if they are in different parts. The chromatic index of such graphs was settled thirty years ago [4], and the value of the chromatic number is obvious, but finding equalized vertex-colorings is not so straightforward (see [5]). For such graphs to have a Hamilton decomposition, clearly they must be regular; to be regular, clearly all parts must have the same size. So this motivates the following definition: Let $K(n, p)$ denote the complete multipartite graph with p parts in which each part contains n vertices. Deciding whether or not $K(n, p)$ has a Hamilton decomposition was settled by Laskar and Auerbach [6] 40 years ago, showing that it exists if and only if $n(p-1)$ is even.

Much more recently, a third family of graphs has drawn wide interest, motivated by the construction of experimental designs in statistics. A block design with two association classes (BDTAC) can be described graph theoretically as follows. Let $K(P, \lambda_1, \lambda_2)$ be the graph in which P is a partition of the vertices, two vertices being joined by λ_1 edges if they are in the same part of P and by λ_2 edges if they are in different parts. The BDTAC is equivalent to a partition of the edges of $K(P, \lambda_1, \lambda_2)$, each element of which is a copy of K_k for some integer k. In the setting of this paper, the natural question is whether or not there exists a Hamilton decomposition of $K(P, \lambda_1, \lambda_2)$, so of particular interest is the regular graph $K(n, p, \lambda_1, \lambda_2)$ where each of the p parts in $K(P, \lambda_1, \lambda_2)$ contains n vertices. This problem was settled by Bahmanian and Rodger 5 years ago [7]. Their method of proof is the main topic in Sect. 2. Continuing the theme of fairness in colorings, the amalgamation proof technique produces a graph H from a given edge-colored graph G, where G is a graph homomorphism of H, such that the edges in H are shared out among the vertices and among color classes in ways that are fair with respect to several notions of balance. The connectivity of color classes is also addressed.

In Sect. 3, the embedding of edge-colored graphs into edge-colored copies of $K(n, p, \lambda_1, \lambda_2)$ is the main focus. This is a great demonstration of the power of amalgamation proofs, but is also motivated by applications in the following sense. Scheduling problems often require prerequisite conditions to be built into the final schedule. For example, when deciding which teachers should teach which classes at what times, some teachers may not be able to teach early in the morning. Hilton [8] developed the notion of an outline schedule where times are compressed into a small number of groups; say early morning, late morning, early afternoon, and last classes. Similarly, subjects being taught, or classes for the same age students could also form such groups. Once this outline schedule has been developed, reversing the amalgamation approach develops the full schedule. This method also allows prerequisites to be built into about a quarter of the entire schedule. Here, we begin with a given edge-colored copy of $K(n, p_1, \lambda_1, \lambda_2)$ and embed it in a copy of $K(n, p_2, \lambda_1, \lambda_2)$ such that each color class induces a Hamilton cycle. In view of the third problem described above, the given copy can be thought of as the given prerequisites in the final Hamilton decomposition that realizes the schedule of the salesman.

2 Amalgamations and Hamilton Decompositions

In 1984, Hilton [9] made a leap forward in the study of Hamilton decompositions. He had the idea of starting with a single vertex, say α, incident with $n(n-1)/2$ loops, n of each of $(n-1)/2$ colors, and attempted to disentangle n vertices from α, one at a time, to end up with the complete graph K_n in which each color class was a Hamilton cycle. The proof was inductive, so was especially powerful in that it allowed one to start midway through the process rather than with a single vertex. All that was needed was for this midway point to satisfy the conditions described in the inductive hypotheses, conditions which actually turn out to be necessary anyway.

It is helpful to think of the single vertex as originally containing the n vertices that eventually appear in the final graph. As each vertex is disentangled from α, one less vertex is still contained in it, so at the ith step one can naturally define the amalgamation function $f_i(\alpha) = n - i$ to be the number of vertices still in α. Inductively, the setup at the ith step is to have: $f_i(\alpha)(f_i(\alpha) - 1)/2$ loops incident with α; one edge between each pair of disentangled vertices; and $n - f_i(\alpha)$ edges between each disentangled vertex and α. Since we hope to end up with K_n then at the ith step, one end of each of $f_i(\alpha) - 1$ loops is detached from α and joined instead of the new vertex being disentangled from α. Also, from each previously disentangled vertex, one of the $n - f_i(\alpha)$ edges joining it to α is detached from α, its new end becoming the disentangled vertex instead of α. So, with these properties in mind, by the time the $(n-1)$th step is completed, it is easy to see our single vertex with loops has been transformed into K_n.

Advantageously, the method is even more flexible than described so far in that it is possible to start with a graph G having p vertices, each vertex, v, containing $f(v) = n$ vertices (or even setups more general than that). If each of the p vertices, v, has $\lambda_1 f(v)(f(v) - 1)/2 = \lambda_1 n(n - 1)/2$ loops on it, and if between each pair of the p vertices, say u and v, there are $\lambda_2 f(u) f(v) = \lambda_2 n^2$ edges, then this graph is the amalgamation (homomorphic image) of $H = K(n, p, \lambda_1, \lambda_2)$: For each of the p parts of $K(n, p, \lambda_1, \lambda_2)$, amalgamate the n vertices into a single vertex to form G. Notice that this includes the classical complete multipartite graphs, when $\lambda_1 = 0$ and $\lambda_2 = 1$.

Of course, the point here is not just to produce K_n or $K(n, p, \lambda_1, \lambda_2)$; we are really trying to produce Hamilton decompositions (or other graph decompositions) of these graphs. The idea is that if the disentangling process can be achieved, then it is much easier to form the amalgamated graph with a suitable edge-coloring (an outline of the final decomposition) than it is to find the final decomposition directly. So attention also needs to be paid to the color of the edges being selected during the disentangling process, both the loops incident with α and the edges joining the previously disentangled vertices to α. It turns out that we now have a lot of control over the disentanglement. Various results appear in the literature, but the following result is a good example of what is possible. Proved in more generality by Bahmanian and Rodger in [7], it ties in nicely with the fairness notions described earlier. Informally, it says that if $D(v)$ is the set of vertices in H disentangled from v in G, then each vertex u in $D(v)$ receives its fair share of the edge ends in G incident with v, and each vertex u in $D(v)$ receives its fair share of the edge ends in G incident with v colored j. That is, $d_H(v) \in \{\lfloor d_G(v)/n \rfloor, \lceil d_G(v)/n \rceil\}$ and $d_{H(j)}(v) \in \{\lfloor d_{G(j)}(v)/n \rfloor, \lceil d_{G(j)}(v)/n \rceil\}$, where $G(j)$ is the subgraph of G induced by the edges colored j. In Theorem 1, ψ plays the role of the amalgamation function, $\ell_G(u)$ is the number of loops in G incident with u, and $m_G(u, v)$ is the number of edges in G joining u to v.

Theorem 1 [7] *Let G be a k-edge-colored graph and let ψ be a function from $V(G)$ into the positive integers such that for each $u \in V(G)$,*

(1) $\psi(u) = 1$ implies $\ell_G(u) = 0$,
(2) $d_{G(j)}(u)/\psi(u)$ is an even integer for all $1 \leq j \leq k$,
(3) $\binom{\psi(u)}{2}$ divides $\ell_G(u)$,
(4) $\psi(u)\psi(v)$ divides $m_G(u, v)$ for each $v \in V(G) \setminus \{u\}$, and
(5) $G(j)$ is connected for $1 \leq j \leq k$.

Then, there exists a detachment H of G in which each $u \in V(G)$ is disentangled into vertices $u_1, \ldots, u_{\psi(u)}$, such that for all $u \in G$:

(i) $m_H(u_i, u_{i'}) = \ell_G(u)/\binom{\psi(u)}{2}$ for all $1 \leq i < i' \leq \psi(u)$ if $\psi(u) \geq 2$,
(ii) $m_H(u_i, v_{i'}) = m_G(u, v)/\psi(u)\psi(v)$ for $v \in V(G) \setminus \{u\}$, $1 \leq i \leq \psi(u)$, and $1 \leq i' \leq \psi(v)$,
(iii) $d_{H(j)}(u_i) = d_{H(j)}(u)/\psi(u)$ for $1 \leq i \leq \psi(u)$ and $1 \leq j \leq k$, and
(iv) Each color class $H(j)$ is connected for $1 \leq j \leq k$.

Condition (2) is critical for proving that connected color classes in G can remain connected during the disentangling process, thus guaranteeing that condition (iv) is satisfied by H. Since we aim to have each color class disentangled into a Hamilton cycle, clearly each vertex v in the amalgamated graph we construct needs to be incident with $2\psi(v)$ edges colored j, for each color j, since each of the $\psi(v)$ vertices inside v needs to be incident with exactly two edges colored j in the disentangled graph.

Not only does this approach give a new proof of Walecki's [3] result, but it also lends itself beautifully to other families of graphs than complete graphs.

Theorem 2 [6, 10] *There exists a Hamilton decomposition of* $\lambda K(n, p)$ *if and only if* $\lambda n(p - 1)$ *is even.*

To see how Theorem 1 is of use in proving Theorem 2, start with p vertices, each joined to each other with λn^2 edges. The edges are then colored with $\lambda n(p - 1)/2$ colors so that each color class is connected and $2n$-regular (the details of how the edge-coloring is accomplished are not included here, but one natural approach is to add Hamilton cycles of K_p, each containing edges of just one color, to complete most of the task). It is easy to see that this edge-colored graph satisfies conditions (1–5) of Theorem 1 with $\psi(v) = n$ for all vertices. So the disentangled graph, H: by condition (ii) H is simple, so it must be that $H = \lambda K(n, p)$; by conditions (iii–iv), each color class of H is 2-regular and connected, so is a Hamilton cycle. This completes the proof.

More recently, the existence of Hamilton decompositions of $K(n, p, \lambda_1, \lambda_2)$ was completely settled in the following theorem.

Theorem 3 [7] *Let* $p > 1$, $\lambda_1 \geq 0$, *and* $\lambda_2 \geq 1$, *with* $\lambda_1 \neq \lambda_2$ *be integers. Then, there exists a Hamilton decomposition of* $K(n, p, \lambda_1, \lambda_2)$ *if and only if*

(ii) $\lambda_1(n - 1) + \lambda_2 n(p - 1)$ *is even, and*
(iii) $\lambda_1 \leq \lambda_2 n(p - 1)$.

It is hopefully not surprising now that the proof of the sufficiency follows the above approaches closely, starting with p vertices, each joined to each other with λn^2 edges, but this time each vertex is also incident with $\lambda_1 n(n - 1)/2$ loops. The edges are then colored so that each color class is connected and $2n$-regular. Once this is done, the result follows essentially immediately from Theorem 1.

The proof of the necessity of Theorem 3 is not included here, but it is worth giving some feel for why condition (iii) is necessary. First note that every Hamilton cycle in $K(n, p, \lambda_1, \lambda_2)$ must use at least p edges joining vertices in different parts in order to be connected. So if we allow λ_1 to grow while holding all other parameters constant, we will eventually run out of the edges joining vertices in different parts. For this reason, an upper bound on λ_1 is to be expected.

3   Embeddings of Edge-Colorings into Hamilton Decompositions

The embedding interest followed the same line as the construction results described in Sect. 2: First studied was embeddings of edge-colored graphs into Hamilton decompositions of K_n (see Theorem 4), then of complete multipartite graphs (see Theorem 5), and then of $K(n, p, \lambda_1, \lambda_2)$ (see Theorems 6 and 7). We now survey this progress, one by one.

In the previous section, Hilton's paper [9] introducing amalgamations as a means of producing graph decompositions was described. One of the great applications he developed was the idea of building prerequisites into the final Hamilton decomposition. In his paper, he proved the following result which completely describes when it is possible to start with a given edge-coloring of K_n and embed it in a Hamilton decomposition of K_m; that is, add $m - n$ new vertices to the given edge-colored K_n, and edges to form a K_m, then color all the added edges so that each color class is a Hamilton cycle. This was truly an amazing result, since typically the given edge-coloring would seemingly need to have much postulated structure or symmetry to make such a result provable. But the amalgamation method is so flexible that he completely solved the problem with the following result.

Theorem 4 [9] *A k-edge-colored K_n (some colors may appear on no edges) can be embedded into a Hamiltonian decomposition of K_m if and only if*

1. *m is odd,*
2. *$k = \lfloor m/2 \rfloor$, and*
3. *Each color class of the given edge-coloring of K_n has at most $m - n$ components, each of which is a path (isolated vertices are considered to be paths of length 0).*

Proof The necessity of these conditions is quite clear: (1–2) follow since in K_m each vertex is incident with exactly two edges of each color; (3) follows because each one of the $m - n$ added vertices can be used to connect just two components in each color class.

Proving the sufficiency clearly demonstrates the power of amalgamations. At first sight, it is not clear at all how to color all the added edges. But we immediately know how to color them in the graph formed by taking any solution to the embedding and amalgamating the added vertices to form a single vertex (in the notation of Sect. 2, the amalgamated vertex is like α, with $f(\alpha) = m - n$). The following shows how to form the amalgamated graph G, even though we do not have a solution (i.e., a Hamilton decomposition of $H = K_m$) in hand.

1. Join each vertex in K_n to the added vertex α with $m - n$ edges.
2. Color the added edges so that each vertex in K_n has degree 2 in each color class. (This is possible since then vertices in K_n would have degree $2k = (n - 1) + (m - n)$.)
3. Add $(m - 1)(m - n - 1)/2$ loops incident with α.
4. To complete the edge-coloring of G, color the loops so that α is incident with exactly $2(m - n)$ edge ends of each color; each loop contributes two edge ends. (This is possible since condition (3) guarantees the number of loops to be added is nonnegative and the number of edges of each color added in the second step is even.)

We can now immediately form a Hamilton decomposition of H from G using Theorem 1 with $\psi(u) = 1$ for all vertices in K_n and $\psi(\alpha) = m - n$. To see this, refer to the various parts of Theorem 1 in turn as follows.

(i) Shows that once the $m - n$ vertices in α are disentangled, the $((m - n)(m - n - 1)/2$ loops on α induce a simple graph, which must be K_{m-n}.

(ii) Shows that once the $m - n$ vertices in α are disentangled, the $m - n$ edges joining each vertex u in K_n to α become single edges joining u to each of the $m - n$ disentangled vertices. So at this stage we know that H is K_m.

(iii) Shows that each vertex in H has degree 2 in each color class.

(iv) Shows, together with what was just shown in (iii), that each color class is a Hamilton cycle.

□

Hilton and Rodger [10] extended Theorem 4 to the complete multipartite graphs. They proved the following result as a corollary of a much more general amalgamation theorem.

Theorem 5 [10] *Suppose that $2t \leq s$. Then, a k-edge-coloring of the complete t-partite graph $K_{a_1,...,a_t}$ can be embedded into a Hamiltonian decomposition of the complete p-partite graph $K(n, p, 0, 1)$ if and only if*

(i) Each color class is a set of vertex-disjoint paths,

(i) $a_i \leq n$ for $1 \leq i \leq t$, and

(i) $p(n - 1)$ is even.

The proof of Theorem 5, while more complicated, follows the approach outlined above for proving Theorem 4. In this case, the given t-partite graph is first embedded greedily into an edge-colored $K(n, t, 0, 1)$ in which each color class is still a set of vertex-disjoint paths; this can be done since we are assuming that $2t \leq s$. The second step introduces one new vertex, an amalgamated vertex playing the role of α in the outline of the proof of Theorem 4 above, but in this case the technique calls for all vertices within the same part to be disentangled before moving on to vertices from other parts still contained in α.

The embedding of edge-colored copies of $K(n, t, \lambda_1, \lambda_2)$ into Hamilton decompositions of $K(n, p, \lambda_1, \lambda_2)$ is really very interesting. Reasonably obvious numerical conditions are sufficient when p is somewhat larger than t (see Theorem 6), but at this stage it appears that there are conditions which depend upon the existence of certain components in a companion bipartite graph to the given edge-colored graph which are necessary for the embedding to exist (see Theorem 7). This structural property is reminiscent of the long-standing unsolved embedding problem for partial idempotent latin squares of order n into idempotent latin squares of order $n + t$ when t is small: When $t \geq n$ numerical conditions do prove to be sufficient (see [11–13]), but for smaller values of t the existence of certain components in a closely related graph can prevent such an embedding (see [11, 14]).

As in other results mentioned so far, the following is a consequence of a more general amalgamation result in [15] which requires some postulations that are unlikely to be necessary in a complete solution to the problem. Nevertheless, the result is sufficiently general to allow the embedding problem to be solved whenever the number of parts, r, being added to the given edge-colored copy of $K(n, t, \lambda_1, \lambda_2)$

is large enough. It is always a little worrying when a result is described in terms of some parameter being sufficiently large. Often that necessary size for the result to work is really very large. However, the good news in this case is that in fact the lower bound on r for the result to be applicable is not really so large, as the following result indicates.

Theorem 6 [15] *Let $n > 1$, $\lambda_1 \geq 0$, $\lambda_2 \geq 1$, $\lambda_1 \neq \lambda_2$, $p = t + r$ and*

$$r \geq \frac{\lambda_1(n-1) + \lambda_2 n(t-1)}{\lambda_2 n(n-1)}. \tag{1}$$

Then, a k-edge-coloring of $K(n, t, \lambda_1, \lambda_2)$ can be embedded into a Hamiltonian decomposition of $K(n, p, \lambda_1, \lambda_2)$ if and only if

1. $k = (\lambda_1(n-1) + \lambda_2 n(p-1))/2$,
2. $\lambda_1 \leq \lambda_2 n(p-1)$,
3. *Every component of $G(j)$ is a path (possibly of length 0) for $1 \leq j \leq k$, and*
4. *$G(j)$ has at most nr components for $1 \leq j \leq k$.*

In the same paper, using the same general amalgamation theorem, it turns out that the case where $r = 1$ (so just one part is being added) can also be completely solved. So now we need to explore the values of r between 1 and $(\lambda_1(n-1) + \lambda_2 n(t-1))/\lambda_2 n(n-1)$. Starting with the smallest values seems enticing! It turns out that even just considering the case where $r = 2$ is particularly challenging. We appear to enter a different world where the structure can play a deciding role in determining whether or not the embedding of the k-edge-coloring of $G = K(n, t, \lambda_1, \lambda_2)$ into a Hamiltonian decomposition of $H = K(n, p = t + 2, \lambda_1, \lambda_2)$ is possible. To see this, it is best to describe the issue in terms of a related bipartite graph, B. Its vertex set is of course partitioned into two sets: $V(G)$ and $C = \{c_j \mid 1 \leq j \leq k\}$. Each $v \in V(G)$ is joined to c_j in B with x edges if and only if $d_{G(j)}(v) = 2 - x$. Recall that in H each color class is a Hamilton cycle, so each vertex has degree 2 in each color class. So B is keeping a track of how many more edges of each color, j, that v needs added during the embedding process. Connectivity is also a critical aspect of the embedding: The added vertices in the $r = 2$ new parts need to be used to connect up all the paths in $G(j)$ for each color j (so $1 \leq j \leq k$). For various reasons, it seems likely, possibly even necessary, that if $d_B(c_j) \equiv 2 \pmod 4$ then at least one of the components (paths) in $G(j)$ must have its end vertices in G, say $v_{j,1}$ and $v_{j,2}$, joined to different new parts in H. Reproducing this during a proof of the sufficiency is managed by forming B^*, a modification of B constructed by disentangling such c_j into two vertices, one having degree 2 being adjacent to $v_{j,1}$ and $v_{j,2}$. As the embedding proceeds, choosing the path for each color, j, which determines $v_{j,1}$ and $v_{j,2}$ seems to be critical, as is described in condition $(*)$ of Theorem 7 below. Let $\Pi = \{\{v_{j,1}, v_{j,2}\} \mid 1 \leq j \leq k\}$ describe this choice. It is conceivable that condition $(*)$ is also a necessary condition. Let C_2 denote the set of vertices in C of degree 2 (mod 4).

Theorem 7 [16] *Let $n > 1$, $\lambda_1 \geq 0$, $\lambda_2 \geq 1$ and $\lambda_1 \neq \lambda_2$. Suppose we are given a k-edge-coloring of $G = K(n, t, \lambda_1, \lambda_2)$, and that*

(*) Π can be chosen such that in the detached graph, B^*, the number of components having an odd number of color vertices of degree divisible by 4 is at most $\lambda_2 n^2$.

Then, the k-edge-coloring of G can be embedded into a Hamiltonian decomposition of $K(n, p = t + 2, \lambda_1, \lambda_2)$ if and only if

(i) Conditions (1–4) of Theorem 6 with $r = 2$ are satisfied, and
(v) $|C_2| \leq 2\lambda_1 \binom{n}{2} + \lambda_2 n^2$.

Apart from amalgamations, there is another aspect of the proof of this result which is of interest here since a 2-edge-coloring of B^* is required that has the colors fairly divided in two ways. An edge-coloring of a graph is said to be *equitable at vertex v* if, for all colors i and j, the number of edges incident with v colored i is within 1 of the number of edges colored j. An edge-coloring of a graph is said to be *evenly equitable at vertex v* if, for all colors i and j, the number of edges incident with v colored i is even and is within 2 of the number of edges colored j. Hilton [17] proved that evenly equitable edge-colorings (i.e., evenly equitable at all vertices) exist whenever all vertices have even degree. Equitable edge-colorings (i.e., equitable at all vertices) are much more problematic (see [18] for example), but de Werra [19] has shown that they always exist for bipartite graphs. To prove Theorem 7, it was critical that these two results of Hilton and de Werra be generalized to require some vertices to be evenly equitably colored and others to be equitably colored. We end with this crucial lemma, which is of interest in its own right.

Lemma 1 [16] Let B be a finite even bipartite graph with bipartition $\{V, C\}$ of its vertex set. For any subset $X \subseteq C$, there exists a 2-edge-coloring $\sigma : E(B) \rightarrow \{1, 2\}$ such that

(i) $d_{B(1)}(v) = d_{B(2)}(v)$ for all $v \in V$,
(ii) $d_{B(1)}(c) = d_{B(2)}(c)$ for all $c \in X$, and
(iii) $|d_{B(1)}(c) - d_{B(2)}(c)| = 2$ for all $c \in C \setminus X$

if and only if

(iv) $|V(D) \cap (C \setminus X)|$ is even for each component D of B.

References

1. McDiarmid, C.J.H.: The solution of a timetabling problem. J. Inst. Math. Appl. **9**, 23–34 (1972)
2. Hajnal, A., Szemerdi, E.: Proof of a Conjecture of P. Erdös, Combinatorial Theory and its Applications, II North-Holland, Amsterdam, , pp. 601–623 (1970)
3. Lucas, E.: Récréations mathématiques, vol. 2, Gauthier-Villars, Paris (1883)
4. Hoffman, D.G., Rodger, C.A.: The chromatic index of complete multipartite graphs. J. Graph Theor. **16**, 159–163 (1992)

5. Lam, P., Shiu, W.C., Tong, C.S., Zhang, Z.F.: On the equitable chromatic number of complete n-partite graphs. Discrete Appl. Math. **113**, 307–310 (2001)
6. Laskar, R., Auerbach, B.: On decomposition of r-partite graphs into edge-disjoint Hamiltonian circuits. Discrete Math. **14**, 265–268 (1976)
7. Bahmanian, M.A., Rodger, C.: Multiply balanced edge colorings of multigraphs. J. Graph Theor. **70**, 297–317 (2012)
8. Hilton, A.J.W.: School timetables, studies on graphs and discrete programming. Ann. Discrete Math. **11**, 177–188 (1981)
9. Hilton, A.J.: Hamiltonian decompositions of complete graphs. J. Comb. Theor. (B) **36**, 125–134 (1984)
10. Hilton, A.J., Rodger, C.A.: Hamiltonian decompositions of complete regular s-partite graphs. Discrete Math. **58**, 63–78 (1986)
11. Andersen, L.D., Hilton, A.J.W., Rodger, C.A.: A solution to the embedding problem for partial idempotent Latin squares. J. London Math. Soc. **26**, 21–27 (1982)
12. Rodger, C.A.: Embedding incomplete idempotent latin squares, Combinatorial Mathematics X. Lecture Notes in Mathematics (Springer), vol. 1036, pp. 355–366 (1983)
13. Rodger, C.A.: Embedding an incomplete latin square in a latin square with a prescribed diagonal. Discrete Math. **51**, 73–89 (1984)
14. Andersen, L.D., Hilton, A.J.W., Rodger, C.A.: Small embeddings of incomplete idempotent Latin squares. Ann. Discrete Math. **17**, 19–31 (1983)
15. Bahmanian, M.A., Rodger, C.: Embedding an edge-colored $K(a^{(p)}; \lambda, \mu)$ into a Hamiltonian decomposition of $K(a^{(p+r)}; \lambda, \mu)$. Graphs Comb. **29**, 747–755 (2012)
16. Demir, M., Rodger, C.A.: Embedding an Edge-Coloring of $K(n^r; \lambda_1, \lambda_2)$ into a Hamiltonian Decomposition of $K(n^{r+2}; \lambda_1, \lambda_2)$, submitted
17. Hilton, A.J.W.: Canonical edge-colourings of locally finite graphs. Combinatorica **2**, 37–51 (1982)
18. Hilton, A.J.W., de Werra, D.: A sufficient condition for equitable edge-colourings of simple graphs. Discrete Math. **128**, 179–201 (1994)
19. de Werra, D.: Equitable colorations of graphs, Rev. Franaise Informat. Recherche Oprationnelle **5**, Sr. R-3, 3–8 (1971)

Chapter 2
On Strong Pseudomonotone and Strong Quasimonotone Maps

Sanjeev Kumar Singh, Avanish Shahi and S. K. Mishra

Abstract We introduce strong pseudomonotone and strong quasimonotone maps of higher order and establish their relationships with strong pseudoconvexity and strong quasiconvexity of higher order, respectively, which yields first-order characterizations of strong pseudoconvex and strong quasiconvex functions of higher order. Moreover, we answer the open problem (converse part of Proposition 6.2) of Karamardian and Schaible (J. Optim. Theory Appl. 66:37–46,1990), for even more generalized functions, namely strongly pseudoconvex functions of higher order.

Keywords Generalized monotone maps · Generalized convexity · First-order conditions

1 Introduction

Minty [9] introduced the concept of monotone maps. Further, in addition to that Karamardian [5] discussed strict monotone and strongly monotone maps. It is well known that every differentiable function is convex if and only if its gradient map is monotone (see [2, 10]). Karamardian [5] stated the relationship between strongly convex functions and strongly monotone maps. In 1976, Karamardian [4] introduced the concept of pseudomonotone maps and showed that a differentiable pseudoconvex function (see [3, 8]) is characterized by pseudomonotonicity of its gradient map and used monotonicity/pseudomonotonicity in establishing several existence theorems for complementarity problems. Further, Karamardian and Schaible [6] introduced strictly pseudomonotone, quasimonotone, strongly monotone, and strongly

S. K. Singh · A. Shahi · S. K. Mishra (✉)
Department of Mathematics, Institute of Science, Banaras Hindu University,
Varanasi 221005, India
e-mail: bhu.skmishra@gmail.com

S. K. Singh
e-mail: sksingh20894@gmail.com

A. Shahi
e-mail: avanishshahi123@gmail.com

© Springer Nature Singapore Pte Ltd. 2018
D. Ghosh et al. (eds.), *Mathematics and Computing*, Springer Proceedings
in Mathematics & Statistics 253, https://doi.org/10.1007/978-981-13-2095-8_2

pseudomonotone maps and showed that for gradient maps, these generalized monotonicity properties are related to generalized convexity properties of the underlying functions.

Lin and Fukushima [7] along with other results for nonlinear programs and mathematical programs with equilibrium constraints introduced strong convexity of order σ and strong monotone maps of order σ. Lin and Fukushima [7] showed that the strong monotonicity of order σ of the gradient map is related to strong convexity of order σ of the function. Arora et al. [1] introduced strongly pseudoconvex functions of order σ and its generalization to characterize solution sets and optimality conditions for optimization problems. Arora et al. [1] have also introduced strongly quasiconvex function of order σ.

It is very natural to see that the concept of strongly monotone maps of order σ due to Lin and Fukushima [7] can be extended to strongly pseudomonotone maps of order σ and strongly quasimonotone maps of order σ can be studied, as Karamardian and Schaible [6] extended the concept of monotone maps to pseudomonotone maps.

In 1990, Karamardian and Schaible [6] left an open problem as the converse of Proposition 6.2 [6], and we have answered that open question positively for a more general function, namely strongly pseudoconvex of order σ, which is also an extension of strongly convex function of order σ given by Lin and Fukushima [7].

2 Preliminaries

2.1 Pseudoconvexity and Quasiconvexity

Definition 1 [2, 6] A differentiable function f on an open convex subset X of $\mathbb{R}^n$ is pseudoconvex on X if, for every pair of distinct points $x, y \in X$, we have

$$\langle \nabla f(y), x - y \rangle \geq 0 \Rightarrow f(x) \geq f(y).$$

Definition 2 [2, 6] A function f is quasiconvex on a convex set X of $\mathbb{R}^n$ if, for all $x, y \in X, \lambda \in [0, 1]$,

$$f(x) \leq f(y) \Rightarrow f(\lambda x + (1 - \lambda)y) \leq f(y).$$

Proposition 1 [2, 6] *A differentiable function f is quasiconvex on an open convex set X of $\mathbb{R}^n$ if and only if, for every pair of points $x, y \in X$, we have*

$$f(x) \leq f(y) \Rightarrow \langle \nabla f(y), x - y \rangle \leq 0.$$

Remark 1 [3] Every pseudoconvex function is quasiconvex, but the converse is not necessarily true.

2.2 Strong Convexity and Strong Monotonicity of Order σ

Definition 3 [7] Let X be a non-empty open and convex subset of $\mathbb{R}^n$. A function $f : X \to \mathbb{R}$ is said to be strongly convex function of order σ if $\exists$ a constant $c > 0$ such that

$$f(\lambda x + (1 - \lambda)y) \leq \lambda f(x) + (1 - \lambda)f(y) - c\lambda(1 - \lambda)\|x - y\|^\sigma,$$

for any $x, y \in X$ and any $\lambda \in [0, 1]$.

Theorem 1 [7] *Let X be a non-empty open and convex subset of $\mathbb{R}^n$. A continuously differentiable function $f : X \to \mathbb{R}$ is strongly convex of order σ on X if and only if $\exists$ a constant $c > 0$ such that*

$$f(x) - f(y) \geq \langle \nabla f(y), x - y \rangle + c\|x - y\|^\sigma, \qquad \forall x, y \in X.$$

Remark 2 [6] For $\sigma = 2$,

$$f(x) - f(y) \geq \langle \nabla f(y), x - y \rangle + c\|x - y\|^2.$$

This function is referred to as strongly convex function in ordinary sense.

Definition 4 [7] Let X be a non-empty open and convex subset of $\mathbb{R}^n$. A mapping $F : X \to \mathbb{R}^n$ is said to be strongly monotone map of order σ if $\exists$ a constant $\beta > 0$ such that

$$\langle F(x) - F(y), x - y \rangle \geq \beta\|x - y\|^\sigma, \qquad \forall x, y \in X.$$

Remark 3 [6] For $\sigma = 2$,

$$\langle F(x) - F(y), x - y \rangle \geq \beta\|x - y\|^2.$$

This map is referred to as strongly monotone map in ordinary sense.

Lin and Fukushima [7] established the relation between strongly convex function of order σ and strongly monotone map of order σ.

Theorem 2 [7] *Let X be a non-empty open and convex subset of $\mathbb{R}^n$. A continuously differentiable function $f : X \to \mathbb{R}$ is strongly convex of order σ if and only if ∇f is strongly monotone of order σ on X.*

3   Strongly Pseudoconvexity of Order σ and Strongly Pseudomonotonicity of Order σ

Definition 5 [1] Let X be a non-empty open and convex subset of $\mathbb{R}^n$. A differentiable function $f : X \to \mathbb{R}$ is said to be strongly pseudoconvex of order σ on X if $\exists \alpha > 0$ such that

$$\langle \nabla f(y), x - y \rangle + \alpha \|x - y\|^\sigma \geq 0 \Rightarrow f(x) \geq f(y), \qquad \forall x \neq y \in X.$$

We introduce strongly pseudomonotone map of order σ.

Definition 6 Let X be a non-empty open and convex subset of $\mathbb{R}^n$. A map $F : X \to \mathbb{R}^n$ is said to be strongly pseudomonotone of order σ on X if $\exists \beta > 0$ such that

$$\langle F(y), x - y \rangle + \beta \|x - y\|^\sigma \geq 0 \Rightarrow \langle F(x), x - y \rangle \geq 0, \qquad \forall x \neq y \in X.$$

We establish the relationship between strong pseudoconvexity and strong pseudomonotonicity of order σ, which is the natural generalization of the strongly pseudoconvex function given by Karamardian and Schaible [6]. Karamardian and Schaible [6] have left an open problem as the converse of the Proposition (6.2), and we prove necessary and sufficient both parts for more general class as strong pseudoconvexity of order σ.

Theorem 3 *Let X be a non-empty open and convex subset of $\mathbb{R}^n$. A continuously differentiable function $f : X \to \mathbb{R}$ is strongly pseudoconvex of order σ if and only if ∇f is strongly pseudomonotone of order σ on X.*

Proof Let f be strongly pseudoconvex of order σ on X, then $\exists \alpha > 0$ such that

$$\langle \nabla f(y), x - y \rangle + \alpha \|x - y\|^\sigma \geq 0 \Rightarrow f(x) \geq f(y), \qquad \forall x \neq y \in X. \quad (1)$$

Since every strongly pseudoconvex function of order σ is quasiconvex function. Therefore,

$$f(\lambda x + (1 - \lambda)y) \leq f(x). \quad (2)$$

By using proposition (1) on Eq. (2),

$$\langle \nabla f(x), (\lambda x + (1 - \lambda)y) - x \rangle \leq 0,$$

$$\langle \nabla f(x), (1 - \lambda)(y - x) \rangle \leq 0,$$

$$\langle \nabla f(x), (x - y) \rangle \geq 0.$$

Therefore, we have

$$\langle \nabla f(y), x - y \rangle + \alpha \|x - y\|^\sigma \geq 0 \Rightarrow \langle \nabla f(x), (x - y) \rangle \geq 0.$$

Thus, ∇f is strongly pseudomonotone of order σ.

Conversely, suppose that ∇f is strongly pseudomonotone of order σ on X and then $\exists\, \beta > 0$ such that

$$\langle \nabla f(y), x - y \rangle + \beta \|x - y\|^\sigma \geq 0 \Rightarrow \langle \nabla f(x), x - y \rangle \geq 0, \qquad \forall x \neq y \in X.$$

Equivalently,

$$\langle \nabla f(x), x - y \rangle < 0 \Rightarrow \langle \nabla f(y), x - y \rangle + \beta \|x - y\|^\sigma < 0. \qquad (3)$$

We want to show that f is strongly pseudoconvex of order σ.

For this, we have to show $\exists\, \alpha > 0$ such that

$$\langle \nabla f(y), x - y \rangle + \alpha \|x - y\|^\sigma \geq 0 \Rightarrow f(x) \geq f(y), \qquad \forall x \neq y \in X. \qquad (4)$$

Suppose on contrary,

$$f(x) < f(y).$$

By the mean value theorem, $\exists\, z = \lambda x + (1 - \lambda) y$, for some $\lambda \in (0, 1)$ such that

$$f(x) - f(y) = \langle \nabla f(z), x - y \rangle, \qquad \forall x \neq y \in X. \qquad (5)$$

$$\langle \nabla f(z), x - y \rangle = \frac{1}{\lambda} \langle \nabla f(z), z - y \rangle < 0.$$

From Eq. (3), we obtain

$$\langle \nabla f(z), z - y \rangle < 0 \Rightarrow \langle \nabla f(y), z - y \rangle + \beta \|z - y\|^\sigma < 0,$$

$$\langle \nabla f(z), z - y \rangle < 0 \Rightarrow \lambda[\langle \nabla f(y), x - y \rangle + \beta \lambda^{\sigma-1} \|x - y\|^\sigma] < 0,$$

$$\langle \nabla f(z), z - y \rangle < 0 \Rightarrow \langle \nabla f(y), x - y \rangle + \beta \lambda^{\sigma-1} \|x - y\|^\sigma < 0,$$

which contradicts that

$$\langle \nabla f(y), x - y \rangle + \alpha \|x - y\|^\sigma \geq 0.$$

So, $f(x) \geq f(y)$ and hence f is strongly pseudoconvex of order σ. $\qquad \square$

Remark 4 [1] Every strongly pseudoconvex function of order σ is pseudoconvex, but the converse is not necessarily true.

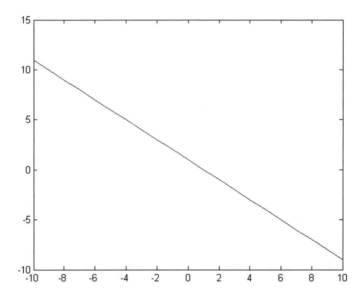

Fig. 1 Strongly pseudomonotone map of order σ.

Remark 5 Every strongly monotone map of order σ is strongly pseudomonotone map of order σ, but the converse is not necessarily true.

Example 1 Let $F : \mathbb{R} \to \mathbb{R}$, defined by $F(x) = 1 - x$, $x \in \mathbb{R}$.

Here, F is strongly pseudomonotone of order σ but not strongly monotone of order σ (Fig. 1).

4 Strongly Quasiconvexity and Strongly Quasimonotonicity of Order σ

Definition 7 [1] Let X be a non-empty open and convex subset of $\mathbb{R}^n$. A differentiable function $f : X \to \mathbb{R}$ is said to be strongly quasiconvex of order σ if $\exists \, \alpha > 0$ such that

$$f(x) \le f(y) \Rightarrow \langle \nabla f(y), x - y \rangle + \alpha \|x - y\|^\sigma \le 0, \qquad \forall x \ne y \in X.$$

We introduce strongly quasimonotone map of order σ.

Definition 8 Let X be a non-empty open and convex subset of $\mathbb{R}^n$. A map $F : X \to \mathbb{R}^n$ is said to be strongly quasimonotone of order σ if $\exists \, \beta > 0$ such that

$$\langle F(y), x - y \rangle > 0 \Rightarrow \langle F(x), x - y \rangle \ge \beta \|x - y\|^\sigma, \qquad \forall x \ne y \in X.$$

Theorem 4 *Let X be a non-empty open and convex subset of $\mathbb{R}^n$. A continuously differentiable function $f : X \to \mathbb{R}$ is strongly quasiconvex of order σ if and only if ∇f is strongly quasimonotone of order σ on X.*

Proof Let f be strongly quasiconvex function of order σ on X, then $\exists \, \alpha > 0$ such that

$$f(x) \le f(y) \Rightarrow \langle \nabla f(y), x - y \rangle + \alpha \|x - y\|^\sigma \le 0, \qquad \forall x \ne y \in X. \quad (6)$$

We have to show that ∇f is strongly quasimonotone of order σ on X.
For this, we have to prove that $\exists \, \beta > 0$ such that

$$\langle \nabla f(y), x - y \rangle > 0 \Rightarrow \langle \nabla f(x), x - y \rangle \ge \beta \|x - y\|^\sigma, \qquad \forall x \ne y \in X.$$

Since every strongly quasiconvex function of order σ is quasiconvex, therefore we have

$$\langle \nabla f(y), x - y \rangle > 0 \Rightarrow f(x) > f(y). \quad (7)$$

As f is strongly quasiconvex function of order σ, then by using Eq. (6), we have

$$f(y) < f(x) \Rightarrow \langle \nabla f(x), y - x \rangle + \alpha \|y - x\|^\sigma \le 0,$$

$$f(y) < f(x) \Rightarrow \langle \nabla f(x), y - x \rangle \le -\alpha \|y - x\|^\sigma,$$

$$f(y) < f(x) \Rightarrow \langle \nabla f(x), x - y \rangle \ge \alpha \|x - y\|^\sigma.$$

Therefore,

$$\langle \nabla f(y), x - y \rangle > 0 \Rightarrow \langle \nabla f(x), x - y \rangle \ge \alpha \|x - y\|^\sigma.$$

So, ∇f is strongly quasimonotone of order σ.
Conversely, let ∇f be strongly quasimonotone of order σ, then $\exists \, \beta > 0$ such that

$$\langle \nabla f(y), x - y \rangle > 0 \Rightarrow \langle \nabla f(x), x - y \rangle \ge \beta \|x - y\|^\sigma, \qquad \forall x \ne y \in X. \quad (8)$$

We have to prove that f is strongly quasiconvex function of order σ.
For this, we have to prove that $\exists \, \beta > 0$ such that

$$f(x) \le f(y) \Rightarrow \langle \nabla f(y), x - y \rangle + \beta \|x - y\|^\sigma \le 0, \qquad \forall x \ne y \in X.$$

Equivalently,

$$\langle \nabla f(y), x - y \rangle + \beta \|x - y\|^\sigma > 0 \Rightarrow f(x) > f(y). \quad (9)$$

Suppose on contrary, $f(x) \leq f(y)$.

By the mean value theorem, $\exists \, z = \lambda x + (1 - \lambda)y$ for some $\lambda \in (0, 1)$ such that

$$f(x) - f(y) = \langle \nabla f(z), x - y \rangle = \frac{1}{\lambda} \langle \nabla f(z), z - y \rangle \leq 0 \Rightarrow \langle \nabla f(z), y - z \rangle > 0, \qquad \forall x \neq y \in X.$$

Since ∇f is strongly quasimonotone of order σ, therefore by using Eq. (8), we obtain

$$\langle \nabla f(z), y - z \rangle > 0 \Rightarrow \langle \nabla f(y), y - z \rangle \geq \beta \|y - z\|^{\sigma},$$

$$\langle \nabla f(z), y - z \rangle > 0 \Rightarrow \lambda \langle \nabla f(y), y - x \rangle \geq \beta \lambda^{\sigma} \|y - x\|^{\sigma},$$

$$\langle \nabla f(z), y - z \rangle > 0 \Rightarrow \langle \nabla f(y), x - y \rangle \leq -\beta \lambda^{\sigma-1} \|x - y\|^{\sigma},$$

$$\langle \nabla f(z), y - z \rangle > 0 \Rightarrow \langle \nabla f(y), x - y \rangle + \alpha \|x - y\|^{\sigma} \leq 0. \qquad (\alpha = \beta \lambda^{\sigma-1})$$

which contradicts to left side inequality of Statement (9).

Hence, $f(x) > f(y)$ and f is strongly quasiconvex function of order σ.   □

Remark 6 Every strongly quasiconvex function of order σ is a quasiconvex function, but the converse is not always true (Fig. 2).

Example 2 $f(x) = 1 - x^3$ on $X = \mathbb{R}$.

Here, f is quasiconvex but not strongly quasiconvex of order σ.

As $f(x) \leq f(y) \Rightarrow \langle \nabla f(y), x - y \rangle \leq 0$.

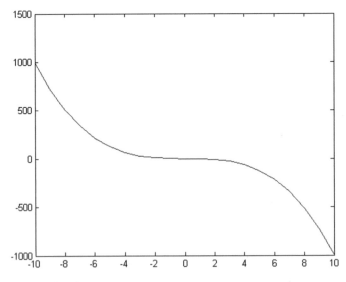

Fig. 2 Quasiconvex function

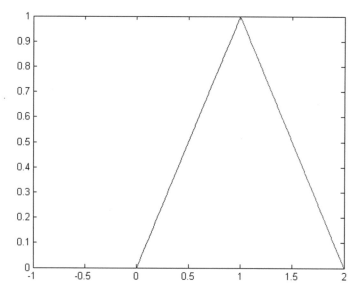

Fig. 3 Strongly quasimonotone map of order σ.

Therefore, f is quasiconvex function. On the other hand, if we take $x = \frac{1}{2}$, $y = 0$ then $f(x) \le f(y)$.

But $\langle \nabla f(y), x - y \rangle + \alpha \|x - y\|^{\sigma} \le 0 \Rightarrow -3y^2(x - y) + \alpha \|x - y\|^{\sigma} \le 0$.

At $x = \frac{1}{2}$, $y = 0$, the above inequality gives $\alpha \le 0$.

But α is positive quantity so this is not applicable for all α.

Hence, f is not strongly quasiconvex of order σ.

Remark 7 As the class of quasifunctions is largest class, a strongly pseudomonotone map of order σ is strongly quasimonotone map of order σ, but the converse is not always true.

Example 3 Define $F : X = [-1, 2] \to \mathbb{R}$, by

$$F(x) = \begin{cases} 0 & \text{for } -1 \le x < 0 \\ x & \text{for } 0 \le x < 1 \\ 2 - x & \text{for } 1 \le x \le 2 \end{cases}$$

Here, F is a strongly quasimonotone map of order σ but not strongly pseudomonotone map of order σ (Fig. 3).

Acknowledgements The first author is financially supported by CSIR-UGC JRF, New Delhi, India, through Reference no.: 1272/(CSIR-UGC NET DEC.2016). The second author is financially supported by UGC-BHU Research Fellowship, through sanction letter no: Ref.No. /Math/Res/Sept.2015/2015-16/918.

References

1. Arora, P., Bhatia, G., Gupta, A.: Characterization of the solution sets and sufficient optimality criteria via higher-order strong convexity. Topics in Nonconvex Optimization, vol. 50, pp. 231–242. Springer Optim Appl, New York (2011)
2. Avriel, M., Diewert, W.E., Schaible, S., Zang, I.: Generalized Concavity. Plenum Publishing Corporation, New York (1988)
3. Cambini, A., Martein, L.: Generalized convexity and optimization. Lecture notes in Economics and Mathematical systems, vol. 616. Springer, Berlin (2009)
4. Karamardian, S.: Complementarity problems over cones with monotone and pseudomonotone maps. J. Optim. Theor. Appl. **18**, 445–454 (1976)
5. Karamardian, S.: The nonlinear complementarity problem with applications, Part 2. J. Optim. Theor. Appl. **4**, 167–181 (1969)
6. Karamardian, S., Schaible, S.: Seven kinds of monotone maps. J. Optim. Theor. Appl. **66**, 37–46 (1990)
7. Lin, G.H., Fukushima, M.: Some exact penalty results for nonlinear programs and Mathematical programs with equilibrium constraints. J. Optim. Theor. Appl. **118**, 67–80 (2003)
8. Mangasarian, O.L.: Nonlinear programming. Corrected Reprint of the 1969 Original, Classical Appl Math, Society for Industrial and Applied Mathematics, SIAM, vol. 10. Philadelphia, PA (1994)
9. Minty, G.J.: On the monotonicity of the gradient of a convex function. Pacific J. Math. **14**, 43–47 (1964)
10. Ortega, J.M., Rheinboldt, W.C.: Interactive Solutions of Nonlinear Equations in Several Variables. Academic Press, New York (1970)

Chapter 3
A Dynamic Non-preemptive Priority Queueing Model with Two Types of Customers

Srinivas R. Chakravarthy

Abstract In this paper, we study a single-server non-preemptive priority queueing model with two types of customers. The customers arrive according to two independent Poisson processes, and the service times are exponential with possibly different parameters. While Type 1 customers, who have non-preemptive priority over Type 2 customers, have a finite waiting room, Type 2 customers have no such restriction. A new dynamic rule based on a predetermined threshold is applied in offering services to lower-priority customers (when higher-priority customers are present) whenever the server becomes free. Using matrix-analytic methods, we analyze the model in steady state and bring out some qualitative and interesting aspects of the model under study. We also compare our model to the classical two-customer non-preemptive priority model to show a marked improvement in the quality of service to customers under the proposed threshold model.

Keywords Queueing · Dynamic non-preemptive priority · Matrix-analytic method · Algorithmic probability

1 Introduction

Preemptive and non-preemptive queueing models have been studied extensively in the literature ever since the classical books on this topic appeared (see, e.g., [4, 7, 15]). Such models have applications in many areas, notably in telecommunications (see, e.g., [14, 15]). Traditional preemptive and non-preemptive queueing models are such that higher-priority customers are first attended before lower-priority customers on a first-come-first-served basis. To avoid excessive delays for lower-priority customers, several modifications to how the preemptive rules are applied have been introduced in the literature. Using the notion of preemptive distance (which is defined as, in the multi-priority queueing model, the difference between the indices of priority

S. R. Chakravarthy (✉)
Departments of Industrial and Manufacturing Engineering & Mathematics,
Kettering University, Flint, MI 48504, USA
e-mail: schakrav@kettering.edu

© Springer Nature Singapore Pte Ltd. 2018
D. Ghosh et al. (eds.), *Mathematics and Computing*, Springer Proceedings
in Mathematics & Statistics 253, https://doi.org/10.1007/978-981-13-2095-8_3

classes), several (see, e.g., [1, 12, 16]) models have been studied, In [6], the authors employ the number of preemptions as a cutoff point for intervening higher-priority customers to yield to lower-priority ones. Using discretion rules such as placing a threshold on the accumulated service effort so as to block further preemptions, a number of models have been studied (see, e.g., [2, 5, 7]). With the help of threshold policy, the authors in [3] introduced policies for preemption based on a certain (a) proportion of service requirements has been met; (b) time units of service has been met; and (c) time remaining for the current service which is less than a pre-specified limit.

All of the papers mentioned above analyzed the queueing models under various assumptions for the arrivals, the services, and the nature of the buffer space (finite or infinite) and derived several system performance measures. Recently, Kim [8] introduced a hysteretic type threshold policy, which depends on the number of (one particular type of) customers present in the system, to determine the priority of two types of customers as well as the rule to switch from one type to another type by preempting the (lower priority) customer in service. More specifically, the author in [8] considers a single-server queue with two types of customers and with $(N, n)-$preemptive priority rule which operates as follows. Whenever the number of Type 1 customers in the system reaches, $N, N \geq 1$, during the time a Type 2 customer is in service, that customer is preempted to provide services to Type 1 (thus getting a priority over Type 2) customers on a first-come-first-served basis and will return to servicing the preempted (Type 2) customer and other Type 2 customers when the number of Type 1 customers is n with $n, 0 \leq n < N$. Under the assumption of Poisson arrivals and general services, the author shows that this new priority discipline enables one to control (within a certain range) the first and second moments of the queue length of high-priority customers and thus the quality of service (QoS) can be improved. It should be pointed out that in this model the author's focus is on the QoS from higher-priority (Type 1) customers' point of view (even though they are already given a higher priority when the upper threshold is reached). Further, the services for Type 2 customers are resumed only when the number of Type 1 customers hits the lower threshold upon completion of a Type 1 service.

It should be pointed out that all the models referenced in the above papers involve preemption in one form or the other causing a disruption in services for one or more types of customers. Our paper focuses on non-preemptive priority queueing model with a new (dynamic) threshold rule such that the lower-priority customers do not have to wait excessively long. Note that in the classical non-preemptive priority queuing model lower-priority customers get pushed out to accommodate higher-priority ones and hence have to wait longer period of time. Also, our model significantly differs from the existing models in the literature including the one considered in [8] by (a) dynamically clearing lower-priority customers as opposed to focusing only on one type of customers through the threshold and (b) focusing on the QoS from lower-priority customers also.

The paper is organized as follows. In Sect. 2, we describe the model under study in more detail and set up the needed notation for understanding the rest of the paper. The steady-state analysis of the model is performed in Sect. 3, and the classical non-preemptive priority queueing model is shown to be the limiting case of the current model in Sect. 4. The comparison of our model and the corresponding classical non-preemptive priority queueing model without threshold is carried out in Sect. 5. Some illustrative examples are presented in Sect. 6, and concluding remarks are outlined in Sect. 7.

2 Model Description and Notation

Two types of customers, say, Type 1 and Type 2, arrive according to two independent Poisson processes with rate λ_1 and λ_2, respectively, to a single-server system. We assume that the service times of Type i customers are exponentially distributed with parameter $\mu_i, i = 1, 2$. Type 1 customers have a waiting area of a finite capacity of size, say, K, while Type 2 customers have no limit in the waiting area. Thus, any arriving Type 1 customers finding the buffer full will be lost. We introduce a new non-preemptive priority rule to offer services to both types of customers as follows. There is a threshold, say, $N, N \geq 1$, such that upon completion of the current service, the server either (a) becomes idle due to no customers waiting in the system; or (b) chooses the customer from the nonempty queue (as only one type of customers is present at that time); or (c) chooses a Type 1 customer to offer service unless the number of Type 2 customers waiting in the system is at least N plus the number of waiting Type 1 customers. That is, the server will offer a service to a Type 2 customer if, say, there are i Type 1 customers and the number of Type 2 customers is at least $N + i$, for $1 \leq i \leq K, N \geq 1$.

For use in sequel, we define a number of auxiliary quantities.

- $\lambda = \lambda_1 + \lambda_2$. This gives the total rate of customers arriving to the system. Note that some of Type 1 customers may be lost due to their buffer being full. So, λ may not always be the effective total arrival rate to the system.
- By e, we will denote a column vector (of dimension $K + 1$) of 1's.
- By e_i, we will denote a unit column vector (of dimension $K + 1$) with 1 in the ith position and 0 elsewhere.
- By I an identity matrix (of dimension $K + 1$).
- Suppose that a is a vector of dimension $K + 1$ with jth element is given by a_j. Then, we denote by $\Delta(a)$ a diagonal matrix of order $K + 1$ with diagonal elements given by $a_j, 1 \leq j \leq K + 1$.
- By $\tilde{\Delta}_i$, we denote a diagonal matrix of order $K + 1$ given by $\tilde{\Delta}_i = \Delta(\sum_{k=1}^{i} e_k)$.
- $\Phi_i, 1 \leq i \leq K$, is a square matrix of order $K + 1$ such that its nonzero entries are 1 and appear in $(i + j, i + j - 1)$th positions, for $1 \leq j \leq K - i + 1$. That is,

$$
\tilde{\Delta}_i = \begin{pmatrix} 1 & & & & & & \\ & 1 & & & & & \\ & & \ddots & & & & \\ & & & 1 & & & \\ & & & & 0 & & \\ & & & & & \ddots & \\ & & & & & & 0 \end{pmatrix}, \quad
\Phi_i = \begin{array}{c} \\ 1 \\ 2 \\ \vdots \\ i \\ i+1 \\ \vdots \\ K+1 \end{array}\!\!
\begin{pmatrix} 1\,2\,\cdots\,i\,\cdots\,K\,K+1 \\ \\ \\ \\ 1 \\ & 1 \\ & & \ddots \\ & & & 1 \end{pmatrix}. \tag{1}
$$

[Note: Here and in the sequel, we will use blank space in matrices or vectors to correspond to the entry being zero unless we need to display 0 for more clarity.]

- The matrix F of dimension $K + 1$ is defined as

$$
F = \begin{pmatrix} -\lambda & \lambda_1 & & & \\ & -\lambda & \lambda_1 & & \\ & & \ddots & \ddots & \\ & & & -\lambda & \lambda_1 \\ & & & & -\lambda_2 \end{pmatrix}. \tag{2}
$$

- Should there be a need to display I or e or e_i of different dimensions other than $K + 1$, we will do so by writing, say, I_m or $e(m)$ or $e_i(m)$ to explicitly identify their dimension given by m, which is different from $K + 1$.
- Finally, we will use the notation "'" appearing as superscript on a vector or a matrix to denote the transpose of a matrix.

3 The Steady-State Analysis

The steady-state analysis of the model described in Sect. 2 will be analyzed in this section. First we define, $N_1(t)$, $N_2(t)$, and $J(t)$, respectively, to be the number of Type 1 customers in the system, the number of Type 2 customers in the system, the status of the server at time t. Note that the status of the server can be either idle ($J(t) = 0$) or busy serving a Type 1 customer ($J(t) = 1$) or busy serving a Type 2 customer ($J(t) = 2$). The process $\{(N_2(t), N_1(t), J(t) : t \geq 0\}$ is a continuous-time Markov chain with state space given by

$$
\begin{aligned}
\Omega = &\{(0, 0, 0)\} \bigcup \{(0, i_1, 1) : 1 \leq i_1 \leq K + 1\} \\
&\bigcup \{(i_2, i_1, r) : 2 - r \leq i_1 \leq K + 2 - r, r = 1, 2, i_2 \geq 1\}.
\end{aligned} \tag{3}
$$

We now define the set of states along with their meanings as follows.

- $* = \{(0, 0, 0)\}$. This corresponds to the system being idle.

- $\underline{0} = \{(0, i_1, 1), 1 \leq i_1 \leq K + 1\}$. This corresponds to the case when there are i_1 Type 1 customers including the one in service and no Type 2 customers in the system.
- $\underline{i_2} = \{(i_2, i_1, r) : 2 - r \leq i_1 \leq K + 2 - r, r = 1, 2, i_2 \geq 1\}$. This set of states corresponds to the case when there are i_2 Type 2 customers, i_1 Type 1 customers in the system, and the server is busy serving a Type r customer. Note that when the server is busy with a Type 1 customer, the number of such customers can be between 1 and $K + 1$, whereas when the server is busy with a Type 2 customer, the number of Type 1 customers will be between 0 and K.

The infinitesimal generator of the Markov chain governing the system is of the form:

$$
Q = \begin{pmatrix}
-\lambda & \lambda_1 e_1' & \lambda_2 h' \\
\mu_1 e_1 & C_1 & C_0 \\
\mu_2 h & \tilde{B}_2 & B_1 & A_0 \\
& & B_2 & B_1 & A_0 \\
& & & \ddots & \ddots & \ddots \\
& & & & B_2 & B_1 & A_0 \\
& & & & & B_2 & E_1 & A_0 \\
& & & & & & E_{2,1} & E_2 & A_0 \\
& & & & & & & E_{3,2} & E_3 & A_0 \\
& & & & & & & & \ddots & \ddots & \ddots \\
& & & & & & & & & E_{K-1,K-2} & E_{K-1} & A_0 \\
& & & & & & & & & & E_{K,K-1} & A_1 & A_0 \\
& & & & & & & & & & & A_2 & A_1 & A_0 \\
& & & & & & & & & & & & A_2 & A_1 & A_0 \\
& & & & & & & & & & & & & \ddots & \ddots & \ddots
\end{pmatrix}, \quad (4)
$$

where

$$
h = e_{K+2}(2K + 2), \quad C_0 = \begin{pmatrix} \lambda_2 I & O \end{pmatrix}, \quad C_1 = F - \mu_1 I + \mu_1 \Phi_1, \quad \tilde{B}_2 = \mu_2 \begin{pmatrix} O \\ \Phi_1 \end{pmatrix},
$$

$$
B_1 = \begin{pmatrix} C_1 & \mu_1 \tilde{\Delta}_1 \\ O & F - \mu_2 I \end{pmatrix}, \quad B_2 = \mu_2 \begin{pmatrix} O & O \\ \Phi_1 & \tilde{\Delta}_1 \end{pmatrix},
$$

$$
(5)
$$

$$
E_i = \begin{pmatrix} F - \mu_1 I + \mu_1 \Phi_{i+1} & \mu_1 \tilde{\Delta}_{i+1} \\ O & F - \mu_2 I \end{pmatrix}, \quad E_{i+1,i} = \mu_2 \begin{pmatrix} O & O \\ \Phi_{i+1} & \tilde{\Delta}_{i+1} \end{pmatrix}, \quad 1 \leq i \leq K - 1,
$$

$$
(6)
$$

$$
A_1 = \begin{pmatrix} F - \mu_1 I & \mu_1 I \\ O & F - \mu_2 I \end{pmatrix}, \quad A_2 = \mu_2 \begin{pmatrix} O & O \\ O & I \end{pmatrix}, \quad A_0 = \lambda_2 I_{2K+2}. \quad (7)
$$

It should be pointed out that C_0 is of dimension $K + 1 \times (2K + 2)$; C_1 is a square matrix of dimension $K + 1$; $\tilde{B}_2$ is of dimension $(2K + 2) \times K + 1$; B_1, B_2, A_0, A_1, A_2, and E_i and $E_{i+1,i}$, $1 \leq i \leq K - 1$, are all square matrices of dimension $(2K + 2)$.

3.1 The Stability Condition

First note that the non-preemptive priority queueing with dynamic priority rule dictated by the threshold parameter, N, under study is governed by a Markov process whose generator [see Eq. (2)] has a modified quasi-birth-and-death (QBD) form. Further, the matrix, $A = A_0 + A_1 + A_2$, is upper triangular and hence is reducible. Thus, we can adopt Theorem 1.4.1 in [10] to our model and obtain the following theorem.

Theorem 1 *The queuing system under study is stable if and only if the following condition is satisfied.*

$$\lambda_2 < \mu_2. \tag{8}$$

Proof Adapting Theorem 1.4.1 in [10], we see that the system under study is stable if and only if

$$\frac{(A_0)_{2K+2,2K+2}}{(A_2)_{2K+2,2K+2}} = \frac{\lambda_2}{\mu_2} < 1.$$

Note: It should be pointed out that the stability condition for the classical two-customer non-preemptive priority queueing model (i.e., our current model without the threshold N) depends not only λ_2 and μ_2 but also on other parameters, namely λ_1, μ_1, and K. We will discuss this in more detail in Sect. 4.

3.2 The Steady-State Probability Vector

The steady-state probability vector, x, of Q satisfying

$$x Q = 0, x e = 1, \tag{9}$$

is partitioned into vectors of smaller dimensions as follows.

$$x = (x^*, u_0, x_1, x_2, \ldots), \quad x_i = (u_i, v_i), \ i \geq 1,$$
$$u_i = (u_{i,1}, u_{i,2}, \ldots, u_{i,K+1}), \ i \geq 0, \quad v_i = (v_{i,0}, v_{i,1}, \ldots, v_{i,K}), \ i \geq 1. \tag{10}$$

Under the stability condition given in (8), the steady-state probability vector x is obtained (see, e.g., [10]) as follows

$$-\lambda x^* + \mu_1 u_{0,1} + \mu_2 v_{1,0} = 0,$$

$$\lambda_1 x^* e_1' + u_0 C_1 + \mu_2 v_1 \Phi_1 = \mathbf{0},$$

$$\lambda_2 u_0 + u_1 C_1 + \mu_2 v_2 \tilde{\Delta}_1 = \mathbf{0},$$

$$\lambda_2 x^* e_1' + \mu_1 u_1 \tilde{\Delta}_1 + v_1 (F - \mu_2 I) + \mu_2 v_2 \tilde{\Delta}_1 = \mathbf{0},$$

$$\lambda_2 u_{i-1} + u_i C_1 + \mu_2 v_{i+1} \Phi_1 = \mathbf{0},$$

$$\lambda_2 v_{i-1} + \mu_1 u_i \tilde{\Delta}_1 + v_i (F - \mu_2 I) + \mu_2 v_{i+1} \tilde{\Delta}_1 = \mathbf{0}, \ 2 \leq i \leq N,$$

$$\lambda_2 u_{i-1} + u_i (F - \mu_1 I + \mu_1 \Phi_{i+1-N}) + \mu_2 v_{i+1} \Phi_{i+1-N} = \mathbf{0},$$

$$\lambda_2 v_{i-1} + \mu_1 u_i \tilde{\Delta}_{i+1-N} + v_i (F - \mu_2 I) + \mu_2 v_{i+1} \tilde{\Delta}_{i+1-N} = \mathbf{0}, \ N+1 \leq i \leq N+K-1,$$

$$x_{N+i} = x_{N+K-1} R^{i+1-K}, \ i \geq K, \tag{11}$$

where the matrix R is the minimal nonnegative solution to the matrix quadratic equation:

$$R^2 A_2 + R A_1 + A_0 = 0, \tag{12}$$

and with the normalizing condition

$$x^* + \sum_{i=0}^{N+K-2} u_i e + \sum_{i=1}^{N+K-2} v_i e + x_{N+K-1} (I - R)^{-1} e = 1. \tag{13}$$

The computation of the steady-state vector, x, can be carried out by exploiting the special structure of the coefficient matrices appearing in (11), and the details are omitted. Once the steady-state vector, x, is obtained, a number of key system performance measures can be obtained. For our focus in this paper, we will consider a few such measures. Two of them will be defined here along with their formulas, and the rest will be presented in appropriate places below. The mean number of Type i customers in the system for the threshold model, denoted by $\mu_{T_i}^{(T)}$, $i = 1, 2$, is given by

$$\mu_{T_1}^{(T)} = \sum_{j=1}^{K+1} j \sum_{i=0}^{\infty} u_{i,j} \text{ and}$$

$$\mu_{T_2}^{(T)} = (N + K - 2) x_{N+K-1} (I - R)^{-1} e + x_{N+K-1} (I - R)^{-2} e + \sum_{i=1}^{N+K-2} i x_i e.$$

3.3 Rate Matrix (R)

Due to special structure of the matrices A_0, A_1, and A_2, the rate matrix, R, also has a special structure of being upper triangular, which can be exploited in its computation.

While logarithmic reduction [9] method for computing R is more efficient, in order to exploit the special structure, especially when K is large, one may want to consider other well-known methods such as (block) Gauss–Seidel iterative method. Since these are well-known and well publicized in the literature, we refer the reader to references such as [9, 13] for details.

3.4 Busy Probabilities at Arbitrary Time

The following theorem displays results, which are intuitively clear, are useful in serving as accuracy checks in numerical computation.

Theorem 2 *The probabilities that the server is busy with Type 1 and Type 2 customers are given by*

$$P_{Busy_1}^{(T)} = \frac{\lambda_1(1 - P_{loss}^{(T)})}{\mu_1}, \tag{14}$$

$$P_{Busy_2}^{(T)} = \frac{\lambda_2}{\mu_2}, \tag{15}$$

where $P_{loss}^{(T)}$ is the probability that a Type 1 customer is lost due to the buffer being full and is given by

$$P_{loss}^{(T)} = \sum_{i=0}^{\infty} [u_{i,K+1} + v_{i+1,K}]. \tag{16}$$

Proof First note that

$$
\begin{aligned}
P_{Busy_1}^{(T)} &= \sum_{i=0}^{\infty} \mathbf{u}_i \mathbf{e}, \\
P_{Busy_2}^{(T)} &= \sum_{i=1}^{\infty} \mathbf{v}_i \mathbf{e}.
\end{aligned}
\tag{17}
$$

From the steady-state equations given in (11), one can easily obtain the following equations.

$$(\lambda_1 + \mu_1) \sum_{i=0}^{\infty} u_{i,1} = \lambda_1 x^* + \mu_1 \sum_{i=0}^{\infty} u_{i,2} + \mu_2 \sum_{i=0}^{N+1} v_{i,1}, \tag{18}$$

$$(\lambda_1 + \mu_1) \sum_{i=0}^{\infty} u_{i,j} = \lambda_1 \sum_{i=0}^{\infty} u_{i,j-1} + \mu_1 \sum_{i=0}^{N+j-1} u_{i,j+1} + \mu_2 \sum_{i=0}^{N+j} v_{i,j}, \ 2 \leq j \leq K, \tag{19}$$

$$\mu_1 \sum_{i=0}^{\infty} u_{i,K+1} = \lambda_1 \sum_{i=0}^{\infty} u_{i,K}, \tag{20}$$

$$\lambda_1 \left(x^* + \sum_{i=1}^{\infty} v_{i,0} \right) = \mu_1 \sum_{i=0}^{\infty} u_{i,1}, \tag{21}$$

$$(\lambda_1 + \mu_2) \sum_{i=1}^{\infty} v_{i,j} = \lambda_1 \sum_{i=1}^{\infty} v_{i,j-1} + \mu_1 \sum_{i=N+j}^{\infty} u_{i,j+1} + \mu_2 \sum_{i=N+j}^{\infty} v_{i,j}, \ 1 \leq j \leq K-1, \tag{22}$$

$$\mu_2 \sum_{i=1}^{\infty} v_{i,K} = \lambda_1 \sum_{i=1}^{\infty} v_{i,K-1} + \mu_1 \sum_{i=N+K}^{\infty} u_{i,K+1} + \mu_2 \sum_{i=N+K+1}^{\infty} v_{i,K}. \tag{23}$$

From Eqs. (18)–(23), through some standard algebraic manipulations, it can easily be verified that

$$\lambda_1 \sum_{i=0}^{\infty} (u_{i,j} + v_{i,j}) = \mu_1 \sum_{i=0}^{\infty} u_{i,j+1}, \ 1 \leq j \leq K-1, \tag{24}$$

$$\lambda_2 (x^* + \boldsymbol{u}_0 \boldsymbol{e}) = \mu_2 \boldsymbol{v}_1 \boldsymbol{e}, \tag{25}$$

$$\lambda_2 (\boldsymbol{u}_i \boldsymbol{e} + \boldsymbol{v}_i \boldsymbol{e}) = \mu_2 \boldsymbol{v}_{i+1} \boldsymbol{e}, \ i \geq 1. \tag{26}$$

The stated result in (14) follows by adding the Eqs. (20), (21), and (24). Similarly, the stated result in (15) is obtained by adding the Eqs. (25) and (26).

3.5   Steady-State Probability at Departure Epoch

In this section, we will derive an expression for the steady-state probability vector at departure epochs. It should be pointed out that due to finite buffer for Type 1 customers, this probability will differ from that of at an arbitrary time.

Suppose that y denotes the steady-state probability vector at departure epoch and that y is partitioned as $y = (y_{0,0}, y_{0,1}, \ldots, y_{0,K}, y_{1,0}, y_{1,1}, \ldots, y_{1,K}, \ldots)$ such that $y_{i,j}$ gives the steady-state probability that at a departure epoch there are j, $0 \leq j \leq K$, Type 1 customers and i, $i \geq 0$, Type 2 customers in the system. The following theorem gives an expression for $y_{i,j}$.

Theorem 3 *The steady-state probability vector y is such that its components are given by*

$$y_{i,j} = \frac{1}{\lambda_2 + \lambda_1 (1 - P_{loss})} \Big[\mu_1 u_{i,j+1} + \mu_2 v_{i+1,j} \Big], \ 0 \leq j \leq K, \ i \geq 0. \tag{27}$$

Proof From the definition of the steady-state probabilities, it is easy to see that

$$y_{i,j} = c\left[\mu_1 u_{i,j+1} + \mu_2 v_{i+1,j}\right], \quad 0 \leq j \leq K, i \geq 0, \tag{28}$$

where c is the normalizing constant. The normalizing constant is obtained as follows.

$$\sum_{i=0}^{\infty}\sum_{j=0}^{K} y_{i,j} = 1 \quad \Rightarrow \quad c\sum_{i=0}^{\infty}\sum_{j=0}^{K}\left[\mu_1 u_{i,j+1} + \mu_2 v_{i+1,j}\right] = 1$$
$$\Rightarrow \quad c\left[\sum_{i=0}^{\infty}\left(u_i e + v_{i+1} e\right)\right] = 1 \quad \Rightarrow \quad c[\lambda_1(1 - P_{loss}) + \lambda_2] = 1,$$

where the last statement follows from Eqs. (14) and (15). Hence, the stated result follows.

In the sequel, we need the following system performance measure defined in terms of the conditional probability at departure epoch to see the qualitative impact of the threshold parameter N. The conditional probability, $P_{Busy_2}^{(T_1>0)}$, that there will be at least one Type 1 customer in the system given that a departure will result in the server offering a service to a Type 2 customer is given by

$$P_{Busy_2}^{(T_1>0)} = \frac{\sum_{j=1}^{K}\sum_{i=N+j}^{\infty} y_{i,j}}{\sum_{i=1}^{\infty} y_{i,0} + \sum_{j=1}^{K}\sum_{i=N+j}^{\infty} y_{i,j}}, \tag{29}$$

which can be obtained in a more computable form using the steady-state probability vectors, u and v. Toward this end, we define

$$(a, b) = x_{N+K-1}(I - R)^{-1}. \tag{30}$$

The simplified and computationally implementable expression for $P_{Busy_2}^{(T_1>0)}$ is given by

$$P_{Busy_2}^{(T_1>0)} = \frac{\mu_1}{\lambda_2 + \lambda_1(1 - P_{loss})}\left[\sum_{j=1}^{K-2}\sum_{i=N+j}^{N+K-2} u_{i,j+1} + \sum_{j=2}^{K+1} a_j - u_{N+K-1,K+1}\right]$$
$$+ \frac{\mu_2}{\lambda_2 + \lambda_1(1 - P_{loss})}\left[\sum_{j=1}^{K-2}\sum_{i=N+j}^{N+K-3} v_{i+1,j} + \sum_{j=2}^{K+1} b_j - \sum_{j=K-1}^{K} v_{N+K-1,j} - v_{N+K,K}\right].$$
$$\tag{31}$$

Note that the above conditional probability is zero in the classical two-customer non-preemptive queueing model and hence will indicate the improvement in the fraction of time the server is paying attention to serving lower-priority customers in our threshold non-preemptive model.

4 Classical Two-Customer Non-preemptive Priority Queueing Model

In this section, we will briefly provide the needed details on the classical non-preemptive priority queueing model with two types of customers (with only higher-priority customers having a finite waiting) so as to compare that model with the model under study here. This is mainly to see the impact of the threshold N on the QoS with respect to Type 2 customers. Also, it is worth pointing out that if we let N approach infinity, our model will reduce to the corresponding classical non-preemptive priority model.

In this case, the state space for this model is same as for the model with threshold N (see Eq. 3), and the generator, $\tilde{Q}$, for the corresponding classical non-preemptive priority queueing model is of the form

$$
\tilde{Q} = \begin{pmatrix}
-\lambda & \lambda_1 e_1' & \lambda_2 h' & & \\
\mu_1 e_1 & C_1 & C_0 & & \\
\mu_2 h & \tilde{B}_2 & B_1 & A_0 & \\
 & & B_2 & B_1 & A_0 \\
 & & & B_2 & B_1 & A_0 \\
 & & & & \ddots & \ddots & \ddots
\end{pmatrix}, \tag{32}
$$

where the entries appearing in (32) are as given in (5)–(7).

4.1 The Steady-State Analysis—Classical Non-preemptive Priority Queueing Model

In this section, we will briefly outline the steady-state analysis starting with the stability condition. In order to derive the stability condition for the classical non-preemptive priority queueing model, we first need the steady-state probability vector of $B = A_0 + B_1 + B_2$, where A_0 is as given in (5) and B_1 and B_2 are as given (7). Toward this end, let $\pi = (\pi_1, \pi_2)$ be the steady-state probability vector of the generator B. That is, π satisfies

$$
\pi B = \mathbf{0}, \ \pi e = 1. \tag{33}
$$

The stability condition for the classical non-preemptive queueing model is given in the following theorem. Before that, we further partition $\pi_r, r = 1, 2$ as $\pi_r = (\pi_{r,1}, \pi_{r,2}, \ldots, \pi_{r,K+1}), r = 1, 2$.

Theorem 4 *The classical two-customer non-preemptive priority queuing system (with only higher-priority customers having a finite waiting room) is stable if and only if the following condition is satisfied.*

$$
\lambda_2 < \mu_2 d, \tag{34}
$$

where d is given by

$$d = \pi_2 e = \left[1 + \frac{\lambda_1}{\mu_1}\left\{1 - \left(\frac{\lambda_1}{\lambda_1 + \mu_2}\right)^K + \sum_{k=1}^{K}\left(\frac{\lambda_1}{\mu_1}\right)^k\left(1 - \left(\frac{\lambda_1}{\lambda_1 + \mu_2}\right)^{K+1-k}\right)\right\}\right]^{-1}. \tag{35}$$

Proof First note that due to the special structure of the matrices A_0, B_1, and B_2 (see Eq. (5)), the steady-state equation given in (34) can be rewritten as

$$\begin{aligned}
\pi_1[F - \mu_1 I + \mu_1 \Phi_1] + \mu_2 \pi_2 \Phi_1 &= \mathbf{0}, \\
\mu_1 \pi_1 \tilde{\Delta}_1 + \pi_2[F - \mu_1 2 + \mu_2 \tilde{\Delta}_1] &= \mathbf{0}, \\
\pi_1 e + \pi_2 e &= 1.
\end{aligned} \tag{36}$$

Noting that (see, e.g., [10]) the necessary and sufficient condition for the classical queue under study in this section to be stable is $\pi A_0 e < \pi B_2 e$, which reduces to

$$\lambda_2 < \mu_2 \pi_2 e. \tag{37}$$

It can easily be verified from (36) that

$$\pi_{1,1} = \frac{\lambda_1}{\mu_1}\pi_{2,1}, \tag{38}$$

$$\pi_{1,j} = \frac{\lambda_1}{\mu_1}\left[\pi_{1,j-1} + \left(\frac{\lambda_1}{\lambda_1 + \mu_2}\right)^{j-1}\right]\pi_{2,1}, \ 2 \leq j \leq K, \tag{39}$$

$$\pi_{1,K+1} = \left(\frac{\lambda_1}{\mu_1}\right)^2\left[\pi_{1,K-1} + \left(\frac{\lambda_1}{\lambda_1 + \mu_2}\right)^{K-1}\right]\pi_{2,1}, \tag{40}$$

$$\pi_{2,j} = \left(\frac{\lambda_1}{\lambda_1 + \mu_2}\right)^{j-1}\pi_{2,1}, \ 1 \leq j \leq K, \tag{41}$$

$$\pi_{2,K+1} = \frac{\lambda_1}{\mu_2}\left(\frac{\lambda_1}{\lambda_1 + \mu_2}\right)^{K-1}\pi_{2,1}. \tag{42}$$

Now adding Eqs. (38) and (39) (over j, $1 \leq j \leq K$), and (40), we get

$$\pi_1 e = \frac{\lambda_1}{\mu_1}\frac{\lambda_1 + \mu_2}{\mu_2}\left[1 - \left(\frac{\lambda_1}{\lambda_1 + \mu_2}\right)^K + \sum_{k=1}^{K}\left(\frac{\lambda_1}{\mu_1}\right)^k\left(1 - \left(\frac{\lambda_1}{\lambda_1 + \mu_2}\right)^{K+1-k}\right)\right]\pi_{2,1}. \tag{43}$$

Adding Eq. (42) to the one obtained by summing over j, $1 \leq j \leq K$ of (41), we get

$$\mu_2 \pi_2 e = (\lambda_1 + \mu_2)\pi_{2,1}. \tag{44}$$

Now the stated result follows immediately from (43) to (44) along with the normalizing equation given in (36).

Note: (1) Note that one can simplify further the expression for d given in (35) by considering three cases: (a) $\lambda_1 = \mu_1$; (b) $\mu_1 = \lambda_1 + \mu_2$; and (c) $\mu_1 \neq \lambda_1 + \mu_2$. The details are omitted.

(2) While the steady-state vector, π, is explicitly given, it is probably more efficient to compute recursively with $\pi_{2,1}$ computed from the normalizing condition.

Under the stability condition given in (34), the steady-state probability vector, say, $\tilde{x}$, of $\tilde{Q}$, is of modified matrix-geometric and is obtained as follows. Once again, we will partition the steady-state vector here like we did for the threshold model. That is, we partition $\tilde{x}$ as

$$\tilde{x} = (\tilde{x}^*, \tilde{u}_0, \tilde{x}_1, \tilde{x}_2, \ldots), \quad \tilde{x}_i = (\tilde{u}_i, \tilde{v}_i), \ i \geq 1,$$
$$\tilde{u}_i = (\tilde{u}_{i,1}, \tilde{u}_{i,2}, \ldots, \tilde{u}_{i,K+1}), \ i \geq 0, \quad \tilde{v}_i = (\tilde{v}_{i,0}, \tilde{v}_{i,1}, \ldots, \tilde{v}_{i,K}), \ i \geq 1.$$

The steady-state probability vector $\tilde{x}$ is obtained by solving the following system of equations.

$$\begin{aligned}
-\lambda\tilde{x}^* + \mu_1\tilde{u}_{0,1} + \mu_2\tilde{v}_{1,0} &= 0, \\
\lambda_1\tilde{x}^*e_1' + \tilde{u}_0 C_1 + \tilde{x}_1\tilde{B}_2 &= \mathbf{0}, \\
\lambda_2\tilde{x}^*e_{K+2}'(2K+2) + \tilde{u}_0 C_0 + \tilde{x}_1[B_1 + \tilde{R}B_2] &= \mathbf{0}, \\
\tilde{x}_i = \tilde{x}_1\tilde{R}^{i-1}, \ i &\geq 1,
\end{aligned} \tag{45}$$

where the matrix $\tilde{R}$ is the minimal nonnegative solution to the matrix quadratic equation:

$$\tilde{R}^2 B_2 + \tilde{R}B_1 + A_0 = 0, \tag{46}$$

and with the normalizing condition

$$\tilde{x}^* + \tilde{u}_0 e + \tilde{x}_1(I - \tilde{R})^{-1}e = 1. \tag{47}$$

The computation of $\tilde{x}$ is done similar to x by exploiting the special structure of the coefficient matrices appearing in (45), and the details are omitted. Like earlier, we display the mean number of Type i customers in the system for the classical model, denoted by $\mu_{T_i}^{(C)}, i = 1, 2$, which is given by

$$\mu_{T_1}^{(C)} = \sum_{j=1}^{K+1} j \sum_{i=0}^{\infty} \tilde{u}_{i,j} \text{ and } \mu_{T_2}^{(C)} = \tilde{x}_1(I - \tilde{R})^{-2}e.$$

The following theorem is very similar to Theorem 2 in that Theorem 5 gives expressions for busy probabilities for the classical non-preemptive priority queueing model.

Theorem 5 *The probabilities that the server is busy with Type 1 and Type 2 customers in the case of classical two-customer non-preemptive priority queueing model are given by*

$$P_{Busy_1}^{(C)} = \frac{\lambda_1(1 - P_{loss}^{(C)})}{\mu_1}, \tag{48}$$

$$P_{Busy_2}^{(C)} = \frac{\lambda_2}{\mu_2}, \tag{49}$$

where

$$P_{loss}^{(C)} = \sum_{i=0}^{\infty}[\tilde{u}_{i,K+1} + \tilde{v}_{i+1,K}]. \tag{50}$$

Proof The proof is very similar to Theorem 2 once the following equations are verified from (45).

$$(\lambda_1 + \mu_1)\sum_{i=0}^{\infty} \tilde{u}_{i,1} = \lambda_1\tilde{x}^* + \mu_1\sum_{i=0}^{\infty} \tilde{u}_{i,2} + \mu_2\sum_{i=0}^{\infty} \tilde{v}_{i,1},$$

$$(\lambda_1 + \mu_1)\sum_{i=0}^{\infty} \tilde{u}_{i,j} = \lambda_1\sum_{i=0}^{\infty} \tilde{u}_{i,j-1} + \mu_1\sum_{i=0}^{\infty} \tilde{u}_{i,j+1} + \mu_2\sum_{i=0}^{\infty} \tilde{v}_{i,j}, \ 2 \le j \le K,$$

$$\mu_1\sum_{i=0}^{\infty} \tilde{u}_{i,K+1} = \lambda_1\sum_{i=0}^{\infty} \tilde{u}_{i,K},$$

$$\lambda_1\left(\tilde{x}^* + \sum_{i=1}^{\infty} \tilde{v}_{i,0}\right) = \mu_1\sum_{i=0}^{\infty} \tilde{u}_{i,1},$$

$$(\lambda_1 + \mu_2)\sum_{i=1}^{\infty} \tilde{v}_{i,j} = \lambda_1\sum_{i=1}^{\infty} \tilde{v}_{i,j-1} \ 1 \le j \le K - 1,$$

$$\mu_2\sum_{i=1}^{\infty} \tilde{v}_{i,K} = \lambda_1\sum_{i=1}^{\infty} \tilde{v}_{i,K-1}. \tag{51}$$

5 Comparison of the Two Models

In this section we, will compare the non-preemptive priority queueing model with the threshold and its corresponding classical model. Toward this end, we first define the traffic intensities of the two models. Let ρ_T and ρ_C denote, respectively, the traffic intensity of the two-customer non-preemptive priority queueing model with threshold and without threshold (i.e., classical). That is,

$$\rho_T = \frac{\lambda_2}{\mu_2}, \quad \rho_C = \frac{\lambda_2}{\mu_2\pi_2 e}, \tag{52}$$

where d is as given in (35). Also, note that the above equation implies

$$\rho_T = \rho_C\pi_2 e. \tag{53}$$

1. Looking at the stability condition (see Theorems 1 and 4 which give expressions for the two models), it is clear that the threshold model can accommodate a larger rate of Type 2 arrivals (assuming all other parameters are fixed) when compared to the classical model. Only when $\lambda_1 \to 0$, which corresponds to essentially not having any higher-priority customers in the model or when $\mu_1 \to \infty$, which assures that Type 1 customers are almost immediately served, we see that the classical non-preemptive priority model will approach to the same level of handling a larger number of Type 2 customers like the threshold non-preemptive priority queueing model without violating the stability condition.

2. When both λ_1 and μ_1 are finite and positive, no matter how fast Type 2 customers are served (i.e., how large μ_2 is), the threshold model can handle more Type 2 customers on the average compared to the classical one. This is due to the fact that $\pi_2 e$ will always be positive.

3. As long as Type 1 customers are allowed to enter into the system and have a finite service time, the threshold model can always accept a larger rate of Type 2 customers (within the allowable level satisfying the stability condition) as compared to the corresponding classical one.

4. Under the assumptions that $\lambda_1 < \mu_1$ (in addition to $\lambda_2 < \mu_2$ which is needed for the stability of the threshold model under study here), it can easily be verified from (35) that $\pi_2 e \to \left(1 - \frac{\lambda_1}{\mu_1}\right)$ as $K \to \infty$. Note that this result is intuitively obvious since in the case when Type 1 customers are admitted without any limit (i.e., they have infinite buffer space like Type 2 customers), the system's stability requires $\lambda_1 < \mu_1$ in addition to $\lambda_2 < \mu_2$.

5. The main purpose of introducing the threshold parameter, N, into the classical non-preemptive priority queueing model is to reduce the average number of Type 2 customers waiting to be processed in the presence of Type 1 customers. We will explore this numerically (due to the complexity of the expressions for this measure for the two models) in the Sect. 6.

6. As $N \to \infty$, the threshold model will approach the classical model. It would be of interest to see an optimal value, say, N^* of N, such that the loss probabilities under the two models are close enough to each other. Again, we will explore this numerically in Sect. 6. This is mainly due to the complexity of the expressions for the loss probability.

6   Numerical Examples

In this section, we will present two representative examples to illustrate the impact of the new type of non-preemptive priority rule. In order to compare the classical non-preemptive priority queueing model to the one studied in this paper in a meaningful way, we need to set the parameters of the model properly taking into account the stability conditions for the two models are different. From Theorems 1 to 4, we note that the stability condition for the classical model depends on λ_1, μ_1, μ_2, and K, whereas for the threshold model with N being positive and finite, it depends only λ_2 and μ_2. Furthermore, as is to be expected, the threshold model can handle a larger load (either through a larger λ_1 or a larger λ_2 or a combination of both) as compared to the classical model. This will be explored further in the examples below.

Example 1 The goal of this example is to find what should be the minimum value of the threshold N, say N^*, such that $|P_{loss}^{(T)} - P_{loss}^{(C)}| < 10^{-3}$ under various scenarios. Toward this end, we fix $\lambda_1 = 1$, $\mu_1 = \mu_2 = 1.1$, vary $K = 1, 2, 3, 5, 10, 15, 20, 50$, and choose λ_2 such that we get a specific value for ρ_C, which is varied over 0.1 through 0.95. Note that in order to properly carry out the comparison, we need to use the same λ_2 value in the threshold model. Thus, ρ_T will be much less than ρ_C (see 53) for the same set of values for the other parameters. Note that in this example, $\lambda_2 = \rho_T$ since we fixed $\mu_2 = 1.0$. In Table 1, we display the values of N^*, and in Tables 2 and 3, respectively, we display the values of $\frac{\mu_{T1}^{(T)}}{\mu_{T1}^{(C)}}$, and $\frac{\mu_{T1}^{(T)}}{\mu_{T1}^{(C)}}$. Note that in obtaining the mean values for Type 1 and Type 2 customers for the threshold model, we set $N = N^*$ so that the comparison of classical and the threshold models makes sense.

From Table 1, we notice that, as expected, for smaller values of ρ_C, one needs a smaller N for all K in the range considered to get the loss probabilities under both models to differ by no more than 10^{-3}. However, as ρ_C becomes larger, one needs a larger N and the value of N appears to increase with K.

A look at the values in Tables 2 and 3 indicates a significant reduction in the mean number of Type 2 customers present in the system, while at the same time, the mean number of Type 1 customers present in the system does not increase appreciably. While the increase is insignificant when K is upto 20, we see relatively significant increase when K is 50. It should be pointed out that one needs to keep in mind the differing values of N when making specific interpretations. However, general observations like the one we made here should be adequate to bring out the qualitative aspects of the threshold model.

Table 1 Optimum N^* values under various scenarios

ρ_C	$K = 1$	$K = 2$	$K = 3$	$K = 5$	$K = 10$	$K = 15$	$K = 20$	$K = 50$
0.1	2	1	1	1	1	1	1	1
0.2	3	3	3	2	2	1	1	1
0.3	4	4	4	4	4	3	2	1
0.4	5	6	6	6	7	7	7	1
0.5	7	8	9	10	12	13	13	1
0.6	10	11	12	14	18	21	23	6
0.7	14	16	18	21	29	34	38	26
0.8	22	25	28	34	47	58	66	64
0.9	41	47	54	65	91	112	130	155
0.95	71	81	92	111	153	188	218	280

Table 2 Ratios of $\mu_{T1}^{(T)}$ over $\mu_{T1}^{(C)}$, under various scenarios at N^*

ρ_C	$K = 1$	$K = 2$	$K = 3$	$K = 5$	$K = 10$	$K = 15$	$K = 20$	$K = 50$
0.1	1.000	1.001	1.002	1.003	1.004	1.005	1.006	1.009
0.2	1.000	1.000	1.001	1.004	1.009	1.021	1.025	1.038
0.3	1.000	1.001	1.001	1.004	1.010	1.024	1.041	1.089
0.4	1.000	1.001	1.001	1.004	1.010	1.021	1.033	1.166
0.5	1.000	1.001	1.001	1.003	1.009	1.018	1.032	1.275
0.6	1.000	1.001	1.001	1.003	1.009	1.017	1.028	1.319
0.7	1.000	1.001	1.001	1.003	1.008	1.016	1.026	1.249
0.8	1.000	1.001	1.001	1.003	1.008	1.014	1.023	1.196
0.9	1.000	1.001	1.001	1.003	1.007	1.013	1.021	1.155
0.95	1.000	1.001	1.001	1.002	1.007	1.013	1.020	1.138

Table 3 Ratios of $\mu_{T2}^{(T)}$ over $\mu_{T2}^{(C)}$, under various scenarios at N^*

ρ_C	$K = 1$	$K = 2$	$K = 3$	$K = 5$	$K = 10$	$K = 15$	$K = 20$	$K = 50$
0.1	0.988	0.911	0.905	0.901	0.902	0.903	0.905	0.909
0.2	0.986	0.978	0.970	0.909	0.888	0.809	0.814	0.831
0.3	0.981	0.968	0.957	0.936	0.898	0.836	0.783	0.761
0.4	0.971	0.972	0.958	0.932	0.904	0.871	0.851	0.695
0.5	0.970	0.964	0.961	0.945	0.915	0.889	0.861	0.626
0.6	0.969	0.956	0.946	0.932	0.904	0.885	0.871	0.647
0.7	0.958	0.947	0.939	0.920	0.899	0.875	0.861	0.740
0.8	0.946	0.930	0.918	0.900	0.870	0.853	0.839	0.768
0.9	0.906	0.885	0.873	0.844	0.804	0.778	0.765	0.721
0.95	0.847	0.817	0.798	0.761	0.705	0.674	0.657	0.621

In summary, this example illustrates the significant advantage in using the type of threshold introduced here to increase the QoS for lower-priority customers without affecting the higher-priority customers. This is an important observation since priority queues occur naturally in practice, and with the classical models, the lower-priority customers get poor QoS.

Example 2 The purpose of this example is to see the impact of N under various scenarios. That is, we look at the non-preemptive priority queueing model with threshold under study in this paper and look at the role played by the parameter N. Toward this end, we fix $\lambda_1 = 1$, $\mu_1 = 1.1$, $\mu_2 = 1$, vary $K = 1, 2, 5, 10, 15, 20$, and choose λ_2 such that we get a specific value for ρ_T, which is taken to be one of four values $\rho_T = 0.1, 0.5, 0.9, 0.95$. Since we fixed $\mu_2 = 1.0$, it is clear (see Eq. (53)) that $\lambda_2 = \rho_T$ in this example.

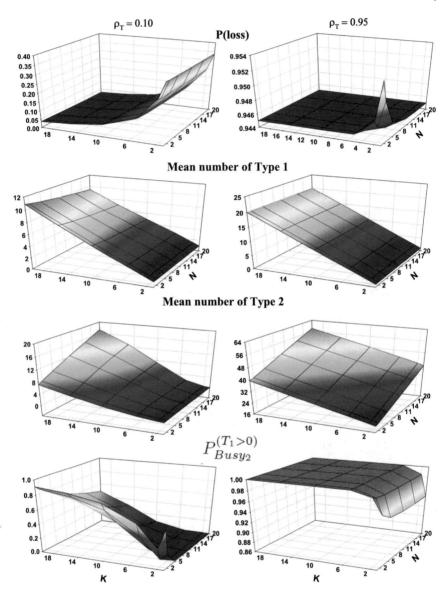

Fig. 1 Selected measures for threshold non-preemptive priority model under various scenarios

In Fig. 1, we display the graphs of four measures, P_{loss}, $\mu_{T1}^{(T)}$, $\mu_{T2}^{(T)}$, and $P_{Busy_2}^{(T_1>0)}$, for the threshold model, for selected values of ρ_T. A brief look at this figure reveals the following key observations.

- For fixed N and for low traffic intensity, we notice that P_{loss} appears to decrease with increasing K. However, in the case of higher traffic intensity, we see such a behavior only for small N. This behavior is as is to be expected since a higher traffic intensity will result in more Type 2 customers arriving to the system resulting in them getting services more frequently. This results in Type 1 customers getting lost more often. Also, notice that in the case of higher traffic intensity, the range for the loss probability is much narrower (in the current case, it varies from 0.945 to 0.952) indicating that one can choose N to be small so as to help increase the quality of service for Type 2 customers.

- For fixed N, we see that $\mu_{T1}^{(T)}$ appears to increase significantly as K increases; however, for fixed K, this measure does not appear to increase significantly as N is increased. This is the case for low as well as high traffic intensity.
- For fixed N we see that $\mu_{T2}^{(T)}$ appears to increase significantly as K increases; similarly, for fixed K this measure appears to increase significantly as N is increased. This is the case for low as well as high traffic intensity.
- With respect to the measure, $P_{Busy_2}^{(T_1>0)}$, we see some interesting observations. These are as follows. First, in the high traffic intensity region, the significant role of K is seen initially as this measure increases and then attains its maximum value. When the traffic intensity is high, the number of Type 2 customers arrives at a faster rate (as μ_2 is fixed and we vary λ_2 to arrive at a specific value for ρ_T). Thus, it is not surprising to see the measure under discussion to be insensitive to N. Secondly, in the low to moderate (the figure contains only for low traffic intensity value due to limiting the number of figures) traffic intensity when N is small, the measure appears to decrease initially and then increase as K is increased. This is somewhat counterintuitive.

7 Concluding Remarks

In this paper, we considered a non-preemptive priority queueing system with two types of customers and introduced a threshold to attend to serving lower-priority customers in the presence of higher priority to increase the quality of service for lower-priority customers. By comparing the current model to the classical two-customer non-preemptive priority queueing model, we showed a marked improvement in the quality of service with the introduction of the new type of threshold parameter. Assuming the buffer size to be finite for Type 1 customers and infinite for Type 2 customers, we studied the model as highly structured QBD process. The model under study can be generalized in a number of ways. For example, we can model the arrivals to follow a versatile point process, namely Markovian arrival process, the service times to be of phase type, and also consider a multi-server system. The results of these and other models will be presented elsewhere.

References

1. Adiri, I., Domb, I.: A single server queueing system working under mixed priority disciplines. Oper. Res. **30**, 97–115 (1982)
2. Avi-Itzhak, B., Brosh, I., Naor, P.: On discretionary priority queueing. Z. Angew. Math. Mech. **6**, 235–242 (1964)
3. Cho, Y.Z., Un, C.K.: Analysis of the M/G/1 queue under acombined preemptive/nonpreemptive priority discipline. IEEE Trans. Commun. **41**, 132–141 (1993)
4. Conway, R.W., Maxwell, W., Miller, L.: Theory of Scheduling. Addison-Wesley, Reading, MA (1967)
5. Drekic, S., Stanford, D.A.: Threshold-based interventions to optimize performance in preemptive priority queues. Queueing Syst.**35**, 289–315 (2000)
6. Drekic, S., Stanford, D.A.: Reducing delay in preemptive repeat priority queues. Oper. Res. **49**, 145–156 (2000)
7. Jaiswal, N.K.: Priority queues. Acadameic Press, USA (1968)
8. Kim, K.: (N, n)-preemptive priority queue. Perform. Eval. **68**, 575–585 (2011)
9. Latouche, G., Ramaswami, V.: Introduction to Matrix Analytic Methods in Stochastic Modeling. SIAM (1999)
10. Neuts, M.F.: Matrix-Geometric Solutions in Stochastic Models: An Algorithmic Approach. The Johns Hopkins University Press, Baltimore, MD. [1994 version is Dover Edition] (1981)
11. Neuts, M.F.: Algorithmic Probability: A Collection of Problems. Chapman and Hall, NY (1995)
12. Paterok, M., Ettl, A.: Sojourn time and waiting time distributions for M/G/1 queues with preemption-distance priorities. Oper. Res. **42**, 1146–1161 (1994)
13. Stewart, W.J.: Introduction to the Numerical Solution of Markov Chains. Princeton University Press, Princeton, NJ (1994)
14. Takagi, H.: Analysis of Polling Systems. MIT, USA (1986)
15. Takagi, H.: Queueing Analysis 1: A Foundation of Performance Evaluation: Vacation and Priority Systems. North-Holland, Amsterdam (1991)
16. Takagi, H., Kodera, Y.: Analysis of preemptive loss priority queues with preemption distance. Queueing Syst. **22**, 367–381 (1996)

Chapter 4
$\mathcal{I}_\theta$-Statistical Convergence of Weight g in Topological Groups

Ekrem Savas

Abstract In this paper, we introduce and study the concept of $\mathcal{I}$-lacunary statistical convergence of weight $g : [0, \infty) \to [0, \infty)$ where $g(x_n) \to \infty$ for any sequence (x_n) in $[0, \infty)$ with $x_n \to \infty$ in topological groups, and finally, we investigate some inclusion relations theorems related to $\mathcal{I}$-lacunary statistical convergence.

Keywords Lacunary sequence · Statistical convergence of weight g Topological groups

1 Introduction

Note that the statistical convergence of a sequence was introduced by Fast [8] and Schoenberg [23]. Later, the concept of statistical convergence has been discussed by Fridy [9], Šalát [14]. More details on statistical convergence and on applications of this concept can be found in Di Maio and Kočinac [13], Das and Savas [5], and Savas [21, 22].

The notion of statistical convergence is related to the density of subsets of the set $\mathbb{N}$ of natural numbers. The density of subset E of $\mathbb{N}$ is defined by

$$\delta(E) = lim_n \frac{1}{n} \sum_{k=1}^{n} \chi_E(k)$$

provided the limit exists, where χ_E is the characteristic functions of E. It is obvious that any finite subset of $\mathbb{N}$ has zero natural density and $\delta(E)^c = 1 - \delta(E)$.

A sequence $x = (x_j)$ is said to be statistically convergent to ξ if for arbitrary $\varepsilon > 0$, the set $E(\varepsilon) = \{n \in \mathbb{N} : |x_j - \xi| \geq \varepsilon\}$ has natural density zero (see [9]). In this case, we write $st - \lim_j x_j = \xi$ and we denote the set of all statistical convergent sequences by S.

E. Savas (✉)
Department of Mathematics, Usak University, Usak, Turkey
e-mail: ekremsavas@yahoo.com

© Springer Nature Singapore Pte Ltd. 2018
D. Ghosh et al. (eds.), *Mathematics and Computing*, Springer Proceedings in Mathematics & Statistics 253, https://doi.org/10.1007/978-981-13-2095-8_4

By a lacunary sequence, we mean an increasing sequence $\theta = (k_p)$ of positive integers such that $k_0 = 0$ and $h_p : k_p - k_{p-1} \to \infty$ as $p \to \infty$. Throughout this paper, the intervals determined by θ will be denoted by $I_p = (k_{p-1}, k_p]$, and the ratio $(k_p)(k_{p-1})^{-1}$ will be abbreviated by q_p.

Also in [10], a new type of convergence called lacunary statistical convergence was introduced as follows: A sequence (x_j) of real numbers is said to be lacunary statistically convergent to ξ (or, S_θ-convergent to ξ) if for any $\varepsilon > 0$,

$$\lim_{p \to \infty} \frac{1}{h_p} |\{j \in I_p : |x_j - \xi| \geq \varepsilon\}| = 0$$

Gadjiev and Orhan (see [11]) has given the order of statistical convergence of a sequence, and also, Colak [4] studied the statistical convergence of order α and strongly p-Cesàro summability of order α.

The (relatively more general) concept of $\mathcal{I}$-convergence was introduced by Kostyrko et al. [12] in a metric space as a generalized form of the concept of statistical convergence, and it is based upon the notion of an ideal of the subset of the set $\mathbb{N}$ of positive integers.

More investigations and more applications of ideals can be found in [6, 7, 15–20].

Recently in [21], we introduce the concepts of $\mathcal{I}$-statistical convergence and $\mathcal{I}$-lacunary statistical convergence in topological groups. Also, Savas [22] extended the above concepts to $\mathcal{I}$-statistical convergence and $\mathcal{I}$-lacunary statistical convergence of order α, $0 < \alpha \leq 1$ in topological groups.

Quite recently in [1], it has been extended to the idea of natural or asymptotic density by taking natural density of weight g where $g : \mathbb{N} \to [0, \infty)$ is a function with $\lim_{n \to \infty} g(n) = \infty$ and $\frac{n}{g(n)} \nrightarrow 0$ as $n \to \infty$.

In a natural way, in this paper we consider new and more general summability methods, namely $\mathcal{I}$-statistical convergence of weight g and $\mathcal{I}_\theta$-statistical convergence of weight g in topological group.

2 Definitions and Notations

In this paper, our study will concern ideal which is given below:

Definition 1 (*see* [12]). A family $\mathcal{I} \subset 2^{\mathbb{N}}$ is said to be an ideal of $\mathbb{N}$ if the following conditions hold:
(a) $P, Q \in \mathcal{I}$ implies $P \cup Q \in \mathcal{I}$,
(b) $P \in \mathcal{I}$, $Q \subset P$ implies $Q \in \mathcal{I}$,

Definition 2 (*see* [12]). A non-empty family $F \subset 2^{\mathbb{N}}$ is said to be an filter of $\mathbb{N}$ if the following conditions hold:
(a) $\phi \notin F$,
(b) $P, Q \in F$ implies $P \cap Q \in F$,
(c) $P \in F$, $P \subset Q$ implies $Q \in F$,

Definition 3 (*see* [12]). A proper ideal $\mathcal{I}$ is said to be admissible if $\{n\} \in \mathcal{I}$ for each $n \in \mathbb{N}$.

Throughout this note, $\mathcal{I}$ will stand for a proper admissible ideal of $\mathbb{N}$.

Definition 4 (*see* [12]) Let $\mathcal{I} \subset 2^{\mathbb{N}}$ be a proper admissible ideal in $\mathbb{N}$. The sequence $x = (x_j)$ of elements of $\mathbb{R}$ is said to be $\mathcal{I}$-convergent to ξ if for each $\epsilon > 0$ the set $K(\epsilon) = \{n \in \mathbb{N} : |x_j - \xi| \geq \epsilon\} \in \mathcal{I}$.

Let $g : \mathbb{N} \to [0, \infty)$ be a function with $\lim\limits_{n \to \infty} g(n) = \infty$. The upper density of weight g was defined in [1] by the formula

$$\overline{d}_g(K) = \lim_{n \to \infty} \sup \frac{K(1, n)}{g(n)}$$

for $K \subset \mathbb{N}$ where as before $K(1, n)$ denotes the cardinality of the set $K \cap [1, n]$. Then, the family

$$\mathcal{I}_g = \{K \subset \mathbb{N} : \overline{d}_g(K) = 0\}$$

forms an ideal. It has been observed in [1] that $\mathbb{N} \in \mathcal{I}_g$ iff. $\frac{n}{g(n)} \to 0$, as $n \to \infty$. So we additionally assume that $n/g(n) \nrightarrow 0$, so that $\mathbb{N} \notin \mathcal{I}_g$ and $\mathcal{I}_g$ is a proper admissible ideal of $\mathbb{N}$. The set of all such weight functions g satisfying the above properties will be denoted by G. Now, we can write the following definition.

Definition 5 A sequence (x_j) of real numbers is said to converge d_g−statistically to ξ if for any given $\varepsilon > 0$, $\overline{d}_g(K(\varepsilon)) = 0$ where $K(\varepsilon)$ is the set defined in Definition 4.

By X, we will note an abelian topological Hausdorff group, written additively, which satisfies the first axiom of countability. For a subset R of X, $s(R)$ will denote the set of all sequences (x_j) such that x_j is in R for $j = 1, 2, \ldots, c(X)$ will denote the set of all convergent sequences. In [2], a sequence (x_j) in X is called to be statistically convergent to an element ξ of X if for each neighbourhood U of 0,

$$\lim_{n \to \infty} \frac{1}{n} |\{j \leq n : x_j - \xi \notin U\}| = 0.$$

The set of all statistically convergent sequences in X is denoted by $st(X)$.

Furthermore, Cakalli [3] considered lacunary statistical convergence in topological groups as follows: A sequence (x_j) is said to be S_θ-convergent to ξ if for each neighbourhood U of 0, $\lim_{p \to \infty} (h_p)^{-1} |j \in I_p : x_j - L \notin U| = 0$. In this case, we define

$$S_\theta(X) = \left\{ (x_j) : \text{ for some } \xi, \; S_\theta - \lim_{j \to \infty} x_j = \xi \right\}.$$

We now introduce the following definitions:

Definition 6 A sequence $x = (x_j)$ in X is said to be statistically convergent of weight g to ξ or $S(\mathcal{I})^g$-convergent of weight g to ξ if for each $\gamma > 0$ and for each neighbourhood U of 0,

$$\{n \in \mathbb{N} : \frac{1}{g(n)} |\{j \leq n : x_j - \xi \notin U\}| \geq \gamma\} \in \mathcal{I}.$$

In this case, we write $x_j \to \xi(S(\mathcal{I})^g)$. The class of all $S(\mathcal{I})^g$-statistically convergent sequences will be denoted by simply $S(\mathcal{I})^g(X)$.

Remark 1 For $\mathcal{I} = \mathcal{I}_{fin} = \{B \subseteq \mathbb{N} : B$ is a finite subset $\}$, $S(\mathcal{I})^g$-convergence coincides with statistical convergence of weight g in topological groups. Further taking $g(n) = n^\alpha$, it reduces to $\mathcal{I}$-statistical convergence of order α in topological groups, which is studied by Savas [22].

Definition 7 Let θ be a lacunary sequence. A sequence $x = (x_j)$ in X is said to be $\mathcal{I}$-lacunary statistically convergent of weight g to ξ or $S_\theta(\mathcal{I})^\alpha$-convergent to ξ if for any $\gamma > 0$ and for each neighbourhood U of 0,

$$\{p \in \mathbb{N} : \frac{1}{g(h_p)} |\{j \in I_p : x_j - \xi \notin U\}| \geq \gamma\} \in \mathcal{I}.$$

In this case, we write

$$S_\theta(\mathcal{I})^g - \lim_{j \to \infty} x_j = \xi \text{ or } x_j \to \xi(S_\theta(\mathcal{I})^g)$$

and define

$$S_\theta(\mathcal{I})^g(X) = \left\{(x_j) : \text{ for some } \xi, S_\theta(I)^g - \lim_{j \to \infty} x_j = \xi\right\}$$

and in particular,

$$S_\theta(\mathcal{I})^g(X)_0 = \left\{(x_j) : S_\theta(\mathcal{I})^g - \lim_{j \to \infty} x_j = 0\right\}.$$

Remark 2 For $\mathcal{I} = \mathcal{I}_{fin}$, $S_\theta(\mathcal{I})^g$-convergence reduces to lacunary statistical convergence of weight g in topological groups, which has not been studied till now. Further, we write in the special case $\theta = 2^r$. Definition 7 reduces to Definition 6.

3 Inclusion Theorems

The following theorem gives inclusion relations

Theorem 1 *Let $g_1, g_2 \in G$ be such that there exist $M > 0$ and $j_0 \in \mathbb{N}$ such that $\frac{g_1(n)}{g_2(n)} \le M$ for all $n \ge j_0$. Then $S(\mathcal{I})^{g_1} \subset S(\mathcal{I})^{g_2}$.*

Proof For any neighbourhood U of 0,

$$\frac{\left|\{j \le n : x_j - \xi \notin U\}\right|}{g_2(n)} = \frac{g_1(n)}{g_2(n)} \cdot \frac{\left|\{j \le n : x_j - \xi \notin U\}\right|}{g_1(n)}$$

$$\le M \cdot \frac{\left|\{j \le n : x_j - \xi \notin U\}\right|}{g_1(n)}.$$

for $n \ge j_0$. Hence for any $\gamma > 0$ and for each neighbourhood U of 0

$$\left\{n \in \mathbb{N} : \frac{\left|\{j \le n : x_j - \xi \notin U\}\right|}{g_2(n)} \ge \gamma\right\}$$

$$\subset \left\{n \in \mathbb{N} : \frac{\left|\{j \le n : x_j - \xi \notin U\}\right|}{g_1(n)} \ge \frac{\gamma}{M}\right\} \cup \{1, 2, \ldots, j_0\}.$$

So we have that $S(\mathcal{I})^{g_1} \subset S(\mathcal{I})^{g_2}$.

Similarly, we can get the following result.

Theorem 2 *Let $g_1, g_2 \in G$ be such that there exist $M > 0$ and $i_0 \in \mathbb{N}$ such that $\frac{g_1(n)}{g_2(n)} \le M$ for all $n \ge i_0$. Then*
  (i) $S_\theta(\mathcal{I})^{g_1}(X) \subset S_\theta(\mathcal{I})^{g_2}(X)$.
  (ii) In particular $S_\theta(\mathcal{I})^{g_1}(X) \subset S_\theta(\mathcal{I})(X)$.

We now record two useful another theorems.

Theorem 3 *For any lacunary sequence θ, $\mathcal{I}$-statistical convergence of weight g implies $\mathcal{I}$-lacunary statistical convergence of weight g if*

$$\liminf_p \frac{g(h_p)}{g(k_p)} > 1.$$

Proof Since $\liminf\limits_p \frac{g(h_p)}{g(k_p)} > 1$, so we get a $H > 1$ such that for sufficiently large p we get

$$\frac{g(h_p)}{g(k_p)} \ge H.$$

Since $x_j \to \xi \left(S(\mathcal{I})^g \right)$, hence for each neighbourhood U of 0 and sufficiently large p we have

$$\frac{1}{g\left(k_p\right)} \left| \left\{ j \leq k_p : x_j - \xi \notin U \right\} \right| \geq \frac{1}{g\left(k_p\right)} \left| \left\{ j \in I_p : x_j - \xi \notin U \right\} \right|$$

$$\geq H \cdot \frac{1}{g\left(h_p\right)} \left| \left\{ j \in I_p : x_j - \xi \notin U \right\} \right|.$$

Then for any $\gamma > 0$, and for each neighbourhood U of 0 we get

$$\left\{ p \in \mathbb{N} : \frac{1}{g\left(h_p\right)} \left| \left\{ j \in I_p : x_j - \xi \notin U \right\} \right| \geq \gamma \right\}$$

$$\subseteq \left\{ p \in \mathbb{N} : \frac{1}{g\left(k_p\right)} \left| \left\{ j \leq k_p : x_j - \xi \notin U \right\} \right| \geq H\gamma \right\} \in \mathcal{I}.$$

This shows that $x_j \to \xi \left(S_\theta(\mathcal{I})^g \right)$.

For the next theorem, we suppose that the lacunary sequence θ fulfils the condition that for any set $C \in F(\mathcal{I}), \bigcup \{n : k_{p-1} < n < k_p, p \in C\} \in F(\mathcal{I})$.

Theorem 4 *For a lacunary sequence θ satisfying the above condition, $\mathcal{I}$-lacunary statistical convergence of weight g implies $\mathcal{I}$-statistical convergence of weight g (where $g(n) \neq n$), if $\sup \sum_{i=1}^{p} \frac{g(h_i)}{g(k_{p-1})} = K(say) < \infty$ where g is also assumed to be monotonically increasing.*

Proof Assume that $x_j \to \xi \left(S_\theta(\mathcal{I})^g \right)$. Take any neighbourhood U of 0. For $\gamma, \gamma_1 > 0$ define the sets

$$P = \left\{ p \in \mathbb{N} : \frac{1}{g\left(h_p\right)} \left| \left\{ j \in I_p : x_j - \xi \notin U \right\} \right| < \gamma \right\}$$

and

$$T = \left\{ n \in \mathbb{N} : \frac{1}{g(n)} \left| \left\{ j \leq n : x_j - \xi \notin U \right\} \right| < \gamma_1 \right\}.$$

From our assumption, it follows that $P \in \mathcal{F}(\mathcal{I})$, the dual filter of $\mathcal{I}$. Also note that

$$A_k = \frac{1}{g(h_k)} \left| \left\{ j \in I_k : x_j - \xi \notin U \right\} \right| < \gamma$$

for all $k \in P$. Let $n \in \mathbb{N}$ be such that $k_{p-1} < n < k_p$ for some $p \in P$. Now

$$\frac{1}{g(n)} \left| \{ j \leq n : x_j - \xi \notin U \} \right|$$

$$\leq \frac{1}{g(k_{p-1})} \left| \{ j \leq k_p : x_j - \xi \notin U \} \right|$$

$$= \frac{1}{g(k_{p-1})} \left| \{ j \in I_1 : x_j - \xi \notin U \} \right| + \cdots + \frac{1}{g(k_{p-1})} \left| \{ j \in I_p : x_j - \xi \notin U \} \right|$$

$$= \frac{g(h_1)}{g(k_{p-1})} \cdot \frac{1}{g(h_1)} \left| \{ j \in I_1 : x_j - \xi \notin U \} \right| +$$

$$\frac{g(h_2)}{g(k_{p-1})} \cdot \frac{1}{g(h_2)} \left| \{ j \in I_2 : x_j - \xi \notin U \} \right| + \cdots +$$

$$\frac{g(h_p)}{g(k_{p-1})} \cdot \frac{1}{g(h_p)} \left| \{ j \in I_p : x_j - \xi \notin U \} \right|$$

$$= \frac{g(h_1)}{g(k_{p-1})} \cdot A_1 + \frac{g(h_2)}{g(k_{p-1})} \cdot A_2 + \cdots + \frac{g(h_p)}{g(k_{p-1})} \cdot A_p$$

By choosing $\delta_1 = \frac{\delta}{K}$ and since $\bigcup \{ n : k_{p-1} < n < k_p, p \in P \} \subset T$ where $P \in \mathcal{F}(\mathcal{I})$, it is obvious that from our assumption on θ that the set T also belongs to $\mathcal{F}(\mathcal{I})$.

Corollary 1 Let $\theta = \{(k_p)\}$ be a lacunary sequence, then $S(\mathcal{I})^\alpha(X) = S_\theta(\mathcal{I})^\alpha(X)$ iff

$$1 < \liminf_p q_p \leq \limsup_p q_p < \infty.$$

Finally, we conclude this paper by proving the following theorem.

Theorem 5 $S(\mathcal{I})^g(X) \subset S_\theta(\mathcal{I})^g(X)$ if $\liminf_n \frac{g(h_p)}{g(n)} > 0$.

Proof Since $\liminf_n \frac{g(h_p)}{g(n)} > 0$, so we can find a $M > 0$ such that for sufficiently large n we have

$$\frac{g(h_p)}{g(n)} \geq M.$$

Since $x_j \to \xi \left(S(\mathcal{I})^g \right)$, hence for any neighbourhood U of 0 and sufficiently large n,

$$\frac{1}{g(n)} \left| \{ j \leq n : x_j - \xi \notin U \} \right| \geq \frac{1}{g(h_p)} \left| \{ j \in I_p : x_j - \xi \notin U \} \right|$$

$$\geq M \frac{1}{g(h_p)} \left| \{ j \in I_p : x_j - \xi \notin U \} \right|.$$

For $\gamma > 0$,

$$\left\{ n \in N : \frac{1}{g(h_p)} \left| \left\{ j \in I_p : x_j - \xi \notin U \right\} \right| \geq \gamma \right\}$$

$$\subset \left\{ n \in N : \frac{1}{g(n)} \left| \left\{ j \in I_p : x_j - \xi \notin U \right\} \right| \geq M\gamma \right\}.$$

Since $\mathcal{I}$ is admissible, the set on the right-hand side belongs to $\mathcal{I}$.

References

1. Balcerzak, M., Das, P., Filipczak, M., Swaczyna, J.: Generalized kinds of density and the associated ideals. Acta Math. Hungar. **147**(1), 97–115 (2015)
2. Çakalli, H.: On statistical convergence in topological groups. Pure Appl. Math. Sci. **43**(1–2), 27–31 (1996)
3. Çakalli, H.: Lacunary statistical convergence in topological groups. Indian J. Pure Appl. Math. **26**(2), 113–119 (1995)
4. Colak, R.: Statistical convergence of order α. Modern Methods in Analysis and Its Applications, 121–129. Anamaya Publisher, New Delhi, India (2010)
5. Das, P., Savaş, E.: On $\mathcal{I}$-convergence of nets in locally solid Riesz spaces. Filomat **27**(1), 84–89 (2013)
6. Das, P., Savaş, E.: On $\mathcal{I}_\lambda$-statistical convergence in locally solid Riesz spaces. Math. Slovaca **65**(6), 1491–1504 (2015)
7. Das, P., Savaş, E.: On $\mathcal{I}$-statistically pre-Cauchy sequences. Taiwanese J. Math. **18**(1), 115–126, FEB (2014)
8. Fast, H.: Sur la convergence statistique. Colloq Math. **2**, 241–244 (1951)
9. Fridy, J.A.: On ststistical convergence. Analysis **5**, 301–313 (1985)
10. Fridy, J.A., Orhan, C.: Lacunary statistical convergence. Pacific J. Math. **160**, 43–51 (1993)
11. Gadjiev, A.D., Orhan, C.: some approximation theorems via statistical convergence. Rocky Mt. J. Math. **32**(1), 508–520 (2002)
12. Kostyrko, P., Šalát, T., Wilczynki, W.: $\mathcal{I}$-convergence. Real Anal. Exch. **26**(2), 669–685 (2000/2001)
13. Maio, G.D., Kocinac, L.D.R.: Statistical convergence in topology. Topology Appl. **156**, 28–45 (2008)
14. Šalát, T.: On statistically convergent sequences of real numbers. Math. Slovaca **30**, 139–150 (1980)
15. Savaş, E., Das, Pratulananda: A generalized statistical convergence via ideals. Appl. Math. Lett. **24**, 826–830 (2011)
16. Savaş, E.: Δ^m-strongly summable sequence spaces in 2-normed spaces defined by ideal convergence and an Orlicz function. Appl. Math. Comput. **217**, 271–276 (2010)
17. Savaş, E.: A sequence spaces in 2-normed space defined by ideal convergence and an Orlicz function. Abst. Appl. Anal. **2011**, Article ID 741382 (2011)
18. Savaş, E.: On some new sequence spaces in 2-normed spaces using Ideal convergence and an Orlicz function. J. Ineq. Appl. Article Number: 482392 (2010). https://doi.org/10.1155/2010/482392
19. Savaş, E.: On generalized double statistical convergence via ideals. In: The Fifth Saudi Science Conference, pp. 16–18 (2012)
20. Savaş, E.: On $\mathcal{I}$-lacunary statistical convergence of order α for sequences of sets. Filomat **29**(6), 1223–1229. 40A35 (2015)

21. Savaş, E.: $\mathcal{I}_\theta$-statistically convergent sequences in topological groups. Mat. Bilten **39**(2), 19–28 (2015)
22. Savaş, E., Savaş Eren, R.E.: $\mathcal{I}_\theta$-statistical convergence of order α in topological groups. Applied mathematics in Tunisia, 141–148, Springer Proc. Math. Stat., 131, Springer, Cham (2015)
23. Schoenberg, I.J.: The integrability methods. Amer. Math. Monthly **66**, 361–375 (1959)

Chapter 5
On the Integral-Balance Solvability of the Nonlinear Mullins Model

Jordan Hristov

Abstract The integral-balance method to the nonlinear Mullins model of thermal grooving has been applied. The successful integral-balance solution utilizing the double-integration techniques has been able after application of the nonlinear Broadbridge transform. The Broadbridge transform converts the Mullins equation into a Dirichlet problem of a nonlinear diffusion equation with a Fujita-type nonlinearity of the diffusion coefficient. The solution is straightforward but needs additional optimization procedure determining the unspecified exponent of the generalized assumed parabolic profile.

Keywords Integral-balance method · Mullins equation · Double-integration method · Approximate solution

1 Introduction

1.1 Mullins Models of Thermal Diffusion Grooving

The thermal grooving on metal surface by mechanisms of evaporation–condensation is modelled by the nonlinear Mullins equation [1–3]

$$\frac{\partial u}{\partial t} = \left[\frac{D(0)}{1 + (\partial u/\partial x)^2} \right] \frac{\partial^2 u}{\partial x^2}, \quad \frac{\partial u(0, t)}{\partial x} = m = const.,$$
$$\frac{\partial u(0, t)}{\partial x} \longrightarrow 0 \quad , x \longrightarrow \infty, \quad u(x, 0) = 0 \tag{1}$$

with initial conditions

$$u_x(0, t) = const., \quad u(x, 0) = 0, \quad u(\infty, 0) = 0, \quad u_x(\infty, t) = 0 \tag{2}$$

J. Hristov (✉)
Department of Chemical Engineering, University of Chemical Technology
and Metallurgy (UCTM), 1756 Sofia, Bulgaria
e-mail: jordan.hristov@mail.bg

© Springer Nature Singapore Pte Ltd. 2018
D. Ghosh et al. (eds.), *Mathematics and Computing*, Springer Proceedings
in Mathematics & Statistics 253, https://doi.org/10.1007/978-981-13-2095-8_5

and subjected to boundary conditions [3]

$$u_x(0, t) = m., \quad u(\infty, t) = u_{xx}(0, t) = u_{xxx}(0, t) = 0. \tag{3}$$

The boundary condition at $x = 0$ actually corresponds to the physical requirement the flux of vapours to be equal to zero at the origin of the groove [1–3]. The model of Mullins considers an axisymmetric groove (about the vertical axis and $x = 0$), and due to this symmetry, we may consider only a half-profile for $x \geq 0$ (see Fig. 1). The model (1)–(3) has been solved and analysed by many authors [1, 2, 4–6]. In addition, when the surface curvature is small (i.e. for $(u_{xx})^2 \ll 1$) a linearization of the model (1) as a fourth-order parabolic equation [3] accounting mainly the groove formation by the surface diffusion mechanisms is possible, namely [1, 3]

$$\frac{\partial u}{\partial t} = -B \frac{\partial^4 u}{\partial x^4}, \quad \frac{\partial u(0, t)}{\partial x} = m, \quad \frac{\partial^3 u(0, t)}{\partial x^3} = 0, \quad 0 < x < \infty, \quad t > 0. \tag{4}$$

Here, the apparent diffusion coefficient $B = D_s \gamma \Omega^2 v / kT$ is a dimensional group involving the coefficient of surface diffusion coefficient D_s, the free surface energy per unit area γ, the molecular volume Ω, and the area v where the surface diffusion occurs.

The linear model (4) was recently solved by a new integral-balance technology named multiple method (MIM) [7] in two recently published articles: the case integer-order time derivative [7] and in a time-fractional (subdiffusion) version (suggested in [8]) [9]. In both cases, the solutions reveal strong subdiffusion behaviour of the process modelled because the groove surface profile evolves in time proportional

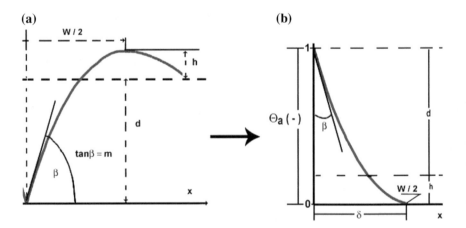

Fig. 1 Schematic groove profile **a** with equivalent Dirichlet diffusion example (by inverting the profile) **b** explaining why the integral-balance method is applied. Adapted from [7] by courtesy of *Thermal Science*

to $t^{1/4}$. Now, the present work addresses an approximate solution of the complete model (1) by the integral-balance method [10–16].

1.2   The Motivation for Doing This Study

The main motivation for this study comes from the elegant work of Broadbridge [1] where the model (1) was solved exactly (see comments in the sequel) as well as a more general class of nonlinear Mullins-type models was defined. Especially to the present author motivation, owning already experience in application of the integral-balance method to nonlinear diffusion problems [12, 13, 17, 18], as well as with solutions of the linearized model (4) [7] and its time-fractional version [9] by MIM, the next challenging tasks were the solution of (1) by the integral method, an attempt never done before. The results of these efforts are presented in this work.

2   The Integral-Balance Method: Necessary Background

The integral-balance method to diffusion models of heat and mass is based on the concept of a finite penetration depth, evolving in time, and a sharp front of the solution [10, 11] propagating with a finite speed. There are two principle integration techniques of the method: simple integration method known as heat-balance integral method (HBIM) of Goodman [10–12, 14–16] and double-integration method (DIM) [12–14, 17–20] (see the sequel). The basic rules of these techniques are explained next.

2.1   Single-Integration Approach

The approach considers (in case of transient diffusion with a constant transport coefficient a) a single-step integration over the penetration depth $\delta(t)$, that is

$$\int_0^\delta \frac{\partial \theta}{\partial t} dx = \int_0^\delta a \frac{\partial^2 \theta}{\partial x^2} dx, \quad \theta(x, t) = 0, t > 0. \tag{5}$$

Physically, the relationship (5) is a simple mass balance over a diffusion layer of finite depth $\delta(t)$, while mathematically it is the zero moment of the diffusion equation. It is worthnoting that the physically based concept of the finite speed (and finite penetration depth) of the diffusant in a semi-infinite medium actually replaces the boundary condition at infinity $\theta(\infty) = 0$ with $\theta(\delta) = 0$ and $\partial \theta(\delta)/\partial x = 0$, known also as Goodmans conditions [10, 11]. This change in the boundary condition forms a sharp front of the solution $\delta(t)$ beyond which the medium is undisturbed. Moreover, it converts the problem defined initially in semi-infinite medium to a two-point problem.

After application of the Leibniz rule, we get from (5) the basic relationship of HBIM

$$\frac{d}{dt}\int_0^\delta \theta(x,t)dx = -a\frac{\partial\theta}{\partial x}(0,t). \tag{6}$$

Replacement of $\theta(x,t)$ by an assumed profile θ_a expressed as a function of the dimensionless space variable x/δ in (6) results in ODE about $\delta(t)$. The principle disadvantage of the single-step integration technique is that the gradient in the right-side of (6) *should be defined through the assumed profile* θ_a.

2.2 Double-Integration Approach

The double-integration method (DIM) in its original version [12–14, 20] employs a two-step integration procedure: first integration from 0 to x and a second one from 0 to δ (see details in the cited references). Here, we will use a modified version [13, 17, 18] (the first formula of (7)) where after application of the Leibniz rule we have (the second formula of (7))

$$\int_0^\delta \int_x^\delta \frac{\partial\theta(x,t)}{\partial t}dxdx = a\theta(0,t) \implies \frac{d}{dt}\int_0^\delta \int_x^\delta \theta(x,t)dxdx = a\theta(0,t). \tag{7}$$

In (7), the first integration is near the front (from x to δ). The approach expressed by (7) is general and applicable to either integer-order [12] and time-fractional models [13, 17–19] (see details in [12–14]) .

3   The Integral-Balance Solution

Prior to applying either the DIM solution to (1), the necessary step is a transform to a more convenient form as a standard nonlinear diffusion equation. Precisely, the Broadbridge transformation (BT) of the nonlinear part of (1) [1] allows the application of the integral-balance method to be successful.

3.1 Transformation to Nonlinear Diffusion Problem

Following Broadbridge [1] and by help of (2) (see also the first BC in (3)), we may apply the substitution $\Theta = u_x/m$. Consequently, we get

$$u = \int_{\infty}^{x} \Theta(z, t) dz. \tag{8}$$

$$\frac{du}{dt} = \frac{d}{dt} \int_{\infty}^{x} \Theta(z, t) dz \tag{9}$$

Applying the Leibniz rule in inverse order (to that used in the previous point) and with the last boundary condition in (2), we may transform (1) as [1]

$$\int_{\infty}^{x} \frac{\partial \Theta}{\partial t} dx = D(\Theta) \frac{\partial \Theta}{\partial x}. \tag{10}$$

Now after differentiation with respect to x, we get a more friendly form

$$\frac{\partial \Theta}{\partial t} = \frac{\partial}{\partial x} \left[D(\Theta) \frac{\partial \Theta}{\partial x} \right], \quad \Theta(x, 0) = 0 \tag{11}$$

with boundary conditions

$$\Theta(0, t) = \frac{u_x(0, t)}{m} = 1, \quad \Theta(\infty, t) = 0. \tag{12}$$

Hence, we got a Dirichlet problem with respect to the variable $\Theta(x, t)$.

Futher, Broadbridge [1] developed an exact self-similar solution in terms of the classical similarity variable $\eta_B = (1/2)(x/\sqrt{D_0 t}) = (\eta/2)$ (η is defined naturally through the solution developed in this work). We will refer further to the Broadbridge solution when specific moments of the solution developed here have to be commented. Now, we go in a different way applying the integral-balance method to (11) and (12).

3.2   Assumed Profile

Prior to applying DIM, we should select the assumed profile $\Theta_a(x/\delta)$. In this work, a parabolic one with unspecified exponent is used, namely

$$\Theta_a = \Theta_s \left(1 - \frac{x}{\delta}\right)^n, \quad \Theta_s = \Theta(0, t) = 1. \tag{13}$$

The profile (13) obeys all boundary conditions at both ends of the penetration layer ($0 \leq x \leq \delta$) for any value of the exponent n [7, 11–15, 18, 19]. This feature offers a flexibility to optimize the numerical value of the exponent as it will demonstrate further in this work. With the boundary conditions (3) and the transform (8), we have

$$\Theta(\delta) = \Theta_x(\delta) = \Theta_{xx}(0, t) = \Theta_{xxx}(0, t) = 0. \tag{14}$$

In accordance with the schematic presentation in Fig. 1a, the penetration depth $\delta(t)$ equals the half-width of the cavity, that is $\delta(t) = w/2$, from the bottom point $x = 0$ up to the inflexion point (because beyond the point of inflection the surface of the next groove begins). The scheme in Fig. 1b is an inverted profile corresponding to the diffusion Dirichlet problem. This was especially done to facilitate understanding of how the integral-balance method, well known from transient diffusion and heat conduction problems, could be applied to the Mullins equation.

3.3 The Nonlinear Diffusion Coefficient $D(\Theta)$

In the transformed model (11), the diffusion coefficient in terms of the variable Θ is transformed as

$$D(u_x) = \frac{D(0)}{1 + (u_x)^2} \implies D(\Theta) = \frac{D(0)}{1 + a\Theta^2}, \quad a = m^2. \tag{15}$$

This is a Fujita-type nonlinearity [21–23] as it was especially commented by Broadbridge [1], and a special transformation $D(\Theta)$ is needed prior application of the integral-balance method. Denoting $D(0) = D_0$, the right-hand side of (11) can be transformed as

$$D(\Theta)\frac{\partial\Theta}{\partial x} = \frac{D_0}{(1 + a\Theta^2)}\frac{\partial\Theta}{\partial x} = D_0\frac{\partial}{\partial x}\left[\frac{arctan(a\Theta)}{\sqrt{a}}\right]. \tag{16}$$

Then, the model (11) takes the form

$$\frac{\partial\Theta}{\partial t} = D_0\frac{\partial^2}{\partial x^2}\left[\frac{arctan(a\Theta)}{\sqrt{a}}\right]. \tag{17}$$

3.4 Penetration Depth

In accordance with the rules of DIM, we have

$$\frac{d}{dt}\int_0^\delta\int_x^\delta \Theta dxdx = \int_0^\delta\int_x^\delta D_0\frac{\partial^2}{\partial x^2}\left[\frac{arctan(a\Theta)}{\sqrt{a}}\right]dxdx \tag{18}$$

The double integration in (18), with account of the boundary conditions (14), yields

$$\frac{1}{(n+1)(n+2)}\frac{d\delta^2}{dt} = D_0\frac{arctan(a)}{\sqrt{a}} = D_0 M(m) \qquad (19)$$

$$M(m) = \frac{arctan(a)}{\sqrt{a}} = \frac{arctan(m^2)}{m} = const. \qquad (20)$$

The initial condition $\Theta(x, 0) = 0$ that corresponds to the physically based $\delta(t = 0) = 0$ results in

$$\delta = \sqrt{D_0 t}\sqrt{M(m)N}, \quad N = (n+1)(n+2). \qquad (21)$$

Therefore, the penetration depth propagates in accordance with the classical (Fickian) diffusion law $\sqrt{t}$ which is in agreement with the result of Broadbridge [1]. In contrast to solution of the linearized model, (4) reveals subdiffusion scaling because $\delta \equiv t^{1/4}$ [7, 9] (see also [8]). Moreover, we may define an effective diffusion coefficient $D_m = D_0 M(m)$ and consequently (21) takes a form mimicking the penetration depth of the linear diffusion problem [10, 11, 15], i.e. as $\delta_m = \sqrt{D_m t}\sqrt{(n+1)(n+2)}$, but the explicit effect of the nonlinearity is lost.

3.5 Approximate Profile (Solution)

With the established relationship about $\delta(t)$, the approximate solution of (11) is

$$\Theta_a = \left(1 - \frac{x}{\sqrt{D_0 t}\sqrt{M(m)N}}\right)^n = \left(1 - \frac{\eta}{\sqrt{M(m)N}}\right)^n, \quad \eta = \frac{x}{\sqrt{D_0 t}} \qquad (22)$$

thus defining in a natural way the Boltzmann similarity variable $\eta = x/\sqrt{D_0 t}$.
   Now remembering that $\partial u/\partial x = \Theta/m \Longrightarrow \partial u_a/\partial x = \Theta/m$, we get

$$\frac{du_a}{dx} = \frac{1}{m}\left(1 - \frac{x}{\sqrt{D_0 t}\sqrt{M(m)N}}\right)^n. \qquad (23)$$

Integration in (23) from 0 to δ yields

$$u_a(x, t) = \int_0^\delta \Theta_a dx = \left\{-\frac{\sqrt{D_0 t}\sqrt{M(m)N}}{m(n+1)}\left(1 - \frac{x}{\delta}\right)^{n+1}\right\}_0^\delta. \qquad (24)$$

Hence, in terms of the original variable $u(x, t)$, precisely the solution about $u_a(x, t)$ is

$$u_a(x, t) = \frac{\sqrt{D_0 t}\sqrt{M(m)N}}{m}\sqrt{\frac{n+2}{(n+1)}}\left(1 - \frac{x}{\delta}\right)^{n+1}. \qquad (25)$$

For $x = 0$ in (25), we get $u_a(0, t)$ which is the groove maximal depth (see the sequel) and the normalized profile following from (25) can be presented as

$$U_a^* = \frac{u_a(x, t)}{u_a(0, t)} = \frac{u_a(x, t)}{\sqrt{D_0 t}} \frac{m}{\sqrt{M(m)}} \sqrt{\frac{n+1}{n+2}} \left(1 - \frac{x}{\delta}\right)^{n+1}. \tag{26}$$

Alternatively, we may scale the groove depth by the natural length scale $\sqrt{D_0 t}$, namely

$$U_a^\circ = \frac{u_a(x, t)}{\sqrt{D_0 t}} = \frac{\sqrt{M(m)}}{m} \sqrt{\frac{n+2}{n+1}} \left(1 - \frac{x}{\delta}\right)^{n+1}. \tag{27}$$

At this moment, we have to mention that the approximate MIM solution of (4) [7, 9] is

$$u_a(x, t)_{MIM} = \frac{m}{n}(Bt)^{1/4} M_4^{1/4} \left(1 - \frac{x}{(Bt)^{1/4} M_4^{1/4}}\right)^n$$

$$= \frac{m}{n}(Bt)^{1/4} M_4^{1/4} \left(1 - \frac{\eta_M}{(Bt)^{1/4} M_4^{1/4}}\right)^n. \tag{28}$$

In (28), $M_4 = \Gamma(n + k + 1)/\Gamma(n + 1)$ and k is the number of the integrations applied by MIM (in the case of the model (4), we have $k = 4$). The similarity variable $\eta_m = x/(Bt)^{1/4}$ is of non-Boltzmann type, and the natural length scale is $(Bt)^{1/4}$ [3, 7, 9]. In this context, it is worth noting that $\eta_m = x/(Bt)^{1/4}$ was used by Mullins [3] as an ansatz allowing transforming the linearized equation (4) into ODE. Hence, the linear problem (4) is easily solvable by MIM, but the solution depends on a nonlinear similarity variable, at the same time as the nonlinear problem (1) results in a solution expressed trough the Boltzmann variable $\eta = x/\sqrt{D_0 t}$ but needs a nonlinear transform (Broadbridge transform, **BT**) at the beginning.

Following (25) at $\eta = \sqrt{M(m)}N$ (corresponding to $x = \delta$), we have $\Theta_a = 0 \Longrightarrow$ $u_a = U_a = 0$. The value of m used in the original study of Mullins [3] was selected as $m = 0.1$. In this case, $M(m) \approx 0.099$ and $\sqrt{M(m)} \approx 0.316$. Therefore, the groove half-width is approximately $\delta = w/2 \approx 0.996\sqrt{D_0 t}\sqrt{(n + 1)(n + 2)}$, where n is still unspecified. The solution of the linearized model (4) in [7] with $m = 0.1$ and $n = 4.555$ (see details in [7, 9]) provides $\delta_4 = w/2 \approx 8.106(Bt)^{1/4}$.

In addition, the condition $x = 0$ defines the maximum of $u(x, t)$ attained by the profile at the groove bottom, denoted as groove depth $G_0(t)$, namely

$$G_0(t) = \frac{\sqrt{D_0 t}\sqrt{M(m)}}{m} \sqrt{\frac{n+2}{n+1}}, \quad G_0^*(t) = \frac{u(0, t)}{\sqrt{D_0 t}} = \frac{\sqrt{M(m)}}{m} \sqrt{\frac{n+2}{n+1}} \tag{29}$$

where $G_0^*(t)$ is the groove depth normalized by the natural length scale $\sqrt{D_0 t}$.

4   Refinement of the Approximate Solution

4.1   Residual Function

When the approximate solution is used as an alternative of the exact one, it is natural that the residual function of the model (1) differs from zero, namely

$$R_u = \left\{ \frac{\partial u_a}{\partial t} - \frac{D_0}{\left[1 + \left(\frac{\partial u_a}{\partial x} \right)^2 \right]} \frac{\partial^2 u_a}{\partial x^2} \right\} \neq 0 \tag{30}$$

or alternatively following (16) and (17) in the forms (31) and (32)

$$R_u = \left[\frac{\partial u_a}{\partial t} - \frac{D_0}{1 + a\Theta^2} \frac{\partial \Theta a}{\partial x} \right] \neq 0 \tag{31}$$

$$R_u = \left\{ \frac{\partial u_a}{\partial t} - D_0 \frac{\partial^2}{\partial x^2} \left[\frac{arctan(a\Theta_a)}{\sqrt{a}} \right] \right\} \neq 0. \tag{32}$$

The refinement of the approximate solution simply means a minimization of R with respect to the exponent n within the range $0 \leq x \leq \delta$ since all other parameters of the model are initially specified. First, let us see what is the behaviour of the residual fiction at the boundaries $x = 0$ and $x = \delta$. For $x = 0$, we have $\Theta_a = 1$, while for $x \longrightarrow \delta$ we get $\Theta_1 \longrightarrow 0$. Now with the assumed profile (13), we have

$$R_{\Theta_2} = n \left(1 - \frac{x}{\delta} \right)^{n-1} \left(\frac{x}{\delta} \right) \frac{1}{\delta} \frac{d\delta}{dt} - D_0 \frac{\partial^2}{\partial x^2} \left[\frac{1}{\sqrt{a}} arctan \left(a \left(1 - \frac{x}{\delta} \right)^n \right) \right]. \tag{33}$$

For in (33), we have $R_{\Theta_2} = 0 - D_0 \frac{\partial^2}{\partial x^2} \left[\frac{1}{\sqrt{a}} arctan(a) \right]$ for any value of n. Further, for $x \longrightarrow \delta$ we have directly $R_{\Theta_2} = 0$ also for any value of n. Moreover, the product $(1/\delta)(d\delta/dt)$ simply reduces to $1/\left(2\sqrt{t} \right)$; that is, the first term in (33) decays in time. Hence, these tests do not provide the needed information about the exponent n and a special attention on the approximation of the diffusion term is needed.

4.2   Approximation of the Diffusion Term

Now, the problem at issue is how to approximate the second term of R_{Θ_2} as a function of x/δ. It is well known that $arctan(y)$ has a convergent series expansion as

$$arctan(y) \approx \sum_{j=0}^{\infty} (-1)^j \frac{y^{2j+1}}{2j+1} \approx y - \frac{y^3}{3} + \frac{y^5}{5} - \frac{y^7}{7} + \cdots \tag{34}$$

with radius of convergence 1, when $-1 \leq y \leq 1$.

In our case $-1 \leq \Theta_a \leq 1$ and $0 \leq a \leq 1$, and with $y = a\Theta_a$, it is possible to obtain a series expansion like (34). However, as a first attempt we will use the linear approximation $arctan(y) \approx (\pi/4) y$ for $-1 \leq y \leq 1$, especially when $y \longrightarrow 1$, which mimics the first term in the series (34) and corresponds to $x/\delta \longrightarrow 0$ when $y = a\Theta_a$. This physically corresponds to a groove profile near its origin $x = 0$, when $m = 1$. Since we always have $(1 - x/\delta) < 1$ and $m < 1 \Longrightarrow a = m^2 \ll 1$, then at least the product $a\Theta_a = a(1 - x/\delta)$ is of order of magnitude 10^{-2}; it is clear from (34) that beyond fourth term all the following will be negligible. Hence, replacing $y = a\Theta_a$ we get $arctan(a\Theta_a) \approx \frac{\pi}{4}(a\Theta_a)$, and with the profile (13), this linear approximation becomes

$$arctan(a\Theta_a) \approx \frac{\pi}{4}(a\Theta_a) \approx \frac{\pi}{4}a\left(1 - \frac{x^n}{\delta}\right). \tag{35}$$

The approximation (35) is valid for $n > 0$ because from the condition $0 \leq y \leq 1$ we should have $(1 - x/\delta)^n < 1$. Therefore, the diffusion term can be approximated as

$$\frac{\partial^2}{\partial x^2}\left[\frac{1}{\sqrt{a}}arctan\left(a\left(1 - \frac{x}{\delta}\right)^n\right)\right] \approx \frac{\pi}{4}\sqrt{a}\left[\frac{n(n-1)}{\delta^2}\left(1 - \frac{x}{\delta}\right)^{n-2}\right]. \tag{36}$$

Now after this approximation the residual function can be presented in two forms

$$R_{\Theta_2} \approx n\left(1 - \frac{x}{\delta}\right)^{n-1}\left(\frac{x}{\delta}\right)\frac{1}{\delta}\frac{d\delta}{dt} - D_0\frac{\pi}{4}\sqrt{a}\left[\frac{n(n-1)}{\delta^2}\left(1 - \frac{x}{\delta}\right)^{n-2}\right] \tag{37}$$

$$R_{\Theta_2} \approx \frac{1}{\delta^2}\left\{n(1-z)^{n-1}z\delta\frac{d\delta}{dt} - D_0\frac{\pi}{4}\sqrt{a}\left[n(n-1)(1-z)^{n-2}\right]\right\}. \tag{38}$$

In (38), the moving boundary domain $0 \leq x \leq \delta$ is transformed into one with fixed boundaries $0 \leq z = x/\delta \leq 1$. The product $\delta\frac{d\delta}{dt}$ is time-independent and therefore

$$R_{\Theta_2} \approx \frac{1}{t}\left\{\frac{nz(1-z)^{n-1}[M(m)(n+1)(n+2)] - \frac{\pi}{2}m[n(n-1)(1-z)^{n-2}}{2[M(m)(n+1)(n+2)]}\right\}. \tag{39}$$

In (39), we have a term (in the waved brackets) which is time-independent but in general R_{Θ_2} decays in time. Now with the new construction of R_{Θ_2} setting $x = 0$ we get $R_{\Theta_2}(x = 0) \approx 0 - \frac{\pi}{2}m(n-1)$ which is obeyed for $n = 1$.

4.3 Optimal Exponents with Linear Approximation of $arctan(a\Theta_a)$

The optimal exponent n can be determined by minimization of the squared error of approximation defined as $E(n, m, t) = \int_0^1 (R_{\Theta_2})^2 dz = \frac{1}{t^2}e(n, m)$, where $e(n, m)$ is

the results of integration of the time-independent term of (38). The procedure is well
described in [12, 13, 15, 18, 19], and we will avoid here huge expressions.

The minimization of $e(n, m)$ for given values of m was performed by Maple.
For two values of m used in the literature: $m = 0.1$ [3] and $m = 0.4$ [1], we have:
$n(m = 1) \approx 0.985$ with $e(n, m) \approx 0.0289$, and $n(m = 0.4) \approx 0.32$ with $e(n, m) \approx$
0.00511, respectively. These values of the exponents do not obey the requirement
$n > 2$. However, numerical tests revealed that the decrease in m, that is for grooves
with small angles β at the origin (see Fig. 1), the values of the optimal exponents
increase and vice versa. As examples supporting this statement, the following results
were obtained: $n(m = 0.01) \approx 2.257$ with $e(n, m) \approx 0.696$, $n(m = 0.02) \approx 1.186$
with $e(n, m) \approx 0.0585$, and $n(m = 0.3) \approx 0.9336$ with $e(n, m) \approx 0.0.00987$.

Therefore, the first attempt to use the approximation (35) provides reasonable data
for small values of m that limits the application of this approach. Nevertheless, if the
number of terms in the series (34) is increased, then we have a rapidly converging
series

$$arctan(a\Theta_a) \approx a \left(1 - \frac{x}{\delta}\right)^n - \frac{a^3}{3} \left(1 - \frac{x}{\delta}\right)^{3n} + \frac{a^5}{5} \left(1 - \frac{x}{\delta}\right)^{5n} + \cdots \quad (40)$$

Then, the approximate diffusion term can be presented as

$$\frac{\partial^2}{\partial x^2} \left[\frac{1}{\sqrt{a}} arctan\left(a\left(1 - \frac{x}{\delta}\right)^n\right)\right] \approx \frac{1}{\sqrt{a}} a \frac{n(n-1)}{\delta^2} \left(1 - \frac{x}{\delta}\right)^{n-2} -$$

$$\frac{a^3}{3} \frac{3n(3n-1)}{\delta^2} \left(1 - \frac{x}{\delta}\right)^{3n-2} + \frac{a^5}{5} \frac{5n(5n-1)}{\delta^2} \left(1 - \frac{x}{\delta}\right)^{5n-1} + \cdots \quad (41)$$

This approach, however, draws a new problem about the reasonable number of terms
in the series (41), which is beyond the scope of this report.

4.4 Brief Notes

To recapitulate the solution results, these are actually the first attempts to solve the
complete Mullins equation by the integral-balance method, especially applying the
double-integration technique. The crucial points are the nonlinear transform of the
diffusion term, after the initial transformation of Broadbridge (BT) and then the
approximation of the diffusion term in the residual function. The new challenging
problem emerging in the determination of the optimal exponent is the approximation
of $arctan(a\Theta_a)$ when Θ_a is the generalized parabolic profile (13), but this figures
new studies beyond the scope of the present communication.

5 Numerical Simulations

Numerical simulations with the approximate solution Θ_a (22), actually of completely normalized profile (26), are shown in Fig. 2a. All these plots correspond to DIM solutions with optimal exponents satisfying the condition $n > 2$ or at least $n \approx 2.0$. The curves reveal a physical adequacy of the solution since larger initial angles (represented by the value of m) result in wider groove openings and vice versa. At this moment of study, these results are, to some extent, qualitative since there are no extensive database of values of m available in the literature. However, the dimensionless presentation shown in Fig. 2 is general since it uses the similarity variable $\eta = \sqrt{D_0 t}$ as independent variable.

As it was mentioned in the previous section, these are results obtained with the linear approximation of $arctan(a\Theta_a)$. The idea to use more terms of the approximating series (34) results in (40) and consequently in (41). Taking into account the small values of m (order of magnitude $10^{-2} \div 10^{-3}$) and the fact that $a = m^2$ (order of magnitude $10^{-4} \div 10^{-6}$) as well as that $(1 - x/\delta) \le 1$ with $n > 2$ we may approximately estimate that, for example, the second term in the series (41) with exponent $3n - 1 > 5$ would have an order of magnitude of about $10^{-6} \div 10^{-10}$. Hence, the linear approximation used in this work is physically reasonable. More terms in (41) could be taken into account when m takes large values of order of magnitude of unity or larger, that is in modelling of grooves with larger openings. This is a good challenge that needs to be proved by modelling and comparison with experimental data on groove shapes, but beyond the scope of this report.

The groove depth evolutions in time presented in Fig. 2b reproduce directly the linear relationship (29) when the natural length scale $\sqrt{D_0 t}$ is used as independent variable: larger values of m result in wider grooves but with slow growths and vice versa.

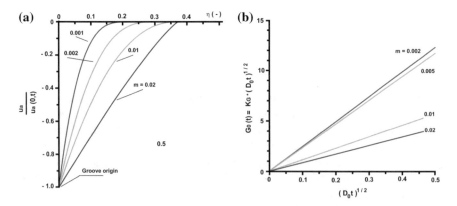

Fig. 2 Numerical simulation with DIM solutions: **a** Dimensionless groove profiles at various m and similarity variable η as independent variable; **b** Groove depth as function of the natural length scale $\sqrt{D_0 t}$

Both groove depth and opening ($w = 2\delta$) are results of a Gaussian diffusion process since $G_0^2 \equiv t$ and $w^2 \equiv t$ (because $\delta^2 \equiv t$) in contrast to the linearized model (4) where $\delta^2 \equiv t^{\frac{1}{4}}$ and the process is a subdiffusive. Precisely, with the approximate profile (13), the mean squared displacement characterizing the diffusion process is $\langle x^2 \rangle \equiv \delta^2$ (see [19]): when $\langle x^2 \rangle \equiv t^\gamma$ with $\gamma = 1$, we have a normal Gaussian process, but when $\gamma < 1$ the process is subdiffusive (see [9, 17]). In the context of the physical process of groove evolution, this should be related to the mechanisms involved: the evaporation–condensation mechanism[(model (1)] is Gaussian, while the surface diffusion mechanism is subdiffusive [7–9].

6   Conclusion

An attempt on the integral-balance method to approximate solution of the nonlinear Mullins model of thermal growing has been reported. The application of the double-integration method (DIM) was successfully applied, but this solution needs two important steps to be done before: (1) application of the Broadbridge transform converting the original model (1) into a Dirichlet problem of nonlinear diffusion equation with Fujita-type nonlinearity and (2) nonlinear transform of the diffusion term, a technique used before [12, 18] allowing application of the assumed parabolic profile.

Principle moment in the process of refinement of the approximate solution is the approximation of the nonlinear diffusion term, and this problem strongly depends on the specific function that should be approximated and doubly differentiated with respect to the space coordinate. The numerical experiments reveal adequate behaviour of the simulated results reasonably modelling groove shapes.

The process of solution developed raises many questions and interesting problem that might be solved in future studies, but the main step to solve the nonlinear Mullins model by the integral-balance method was already done.

References

1. Broadbridge, P.: Exact solvability of the Mullins nonlinear diffusion model of groove development. J. Math. Phys. **30**, 1648–1651 (1989)
2. Broadbridge, P.: Exact solution of a degenerate fully nonlinear diffusion equation. Z. Angw. Math. Phys. **55**, 34–538 (2004)
3. Mullins, W.W.: Theory of thermal grooving. J. Appl. Phys. **28**, 333–339 (1957)
4. Kitada, A.: On properties of a classical solution of nonlinear mass transport equation. J. Math. Phys. **27**, 1391–1392 (1986)
5. Martin, P.A.: Thermal grooving by surface diffusion: Mullins revisited and extended to multiple grooves. Q. J. Appl. Math. **67**, 125–36 (2009)
6. Robertson, W.M.: Grain-boundary growing by surface diffusion for finite slopes. J. Appl. Phys. **42**, 463–467 (1971)

7. Hristov, J.: Multiple integral-balance method: basic idea and an example with Mullinss model of thermal grooving. Therm. Sci. **21**, 1555–1560 (2017)
8. Abu Hamed, M., Nepomnyashchy, A.A.: Groove growth by surface subdiffusion. Physica D: Nonlinear Phenom. **298299**, 42–47 (2015)
9. Hristov, J.: Fourth-order fractional diffusion model of thermal grooving: integral approach to approximate closed form solution of the Mullins model. Math. Model Natur. Phenom. **13**, 1–6 (2018)
10. Goodman, T.R.: The heat balance integral and its application to problems involving a change of phase. Trans. ASME **80**, 335–342 (1958)
11. Hristov, J.: The heat-balance integral method by a parabolic profile with unspecified exponent: analysis and Benchmark exercises. Therm. Sci. **13**, 27–48 (2009)
12. Hristov, J.: Integral solutions to transient nonlinear heat (mass) diffusion with a power-law diffusivity: a semi-infinite medium with fixed boundary conditions. Heat Mass Transf. **52**, 635–655 (2016)
13. Hristov, J.: Double integral-balance method to the fractional subdiffusion equation: approximate solutions, optimization problems to be resolved and numerical simulations. J. Vib. Control **23**, 2795–2818 (2017)
14. Mitchell, S.L., Myers, T.G.: Application of standard and rened heat balance integral methods to one-dimensional Stefan problems. SIAM Rev. **52**, 57–86 (2010)
15. Myers, J.G.: Optimizing the exponent in the heat balance and refined integral methods. Int. Commun. Heat Mass Transf. **36**, 143–147 (2009)
16. Sahu, S.K., Das, P.K., Bhattacharyya, S.: A comprehensive analysis of conduction-controlled rewetting by the heat balance integral method. Int. J. Heat Mass Transf. **49**, 4978–4986 (2006)
17. Hristov, J.: Approximate solutions to time-fractional models by integral balance approach, Chapter 5. In: Cattani, C., Srivastava, H.M., Yang, X.-J. (eds.) Fractional Dynamics, pp. 78–109, De Gruyter Open (2015)
18. Hristov, J.: Integral-balance solution to nonlinear subdiffusion equation, Chapter 3, In: Bhalekar, S. (ed.) Frontiers in Fractional Calculus, pp. 57–88. Bentham Science Publishers (2017)
19. Hristov, J.: Subdiffusion model with time-dependent diffusion coefficient: integral-balance solution and analysis. Therm. Sci. **21**, 69–80 (2017)
20. Volkov, V.N., Li-Orlov, V.K.: A refinement of the integral method in solving the heat conduction equation. Heat Transf. Sov. Res. **2**, 41–47 (1970)
21. Fujita, H.: The exact pattern of a concentration-dependent diffusion in a semi-infinite medium, Part II. Text. Res. J. **22**, 823–827 (1952)
22. Fujita, H.: The exact pattern of a concentration-dependent diffusion in a semi-infinite medium, Part 1. Text. Res. J. **22**, 757–760 (1952)
23. Fujita, H.: The exact pattern of a concentration-dependent diffusion in a semi-infinite medium, Part III. Text. Res. J. **24**, 234–240 (1954)

Chapter 6
Optimal Control of Rigidity Parameter of Elastic Inclusions in Composite Plate with a Crack

Nyurgun Lazarev and Natalia Neustroeva

Abstract Equilibrium problems for a family of composite plates with a crack passing along the boundary of an elastic inclusion are considered. We assume that the Signorini-type condition for nonpenetration of the opposite crack faces is fulfilled. It is shown that there exists a solution of the optimal control problem with the cost functional given with the help of an arbitrary continuous functional in the solution space.

Keywords Timoshenko plate · Rigid inclusion · Crack · Nonpenetration conditions · Variational inequality · Derivative of energy functional
Shape control

1 Introduction

It is well known that the difference between the coefficients of thermal expansion and moduli elasticity for heterogeneous materials often leads to initiation of cracks (delamination) and ruptures at the boundary interface of different materials. In this regard, it is important to analyze high-level mathematical models of elastic bodies with delaminated inclusions and to investigate dependence of solutions on the variation of physical parameters of inclusions. We consider two types of inclusions: For the first type, we have inclusions which are described by the Timoshenko model, and the second type of inclusions corresponds to the Kirchhoff–Love model. Optimal control problem considered in this work consists in finding the best rigidity parameter of an elastic inclusion. The cost functional is defined with the help of an arbitrary continuous functional in the solution space.

The main difficulty in studying this problem is due to the presence of the nonlinear boundary conditions of inequality type. Since the beginning of 1990s, a crack theory

N. Lazarev (✉) · N. Neustroeva
North-Eastern Federal University, Yakutsk 677891, Russia
e-mail: nyurgun@ngs.ru

N. Lazarev
Lavrentyev Institute of Hydrodynamics SB RAS, Novosibirsk 630090, Russia

© Springer Nature Singapore Pte Ltd. 2018
D. Ghosh et al. (eds.), *Mathematics and Computing*, Springer Proceedings
in Mathematics & Statistics 253, https://doi.org/10.1007/978-981-13-2095-8_6

with nonpenetration conditions at the crack faces has been under active study (see, e.g., [3–7, 12, 16, 17]). Some of these works are devoted to the investigation of various nonlinear mathematical models of crack theory. We refer the reader to [7, 15, 19, 21] for results concerning the shape sensitivity analysis to nonlinear problems in domains with cuts. The fictitious domain and smooth domain methods were proposed in [1, 2]. Invariant integrals in the framework of nonlinear elasticity problems with Signorini-type conditions were constructed in [2, 7, 24]. The problems concerning equilibrium models for elastic bodies with rigid inclusions [7–11, 13, 17, 18, 22, 23] or elastic inclusions [14] were studied. It is worth mentioning that these problems belong to the class of free boundary value problems.

2 Equilibrium Problems

We formulate the two types of variational problems. These both types of problems are formulated with respect to the identical geometrical objects. Let us consider a bounded domain $\Omega \subset \mathbf{R}^2$ with a boundary $\Gamma \in C^{0,1}$. Let a subdomain ω be strictly contained in Ω, i.e., $\overline{\omega} \cap \Gamma = \emptyset$, and let a boundary $\partial\omega$ be sufficiently smooth. Assume that $\partial\omega$ consists of two disjoint curves γ_c and $\partial\omega \setminus \gamma_c$, meas $\partial\omega \setminus \gamma_c > 0$. The outward pointing unit normal to $\partial\omega$ is denoted by $\nu = (\nu_1, \nu_2)$.

We require that the curve γ_c can be extended up to the outer boundary Γ in such a way that Ω is divided into two subdomains Ω_1, Ω_2 with the Lipschitz boundaries. The latter condition is sufficient to fulfill the Korn and Poincare inequalities in the domain $\Omega_c = \Omega \setminus \overline{\gamma}_c$ [7].

For simplicity, suppose the plate has a uniform thickness $2h = 2$. Let us assign a three-dimensional Cartesian space $\{x_1, x_2, z\}$ with the set $\{\Omega_c\} \times \{0\} \subset \mathbf{R}^3$ corresponding to the middle plane of the plate. The curve γ_c defines a crack (a cut) in the plate. This means that the cylindrical surface of the crack may be defined by the relations $x = (x_1, x_2) \in \gamma_c$, $-1 \le z \le 1$ where $|z|$ is the distance to the middle plane. Following our arguments, an elastic inclusion is specified by the set $\omega \times [-1, 1]$; i.e., the boundary of the elastic inclusion is defined by the cylindrical surface $\partial\omega \times [-1, 1]$. An unaltered part of the plate corresponds to the domain $\Omega_c \setminus \overline{\omega}$.

Denote by $\chi = (W, w)$ the displacement vector of the mid-surface points ($x \in \Omega_c$), by $W = (w_1, w_2)$ the displacements in the plane $\{x_1, x_2\}$, and by w the displacements along the axis z. The angles of rotation of a normal fiber are denoted by $\psi = \psi(x) = (\psi_1, \psi_2)$, ($x \in \Omega_c$).

In accordance with the direction of the outer normal ν to $\partial\omega$, it is possible to speak about a positive face $\partial\omega^+$ and a negative face $\partial\omega^-$ of the curve $\partial\omega$. If the trace of a function v is chosen on the positive (from the side of the domain $\Omega \setminus \overline{\omega}$) face $\partial\omega^+$, we use the notation $v^+ = v|_{\partial\omega^+}$, and if it is chosen on the negative face, then $v^- = v|_{\partial\omega^-}$. In addition, the jump $[v]$ of the function v on the curve γ_c can be found by the formula $[v] = v|_{\gamma_c^+} - v|_{\gamma_c^-}$.

Assume that deformation of the unaltered part (which corresponds to the set $\Omega_c \setminus \overline{\omega}$) is described by the Timoshenko model. The corresponding formulas for strains and other mechanical values have the form [20]:

$$\varepsilon_{ij}(\psi) = \frac{1}{2}\left(\frac{\partial \psi_j}{\partial x_i} + \frac{\partial \psi_i}{\partial x_j}\right), \quad \varepsilon_{ij}(W) = \frac{1}{2}\left(\frac{\partial w_j}{\partial x_i} + \frac{\partial w_i}{\partial x_j}\right). \tag{1}$$

The tensors of moments $m(\psi) = \{m_{ij}(\psi)\}$ and stresses $\sigma(W) = \{\sigma_{ij}(W)\}$ are expressed by the formulas (summation is performed over repeated indices)

$$m_{ij}(\psi) = b_{ijkl}\varepsilon_{kl}(\psi), \quad \sigma_{ij}(W) = 3b_{ijkl}\varepsilon_{kl}(W), \quad i, j, k, l = 1, 2, \tag{2}$$

with nonzero components of elasticity tensor $B = \{b_{ijkl}\}$ specified by the relations

$$a_{iiii} = D, \quad a_{iijj} = D\kappa, \quad a_{ijij} = a_{ijji} = D(1 - \kappa)/2, \quad , i \neq j, \quad i, j = 1, 2, \tag{3}$$

where D and κ are the constants: D is a cylindrical rigidity of the plate, κ is the Poisson ratio, $0 < \kappa < 1/2$. The transverse forces in the Timoshenko-type model are defined by the expressions

$$q_i(w, \psi) = L(w_{,i} + \psi_i), \quad i = 1, 2, \quad \left(v_{,i} = \frac{\partial v}{\partial x_i}\right),$$

where $L > 0$ is a constant coefficient describing elastic plate characteristics with respect to transverse shear [20].

Next, we describe the mathematical models corresponding to elastic inclusion which refers to the domain ω. There are two types of inclusions. For the first type, we have the same relations (1)–(3) with some other constant coefficients D', κ', $B' = \{b'_{ijkl}\}$. For the transverse forces, we accept the formulas

$$q_i(w, \psi) = \frac{L'}{\lambda}(w_{,i} + \psi_i), \quad i = 1, 2,$$

where $L' > 0$ is a constant value, and $\lambda \in (0, 1]$.

The second type of elastic inclusion is described by the Kirchhoff–Love model, so that the following relations are fulfilled in the domain ω:

$$m_{ij} = -b'_{ijkl}w_{,kl} .$$

$$\sigma_{ij}(W) = 3b'_{ijkl}\varepsilon_{kl}(W), \quad i, j, k, l = 1, 2.$$

As the next step, we want to formulate the corresponding variational problems. For the first type of inclusions, we formulate a family of variational problems. In order to define a potential energy functional, introduce bilinear forms $B(Q, \cdot, \cdot), b(Q, \cdot, \cdot)$ determined by the equalities

$$B(Q, \overline{\eta}, \eta) = \int_Q \sigma_{ij}(\overline{W})\, \varepsilon_{ij}(W) + m_{ij}(\overline{\psi})\, \varepsilon_{ij}(\psi),$$

$$b(Q, \overline{\eta}, \eta) = \int_Q (\overline{w}_{,i} + \overline{\psi}_i)(w_{,i} + \psi_i),$$

where $Q \subset \Omega_c$, $\eta = (W, w, \psi)$, $\overline{\eta} = (\overline{W}, \overline{w}, \overline{\psi})$.

The potential energy functional of the plate has the following representation [20]:

$$\Pi^\lambda(\Omega_c, \eta) = \frac{1}{2} B(\Omega_c, \eta, \eta) + \frac{1}{2} \Lambda(\lambda) b(\Omega_c, \eta, \eta) - \int_{\Omega_c} F\eta, \quad \eta = (W, w, \psi),$$

where the vector $F = (f_1, f_2, f_3, f_4, f_5) \in L^2(\Omega_c)^5$ describes the body forces [20],

$$\Lambda(\lambda) = \begin{cases} L, & x \in \Omega_c \backslash \overline{\omega}, \\ \frac{L'}{\lambda}, & x \in \omega. \end{cases}$$

In what follows, we suppose that $f_4 = f_5 = 0$. Introduce the Sobolev spaces

$$H^{1,0}(\Omega_c) = \left\{ v \in H^1(\Omega_c) \mid v = 0 \text{ on } \Gamma \right\}, \quad H(\Omega_c) = H^{1,0}(\Omega_c)^5.$$

Note that the following inequality holds (with some fixed value λ)

$$B(\Omega_c, \eta, \eta) + \Lambda(\lambda) b(\Omega_c, \eta, \eta) \geq c \|\eta\|_{H(\Omega_c)}^2 \quad \forall \eta \in H(\Omega_c), \tag{4}$$

where the constant $c > 0$ is independent of η [16]. This estimate ensures that the bilinear form $B(\Omega_c, \eta, \eta)$ defines a norm equivalent to the standard norm on $H(\Omega_c)$.

The condition of mutual nonpenetration of opposite faces of the crack is given by

$$[W]\nu \geq |[\psi]\nu| \quad \text{on} \quad \gamma_c. \tag{5}$$

The derivation and justification of the condition (5) can be found in [16]. Introduce the set of admissible functions

$$K_1 = \{ \eta = (W, w, \psi) \in H(\Omega_c) \mid [W]\nu \geq |[\psi]\nu| \text{ on } \gamma_c \}.$$

Now, we can formulate a family of the equilibrium problems for the plate with a crack on the boundary of the elastic inclusion. We fix the parameter $\lambda \in (0, 1]$ and set the minimization problem

$$\inf_{\eta \in K_1} \Pi^\lambda(\Omega_c, \eta). \tag{6}$$

Using the same reasoning as in the paper [16], it is possible to prove the existence of a unique solution ξ^λ to the problem (6). Besides, it can be shown that the problem (6) is equivalent to the following variational inequality [16]

$$\xi^\lambda \in K_1,$$
$$B(\Omega_c, \xi^\lambda, \eta - \xi^\lambda) + \Lambda(\lambda)b(\Omega_c, \xi^\lambda, \eta - \xi^\lambda) \geq \int_{\Omega_c} F(\eta - \xi^\lambda) \quad \forall \eta \in K_1. \quad (7)$$

Next, let us formulate a variational problem for plate with a inclusion of the second type, i.e., if deformation of the elastic inclusion is described by the Kirchhoff–Love model. We start with the introduction of the conditions describing the Kirchhoff–Love hypothesis of straight-line normals

$$w_{,i} + \psi_i = 0 \text{ in } \omega, \quad i = 1, 2.$$

Therefore, the set of admissible functions K_2 is defined by the following relation

$$K_2 = \{ \eta = (W, w, \psi) \in H(\Omega_c) \mid w_{,i} + \psi_i = 0 \text{ in } \omega, \ i = 1, 2; \ [W]v \geq |[\psi]v| \text{ on } \gamma_c \}$$

For this case, we can represent the potential energy of the plate in the following form:

$$\Pi^K(\eta) = \frac{1}{2}B(\Omega_c, \eta, \eta) + \frac{L}{2}b(\Omega_c \setminus \overline{\omega}, \eta, \eta) - \int_{\Omega_c} F\eta.$$

The variational setting of problem is as follows. In the domain Ω_c, we have to find function $\xi^K \in K_2$ such that

$$\Pi^K(\xi^K) = \min_{\eta \in K_2} \Pi^K(\eta), \quad (8)$$

As we have to find the solution ξ^K in the space $H(\Omega_c)$, we assume that the gluing conditions are satisfied on the interface between the media outside the crack:

$$[\xi^K] = (0, 0, 0) \text{ on } \partial\omega \setminus \gamma_c.$$

The convexity, weak semi-continuity, and coercivity of the functional $\Pi^K(\eta)$ in the space $H(\Omega_c)$ can be established similar to that made in [16]. The properties of $\Pi^K(\eta)$ and the convexity and closedness of the set K_2 guarantee the existence and uniqueness of the solution $\xi^K = (U^K, u^K, \phi^K)$ of problem (8). Besides, the problem (8) is equivalent to the following variational inequality

$$\xi^K = (U^K, u^K, \phi^K) \in K_2, \quad B(\Omega_c, \xi^K, \eta - \xi^K) + Lb(\Omega \setminus \overline{\omega}, \xi^K, \eta - \xi^K) \geq$$
$$\geq \int_{\Omega_\gamma} F(\eta - \xi^K) \quad \forall \eta = (W, w, \psi) \in K_2. \quad (9)$$

3   Optimal Control Problem

In this section, we prove the main result of the paper which provides an existence of the optimal rigidity parameter $\lambda^* \in [0, 1]$ for the elastic inclusion. Here, the limiting case $\lambda = 0$ corresponds to the inclusion with an infinite shear rigidity or Kirchhoff–Love's inclusion. We define the cost functional $J : [0, 1] \to \mathbf{R}$ of an optimal control problem with the use of the following relation

$$
J(\lambda) = \begin{cases} G(\xi^\lambda), & \lambda \in (0, 1], \\ G(\xi^K), & \lambda = 1, \end{cases}
$$

where $G(\eta) : H(\Omega_\gamma) \to \mathbf{R}$ is an arbitrary continuous functional.

As examples of such functionals having physical sense, we can give the following functionals. The functional $G_1(\eta) = \int_{\gamma_c} |[\chi]| \, (\eta = (W, w, \psi), \chi = (W, w))$ characterizes the opening of the crack. The functional $G_2(\eta) = \|\eta - \eta_0\|_{H(\Omega_c)}$ characterizes the deviation of the displacement vector from a given function χ_0 by η_0. Consider the optimal control problem:

$$
\text{Find} \quad \lambda^* \in [0, 1] \quad \text{such that} \quad J(\lambda^*) = \sup_{\lambda \in [0,1]} J(\lambda). \tag{10}
$$

Theorem 1 *There exists a solution of the optimal control problem* (10).

Proof Consider a maximizing sequence $\lambda_n \in [0, 1]$. In view of evidence, we can exclude the simple situations corresponding to the following case: $\lambda_n = \hat{\lambda}$ for all $n > n_0$. Therefore, we have to deal with the following two cases:

1. $\lambda_n \to \alpha$, $\lambda_n \in (0, 1]$, $\alpha \in (0, 1]$,
2. $\lambda_n \to 0$, $\lambda_n \in (0, 1)$.

We start from the first case. For each fixed λ_n, there exists a solution $\xi^n = \xi^{\lambda_n}$, $n = 1, 2, \ldots$ of the variational inequality like (7), i.e.,

$$
\xi^n \in K_1, \quad B(\Omega_c, \xi^n, \eta - \xi^n) + \Lambda(\lambda_n) b(\Omega_c, \xi^n, \eta - \xi^n) \geq \int_{\Omega_c} F(\eta - \xi^n) \quad \forall \eta \in K_1. \tag{11}
$$

By substituting $\eta = 2\xi^\lambda$ and $\eta = 0$ into the variational inequalities (7), we get

$$
B(\Omega_c, \xi^\lambda, \xi^\lambda) + \Lambda(\lambda) b(\Omega_c, \xi^\lambda, \xi^\lambda) = \int_{\Omega_c} F\xi^\lambda. \tag{12}
$$

Taking into account the inequality (4), we can derive from the last equality that

$$\|\xi^\lambda\|^2_{H(\Omega_c)} \leq B(\Omega_c, \xi^\lambda, \xi^\lambda) + \Lambda(1)b(\Omega_c, \xi^\lambda, \xi^\lambda)$$
$$\leq B(\Omega_c, \xi^\lambda, \xi^\lambda) + \Lambda(\lambda)b(\Omega_c, \xi^\lambda, \xi^\lambda) = \int_{\Omega_c} F\xi^\lambda.$$

From this, we get the following uniform estimation

$$\|\xi^n\|_{H(\Omega_c)} \leq C. \tag{13}$$

Choosing a subsequence, if necessary, we can assume that as $n \to \infty$

$$\xi^n \to \tilde{\xi} \quad \text{weakly in} \quad H(\Omega_c),$$
$$\frac{(u^n_{,i} + \phi^n_i)}{\lambda_n} \to \frac{(\tilde{u}_{,i} + \tilde{\phi}_i)}{\alpha} \quad \text{weakly in} \quad L^2(\omega). \tag{14}$$

$$(\lambda_n)^{-1/2}(u^n_{,i} + \phi^n_i) \to \alpha^{-1/2}(\tilde{u}_{,i} + \tilde{\phi}_i) \quad \text{weakly in} \quad L^2(\omega). \tag{15}$$

Using the strong convergence of $U^n \to \tilde{U}$, $\phi^n \to \tilde{\phi}$ in $H^1(\Omega_c)^2$ as $n \to \infty$, it can be easily shown that $\tilde{\xi} \in K_1$. In view of (14), (15), we pass to the limit as $n \to \infty$ in (11), which yields

$$\tilde{\xi} \in K_1, \quad B(\Omega_c, \tilde{\xi}, \eta - \tilde{\xi}) + \Lambda(\alpha)b(\Omega_c, \tilde{\xi}, \eta - \tilde{\xi}) \geq \int_{\Omega_c} F(\eta - \tilde{\xi}) \quad \forall \eta \in K_1.$$

By the arbitrariness of η, this inequality means that the last inequality is variational and $\tilde{\xi} = \xi^\alpha$. Now, we will prove that $\xi^n \to \xi^\alpha$ is strong in $H(\Omega_c)$. The weak convergence $\xi^n \to \xi^\alpha$ as $n \to \infty$ implies that

$$\lim_{n\to\infty} \int_{\Omega_c} F\xi^n = \int_{\Omega_c} F\xi^\alpha$$

Consequently, the limit of the right side of (12) exists and is equal to

$$\lim_{n\to\infty} \left(B(\Omega_c, \xi^n, \xi^n) + \Lambda(\lambda_n)b(\Omega_c, \xi^n, \xi^n) \right)$$
$$= \lim_{n\to\infty} \left(B(\Omega_c, \xi^n, \xi^n) + \Lambda(\alpha)b(\Omega_c, \xi^n, \xi^n) \right) = \lim_{n\to\infty} \int_{\Omega_c} F\xi^n = \int_{\Omega_c} F\xi^\alpha.$$

On the other hand, by (12), we derive

$$\lim_{n\to\infty} \left(B(\Omega_c, \xi^n, \xi^n) + \Lambda(\alpha)b(\Omega_c, \xi^n, \xi^n) \right)$$
$$= B(\Omega_c, \xi^\alpha, \xi^\alpha) + \Lambda(\alpha)b(\Omega_c, \xi^\alpha, \xi^\alpha). \tag{16}$$

We should recall that by the estimation (4), the bilinear form

$$B(\Omega_c, \cdot, \cdot) + \Lambda(\alpha) b(\Omega_c, \cdot, \cdot)$$

determines an equivalent norm in the space $H(\Omega_c)$. This fact and the relation (16) allow us to obtain as $n \to \infty$

$$\|\xi^n\|_{H(\Omega_c)} \to \|\xi^\alpha\|_{H(\Omega_c)}.$$

Next, based on this convergence of norms and the weak convergence $\xi^n \to \xi^\alpha$ in $H(\Omega_c)$, we get the desired strong convergence $\xi^n \to \xi^\alpha$ in $H(\Omega_c)$. Thus, we have relations

$$\sup_{\lambda \in [0,1]} J(\lambda) = \lim_{n \to \infty} J(\lambda_n) = \lim_{n \to \infty} G(\xi^n) = G(\xi^\alpha) = J(\alpha),$$

which prove the statement for the first case.

Let us consider the second case. We suppose that the maximizing sequence λ_n converges to 0. Analogously, from (13), we can conclude that there is a subsequence (retain notation) such that ξ^n converges weakly in $H(\Omega_c)$ to some $\tilde{\xi}$. Next, we can represent (12) in the following form

$$B(\Omega_c, \xi^\lambda, \xi^\lambda) + Lb(\Omega_c \setminus \overline{\omega}, \xi^\lambda, \xi^\lambda) + \frac{L'}{\lambda} b(\omega, \xi^\lambda, \xi^\lambda) = \int_{\Omega_c} F\xi^\lambda$$

and conclude that

$$\|(u^n_{,i} + \phi^n_i)\|^2_{L^2(\omega)} \leq C\lambda_n, \quad (\tilde{u}_{,i} + \tilde{\phi}_i) = 0 \quad \text{a.e. in } \omega \tag{17}$$

with some positive constant C. Therefore, in view of the last equation in (17), we can obtain that $\tilde{\xi} \in K_2$. Now, we can substitute some fixed element $\eta \in K_2$ as the test function in (11) and pass to the limit as $n \to \infty$. As a result, we arrive at the relation

$$\tilde{\xi} \in K_2, \quad B(\Omega_c, \tilde{\xi}, \eta - \tilde{\xi}) + Lb(\Omega_c \setminus \overline{\omega}, \tilde{\xi}, \eta - \tilde{\xi}) \geq \int_{\Omega_c} F(\eta - \tilde{\xi}) \quad \forall \eta \in K_2.$$

The arbitrariness of the test function η means that the last inequality is variational and $\tilde{\xi} = \xi^K$.

In the next step, we prove the strong convergence $\xi^n \to \xi^K$ as $n \to \infty$. To this end, we rewrite (12) for the parameters λ_n as follows

$$B(\Omega_c, \xi^n, \xi^n) + Lb(\Omega_c \setminus \overline{\omega}, \xi^n, \xi^n) + \frac{L'}{\lambda_n} b(\omega, \xi^n, \xi^n) = \int_{\Omega_c} F\xi^n.$$

From this, using the weak lower semi-continuity of the bilinear forms $B(\Omega_c, \cdot, \cdot)$, $b(\Omega_c, \cdot, \cdot)$, we can deduce

$$
\begin{aligned}
\lim_{n\to\infty} \sup \frac{L'}{\lambda_n} b(\omega, \xi^n, \xi^n) &\leq \lim_{n\to\infty} \sup \left(-B(\Omega_c, \xi^n, \xi^n) - Lb(\Omega_c\setminus\overline{\omega}, \xi^n, \xi^n) + \int_{\Omega_c} F\xi^n \right) \\
&\leq \left(-B(\Omega_c, \xi^K, \xi^K) - Lb(\Omega_c\setminus\overline{\omega}, \xi^K, \xi^K) + \int_{\Omega_c} F\xi^K \right) = 0.
\end{aligned}
\tag{18}
$$

The last equality to zero in (18) is provided by the following identity

$$
B(\Omega_c, \xi^K, \xi^K) + Lb(\Omega_c\setminus\overline{\omega}, \xi^K, \xi^K) = \int_{\Omega_c} F\xi^K,
\tag{19}
$$

which can be obtained from the variational inequality (9) by substituting $\eta = 0$, $\eta = 2\xi^K$. Therefore, we get

$$
\lim_{n\to\infty} \sup \frac{L'}{\lambda_n} b(\omega, \xi^n, \xi^n) = \lim_{n\to\infty} L'b(\omega, \xi^n, \xi^n) = 0.
$$

Consequently, we have

$$
\begin{aligned}
&\lim_{n\to\infty} \left(B(\Omega_c, \xi^n, \xi^n) + Lb(\Omega_c\setminus\overline{\omega}, \xi^n, \xi^n) + L'b(\omega, \xi^n, \xi^n) \right) \\
&= \lim_{n\to\infty} \left(\int_{\Omega_c} F\xi^n - \frac{L'}{\lambda_n} b(\omega, \xi^n, \xi^n) + L'b(\omega, \xi^n, \xi^n) \right) = \int_{\Omega_c} F\xi^K.
\end{aligned}
\tag{20}
$$

Finally, taking into account the identity $b(\omega, \xi^K, \xi^K) = 0$ and relations (19), (20), we get

$$
\begin{aligned}
&\lim_{n\to\infty} \left(B(\Omega_c, \xi^n, \xi^n) + Lb(\Omega_c\setminus\overline{\omega}, \xi^n, \xi^n) + L'b(\omega, \xi^n, \xi^n) \right) \\
&= \lim_{n\to\infty} \left(\int_{\Omega_c} F\xi^n - \frac{L'}{\lambda_n} b(\omega, \xi^n, \xi^n) + L'b(\omega, \xi^n, \xi^n) \right) = \int_{\Omega_c} F\xi^K \\
&= B(\Omega_c, \xi^K, \xi^K) + Lb(\Omega_c\setminus\overline{\omega}, \xi^K, \xi^K) + L'b(\omega, \xi^K, \xi^K).
\end{aligned}
$$

This means that we have the convergence of norms

$$
\|\xi^n\|_{H(\Omega_c)} \to \|\xi^K\|_{H(\Omega_c)}
$$

as $n \to \infty$, which together with the weak convergence $\xi^n \to \xi^K$ in $H(\Omega_c)$ provides the desired strong convergence $\xi^n \to \xi^K$ as $n \to \infty$ in $H(\Omega_c)$. At last, we have relations

$$\sup_{\lambda \in [0,1]} J(\lambda) = \lim_{n \to \infty} J(\lambda_n) = \lim_{n \to \infty} G(\xi^n) = G(\xi^K) = J(0).$$

Thus, we have established the existence of solutions of (10) for all possible cases. Theorem is proved.

References

1. Alekseev, G.V., Khludnev, A.M.: Crack in elastic body crossing the external boundary at zero angle. Vestnik Q. J. Novosibirsk State Univ. Ser.: Math, Mech. inform. **9**(2), 15–29 (2009)
2. Andersson, L.-E., Khludnev, A.M.: On crack crossing an external boundary. Fictitious domain method and invariant integrals. Siberian. J Ind. Math. **11**(3), 15–29 (2008)
3. Hömberg, D., Khludnev, A.M.: On safe crack shapes in elastic bodies. Eur. J. Mech. A/Solids **21**(6), 991–998 (2002)
4. Itou, H., Khludnev, A.M.: On delaminated thin Timoshenko inclusions inside elastic bodies. Math. Methods Appl. Sci. **39**(17), 4980–4993 (2016)
5. Itou, H., Khludnev, A.M., Rudoy, E.M., Tani, A.: Asymptotic behaviour at a tip of a rigid line inclusion in linearized elasticity. Z. Angew. Math. Mech. **92**, 716–730 (2012)
6. Itou, H., Kovtunenko, V.A., Tani, A.: The interface crack with Coulomb friction between two bonded dissimilar elastic media. Appl. Math. **56**(1), 69–97 (2011)
7. Khludnev, A.M.: Elasticity Problems in Nonsmooth Domains. Fizmatlit, Moscow (2010)
8. Khludnev, A.M.: Problem of a crack on the boundary of a rigid inclusion in an elastic plate. Mech. Solids **45**(5), 733–742 (2010)
9. Khludnev, A.M.: Optimal control of crack growth in elastic body with inclusions. Eur. J. Mech. A/Solids. **29**(3), 392–399 (2010)
10. Khludnev, A.M.: Thin rigid inclusions with delaminations in elastic plates. Eur. J. Mech. A/Solids. **32**(1), 69–75 (2012)
11. Khludnev, A.M.: Shape control of thin rigid inclusions and cracks in elastic bodies. Arch. Appl. Mech. **83**(10), 1493–1509 (2013)
12. Khludnev, A.M., Kovtunenko, V.A.: Analysis of Cracks in Solids. WIT Press, Southampton-Boston (2000)
13. Khludnev, A.M., Negri, M.: Optimal rigid inclusion shapes in elastic bodies with cracks. Z. Angew. Math. Phys. **64**(1), 179–191 (2013)
14. Khludnev, A.M., Popova, T.S.: Junction problem for Euler-Bernoulli and Timoshenko elastic inclusions in elastic bodies. Q. Appl. Math. **74**(4), 705–718 (2016)
15. Kovtunenko, V.A.: Primal-dual methods of shape sensitivity analysis for curvilinear cracks with nonpenetration. IMA J. Appl. Math. **71**, 635–657 (2006)
16. Lazarev, N.P.: An iterative penalty method for a monlinear problem of equilibrium of a Timoshenko-type plate with a crack. Num. Anal. Appl. **4**(4), 309–318 (2011)
17. Lazarev, N.P.: An equilibrium problem for the Timoshenko-type plate containing a crack on the boundary of a rigid inclusion. J. Siberian Fed. Univ. Math. Phys. **6**(1), 53–62 (2013)
18. Lazarev, N.P.: Optimal control of the thickness of a rigid inclusion in equilibrium problems for inhomogeneous two-dimensional bodies with a crack. Z. Angew. Math. Mech. **96**(4), 509–518 (2016)
19. Lazarev, N.P., Rudoy, E.M.: Shape sensitivity analysis of Timoshenko's plate with a crack under the nonpenetration condition. Z. Angew. Math. Mech. **94**(9), 730–739 (2014)
20. Pelekh, B.L.: Theory of Shells with Finite Shear Modulus. Nauk. Dumka, Kiev (1973)
21. Rudoy, E.M.: Shape derivative of the energy functional in a problem for a thin rigid inclusion in an elastic body. Z. Angew. Math. Phys. **66**(4), 1923–1937 (2014)
22. Shcherbakov, V.V.: On an optimal control problem for the shape of thin inclusions in elastic bodies. J. Appl. Ind. Math. **7**(3), 435–443 (2013)

23. Shcherbakov, V.V.: Existence of an optimal shape of the thin rigid inclusions in the Kirchhoff-Love plate. J. Appl. Indust. Math. **8**(1), 97–105 (2014)
24. Shcherbakov, V.V.: The Griffith formula and J-integral for elastic bodies with Timoshenko inclusions. Z. Angew. Math. Mech. **96**(11), 1306–1317 (2016)

Chapter 7
Convergence of Generalized Mann Type of Iterates to Common Fixed Point

T. Som, Amalendu Choudhury, D. R. Sahu and Ajeet Kumar

Abstract The present paper deals with the convergence of two modified Mann type of iteration schemes for a single and a finite family of mappings to the fixed and common fixed point, respectively, of a single and a finite family of quasi-nonexpansive mappings on a uniformly convex Banach space. An example is added in support of our main result. The results obtained generalize the earlier results of Rhoades (J Math Anal Appl 56:741–750, [6]), Som et al. (Proc Nat Acad Sci (India) 70(A)(II):185–189, [8]), and others in turn.

Keywords Quasi-nonexpansive map · Generalized Mann iterates · Convergence Fixed point

1 Introduction

The present paper deals with the generalization of Mann type of iteration scheme [4] for a single mapping to two different iteration schemes involving firstly a single map and secondly a finite family of mappings, respectively, and then studies the convergence of such an iteration scheme for quasi-nonexpansive self-mappings of a

T. Som (✉)
Department of Mathematical Sciences, Indian Institute of Technology (BHU),
Varanasi 221005, India
e-mail: tsom.apm@itbhu.ac.in

A. Choudhury
Department of Mathematics and Statistics, Haflong Government College,
Haflong, Dima Hasao 788819, Assam, India
e-mail: amalendu_choudhury@yahoo.com

D. R. Sahu · A. Kumar
Department of Mathematics, Banaras Hindu University, Varanasi 221005, India
e-mail: drsahudr@gmail.com

A. Kumar
e-mail: ajeetbhu09@gmail.com

© Springer Nature Singapore Pte Ltd. 2018
D. Ghosh et al. (eds.), *Mathematics and Computing*, Springer Proceedings
in Mathematics & Statistics 253, https://doi.org/10.1007/978-981-13-2095-8_7

convex subset of a uniformly convex Banach space to the fixed and common fixed point, which mainly generalize a fixed point result of Rhoades [6].

Some preliminary definitions and earlier results of other authors noted in [1, 2, 4–7] and an extension of Mann type of iteration scheme defined for a finite sequence of mappings are the following:

Definition 1 A mapping T from a Banach space X into itself is said to be nonexpansive if T satisfies

$$\|Tx - Ty\| \leq \|x - y\| \text{ for all } x, y \in X.$$

In the setting of a Banach space, Dotson [1] introduced a new class of mappings, called quasi-nonexpansive, in the following manner:

Definition 2 [1] Let X be a Banach space, and let C be a convex subset of X. A self-mapping T of C is said to be quasi-nonexpansive, provided T has a fixed point, say p, in C, if

$$\|Tx - p\| \leq \|x - p\|$$

is true for all $x \in C$.

Definition 3 The modulus of convexity of a Banach space E is a function $\delta : (0, 2] \to (0, 1]$ defined by

$$\delta(\epsilon) = \inf\{\|x - y\| : x, y \in E, \|x\| = \|y\| = 1, \|x - y\| \geq \epsilon\}.$$

It is well known [4] that if E is uniformly convex then δ is strictly increasing, $\lim_{\epsilon \to 0} \delta(\epsilon) = 0$ and $\delta(2) = 1$. Let η be the inverse of δ, then we note that $\eta(t) < 2$ for $t < 1$.

Lemma 1 [2] *Let E be a uniformly convex Banach space and B_r be the closed ball in E centered at the origin with radius $r > 0$. If $x_1, x_2, x_3 \in B_r$,*
$$\|x_1 - x_2\| \geq \|x_2 - x_3\| \geq d > 0 \text{ and } \|x_2\| \geq \left(1 - \frac{1}{2}\delta\left(\frac{d}{r}\right)r\right)$$
then
$$\|x_1 - x_3\| \leq \eta\left(1 - \frac{1}{2}\delta\left(\frac{d}{r}\right)\right)\|x_1 - x_2\|.$$

Petryshyn and Williamson [5] proved the following result on the convergence of iterates of a quasi-nonexpansive mapping.

Theorem 1 *Let C be a closed subset of a Banach space X and $T : C \to X$ be a quasi-nonexpansive mapping. Suppose there exists a point x_0 in C such that $x_n = T^n x_0 \in C, n \in \mathbb{N}$. Then, the sequence $\{x_n\}$ converges to a fixed point of T in C if and only if $\lim_{n \to \infty} D(x_n, F(T)) = 0$, where $F(T)$ is the fixed point set of T.*

Definition 4 [4] For a self-mapping T of C and $x_0 \in C$, the Mann type of iteration is defined as

$$x_{n+1} = (1 - t_n)x_n + t_n T x_n, \tag{1}$$

where $t_n \in (\alpha, \beta)$, $0 < \alpha < \beta < 1$, $n = 0, 1, 2, \ldots$

Theorem 2 [7] *Let X be a uniformly convex Banach space, C a closed convex subset of X, and T a quasi-nonexpansive mapping of C into itself. Let $\phi : [0, \infty) \to [0, \infty)$ be a nondecreasing function with $\phi(0) = 0$ and $\phi(t) > 0$ for $t \in (0, \infty)$. If T satisfies*

$$\|x - Tx\| > \phi(D(x, F(T)))$$

for all $x \in C$, then for arbitrary $x_0 \in C$, the sequence of Mann type of iterates given in (1) converges to a member of $F(T)$.

In a strictly convex Banach space, Rhoades [6] proved the following theorem:

Theorem 3 [6] *Let X be a strictly convex Banach space and C a closed convex subset of X. Let $T : C \to C$ be continuous, quasi-nonexpansive mapping and $T(C)$ be a subset of a compact set K of X. Then, the Mann iterates given by (1) converge strongly to a fixed point of T.*

Som et al. [8] generalized the Mann type of iteration as in (1) in the following manner:

Definition 5 Let $\{T_k\}_{k=1}^N$ be a finite family of self-mappings of a convex subset C of a Banach space X, and let $\{t_k\}_{k=1}^N$ be a finite sequence in $(0, 1]$. For $x_0 \in C$, we define the modified Mann type of iteration as

$$
\begin{cases}
x_{iN+1} = t_1 x_{iN} + (1 - t_1) T_1 x_{iN}; \\
x_{iN+2} = t_2 x_{iN+1} + (1 - t_2) T_2 x_{iN+1}; \\
\ldots \quad \ldots \quad \ldots \quad \ldots \quad \ldots \\
x_{iN+k} = t_k x_{iN+k-1} + (1 - t_k) T_k x_{iN+k-1}; \\
\ldots \quad \ldots \quad \ldots \quad \ldots \quad \ldots \\
x_{iN+N} = t_N x_{iN+N-1} + (1 - t_N) T_N x_{iN+N-1};
\end{cases}
\tag{2}
$$

for $i = 0, 1, 2, \ldots$.

For a finite family $\{T_k\}_{k=1}^N$ of quasi-nonexpansive mappings, Som et al. [8] proved the following result on convergence of Mann type of iterates to common fixed point of a finite family of mappings.

Theorem 4 [8] *Let C be a nonempty convex subset of a uniformly convex Banach space. Let $\{T_k\}_{k=1}^N$ be a finite sequence of quasi-nonexpansive mappings of C into itself. Let the graph of each T_k be closed and one of $T_k(C)$, $k = 1, 2, \ldots, N$ be compact. If the family $\{T_k\}_{k=1}^N$ has a common fixed point in C, then the modified Mann type of iterates given by (2) converge to the common fixed point of the family.*

2 Main Results

As a particular case of Definition 5, we note the following as the modification of Mann type of iteration (1) for a single map:

Definition 6 Let T be a self-mapping of a convex subset C of a Banach space X and $\{t_k\}_{k=1}^N$ be a finite sequence in $(0, 1]$. For $x_0 \in C$, we define the modified Mann type of iteration as

$$
\begin{cases}
x_{iN+1} = t_1 x_{iN} + (1 - t_1) T x_{iN}; \\
x_{iN+2} = t_2 x_{iN+1} + (1 - t_2) T x_{iN+1}; \\
\ldots \quad \ldots \quad \ldots \quad \ldots \quad \ldots \\
x_{iN+k} = t_k x_{iN+k-1} + (1 - t_k) T x_{iN+k-1}; \\
\ldots \quad \ldots \quad \ldots \quad \ldots \quad \ldots \\
x_{iN+N} = t_N x_{iN+N-1} + (1 - t_N) T x_{iN+N-1};
\end{cases}
\tag{3}
$$

for $i = 0, 1, 2, \ldots$.

Using the iteration scheme (3), we have our first result on convergence to the fixed point of a single quasi-nonexpansive mapping, which generalizes Theorem 3 in respect of the iteration scheme.

Theorem 5 *Let C be a nonempty convex subset of a uniformly convex Banach space. Let T be a quasi-nonexpansive mapping of C into itself. Let the graph of T be closed and $T(C)$ be compact. If T has a fixed point in C, then the modified Mann type of iterates given by (3) converge to a fixed point of T.*

Proof The proof is similar to that of Theorem 4 [8], so we omit it.

As the next generalization of Mann iteration scheme for single mapping defined in (1) and also of the generalized Mann iteration scheme for n-mappings of Som et al. [8], we further modify it for a finite family of mappings in a different way and define it in the following manner.

Definition 7 Let $\{T_k\}_{k=1}^{N+1}$ be a finite family of self-mappings of a convex subset C of a Banach space X, and let $\{t_k\}_{k=1}^N$ be a finite sequence in $(0, 1]$. For $x_0 \in C$, we define the modified Mann type of iteration as

$$
\begin{cases}
x_{iN+1} = t_1 T_1 x_{iN} + (1 - t_1) T_2 x_{iN}; \\
x_{iN+2} = t_2 T_2 x_{iN+1} + (1 - t_2) T_3 x_{iN+1}; \\
\ldots \quad \ldots \quad \ldots \quad \ldots \quad \ldots \\
\ldots \quad \ldots \quad \ldots \quad \ldots \quad \ldots \\
x_{iN+k} = t_k T_k x_{iN+k-1} + (1 - t_k) T_{k+1} x_{iN+k-1}; \\
\ldots \quad \ldots \quad \ldots \quad \ldots \quad \ldots \\
x_{iN+N} = t_N T_N x_{iN+N-1} + (1 - t_N) T_{N+1} x_{iN+N-1};
\end{cases}
\tag{4}
$$

for $i = 0, 1, 2, \ldots$.

For such a finite family $\{T_k\}_{k=1}^{N+1}$ of quasi-nonexpansive mappings, we have the following result generalizing all such previous results established by other authors.

Theorem 6 *Let C be a nonempty convex subset of a uniformly convex Banach space. Let $\{T_k\}_{k=1}^{N+1}$ be a finite family of quasi-nonexpansive mappings of C into itself. Let the graph of each T_k be closed and one of $T_k(C)$, $k = 1, 2, \ldots, N + 1$ be compact. If the family of mappings $\{T_k\}_{k=1}^{N+1}$ has a common fixed point in C, then the modified Mann type of iterates given by (4) converge to the common fixed point of the family.*

Proof First, we show that

$$\|T_k x_{iN+k-1} - T_{k+1} x_{iN+k-1}\| \to 0 \text{ as } iN \to \infty$$

for each $k = 1, 2, \ldots, N + 1$.

If possible, let for a given $\epsilon > 0$, there exists a subsequence $\{iN_j\}$ of $\{iN\}$ such that

$$\|T_k x_{iN_j+k-1} - T_{k+1} x_{iN_j+k-1}\| \geq \epsilon. \tag{5}$$

Let $u \in C$ be a common fixed point of the family of mappings $\{T_k\}_{k=1}^{N+1}$. Then by quasi-nonexpansiveness of T_k and T_{k+1}, we get for each $k = 1, 2, \ldots, N + 1$,

$$
\begin{aligned}
\|x_{iN+k} - u\| &= \|(t_k T_k x_{iN+k-1} + (1 - t_k) T_{k+1} x_{iN+k-1}) - (t_k T_k u + (1 - t_k) T_{k+1} u)\| \\
&\leq t_k \|T_k x_{iN+k-1} - u\| + (1 - t_k)\|T_{k+1} x_{iN+k-1} - u\| \\
&\leq t_k \|x_{iN+k-1} - u\| + (1 - t_k)\|x_{iN+k-1} - u\| \\
&= \|x_{iN+k-1} - u\|.
\end{aligned}
$$

Thus, $\{\|x_{iN+k} - u\|\}$ is a decreasing sequence of nonnegative reals, and therefore, it is convergent. From (5), we have,

$$\|(T_k(x_{iN_j+k-1}) - T_k u) - (T_{k+1}(x_{iN_j+k-1}) - T_{k+1} u)\| \geq \epsilon.$$

Then for this ϵ, by uniform convexity of the space, there exists a δ, $0 < \delta < 1$, such that

$$
\begin{aligned}
\|x_{iN_j+k} - u\| &= \|(t_k T_k x_{iN_j+k-1} + (1 - t_k) T_{k+1} x_{iN_j+k-1}) - (t_k T_k u + (1 - t_k) T_{k+1} u)\| \\
&\leq \delta \max\{\|T_k x_{iN_j+k-1} - u\|, \|T_{k+1} x_{iN_j+k-1} - u\|\}.
\end{aligned}
$$

Since T_{k+1} is quasi-nonexpansive, we get,

$$
\begin{aligned}
\|x_{iN_j+k} - u\| &\leq \delta \|T_{k+1} x_{iN_j+k-1} - u\| \\
&\leq \delta \|x_{iN_j+k-1} - u\| \\
&\leq \delta^{j+k} \|x_0 - u\| \to 0 \text{ as } j \to \infty.
\end{aligned}
$$

Therefore, $\|x_{iN+k} - u\| \to 0$ as $iN \to \infty$.
Now,

$$\|T_k x_{iN_j+k-1} - T_{k+1} x_{iN_j+k-1}\| \le \|T_k x_{iN_j+k-1} - u\| + \|u - T_{k+1} x_{iN_j+k-1}\|$$
$$\le 2\|x_{iN_j+k-1} - u\| \to 0, \text{ as } iN_j \to \infty.$$

which contradicts (5) and therefore, for $k = 1, 2, \ldots, N + 1$,

$$\|T_k x_{iN_j+k-1} - T_{k+1} x_{iN_j+k-1}\| \to 0, \quad \text{as } iN \to \infty. \tag{6}$$

Let $T_{k+1}(C)$ be compact. Then by compactness, there is a subsequence $\{T_{k+1} x_{iN_j+k-1}\}$ which is convergent in C.
Let

$$\lim_{j \to \infty} T_{k+1} x_{iN_j+k-1} = z \in C.$$

Then from (6), we have $\lim_{j \to \infty} T_k x_{iN_j+k-1} = z$. Now

$$\|z - T_{k+1}z\| \le \|z - T_k x_{iN_j+k-1}\| + \|T_k x_{iN_j+k-1} - T_{k+1} x_{iN_j+k-1}\|$$
$$+ \|T_{k+1} x_{iN_j+k-1} - T_{k+1}z\| \tag{7}$$

Since T_{k+1} has a closed graph, therefore $\|T_{k+1} x_{iN_j+k-1} - T_{k+1}z\| \to 0$ as $j \to \infty$ as such right-hand side of (7) tends to zero as $j \to \infty$. Hence,

$$T_{k+1}z = z \text{ for } k = 0, 1, 2, \ldots, N.$$

Thus, z is a common fixed point of $\{T_k\}_{k=1}^{N+1}$, as such the sequence $\{\|x_{iN+k} - z\|\}$ is a decreasing sequence. But

$$\|x_{iN_j+k} - z\| \le \|(t_k T_k x_{iN_j+k-1} + (1 - t_k) T_{k+1} x_{iN_j+k-1}) - T_k x_{iN_j+k-1}\|$$
$$+ \|T_k x_{iN_j+k-1} - z\|$$
$$\le (1 - t_k)\|T_{k+1} x_{iN_j+k-1} - T_k x_{iN_j+k-1}\| + \|T_k x_{iN_j+k-1} - z\|$$

which tends to 0 as $j \to \infty$.
That is, the sequence $\{\|x_{iN+k} - z\|\}$ has a subsequence converging to 0 and therefore

$$\|x_{iN+k} - z\| \to 0$$

as $iN \to \infty$, i.e., $x_{iN+k} \to z$ as $iN \to \infty$. This completes the proof of the theorem.

3 Numerical Example

The following example is in support of Theorem 6.

Example 1 Let $X = \mathbb{R}$ and $C = [0, 1]$. Define a mapping $T_i : C \to C$ by

$$T_i x = x^{i+1} \text{ for all } x \in C.$$

It is clear that each T_i is a nonlinear continuous self-mapping on C with unique fixed point $p = 0$. Moreover,

$$\begin{aligned}
|T_i x - p| = |x^{i+1} - 0| &= |x^i||x - 0| \\
&\leq |x - p| \\
&\leq |x - p| \text{ for all } x \in C,
\end{aligned}$$

i.e., T_i is quasi-nonexpansive mapping. However, T_1 is not nonexpansive. Indeed, for $x = 14/30$ and $y = 17/30$, we have

$$\begin{aligned}
|T_1 x - T_1 y| = |(14/30)^2 - (17/30)^2| \\
= 0.103333 > 1/10 = |14/30 - 17/30| = |x - y|.
\end{aligned}$$

Similarly, T_2 is not nonexpansive. Indeed, for $x = 19/30$ and $y = 20/30$, we have

$$\begin{aligned}
|T_2 x - T_2 y| = |(19/30)^3 - (20/30)^3| \\
= 0.042259 > 1/30 = |19/30 - 20/30| = |x - y|.
\end{aligned}$$

Finally, T_3 is not nonexpansive. Indeed, for $x = 19/30$ and $y = 20/30$, we have

$$\begin{aligned}
|T_3 x - T_3 y| = |(19/30)^4 - (20/30)^4| \\
= 0.036640 > 1/30 = |19/30 - 20/30| = |x - y|.
\end{aligned}$$

Here $N = 2$. Hence, the Mann type of iteration (4) reduces to

$$\begin{cases}
x_{i(N-1)+1} = t_1 T_1 x_{i(N-1)} + (1 - t_1) T_2 x_{i(N-1)}; \\
x_{i(N-1)+2} = t_2 T_2 x_{i(N-1)+1} + (1 - t_2) T_3 x_{i(N-1)+1};
\end{cases} \tag{8}$$

for all $i = 0, 1, 2, \ldots$.

Thus, all the assumptions of Theorem 6 are satisfied. Hence, from Theorem 6, it follows that the sequence generated by (8) converges to the common fixed point of the family $\{T_k\}_{k=1}^{N+1}$. It is clear that the sequence $\{x_n\}$ generated by the proposed iterative scheme converges to $\{0\}$. For different initial values $x_0 = 0.99, .999, .9999, .999999$ and $x_0 = 0.999999999$ and $t_1 = t_2 = .5$, the convergence of sequence $\{x_n\}$ is shown in Fig. 1.

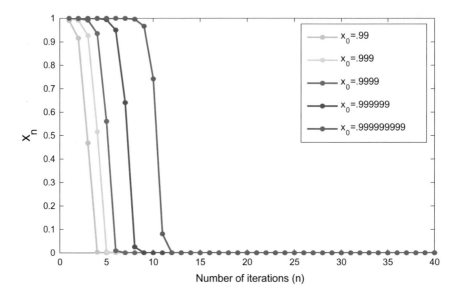

Fig. 1 Convergence of iterative method (4)

4 Conclusion

Our definition 7 is a more generalized version of the iteration scheme involving $N + 1$ mappings, and the Theorems 5 and 6 generalize the earlier results of Petryshyn and Williamson [5], Senter and Dotson [7], and Rhoades [6] not only in the sense of iteration scheme but also in the sense of mapping, which was considered to be continuous in the result of Rhoades [6]. In our case, the mappings considered need not be continuous.

References

1. Dotson Jr., W.G.: Fixed points of quasi nonexpansive mappings. J. Aust. Math. Soc. **13**, 167–170 (1972)
2. Goebel, K., Kirk, W.A., Shimi, T.N.: A fixed point theorem in uniformly convex spaces. Bol. Un. Mat. Ital. **7**(4), 67–75 (1973)
3. Iseki, K.: On common fixed point of mappings. Bull. Aus. Math. Soc. **10**, 75–87 (1974)
4. Mann, W.R.: Mean value methods in iteration. Proc. Amer. Math. Soc. **4**, 506–510 (1953)
5. Petryshyn, W.V., Williamson, T.E.: A necessary and sufficient condition for the convergence of iterates for quasi non-expansive mappings. Bull. Amer. Math. Soc. **78**, 1027–1031 (1972)
6. Rhoades, B.E.: Comments on two fixed point iteration methods. J. Math. Anal. Appl. **56**, 741–750 (1976)
7. Sentor, H.F., Dotson, W.G.: Approximating fixed points of non expansive mappings. Proc. Amer. Math. Soc. **44**, 375–379 (1974)
8. Som, T., Das S.: Convergence of modified Mann type of iterates and fixed point. Proc. Nat. Acad. Sci. (India) **70**(A)(II), 185–189 (2000)

Chapter 8
Geometric Degree Reduction of Bézier Curves

Abedallah Rababah and Salisu Ibrahim

Abstract We consider the weighted-multi-degree reduction of Bézier curves. Based on the fact that exact degree reduction is not possible, therefore approximative process to reduce a given Bézier curve of high degree n to a Bézier curve of lower degree m, $m < n$ is needed. The weight function is used to better representing the approximative curve at some parts that need more details, and the error is greater than other parts. The L_2 norm is used in the degree reduction process. Numerical results and comparisons are supported by examples. The numerical results obtained from the new method yield minimum approximation error, improve the approximation in some parts of the curve, and show up possible applications in science and engineering.

Keywords Bézier curves · Multiple degree reduction · Geometric continuity

1 Introduction

The problem of degree reduction of Bézier curve is to approximate an original Bézier curve of degree n with another Bézier curve of degree m, $m < n$ under the satisfaction of boundary conditions and minimum error conditions. Degree reduction is important in different fields of science, medical physics, network design, engineering, and industrial applications. So many scientists had tried several times to find a solution to degree reduction. The approach to the problem of degree reduction leads to solving a nonlinear problem. This requires numerical methods. In 1999, Lutterkort et al. proved in [1] that degree reduction of Bézier curves in the L_2 norm equals best

A. Rababah (✉)
Department of Mathematical Sciences, United Arab Emirates University, Al Ain, UAE
e-mail: rababah@just.edu.jo

S. Ibrahim
Department of Mathematics Northwest University, Kano 3220, Nigeria
e-mail: salisuibrahim@nwu.edu.ng

© Springer Nature Singapore Pte Ltd. 2018
D. Ghosh et al. (eds.), *Mathematics and Computing*, Springer Proceedings
in Mathematics & Statistics 253, https://doi.org/10.1007/978-981-13-2095-8_8

Euclidean approximation of Bézier points; see also [5]. These results are generalized to the constrained case by Ahn et al. in [2], and the discrete cases have been studied in [3]. In 2007, Rababah et al. used in [4] the idea of basis transformation between Bernstein and Jacobi basis to ascertain multi-degree reduction of Bézier curves. The existing methods to find degree reduction have many issues including: accumulate round-off errors, stability issues, complexity, accuracy, losing conjugacy, requiring the search direction to be set to the steepest descent direction frequently, experiencing ill-conditioned systems, leading to a singularity, and the most challenging difficulty is in applying the methods (difficulty and indirect). A. Rababah and S. Mann presented also in [5] linear G^1, G^2, and G^3-multiple degree reduction methods for Bézier curves. The weighted G^0- and G^1, weighted G^1, and weighted G^2-multiple degree reduction methods for Bézier curves are studied by Rababah and Ibrahim in [6–8] respectively. Woźny and lewanowrez degree reduced Bézier curves using dual Bernstein basis in [9]. Due to the new development in digital technology, [10] use the approach of Bézier curve for Automated Offline Signature Verification with Intrusion Identification. The research on Bézier curves has extended to the area of Medical Image Visual Appearance Improvement Using Bihistogram Bézier curves Contrast Enhancement in [11]. In all existing degree reducing methods, the conditions and free parameters were applied at the end points.

The main contribution of this paper is to introduce the weight with the problem of degree reduction of Bézier curves. So that it gives more weight to the center of the curve. It is appropriate to consider degree reduction with the weight function $w(t) = 2t(1 - t), t \in [0, 1]$. The result obtained carries all general advantages such as better approximation at the center of the curves, minimum error, simplicity in design, and implementation over existing results.

2 Preliminaries

Definition 1 A Bézier curve $P_n(t)$ of degree n is defined algebraically as follows:

$$P_n(t) = \sum_{i=0}^{n} p_i B_i^n(t) \quad 0 \le t \le 1, \tag{1}$$

where

$$B_i^n(t) = \binom{n}{i}(1 - t)^{n-i} t^i, \quad i = 0, 1, \ldots, n,$$

are the Bernstein polynomials of degree n, and $p_0, p_1, \ldots, p_n$ are called the Bézier control points of the Bézier curve.

Multiplication of two Bernstein polynomials with the weight function $2t(1 - t)$ is given by

$$B_i^m(t)B_j^n(t)2t(1-t) = \frac{2\binom{m}{i}\binom{n}{j}}{\binom{m+n+2}{i+j+1}}B_{i+j+1}^{m+n+2}(t).$$

We define the Gram matrix $G_{m,n}$ as the $(m+1) \times (n+1)$-matrix, whose elements are given by

$$g_{ij} = \int_0^1 B_i^m(t)B_j^n(t)2t(1-t)dt = \frac{2\binom{m}{i}\binom{n}{j}}{(m+n+3)\binom{m+n+2}{i+j+1}}, \qquad (2)$$
$$i = 0,\ldots,m, \quad j = 0,1,\ldots,n.$$

The matrix $G_{m,m}$ is real, symmetric, and positive definite [5].

Geometric continuity describes the continuity of two curves with some geometric properties. It is independent of their parametrization and denoted by G^k. Geometric continuity produces additional free parameters; see [5, 12] that are used to minimize the error.

Definition 2 Bézier curves P_n and R_m are said to be G^k-continuous at $t = 0, 1$ if there exists a strictly increasing parametrization $s(t) : [0, 1] \rightarrow [0, 1]$ with $s(0) = 0, s(1) = 1$, and

$$R_m^{(i)}(t) = P_n^{(i)}(s(t)), \quad t = 0,1, \quad i = 0,1,\ldots,k. \qquad (3)$$

3 Degree Reduction of Bézier Curves

Degree reduction can be defined as a method of approximating a given Bézier curve of degree n by a Bézier curve of degree m, $m < n$. Degree reduction is approximative process in nature, and exact degree reduction is ordinarily not possible. In this paper, our goal is to find a Bézier curve $R_m(t)$ of degree m with control points $\{r_i\}_{i=0}^m$ that approximates a given Bézier curve $P_n(t)$ of degree n with control points $\{p_i\}_{i=0}^n$, where $m < n$. The Bézier curve R_m has to satisfy the following two conditions:

(1) P_n and R_m are G^k-continuous at the end points for k = 0,1, and
(2) the L_2-error between P_n and R_m is minimum.

We can write the two Bézier curves $P_n(t)$ and $R_m(t)$ in matrix form as.

$$P_n(t) = \sum_{i=0}^n p_i B_i^n(t) =: B_n P_n, \quad 0 \leq t \leq 1, \qquad (4)$$

$$R_m(t) = \sum_{i=0}^{m} r_i B_i^m(t) =: B_m R_m, \quad 0 \leq t \leq 1, \tag{5}$$

where B_n, B_m are the row matrices containing the Bernstein polynomials of degree n, m, respectively, and P_n and R_m are the column matrices containing the Bézier points of degrees n and m, respectively.

In this paper, we use the weighted L_2-norm to measure distance between the Bézier curves $P_n(t)$ and $R_m(t)$; therefore, the error term becomes

$$\varepsilon = \int_0^1 ||B_n P_n - B_m R_m||^2 2t(1-t)dt$$

$$= \int_0^1 ||B_n P_n - B_m^c R_m^c - B_m^f R_m^f||^2 \cdot 2t(1-t)dt. \tag{6}$$

The linear system is constructed and solved for each of the conditions of the G^1- and G^2-degree reductions. The control points of the Bézier curve are expanded into their x and y components. Therefore, the variables of our system of equations are $r_k^x, r_k^y, k = 2, \ldots, m-2$, δ_0 and δ_1 and $r_k^x, r_k^y, k = 3, \ldots, m-3$, η_0 and η_1 for G^1- and G^2-degree reductions respectively.

The unknowns have the following solution form; see [5]

$$R_m^F = (G_{m,m}^F)^{-1}\left(G_{m,n}^{PC} P_n^C - G_{m,m}^C R_m^C\right). \tag{7}$$

4   Applications

This section provides two examples to support and validates the theoretical results of the discussed methods.

Example 4.1 Given a Bézier curve $P_n(t)$ of degree 12 with control points;
$P_0 = (0.224, 0.213)$, $P_1 = (0.248, 0.327)$, $P_2 = (0.079, 0.377)$, $P_3 = (0.004, 0.497)$, $P_4 = (0.544, 0.587)$,
$P_5 = (0.068, 0.511)$, $P_6 = (0.529, 0.131)$, $P_7 = (-0.274, 0.516)$, $P_8 = (0.248, 0.531)$, $P_9 = (0.194, 0.383)$,
$P_{10} = (0.202, 0.357)$, $P_{11} = (0.494, 0.306)$, $P_{12} = (0.193, 0.141)$.
This curve is reduced to Bézier curve $R_m(t)$ of degree 8.

Figure 1 depicts the original curve in solid-black, weighted G^1- and G^2-degree reduction in dashed-green and dashed-red curve.

Figure 2 depicts the error plots in long thick blue, and dashed-orange curves represent weighted G^2- and G^1-degree reduction respectively.

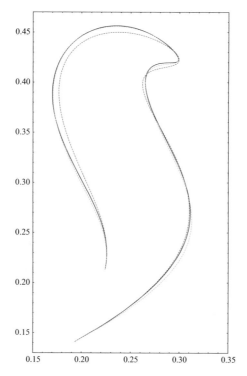

Fig. 1 Curves of degree 12 reduced to degree 8 with weighted G^1 and G^2 in (dashed-green and dashed-red) and original curve in (black)

Example 4.2 Given a Bézier curve $P_n(t)$ of degree 15 with control points; see [13].
$\mathbf{P}_0 = (0, 0)$, $\mathbf{P}_1 = (1.5, -2)$, $\mathbf{P}_2 = (4.5, -1)$, $\mathbf{P}_3 = (9, 0)$, $\mathbf{P}_4 = (4.5, 1.5)$,
$\mathbf{P}_5 = (2.5, 3)$, $\mathbf{P}_6 = (0, 5)$, $\mathbf{P}_7 = (-4, 8.5)$, $\mathbf{P}_8 = (3, 9.5)$, $\mathbf{P}_9 = (4.4, 10.5)$,
$\mathbf{P}_{10} = (6, 12)$, $\mathbf{P}_{11} = (8, 11)$, $\mathbf{P}_{12} = (9, 10)$, $\mathbf{P}_{13} = (9.5, 5)$, $\mathbf{P}_{14} = (7, 6)$,
$\mathbf{P}_{15} = (5, 7)$.
This curve is reduced to Bézier curve $R_m(t)$ of degree 8.

Figure 3 depicts the original curve in solid-black, weighted G^1 and G^2-degree reduction in dashed-green and dashed-red curve.

Figure 4 shows the curves with polygon are reduced to degree 8 with weighted G^1 and G^2-degree reduction in dashed-green and dashed-red curve and original curve in (black).

Figure 5 depicts the error plots in long thick blue, and dashed-orange curves represent weighted G^2- and G^1-degree reduction respectively. Figure 6 depict the figure from existing method; see [13].

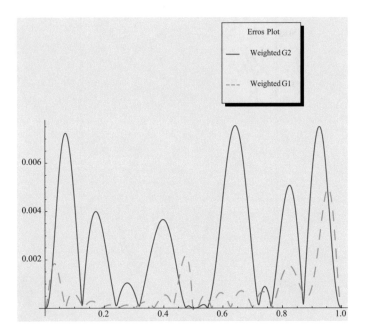

Fig. 2 Error Plots

Fig. 3 Curves of degree 15 reduced to degree 8 with weighted G^1 and G^2 in (dashed-green and dashed-red) and original curve in (black)

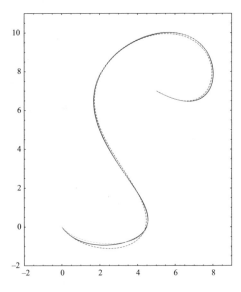

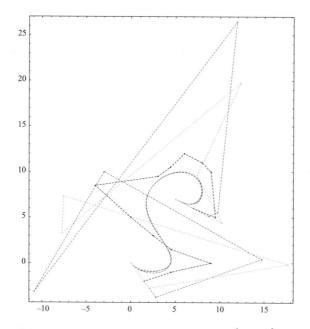

Fig. 4 Polygon of degree 15 reduced to degree 8 with weighted G^1 and G^2 in (dashed-green and dashed-red) and original curve in (black)

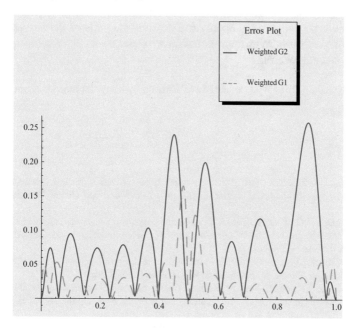

Fig. 5 Error plots

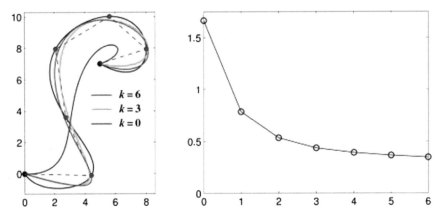

Fig. 6 Figure from existing method; see [13]

5 Conclusion

This paper investigates weighted-multi-degree reduction of Bézier curves. Explicit formula for weighted G^1 and G^2 method is used to reduce a given Bézier curve of high degree n to a Bézier curve of lower degree m, $m < n$, and these are achieved with the help of mathematica 9. Finally our numerical results show that the weight function helps to improve the approximation in some parts of the curves, and our new method yields minimum approximation error and shows up possible application in science and engineering.

Acknowledgements The authors would like to thank the reviewers for helpful comments.

References

1. Lutterkort, D., Peters, J., Reif, U.: Polynomial degree reduction in the L_2-norm equals best Euclidean approximation of Bézier coefficients. Comput. Aided Geom. Des. **16**, 607–612 (1999)
2. Ahn, Y., Lee, B.G., Park, Y., Yoo, J.: Constrained polynomial degree reduction in the L_2-norm equals best weighted Euclidean approximation of Bézier coefficients. Comput. Aided Geom. Des. **21**, 181–191 (2004)
3. Ait-Haddou, R.: Polynomial degree reduction in the discrete L_2-norm equals best Euclidean approximation of h-Bézier coefficients. BIT Numer, Math (2016)
4. Rababah, A., Lee, B.G., Yoo, J.: Multiple degree reduction and elevation of Bézier curves using Jacobi-Bernstein basis transformations. Num. Funct. Anal. Optim. **28**(9–10), 1179–1196 (2007)
5. Rababah, A., Mann, S.: Linear methods for G^1, G^2, and G^3–multi-degree reduction of Bézier curves. Comput.-Aided Des. **45**(2), 405–14 (2013)
6. Rababah, A., Ibrahim, S.: Weighted G^1-multi-degree reduction of Bézier curves. Int. J. Adv. Comput. Sci. Appl. **7**(2), 540–545 (2016)

7. Rababah, A., Ibrahim, S.: Weighted degree reduction of Bézier curves with G^2-continuity. Int. J. Adv. Appl. Sci. **3**(3), 13–18 (2016)
8. Rababah, A., Ibrahim, S.: Weighted G^0- and G^1 multi-degree reduction of Bézier curves. In: AIP Conference Proceedings 1738, 05, vol. 7, issue 2. p. 0005 (2016). https://doi.org/10.1063/1.4951820
9. Woźny, P., Lewanowicz, S.: Multi-degree reduction of Bézier curves with constraints, using dual Bernstein basis polynomials. Comput. Aided Geom. Des. **26**, 566–579 (2009)
10. Vijayaragavan, A., Visumathi, J., Shunmuganathan, K.L.: Cubic Bézier curve approach for automated offline signature verification with intrusion identification. Math. Prob. Eng. **2014**(Article ID 928039), 8 pp. (2014)
11. Gan, H.-S., Swee, T.T., Abdul Karim, A.H., Amir Sayuti, K., Abdul Kadir, M.R., Tham, W.-T., Wong, L.-X., Chaudhary, K.T., Ali, J., Yupapin, P.P.: Medical image visual appearance improvement using bihistogram Bezier curve contrast enhancement: data from the osteoarthritis initiative. Sci. World J. **2014**(Article ID 294104), 13 pp. (2014)
12. Lu, L., Wang, G.: Optimal multi-degree reduction of Bézier curves with G^2-continuity. Comput. Aided Geom. Des. **23**, 673–683 (2006)
13. Lu, L.: Sample-based polynomial approximation of rational Bézier curves. J. Comput. Appl. Math. **235**, 1557–1563 (2011)

Chapter 9
Cybersecurity: A Survey of Vulnerability Analysis and Attack Graphs

Rachid Ait Maalem Lahcen, Ram Mohapatra and Manish Kumar

Abstract The network infrastructure is the most critical technical asset of any organization. This network architecture must be useful, efficient, and secure. However, their cybersecurity challenges are immense as the number of attacks is increasing. Consequently, there is a need to have efficient tools to assess the risks, know the vulnerabilities, and find the solutions before the attackers exploit them. The challenges remain in integrating the vulnerability analysis tools in a holistic process that cyber defenders can use to detect an intrusion and respond quickly. Attack graphs showed great importance in analyzing security. In this paper, we present a survey of raised and related topics to the field of vulnerability analysis and attack graphs.

Keywords Attack graphs · Cybersecurity · Cyber situational awareness
Vulnerability analysis

1 Introduction

Enterprise networks continue to struggle with maintenance of network performance, availability, and security [1]. For instance, the Identity Theft Resources Center [2] had recorded 1339 US data breaches in 2017, exposing more than 174,402,528 confidential records. In cumulative view, between January 1, 2005, and December 27, 2017, number of breaches is 8190 with 1,057,771,011 of exposed records. Based on The Federal Bureau Investigation's (FBI) Internet Crime Complaint Center [3] receives an average of 280,000 complaints each year, or an average of 800 complaints a day, and in 2016 there was a total loss of $1.33 Billion. It is also widely recognized

R. Ait Maalem Lahcen · R. Mohapatra (✉)
Department of Mathematics, 4000 Central Florida Blvd., Orlando, FL 32816, USA
e-mail: ram.mohapatra@ucf.edu

R. Ait Maalem Lahcen
e-mail: rachid@ucf.edu

M. Kumar
Department of Mathematics, Birla Institute of Technology and Science-Pilani,
Hyderabad Campus, Hyderabad 500078, Telangana, India

© Springer Nature Singapore Pte Ltd. 2018   97
D. Ghosh et al. (eds.), *Mathematics and Computing*, Springer Proceedings
in Mathematics & Statistics 253, https://doi.org/10.1007/978-981-13-2095-8_9

that the time it takes for an organization to realize that they have been successfully attacked is measured in hundreds of days, and not in hours. Organizations in Europe, Middle East and Africa (EMEA) report [4] that it took three times longer to detect a compromise in the region, it was 469 days versus a global average of 146 days. Many organizations in EMEA were re-compromised within months of an initial breach. Consequently, it is crucial for any institution to analyze the security of its network from every access point. The attackers may have one or various motives, and they are determined to breach the systems. Once they enter one access point, they will try to penetrate every level in the network. Hence, the motivation of the defender to protect the systems cannot stop at the administrative duties. The defender must possess tools that can analyze enterprise network to discover vulnerabilities before the attackers do. One of the most effective methods is to search for all possible multi-access points, the various possible attack paths by building attack graphs and simulate the attacks [5]. The scenario graph demonstrates every possible path to break into a network security [5]. Consequently, network attack graph depicts all possible penetration scenarios. Attack graphs give an overview of potential scenarios that can lead to an unauthorized intrusion [6]. The challenge in security of zero-day exploits will always be a challenge since attackers develop exploits for those vulnerabilities that have not yet been disclosed. Hence, it is necessary to explore unexpected attackers behavior and not be limited by predefined information [7]. Since we see cyber situational awareness to be an important framework in which attack graphs can be implemented, we'll address it first. Some related and interesting work can be found in [3, 6–16].

2   Cyber Situational Awareness

Cyber situational awareness (CSA) is important for an effective cybersecurity analysis and incident response. However, it hasn't been well studied [11]. Several open-source tools and products were developed to tackle cyber problems, with US Government being a primary client. However, those tools have not improved CSA of cybersecurity analysts. Braford et al. in chapter 1 of "Cyber SA: Situational Awareness for Cyber Defense" discuss that aspects of situational awareness (SA) consist of [12]:

- Situation perception that includes recognition and identification of the type of attack, the target.
- Impact assessment that includes assessment of current and of future damage.
- Situation tracking is important to be aware of its progress.
- Awareness of intent and threat hunting techniques.
- Backtracking and forensics to analyze reasons and methods that caused a situation attack.
- Evaluation of the collected SA information.
- Lessons learned about current situation and how it'll evolve in the future.

Therefore, CSA can be summarized in three major steps [12]:

1. **Recognition** which provides basis for better SA.
2. **Comprehension** in which knowledge and data apply context to make sure the information is meaningful to the specific circumstances.
3. **Projection** that is used to make educated and informed assessment about future attacks and mitigate their threats.

The diagram in Fig. 1 shows how the situation can evolve in a nonlinear way.

This is equivalent to sensemaking in [8] that includes learning new areas, solving not so well-defined problems, acquiring SA, and participating in knowledge sharing; as those steps should lead to deeper understanding. CSA requires time to develop, and one should work on building a model that better prepares for future attacks. It is clear that cyber defenders ought to deal with uncertainty as it is not possible for them to be aware of everything running within every computer inside the network. There is also no efficient mechanism to digest the logs even if every device can be logged. To summarize, one should find answers to these questions in CSA [17]:

- Is there an intrusion?
- Where is the intruder?
- How does the situation evolve?
- What is the impact of the attack on the network?
- How to assess a damage?
- What behavior is expected from the attackers?
- What strategies they may take?

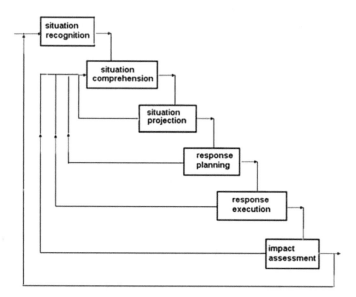

Fig. 1 A nonlinear SA process [17]

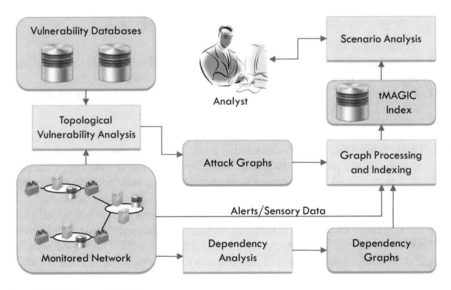

Fig. 2 CSA framework [17]

- Can we predict future scenarios of the current situation ?
- How did the intruder manage to make it happen?
- What was the target or goal of the intruder?

Figure 2 depicts CSA framework in which vulnerability analysis is conducted by a topological approach allowing to generate attack graphs by encoding probabilistic knowledge of the attackers' behavior. They merged multiple attack types to a compact data structure and define an index structure on top of it to classify multiple alerts and data from sensors. A dependency analysis is performed to generate dependency graphs. Consequently, attack graph scenario is made from joining dependency graphs and attack graphs. Scenario graphs show ways in which an intruder can exploit known vulnerabilities and affect the system. The authors also proposed an algorithm for both detection and prediction, and it scaled well with large graphs [17].

3   Attack Graphs

Computer networks may have vulnerabilities that can be exploited in ways that serve the goals of the intruder. Although a successful attack may require multiple steps in various order, the usual network attack consists of these stages:

1. **Reconnaissance** in which attackers gather information about a target to use in the next step. Some of the techniques used are social engineering, physical reconnaissance, and dumpster diving. Reconnaissance can be active or passive depending on whether the interaction happened with the system or not.

2. **Scanning** is the next step to discover running services on a target computer or network. It is a development of active reconnaissance since the attacker engages with system to learn about its vulnerabilities.
3. **Gaining Access** is a logical next step after attempting to exploit identified vulnerabilities.
4. **Maintaining Access** is possible with the intruder planting own Trojan software, packet analyzer, or additional backdoor network access codes.
5. **Covering tracks** or a hiding stage in which the intruder tries to cover-up the crime. This stage may include cleaning logs, hidden background programs, and installing codes to conceal malicious software from legitimate users.

A case example with an attack graph is given by J. Li, X. Ou, and R. Rajagopalan in chapter 4 of [12]. In this example, attack paths are found after configuration analysis. Figure 3 shows it.

The intruder breaches web server (a critical attack vector, i.e. used in Equifax breach in 2017) from a remote location by exploitation of CVE-2002-0392 vulnerability and gains local access on the server. Then attempts to alter data on file server in order to exploit vulnerabilities to get access on the machine. The intruder installs a Trojan-horse program, and wait for a user on workstation to run it, and gain control of the station. Details of this scenario graph can be found in [12]. Although this attack graph, or any other attack graph of similar size, may look simple, it could still involve complicated computations of the likelihood that an attack can be successful. Figure 4 shows a simple example of an attack graph found in [16]. The oval nodes being the exploit nodes and the conditional nodes being the text nodes.

The complexity of attack graphs topology creates many shared dependencies. For instance, node c_{10} can be reached by an intruder from exploiting e_4 or e_5 that fully depend upon c_7. Hence, the paths to e_4 and e_5 are not independent. Furthermore, one

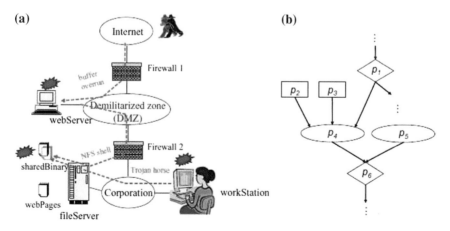

Fig. 3 A case example of an attack scenario and attack graph (WebSevrer p_1, NFS protocol p_2, WebServer p_3, File server p_4) [12]

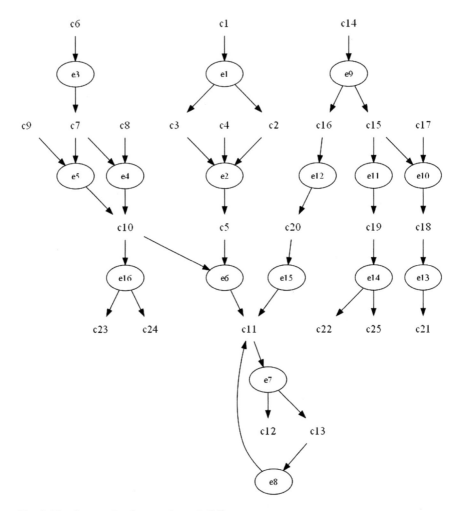

Fig. 4 Simple example of an attack graph [16]

cannot assume independence in attack graphs, and should measure the probability of possible multistep attacks. Yun et al. in [16] presented a method for security risk assessment that combines the attack graphs and the Common Vulnerability Scoring System (CVSS) in order to address incorrect probability computing caused by conjoint dependencies in nodes. Briefly, CVSS helps to identify the principal characteristics of a vulnerability and scores its severity. CVSS is formed by three metric groups stated in [10]:

1. Base including exploit ability metrics and impact metrics. This includes the ease to exploit the vulnerability component, and the consequence or the impacted component. Vulnerability characteristics that are constant across user and environment.

2. Temporal represents the characteristics of a changing vulnerability, yet not across user environments.
3. Environmental represent metrics characteristics of a vulnerability related to a particular user's environment.

The algorithm in [16] calculated either accurately or approximately the probability of nodes depending on their depth, a setting number, and a formed theorem. This algorithm solved the problem of probabilistic incorrect computing. It was experimented in a 5–20 hosts in a simulation and showed some effectiveness over HOMER's algorithm [18]. Wang, Du, and Yang presented an automated method that generates and analyzes attack graphs in [13]. They formed it using symbolic model checking algorithms and tested it on a small network example. They tested it on a small operational network using applied Network Security Planning Architecture and found a faulty firewall. Shahriari, Ganjisaffar, Jalili, and Habibi modeled networks' topologies, their configurations and vulnerabilities in [19]. A framework that is similar to MulVAL which we'll address later in the paper. They implemented an expert system based on a framework for automatic topological multihost vulnerability analysis. A methodology that explores all paths of attacks and combats unauthorized access by an attacker. The output of the expert system is accessed by the network administrator from the user interface which allows to control the inference engine. The latter processes logical inferences based on the knowledge base input, which collects facts and inference rules. Knowledge base component gets input from the host vulnerability extractor that takes information from vulnerability databases and host scanners. The expert system performed vulnerability analysis of a network with 1600 hosts in reasonable time (31 s) [19].

Noel and Jajodia applied adjacency matrix clustering to network attack graphs in order to correlate attacks and predict them [20]. Reachability across the network is found by self-multiplying the clustered adjacency matrices to find number of steps to an attack. The reachability analysis summaries how changing a network configuration can affect the attack graph. The graphical technique matches columns and rows of the clustered adjacency matrix to show multiple step attacks. This allows to identify impact depending on the number of steps to victim machines and identify the sources of the attack. The adjacency matrix brings simplicity to their approach since a single matrix element represents each graph edge. Graph vertices are implicitly represented as matrix rows and columns. The adjacency matrix avoids the typical crowded edge representation of small and large graphs. Their clustering algorithm is advantageous because it scales linearly with network size, it is parameter-free and completely automatic. Yang et al. experiment in [15] show that the built hierarchical architecture constructed is good for assessing the potential security risks of four levels: network, hosts, services, and vulnerabilities. The vulnerability attack link generated algorithm proposed in their paper could help system administrators mitigate the potential security risks in the computer system. This algorithm is composed of two subalgorithms: (1) host access link generated algorithm and (2)vulnerability attack link generated algorithm, details can be found in [15]. Abraham and Nair propose in [7] a stochastic approach for security evaluation based on attack graphs, taking into account CVSS scoring. They used MulVAL (developed by Kansas State University) to generate logical attack graphs in a polynomial time. A simulation of the Absorbing Markov chain is conducted on the attack graph generated for the network. They used a realistic network to analyze and capture security properties and optimize the application of patches. The proposed model can assist to harden the system by identifying its critical parts and predicting the total security variation over time [7].

Lippmann and Ingols, in 2005, surveyed attack graphs papers that focused on three goals [21]:

1. Papers construct attack graphs to analyze network security.
2. Papers about formal languages that are complex or simple to describe states in attack graphs. Those languages would typically define preconditions for a successful intrusion, and postconditions or changes in network state after an intrusion.
3. Papers describe attack graphs used with intrusion detection systems (IDS) to group alerts.

They found that most of algorithms were tested on small networks with fewer than 20 hosts. Consequently, we find that, after 2005, several papers tackled scalability problem and attempted larger networks but not the desirable to enterprise networks with over 10,000 hosts. Nevertheless, research using attack graphs has achieved a number of good prototypes that are summarized in Table 1.

Cauldron is a commercialized TVA that was developed by George Mason University, hence, it applies the concept shown in Fig. 5. In this paper, we limit the survey to TVA discussed below. FireMon is the commercialized NETSPA, adopted by FireMon, LLC. We also limit discussion to NETSPA. Another commercial toolkit is Skybox View by Sktbox Security Inc.; it has a polynomial complexity $O(n^3)$.

Table 1 Attack graphs toolkits [7]

Toolkit name	Complexity	Open source	Developer
MulVAL	$O(n^2)$ $O(n^3)$	Yes	Kansas State University
TVA	$O(n^2)$	No	George Mason University
NETSPA	$O(nlogn)$	No	MIT
Cauldron	$O(n^2)$	No	Commercial
Firemon	$O(nlogn)$	No	Commercial

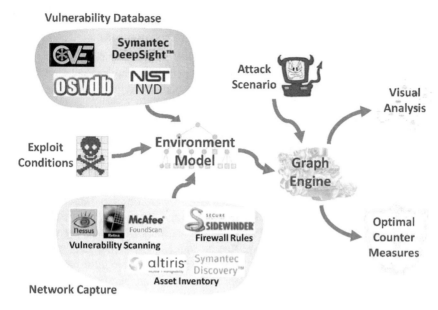

Fig. 5 Topological vulnerability analysis [12]

3.1 Topological Vulnerability Analysis

Jajodia and Noel discuss Topological Vulnerability Analysis (TVA) in [12]. TVA tries to discover the paths through a network that an intruder may follow. Figure 5 shows the concept of TVA architecture.

Network Capture builds a model of the network, *Vulnerability Database* represents a comprehensive repository of reported vulnerabilities and the record listing of the affected software or hardware, and the *Exploit Conditions* conceals how each vulnerability may be exploited and the consequence of the breach (preconditions and postconditions). All inputs from network capture are used to set up an *Environment Model* for multistep attack graph simulation. The *Graph Engine* generates all possible attack path scenarios after analyzing vulnerability dependencies, coordinating the before and after exploitation conditions. The TVA outputs *Visual Analysis* of attack graphs and calculate *Optimal Counter Measures*. TVA attack graphs can support intrusion detection system. TVA matches the network model against a database of reported vulnerabilities from the examples included in Fig. 5 [12]. Although TVA has some technical challenges like entering the exploits information by hand, it can be used to determine safe network configurations with respect to the goal of maximizing available network services. It also has potential application to identify possible attack responses and improve intrusion detection systems.

3.2 A Network Security Planning Architecture

A Network Security Planning Architecture (NETSPA) generates attack graphs from a network topology and graphs of all potential paths that can be exploited for a user-defined network. These graphs and their associated statistics, such as number of hosts compromised and attacker privilege levels, allow a network administrator to determine likely intrusion paths and extrapolate this data to determine the current and future security of the network given past software vulnerability frequencies. As the attack graphs are displayed in near real time, an administrator can change the network topology slightly, recompute the graphs for the new topology, and compare the graphs produced from different configurations. This allows an administrator to weigh network security against other factors, such as hardware costs and ease of maintenance. Finally, NETSPA imports information from several existing security and network planning tools. Existing network configuration information can be obtained through the use of tools such as nmap, Nessus, and NetViz. Online databases such as ICAT and the Nessus vulnerability plug-ins provide valuable information about attack requirements and effects [22]. Construction of an attack graph requires several pieces of information about the type of attacker, underlying network topology, number of attacks available to the attacker, and their types. Figure 6 illustrates these input components.

Only three inputs (the attack model, network and host vulnerabilities, and network topology) are essential to the creation of a useful attack graph. The attack model defines the state transition relation of an attack by stating its requirements and effects of executing an attack. The network topology limits the physical paths that an attacker

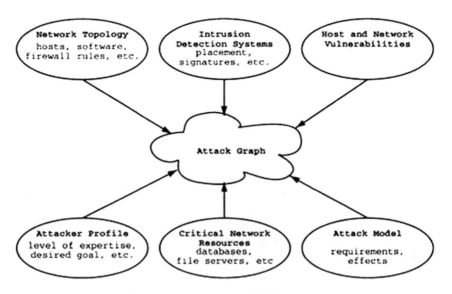

Fig. 6 Necessary information to create an attack graph [22]

can traverse within the network, subject to network connectivity and firewall rules. The host and network vulnerabilities and configurations define the possible set of initial actions that the attacker can take against the network using attacks from the attack model. The other three inputs to the attack graph (attacker profiles, intrusion detection systems, and critical network resources) are not required to generate an attack graph; however, they increase the utility of the constructed graph. The attacker profile defines the starting state of the intruder, as well as the methodology that he uses in choosing the next attack to execute. This enables the administrator to optimize a network's security against novice outside attackers, while accepting the possibility of an insider attack. A list of critical network resources also allow the security administrator to prune the complete attack graph to only those states which are judged critical, such as not allowing attacker access to a central billing database. Finally, the placement and type of the intrusion detection systems allow the graph generator to determine which paths are visible.

NETSPA was created to fill a void in existing security software. The primary design goal of NETSPA was to create a system that could automatically compute complete attack graphs for real, user-specified networks. This, in turn, leads to three separate subgoals: Allow a user to easily define a network and its resulting config-uration, enable quick modeling of realistic actions, and efficiently compute worst-case attack graphs with sufficient meta-information to be easily useful to the user. Secondary to the notion of attack graphs was that of simplicity and information reuse, most notably in the action specifications. The worst-case graphs generated by NETSPA illustrate all possible cyber-attack paths. They do not model physical attacks or human engineering attacks. Graph generation does not take into account the skill or predisposition of the attacker. It also assumes that attempts at "security by obscurity," such as passing SMTP traffic through the firewall on a non-standard port, fail. In addition, the model of an IDS is assumed to be "best-case." A host-based IDS always detects an attack launched against it, while a network-based IDS always detects attacks that are visible on the network if it has a signature for the attack. NETSPA is divided into several modules to achieve its goals, each component and resulting connectivity is shown in Fig. 7. As seen in the upper left of the figure and illustrated, the software database is the repository of software information used by NETSPA to make software names consistent. The software database is used by both the action database and the input filters to the network model to create a consistent software naming scheme among network configuration and action definitions. The action database shown in the middle left of figure contains information about every possible action that an attacker can execute against a user-defined network. The cre-ation of this network is aided by the user input filters, shown in the upper right of Fig. 7, which populate the network model with network configuration information. This network model is then used to create an initial network state, which is provided, along with the database of possible actions and set of existing trust relationships, to the computation engine. The computation engine then creates a worst-case attack graph for the specific set of inputs [22].

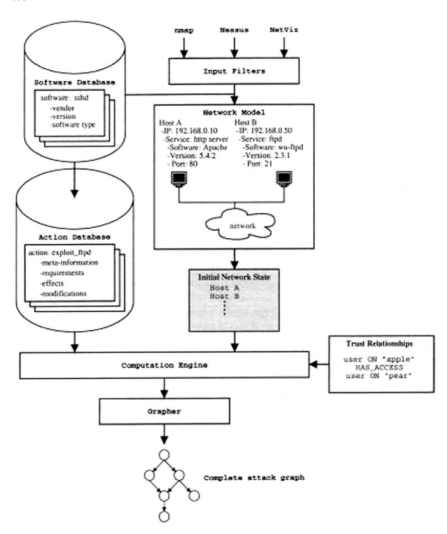

Fig. 7 NETSPA component diagram [22]

3.3 Multihost, Multistage, Vulnerability Analysis

Multihost, multistage, vulnerability analysis (MulVAL) project was developed at
Kansas State University as a research tool to better manage the configuration of an
enterprise network. Xinming Ou, Govindavajhala, and Appel discuss that MulVAL
uses datalog as the artificial language for the elements in the analysis [23]. The inputs
to MulVALs analysis are reported vulnerabilities or advisors, host configuration,
network configuration, the network users or principals, and policies like access levels.
Figure 8 shows MulVAL framework.

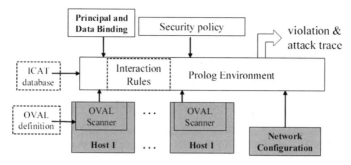

Fig. 8 The MulVAL framework [23]

The reasoning engine in MulVAL can handle the network size and perform analysis for thousands of machines. For scalability, MulVaL was tested on up to 2000 hosts. The scanners can execute in parallel on multiple machines. The analysis engine then operates on the data collected from all hosts. The OVAL scanner collects machine configuration information and compares the configuration with formal advisories to assess for vulnerabilities existence on a system. However, when a new advisory comes, the scanning will have to be repeated on each host which is not the most desirable technique. OVAL language is an XML-based language for specifying machine configuration tests. MulVAL runs efficiently for networks with thousands of hosts, and it has found security problems in a real network [23]. MulVaL is an open source and that gives an advantage to academic researchers.

4 Conclusion

Predicting total security on a given time is still a challenging task, and blocking sophisticated threats or advanced malware attacks is still less effective [24]. Attack graphs representation approaches had several developments since 1996 from enumeration approach to hybrid condition with exploit oriented approach and vulnerability oriented approach [25]. Good strides in addressing scalability and network vulnerability analysis were made, yet, there is still need to address complex large enterprise and multiple stage attacks. Those complex networks demand automatic expert system to analyze network topology, show exploitation scenarios, and rank relevant subgraphs to determine security measures that need to be deployed first. In addition, future research should improve the application of graph attacks algorithms by decreasing their complexity. Finally, there is also a need for research designs of security systems to better integrate and automate cyber situational awareness [26–33].

References

1. Filkins, B.: Network Security Infrastructure and Best Practices: A SANS Survey. SANS Institute, Washington (2017)
2. Identity Theft Resource Center: ITRC Data Breach Report (2017)
3. Smith, S.S.: Internet Crime Report **2016**, 29920 (2016)
4. Hall, T., Hau, B., Penrose, M., Bevilacqua, M.: Mandiant M-Trends 2016 EMEA Edition, pp. 1–18 (2016)
5. Liu, Z., Li, S., He, J., Xie, D., Deng, Z.: Complex network security analysis based on attack graph model. In: 2012 Second International Conference on Instrumentation, Measurement, Computer, Communication and Control, pp. 183–186 (2012)
6. Sheyner, O., Wing, J.: Tools for generating and analyzing attack graphs. In: 2nd International Symposium on Formal Methods for Components and Objects (FMCO'03), vol. 3188, pp. 344–371 (2004)
7. Abraham, S., Nair, S.: A Predictive Framework for Cyber Security Analytics Using Attack Graphs, pp. 1–17 (2015)
8. Pirolli, P., Russell, D.M.: Introduction to this special issue on sensemaking. Hum.-Comput. Interact. **26**, 1–8 (2011)
9. Seuwou, P., Banissi, E., Ubakanma, G., Sharif, M.S., Healey, A.: Actor-network theory as a framework to analyse technology acceptance model's external variables: the case of autonomous vehicles. Commun. Comput. Inf. Sci. **630**, 305–320 (2016)
10. Singhal, A., Ou, X.: Security Risk Analysis of Enterprise Networks Using Probabilistic Attack Graphs, pp. 1–22 (2011). https://doi.org/10.6028/nist.ir.7788
11. Stevens-Adams, S., Carbajal, A., Silva, A., Nauer, K., Anderson, B., Reed, T., Forsythe, C.: Enhanced Training for Cyber Situational Awareness. Lecture Notes in Computer Science (Including Subseries Lecture Notes in Artificial Intelligence and Lecture Notes in Bioinformatics) (LNAI), vol. 8027, pp. 90–99 (2013)
12. Jajodia, S., Peng, L., Swarup, V., Wang, C.: Cyber Situational Awareness Testing, vol. 2016, pp. 209–233. Springer (2016)
13. Wang, C., Du, N., Yang, H.: Generation and analysis of attack graphs. Procedia Eng. **29**, 4053–4057 (2012)
14. Wang, X., Liao, Y.: A replication detection scheme for sensor networks. Procedia Eng. **29**, 21–26 (2012)
15. Yang, J., Liang, L., Yang, Y. and Zhu, G.: A hierarchical network security risk assessment method based on vulnerability attack link generated. In: 2012 4th International Symposium on Information Science and Engineering (ISISE 2012), vol. 1, pp. 113–118 (2012)
16. Ye, Y., Xu, X.S., Qi, Z.C.: A probabilistic computing approach of attack graph-based nodes in large-scale network. Procedia Environ. Sci. **10**, 3–8 (2011)
17. Pino, R.E.: Cybersecurity Systems for Human Cognition Augmentation. Springer, New York (2014)
18. Homer, J., Ou, X., Schmidt, D.: A sound and practical approach to quantifying security risk in enterprise networks. Technical Report, pp. 1–15. Kansas State University (2009)
19. Hamid, R.S., Yasser, G., Rasool, J.: Topological analysis of multi-phase attacks using expert systems. J. Inf. Sci. Eng. **767**, 743–767 (2008)
20. Noel, S., Jajodia, S.: Understanding complex network attack graphs through clustered adjacency matrices. Proceedings-Annual Computer Security Applications Conference (ACSAC) **2005**, 160–169 (2005)
21. Lippmann, R.P., Ingols, K.W.: An annotated review of past papers on attack graphs. No. PR-IA-1 (2005)
22. Artz, M.L.: NetSPA: a network security planning architecture. Netw. Secur. **2001**, 1–97 (2002)
23. Ou, X., Govindavajhala, S., Appel, A.W.: MulVAL: a logic-based network security analyzer. In: Proceedings of the 14th conference on USENIX Security Symposium, vol. 14 (2005)
24. Oltsik, J.: Integrated Network Security Architecture: Threat-Focused Next-generation Firewall. The Enterprise Strategy Group, Inc. (2014)

25. Mell, P., Harang, R.: Minimizing attack graph data structures. In: ICSEA 2015: Tenth International Conference on Software Engineering Advances. Barcelona (2015)
26. Bacic, E., Froh, M., Henderson, G.: Mulval extensions for dynamic asset protection (2006)
27. Frigault, M., Wang, L.: Measuring network security using bayesian network-based attack graphs. In: Proceedings—International Computer Software and Applications Conference, pp. 698–703 (2008)
28. Kaynar, K.: A taxonomy for attack graph generation and usage in network security. J. Inf. Secur. Appl. **29**, 27–56 (2016)
29. Long, X., Wu, X.: Motion segmentation based on edge detection. Procedia Eng. **29**, 74–78 (2012)
30. Ma, J.C., Wang, Y.J., Sun, J.Y., Chen, S.: A minimum cost of network hardening model based on attack graphs. Procedia Eng. **15**, 3227–3233 (2011)
31. Mourad, A., Soeanu, A., Laverdière, M.A., Debbabi, M.: New aspect-oriented constructs for security hardening concerns. Comput. Secur. **28**, 341–358 (2009)
32. Ou, X., Govindavajhala, S., Appel, A.W: Policy-based multihost multistage vulnerability analysis (2005)
33. Dimitrios, P., Sarandis, M., Christos, D.: Expanding topological vulnerability analysis to intrusion detection through the incident response intelligence system. Inf. Manage. Comput. Secur. **4** (2010)

Chapter 10
A Solid Transportation Problem with Additional Constraints Using Gaussian Type-2 Fuzzy Environments

Sharmistha Halder (Jana), Debasis Giri, Barun Das, Goutam Panigrahi, Biswapati Jana and Manoranjan Maiti

Abstract This paper deals with nonlinear transportation problem where one part of unit transportation cost varies with distance from some origin, and the problems consist one more impurity restriction. Moreover, the fixed unit transportation costs are imprecise ones. In model I, some parameters (i.e. production cost, transport cost, supply, demand and unit of impurity at demand point) are considered as Gaussian type-2 fuzzy variable, while model II considered only the supply and demand which are deterministic. The type-2 fuzzy variables are transformed into type-I fuzzy variables with the help of CV-based reduction method. Genetic algorithm (GA) has been applied to solve the proposed models. Finally, an illustration is presented numerically to demonstrate the experimental results.

S. Halder (Jana)
Department of Mathematics, Midnapore College [Autonomous],
Midnapore 721101, India
e-mail: sharmistha792010@gmail.com

D. Giri (✉)
Department of Computer Science and Engineering,
Haldia Institute of Technology, Haldia, East Midnapore 721657, India
e-mail: debasis_giri@hotmail.com

B. Das
Department of Mathematics, Sidho Kanho Birsha University,
Purulia 723104, West Bengal, India
e-mail: bdasskbu@gmail.com

G. Panigrahi
Department of Mathematics, National Institute of Technology,
Durgapur 713209, West Bengal, India
e-mail: panigrahi_goutam@rediffmail.com

B. Jana
Department of Computer Science, Vidyasagar University,
Midnapore 721102, West Bengal, India
e-mail: biswapatijana@gmail.com

M. Maiti
Department of Applied Mathematics with Oceanology and Computer Programming,
Vidyasagar University, Midnapore 721102, West Bengal, India
e-mail: mmaiti2005@yahoo.co.in

© Springer Nature Singapore Pte Ltd. 2018
D. Ghosh et al. (eds.), *Mathematics and Computing*, Springer Proceedings
in Mathematics & Statistics 253, https://doi.org/10.1007/978-981-13-2095-8_10

Keywords Nonlinear solid transportation problem · Impurity constraints · Critical value · Gaussian type-2 fuzzy variables · Genetic algorithm · Reduction method

1 Introduction

The traditional transportation issue is one of the subclasses of nonlinear programming problem in which all the constrains are of equal type or of in-equal type. In traditional shape, the issue limits the aggregate of transporting an item which is accessible at a few sources and are required at different goals. The unit cost, i.e., the cost of transporting one unit from a specific supply point to a specific request point, the amount accessible at the supply focuses and the amount required at the request focuses are the parameters of the transportation issues. In reality circumstances, the transportation issue typically includes nonlinear, noncommensurable, numerous and clashing target capacities. This sort of issue is called nonlinear multi-target transportation issue. A few creators apply a distance function to present a numerical model of the nonlinear multi-target transportation issue. In this case, we propose the single objective function of transportation problem which become non-linear. In the wake of presenting the idea of fuzzy set theory by Zadeh [1] in 1965, Zimmermann [2] connected the fuzzy programming technique with some reasonable enrolment capacities to tackle multi-objective linear programming issues. The outcome acquired by fuzzy linear programming leads to effective arrangements, as well.

The standard/usual transportation issue [3] is a well-known improvement issue in operational research, in which two sorts of imperatives (source and goals) are mulled over. In any case, in genuine circumstances, it deals with other constraints such as the type of products, mode of transport and distance of path travels. As a result, the conventional transportation problems (2D-TPs) with conveyance constraints turn into the solid transportation problems (STPs/3D-TPs). The STP was first proposed by Schell [4]. As a speculation of normal TP, STP was presented by Haley [5] in 1962. In current years, STP has received much attention, many models and algorithms under both crisp and uncertain environment have been developed. There are many researchers who have worked in this area such as Jimenez et al. [6], Yang et al. [7], Liu et al. [8], Kocken et al. [9].

Ordinarily, separations of the courses between the sources to goals are not considered in TPs as the distance of the route stays unaltered and does not cause any effect in the minimization of cost/time. In true issue, these might be distinctive courses/ways for travel between a origin and a goal. Amongst these paths, the distance between the sources to goals is different. Per unit transportation costs and fixed charges along the routes are also different. Hence, choice of routes plays a major roll in maximization the profit in a TP. Thus, in a transportation problem, if different routes are considered besides different vehicles, then the three-dimensional transportation problem (3D-TP) is transformed into four-dimensional transportation problem (4D-TP). In an STP, when in excess of one items are put away at various steps and are transported to various goals utilizing diverse kinds of conveyances, the issue diminishes to a multi-item STP (MISTP). Many researchers such as Ojha et al. [10], Kundu et al. [11], Giri et al. [12] and others worked on MISTP. Similarly when in a 4D-TP, more than one item is considered as it becomes a multi-item 4D-TP (MI4D-TP).

Type-2 fuzzy sets are used due to its flexibility and degrees of freedom, and it is treated as three dimension. So, type-2 sets are more efficient for modelling uncertain problem accurately than type-1 fuzzy variable. Dubois and Prade [13] and Mizumoto and Tanaka [14] investigate the logical tasks of fuzzy type-2. Afterwards, huge lists of hypothetical research take a shot at the property of type-2 fuzzy variables [6, 13, 15], and its practical application has been developed [16, 17].

The present paper essentially researches the accompanying things:

- A computationally effective defuzzification procedure of type-2 fuzzy variables are introduced.
- In spite of the fact that TPs with type-1 fuzzy parameters are talked about by numerous specialists, transportation issues of type-2 fuzzy variable are composed and comprehended.
- Chance-constrained programming model with type-2 fuzzy variables is formulated.

Here, we have presented profit maximization STP with Gaussian type-2 fuzzy variables. A few sort of conveyances are utilized for transportation of merchandise from source to destinations. Here the transportation system is planned regarding a dealer who buys the source amounts at various starting points and sells the transported amount at different destinations as per the demands at destinations. Purchasing costs and selling price at different origins and destinations are different. Transportation costs, demands at destination, conveyance procurement cost and capacities are also Gaussian type-2 fuzzy variables.

2   Mathematical Model Formulation

2.1   Notations

The following notations are used

Index sets

- i: index for source for all i = 1, 2, ..., M.
- j: index for destination for all i = 1, 2, ..., N.

Decision variables

- w_{ij} units transported from ith origin to jth destination.
- (x_i, y_i) position of the ith origin. Objective functions z1 total transportation cost from ith origin to jth destinations.

Objective functions:

z1 total transportation cost from ith origin to jth destinations.

Parameters

- h_{ij} creation cost per unit conveyed from ith source to jth destination.
- c_{ij} transportation cost per unit conveyed from warehouses ith to markets jth add up to availability supply for each source (or origin) i.
- A_i Add up to accessible supply for each source (or origin) i.
- B_j Add up to request of every goal (destination) j. (p_j, q_j) position of the jth destination.
- d_{ij} distance of per unit delivered from ith warehouses to jth markets.

2.2 Model Formulation

Objective functions:

Give us a chance to consider a transportation problem with M origins $O_i(i = 1, 2, ..., M)$ and N destinations $D_j(j = 1, 2, ..., N$, in which the position (x_i, y_i) of origins to be decided with respect to the position of destination $p_j, q_j)$ in of that the units of transportation w_{ij} from ith origin to jth destination. Also to be decided the first part of the objective function is the cost associated with the amount to be transported and second part is associated with the distance from origin to destinations. Hence, the objective function of the nonlinear solid transportation problem is as follows:

$$z_1 = Min\left\{\sum_{i=1}^{M}\sum_{j=1}^{N}\widetilde{h}_{ij}w_{ij} + \sum_{i=1}^{M}\sum_{j=1}^{N}\widetilde{c}_{ij}d_{ij}y_{ij}\right\}$$

$$where \ y_{ij} = \begin{cases} 1, & if \quad w_{ij} \neq 0; \\ 0, & if \quad w_{ij} = 0 \end{cases} \tag{1}$$

$$where \quad d_{ij} = \sqrt{(x_i - p_j)^2 + (y_i - q_j)^2}$$

Constraints:

For the ith origin O_i to the total amount shipment $\sum_{j=1}^{N} w_{ij}$ cannot exceed its availability A_i. That is, we must require

$$\sum_{j=1}^{N} w_{ij} \leq \widetilde{A}_i \qquad i = 1, 2, \ldots, M. \tag{2}$$

The aggregate incoming shipment at jth destination is $\sum_{i=1}^{M} w_{ij}$ should satisfied its requirement or demand. That is, we must require

$$\sum_{i=1}^{M} w_{ij} \geq \tilde{B}_j \qquad j = 1, 2, \ldots, N. \tag{3}$$

Consider one unit of the item at the ith supply point contains f_i units of polluting influence. The total impurity at origin i is $\sum_{i=1}^{m} f_i w_{ij}$. Request point j cannot get more than g_j units of impurity. That is, we should require

$$\sum_{i=1}^{M} f_i w_{ij} \leq \tilde{g}_j \qquad j = 1, 2, \ldots, N. \tag{4}$$

Non-negativity constraints on decision factors: $w_{ij} \geq 0\ \forall i, j$

2.3   Defuzzification of Gaussian Type-2 Fuzzy Variables

$$Min \bar{f}$$

$$s.t \quad Cr\left\{ \sum_{i=1}^{m}\sum_{j=1}^{n} \tilde{h}_{ij} w_{ij} + \sum_{i=1}^{m}\sum_{j=1}^{n} \tilde{c}_{ij} d_{ij} y_{ij} \geq \bar{f} \right\} \geq \alpha$$

$$Cr\left\{ \sum_{j=1}^{n} w_{ij} \leq \tilde{A}_i \right\} \geq \alpha_i \qquad i = 1, 2, \ldots M,$$

$$Cr\left\{ \sum_{i=1}^{m} w_{ij} \geq \tilde{B}_j \right\} \geq \beta_j \qquad j = 1, 2, \ldots N,$$

$$Cr\left\{ \sum_{i=1}^{m} f_i w_{ij} \leq \tilde{g}_j \right\} \geq \gamma_k \qquad j = 1, 2, \ldots N, x_{ijk} \geq 0, \forall i, j, k.$$

$$\tag{5}$$

Here $Min f_1$ indicate the minimum value and the objective function accomplish with generalized credibility $\alpha(0 < \alpha \leq 1).\alpha_i, \beta_j, \gamma_k (0 < \alpha_i, \beta_j, \gamma_k \leq 1)$ which are the present generalized credibility satisfaction level of the origin and end point restriction respectively for all i, j, k.

Case i:

When $\alpha \epsilon (0, 0.25]$, then the parametric problem of the model representation (5) as:

Minf̄

$$s.t \quad \sum_{i=1}^{M}\sum_{j=1}^{N}\left((\mu_{\tilde{h}_{ij}} - \sigma_{\tilde{h}_{ij}}\sqrt{2\ln(1+(1-4\alpha)\theta_{r,\tilde{h}_{ij}})-2\ln 2\alpha})w_{ij}\right.$$

$$+(\mu_{\tilde{c}_{ij}} - \sigma_{\tilde{c}_{ij}}\sqrt{2\ln(1+(1-4\alpha)\theta_{r,\tilde{c}_{ij}})-2\ln 2\alpha})y_{ij}$$

$$and \quad \sum_{j=1}^{N}w_{ij} \le (\mu_{\tilde{A}_i} - \sigma_{\tilde{A}_i}\sqrt{2\ln(1+(1-4\alpha_i)\theta_{r,\tilde{A}_i})-2\ln 2\alpha_i}), \qquad i=1,2,3,\ldots M$$

$$\sum_{i=1}^{M}w_{ij} \ge (\mu_{\tilde{B}_j} - \sigma_{\tilde{B}_j}\sqrt{2\ln(1+(1-4\beta_j)\theta_{r,\tilde{B}_j})-2\ln 2\beta_j}), \qquad j=1,2,3,\ldots N$$

$$\sum_{i=1}^{M}f_i w_{ij} \le (\mu_{\tilde{g}_j} - \sigma_{\tilde{g}_j}\sqrt{2\ln(1+(1-4\gamma_k)\theta_{r,\tilde{g}_j})-2\ln 2\gamma_k}), \qquad j=1,2,3,\ldots N$$

Case ii:

When $\alpha \epsilon (2.5, 0.5]$, then the parametric problem of the model representation (5) as:

Minf̄

$$s.t \quad \sum_{i=1}^{M}\sum_{j=1}^{N}\left((\mu_{\tilde{h}_{ij}} - \sigma_{\tilde{h}_{ij}}\sqrt{2\ln(1+(4\alpha-1)\theta_{r,\tilde{h}_{ij}})-2\ln(2\alpha+(4\alpha-1)\theta_{1,h_{ij}})})w_{ij}\right.$$

$$+(\mu_{\tilde{c}_{ij}} - \sigma_{\tilde{c}_{ij}}\sqrt{2\ln(1+(4\alpha-1)\theta_{r,\tilde{c}_{ij}})-2\ln(2\alpha+(4\alpha-1)\theta_{1,c_{ij}})})y_{ij}$$

$$and \quad \sum_{j=1}^{N}w_{ij} \le (\mu_{\tilde{A}_i} - \sigma_{\tilde{A}_i}\sqrt{2\ln(1+(4\alpha_i-1)\theta_{r,\tilde{A}_i})-2\ln(2\alpha_i+(4\alpha_i-1)\theta_{1,A_i})}), \qquad i=1,2,3,\ldots M$$

$$\sum_{i=1}^{M}w_{ij} \ge (\mu_{\tilde{B}_j} - \sigma_{\tilde{B}_j}\sqrt{2\ln(1+(4\beta_j-1)\theta_{r,\tilde{B}_j})-2\ln(2\beta_j+(4\beta_j-1)\theta_{r,B_j})}), \qquad j=1,2,3,\ldots N$$

$$\sum_{i=1}^{M}f_i w_{ij} \le (\mu_{\tilde{g}_j} - \sigma_{\tilde{g}_j}\sqrt{2\ln(1+(4\gamma_k-1)\theta_{r,\tilde{g}_j})-2\ln(2\gamma_k+(4\gamma_k-1)\theta_{1,g_j})}), \qquad j=1,2,3,\ldots N$$

Case iii:

When $\alpha \epsilon (0.5, 7.5]$, then the parametric problem of the model representation (5) as:

Minf̄

$$s.t \quad \sum_{i=1}^{M}\sum_{j=1}^{N}\left((\mu_{\tilde{h}_{ij}} + \sigma_{\tilde{h}_{ij}}\sqrt{2\ln(1+(3-4\alpha)\theta_{r,\tilde{h}_{ij}})-2\ln(2(1-\alpha)+(3-4\alpha)\theta_{1,\tilde{h}_{ij}})})w_{ij}\right.$$

$$+(\mu_{\tilde{c}_{ij}} + \sigma_{\tilde{c}_{ij}}\sqrt{2\ln(1+(3-4\alpha)\theta_{r,\tilde{c}_{ij}})-2\ln(2(1-\alpha)+(3-4\alpha)\theta_{1,\tilde{c}_{ij}})})y_{ij}$$

$$and \quad \sum_{j=1}^{N}w_{ij} \le (\mu_{\tilde{A}_i} + \sigma_{\tilde{A}_i}\sqrt{2\ln(1+(3-4\alpha_i)\theta_{r,\tilde{A}_i})-2\ln(2(1-\alpha_i)+(3-4\alpha_i)\theta_{1,\tilde{A}_i})}), \qquad i=1,2,3,\ldots M$$

$$\sum_{i=1}^{M}\sum_{k=1}^{K}w_{ij} \ge (\mu_{\tilde{B}_j} + \sigma_{\tilde{B}_j}\sqrt{2\ln(1+(3-4\beta_j)\theta_{r,\tilde{B}_j})-2\ln(2(1-\beta_j)+(3-4\alpha)\theta_{1,\tilde{B}_j})}), \qquad j=1,2,3,\ldots N$$

$$\sum_{i=1}^{M}f_i w_{ij} \le (\mu_{\tilde{g}_j} + \sigma_{\tilde{g}_j}\sqrt{2\ln(1+(3-4\gamma_k)\theta_{r,\tilde{g}_j})-2\ln(2(1-\gamma_k)+(3-4\gamma_k)\theta_{1,\tilde{g}_j})}), \qquad k=1,2,3,\ldots K$$

Case iv:

When $\alpha \in (0.75, 1]$, then the parametric problem of the model representation (5) as:

$$Min \bar{f}$$

$$s.t \quad \sum_{i=1}^{M}\sum_{j=1}^{N}\left((\mu_{\tilde{h}_{ij}} + \sigma_{\tilde{h}_{ij}}\sqrt{2\ln(1 + (4\alpha - 3)\theta_{r,\tilde{h}_{ij}}) - 2\ln(2(\alpha - 1))}w_{ij}\right.$$

$$\left. + (\mu_{\tilde{c}_{ij}} + \sigma_{\tilde{c}_{ij}}\sqrt{2\ln(1 + (4\alpha - 3)\theta_{r,\tilde{c}_{ij}}) - 2\ln(2(1 - \alpha))}y_{ij}\right)$$

$$and \quad \sum_{j=1}^{N}w_{ij} \leq (\mu_{\tilde{A}_i} + \sigma_{\tilde{A}_i}\sqrt{2\ln(1 + (4\alpha_i - 3)\theta_{r,\tilde{A}_i}) - 2\ln(2(\alpha_i - 1))}, \qquad i = 1, 2, 3, \ldots M$$

$$\sum_{i=1}^{M}w_{ij} \geq (\mu_{\tilde{B}_j} + \sigma_{\tilde{B}_j}\sqrt{2\ln(1 + (4\beta_j - 3)\theta_{r,\tilde{B}_j}) - 2\ln(2(1 - \beta_j))}, \qquad j = 1, 2, 3, \ldots N$$

$$\sum_{i=1}^{M}f_i w_{ij} \leq (\mu_{\tilde{g}_j} + \sigma_{\tilde{g}_j}\sqrt{2\ln(1 + (4\gamma_k - 3)\theta_{r,\tilde{g}_j}) - 2\ln(2(\gamma_k - 1))}, \qquad j = 1, 2, 3, \ldots N$$

2.4   Model 2: Production Cost, Unit Transportation Cost, Impurity at Demand Point are treated as Gaussian Type-2 Fuzzy Variables and Source, Demands are Crisp

$$Min f_1 = \left\{\sum_{i=1}^{M}\sum_{j=1}^{N}\tilde{h}_{ij}w_{ij} + \sum_{i=1}^{M}\sum_{j=1}^{N}\tilde{c}_{ij}d_{ij}y_{ij}\right\}$$

$$where \quad d_{ij} = \sqrt{(x_i - p_j)^2 + (y_i - q_j)^2}$$

$$\sum_{j=1}^{N}w_{ij} \leq \tilde{A}_i \qquad i = 1, 2, \ldots, M.$$

$$\sum_{i=1}^{m}w_{ij} \geq \tilde{B}_j \qquad j = 1, 2, \ldots, N.$$

$$\sum_{i=1}^{m}f_i w_{ij} \leq \tilde{g}_j \qquad j = 1, 2, \ldots, N.$$

Case i:

When $\alpha \in (0, 0.25]$, then the parametric problem of the model representation (5) as:

$$Min \, TP = \sum_{i=1}^{M}\sum_{j=1}^{N}\left((\mu_{\tilde{h}_{ij}} - \sigma_{\tilde{h}_{ij}}\sqrt{2\ln(1 + (1 - 4\alpha)\theta_{r,\tilde{h}_{ij}}) - 2\ln 2\alpha}w_{ij}\right.$$

$$\left. + (\mu_{\tilde{c}_{ij}} - \sigma_{\tilde{c}_{ij}}\sqrt{2\ln(1 + (1 - 4\alpha)\theta_{r,\tilde{c}_{ij}}) - 2\ln 2\alpha}y_{ij}\right)$$

$$s.t \quad (11)-(13) \tag{6}$$

Case ii:

When $\alpha \in (2.5, 0.5]$, then the parametric problem of the model representation (5) as:

$$
Min\ TP = \sum_{i=1}^{M}\sum_{j=1}^{N}\Big((\mu_{\tilde{h}_{ij}} - \sigma_{\tilde{h}_{ij}}\sqrt{2\ln(1 + (4\alpha - 1)\theta_{r,\tilde{h}_{ij}}) - 2\ln(2\alpha + (4\alpha - 1)\theta_{1,h_{ij}})})w_{ij}
$$
$$
+ (\mu_{\tilde{c}_{ij}} - \sigma_{\tilde{c}_{ij}}\sqrt{2\ln(1 + (4\alpha - 1)\theta_{r,\tilde{c}_{ij}}) - 2\ln(2\alpha + (4\alpha - 1)\theta_{1,c_{ij}})})y_{ij}
$$

$$
s.t \quad (11)-(13) \tag{7}
$$

Case iii:

When $\alpha \in (0.5, 7.5]$, then the parametric problem of the model representation (5) as:

$$
Min\ TP = \sum_{i=1}^{M}\sum_{j=1}^{N}\Big((\mu_{\tilde{h}_{ij}} + \sigma_{\tilde{h}_{ij}}\sqrt{2\ln(1 + (3 - 4\alpha)\theta_{r,\tilde{h}_{ij}}) - 2\ln(2(1 - \alpha) + (3 - 4\alpha)\theta_{1,\tilde{h}_{ij}})})w_{ij}
$$
$$
- (\mu_{\tilde{c}_{ij}} + \sigma_{\tilde{c}_{ij}}\sqrt{2\ln(1 + (3 - 4\alpha)\theta_{r,\tilde{c}_{ij}}) - 2\ln(2(1 - \alpha) + (3 - 4\alpha)\theta_{1,\tilde{c}_{ij}})})y_{ij}
$$

$$
s.t \quad (11)-(13) \tag{8}
$$

Case iv:

When $\alpha \in (0.75, 1]$, then the parametric problem of the model representation (5) as:

$$
Min\ TP = \sum_{i=1}^{M}\sum_{j=1}^{N}\Big((\mu_{\tilde{h}_{ij}} + \sigma_{\tilde{h}_{ij}}\sqrt{2\ln(1 + (4\alpha - 3)\theta_{r,\tilde{h}_{ij}}) - 2\ln(2(\alpha - 1))})w_{ij}
$$
$$
- (\mu_{\tilde{c}_{ij}} + \sigma_{\tilde{c}_{ij}}\sqrt{2\ln(1 + (4\alpha - 3)\theta_{r,\tilde{c}_{ij}}) - 2\ln(2(1 - \alpha))})y_{ij}
$$

$$
s.t \quad (11)-(13) \tag{9}
$$

3   Solution Procedures

Genetic algorithm (GA) has been utilized to take care of the issue of given model. GA is utilized to find optimization through heuristic inquiry process that corresponds related regular choice (natural selection). Here population is as an arrangement of feasible solutions of proposed issue. Genotype is called as considered member of population, a chromosome, a string or a permutation. A GA performed three different operations—reproduction, crossover and mutation.

3.1   Parameters

The different parameters are considered to solve the problem through GA as follows.
(*MAXGEN*)-number of generation (set 5000)
(*POPSIZE*)-size of population (set 100)
(*PXOVER*)- probability of crossover (set 0.6)
(*PMU*)-probability of mutation (set 0.2).

3.2   Representation of Chromosome

The variables in this proposed models are nonlinear. So, a real number is used to represent the chromosome to solve the proposed model. Many nonlinear real problems used binary vectors but those were not effective.

3.3   Reproduction

To evaluate the chromosome, parents are randomly selected. The boundaries, dependent variables, independent variables are determined from all (here 16) variables to initialize the population.

3.4   Crossover

The main operator of GA is crossover. It is used to exchange the parent's characteristics and communicate to the children. It may happen in two steps:

(i) Selection for crossover: A random number r is generated for each solution of $P^1(T)$ from the range [0...1]. The solution is considered for crossover, if $r < p_c$, where p_c is crossover probability.

(ii) Crossover process: After selection some solution, crossover has been applied. The random number c has been taken from the range [0...1] for the pair of solutions Y_1, Y_2. Y_{11} and Y_{21} are calculated using Y_1, Y_2 as follows: where $Y_{11} = cY_1 + (1 - c)Y_2$, $Y_{21} = cY_2 + (1 - c)Y_1$, where Y_{11}, Y_{21} must meet the problem constraints.

3.5   Mutation

To recover any loss of some important characteristics, we need to perform mutation operation. It is also used for maintaining population diversity. It is done in two steps:

(i) Mutation Selection: A random number r is generated for each solution of $P^1(T)$ from the range [0...1]. The solution is considered for mutation, if $r < p_m$, where p_m is the mutation probability.

(ii) Mutation process: A random number r is selected with in the range [1...K]. Then by replacement of x_r within rth component of X they are random number. We get a solution $X = (x_1, x_2, \ldots x_k)$, which is a solution through mutation.

3.6   Evaluation

The evaluation function used to solve this problem is
$$eval(V_i) = objective\ function\ value$$
Through Roulette wheel selection chromosome. Here better chromosome has been chosen from the population to create the new chromosomes. Presently, new enhanced better chromosomes are produced through arithmetic crossover and mutation. The steps of the proposed algorithm are given below:

Step-1: Begin
t=0; Where t is considered as number of iteration.
Step-2: Population(t) is initialized.
Step-3: Population(t) is evaluated.
Step-4: while(condition is true)
{
Population(t) is selected from Population(t-1).
Perform crossover on Population(t)
Perform mutation on Population(t)
evaluate Population(t)
}
Step-5: Optimization Result Printed
Step-6: end.

4   Numerical Experiments

To present the relevancy and utility of the proposed model, a numerical illustration with three sources and three destination and three convenances are considered in these models. The model described above is coded in GA to solve the minimization solid transportation problem (Tables 1, 2, 3 and 4).

Table 1 Gaussian T2 fuzzy unit transportation costs

Product	c(11)	c(12)	c(21)	c(22)
	(10, 1.0; 0.8, 1.8)	(12, 1.2; 0.9, 1.5)	(9, 1.0; 1.1, 1.5)	(11, 1.0; 1.1, 1.5)

Table 2 Solid transportation problem parameters

i	Source	j	Demand	j	Max impurity received
1	(30, 1.5; 0.8, 1.0)	1	(23, 2.1; 0.5, 0.8)	1	(28, 2.1; 1.2, 1.6)
2	(31, 1.2; 0.1, 1.0)	2	(21, 1.1; 0.5, 0.8)	2	(35, 2.1; 1.0, 1.6)

Table 3 Value of h_{ij}

Product	h(11)	h(12)	h(21)	h(22)
	(4.5, 1.0; 0.8, 1.8)	(3.1, 1.2; 0.9, 1.5)	(2.23, 1.0; 1.1, 1.5)	(5.23, 1.0; 1.1, 1.5)

Table 4 Value of d_{ij} and impurity

Distance	Unknown location	Impurity
$d_{11} = 3.17, d_{12} = 0.1$	$x1 = 5.29, x2 = 4.2$	f1 = 7
$d_{21} = 1.66, d_{22} = 1.64$	$y1 = 8.1, y2 = 9.2$	f2 = 5

5   Discussion

We obtained different solutions from the experiment which are distinct with different degrees. The performance of this model has been shown through the experimental result in Table 5. The obtained results demonstrate the applicability and managerial insight of the proposed scheme. The proposed algorithm is very effective for searching better solution, and we achieve Pareto optimal solutions for managerial decision.

Table 5 Different models results (optimum)

α	Model	Amount	Transportation Cost
0.95	1	40.189	202.337
	2	31.421	190.660
0.90	1	42.09	229.418
	2	32.75	192.658
0.85	1	43.189	236.972
	2	34.56	192.880
0.80	1	43.89	241.672
	2	32.63	190.186

Here, we observed that cost of Model 1 is greater than the cost in Model 2. GA has been used to shown the crossover of results. It is possible to get the result through the variations of population size, iteration, crossover and mutation.

6 Comparison with Earlier Work

It has been observed that few development has been done using STP with cost minimization. Most work has been developed by considering profit maximization. Here we have investigated the problem in the angle of cost minimization. These two approaches are opposite angle, and it is hard for comparison between them. This proposed scheme is the new development using cost minimization solid transportation problem with Gaussian type-2 fuzzy environments. So, this is a another innovative examination towards the field of transportation according as far as anyone is concerned.

7 Conclusions and Future Scope

A new cost minimization STP with most parameters is considered as Gaussian type-2 fuzzy environments. Here, the parameters are supply, demand, production cost, transport cost and impurity at demand point. The GA has been used to solve the proposed model and achieve good results. The main contributions are mentioned below:

- This is the new attempt in STP with cost minimization.
- Gaussian type-2 has been used to get accurate result which is more precise than type-1.
- A new concept has been developed using these models. One can apply using time minimization budget constraint, damage item, discount of price, festival offer, etc.
- This model can be solved through different environments like rough, fuzzy rough, intuitionists fuzzy environment.

References

1. Zadeh, L.A.: Fuzzy sets. Inf. Control. **8**, 338–353 (1965)
2. Zimmermann, H.J.: Fuzzy programming and linear programming with several objective functions. Fuzzy Sets Syst. **1**, 4555 (1978)
3. Hitchcock, F.L.: The distribution of a product from several sources to numerous localities. J. Math. Phys. **20**(1), 224–230 (1941)
4. Shell, E.: Distribution of a product by several properties. In: Directorate of Management Analysis. Proceedings of the Second Symposium in Linear Programming, vol. 2 (1955)

5. Haley, K.B.: New methods in mathematical programming-The solid transportation problem. Oper. Res. **10**(4), 448–463 (1962)
6. Jimenez, F., Verdegay, J.L.: Solving fuzzy solid transportation problems by an evolutionary algorithm based parametric approach. Eur. J. Oper. Res. **117**, 485–510 (1999)
7. Yang, L., Liu, L.: Fuzzy fixed charge solid transpotation problem and algorithm. Appl. Soft Comput. **7**, 879–889 (2007)
8. Liu, P., Yang, L., Wang, L., Li, S.: A solid transportation problem with type-2 fuzzy variables. Appl. Soft Comput. **24**, 543–558 (2014)
9. Kocken, H.G., Sivri, M.: A simple parametric method to generate all optimal solutions of fuzzy solid transportation problem. Appl. Math. Model. **40**(7–8), 4612–4624 (2016)
10. Ojha, A., Das, B., Mondal, S.K., Maiti, M.: A multi-item transportation problem with fuzzy tolerance. Appl. Soft Comput. **13**(8), 3703–3712 (2013)
11. Kundu, P., Kar, S., Maiti, M.: A fixed charge transportation problem with type-2 fuzzy variables. Inf. Sci. **255**, 170–186 (2014)
12. Giri, P.K., Maiti, M.K., Maiti, M.: Fully fuzzy fixed charge multi-item solid transportation problem. Appl. Soft Comput. **27**, 77–91 (2015)
13. Dubois, D., Prade, H.: Fuzzy Sets and Systems: Theory and Applications. Academic Press, New York (1980)
14. Mizumoto, M., Tanaka, K.: Fuzzy sets of type-2 under algebraic product and algebraic sum. Fuzzy Sets Syst. **5**(3), 277–280 (1981)
15. Gray, P.: Exact solution of the fixed charge transportation problem. Oper. Res. **19**(6), 1529–1538 (1971)
16. Bit, A.K., Biswal, M.P., Alam, S.S.: Fuzzy programming approach to multiobjective solid transportation problem. Fuzzy Sets Sys. **57**(2), 183–194 (1993)
17. Greenfield, S., Chiclana, F., John, R.I., Coupland, S.: The sampling method of defuzzification for type-2 fuzzy sets: experimental evaluation. Inf. Sci. **189**, 77–92 (2012)

Chapter 11
Complements to Voronovskaya's Formula

Margareta Heilmann, Fadel Nasaireh and Ioan Raşa

Abstract We generalize some known results concerning Voronovskaya-type formulas for the composition of two linear operators acting on an arbitrary Banach space.

Keywords Voronovskaya-type formula · Composition of operators · Bernstein operator · Inverse of Bernstein operator

MSC (2010): 41A36 · 41A35

1 Introduction

Voronovskaya-type formulas are usually established for sequences of positive linear operators. They are important tools in approximation theory, used in order to investigate the rate of convergence and saturation properties. The classical Voronovskaya formula is related to the Bernstein operators $B_n : C[0, 1] \longrightarrow C[0, 1]$,

$$B_n f(x) = \sum_{k=0}^{n} \binom{n}{k} x^k (1 - x)^{n-k} f\left(\frac{k}{n}\right), \ f \in C[0, 1], \ x \in [0, 1],$$

M. Heilmann (✉)
School of Mathematics and Natural Sciences, University of Wuppertal,
Gaußstraße 20, 42119 Wuppertal, Germany
e-mail: heilmann@math.uni-wuppertal.de

F. Nasaireh · I. Raşa
Department of Mathematics, Technical University,
Str. Memorandumului 28, 400114 Cluj-Napoca, Romania
e-mail: fadelnasierh@gmail.com

I. Raşa
e-mail: Ioan.Rasa@math.utcluj.ro

© Springer Nature Singapore Pte Ltd. 2018
D. Ghosh et al. (eds.), *Mathematics and Computing*, Springer Proceedings
in Mathematics & Statistics 253, https://doi.org/10.1007/978-981-13-2095-8_11

and reads as follows:

$$\lim_{n \to \infty} n \left(B_n f(x) - f(x) \right) = K f(x),$$

where $K f(x) = \frac{x(1-x)}{2} f''(x)$, $f \in C^2[0, 1]$, $x \in [0, 1]$, the convergence being uniform on $[0, 1]$.

If f has continuous derivatives of higher degree, the above Voronovskaya formula can be extended; see, e.g., [4] and the references therein.

The investigation of Voronovskaya-type formulas for the composition of two linear operators P_n and Q_n, acting on an arbitrary Banach space X, was initiated in [4].

The main aim of this article is to generalize the results of [4]. This is done in Sect. 2.

In particular, in [4] was investigated the case when $P_n Q_n$ acts as the identity on some linear subspace of X. We are concerned with this case in Sects. 3 and 4, where the operators B_n and B_n^{-1} are considered, acting on polynomials.

In Sects. 3 and 4, we use the eigenstructure of the operators B_n. Similar results can be obtained for several other operators, for which the eigenstructure is known; this will be done elsewhere.

A problem is mentioned in Sect. 4.

2 Voronovskaya's Formula for Composition of Operators

Let X be a Banach space. For a given $m \in \mathbb{N}$, consider the linear subspaces $Y_m \subseteq Y_{m-1} \subseteq \cdots \subseteq Y_1 \subseteq Z \subseteq Y_0 = X$.

Let $P_n : X \longrightarrow X$, $Q_n : Z \longrightarrow X$, $n \in \mathbb{N}$, be linear operators. Suppose that each operator P_n is bounded and

$$\lim_{n \to \infty} P_n x = x, \ x \in X. \tag{1}$$

For $l = 0, 1, \ldots, m$, consider the linear operators $K_l : Y_l \longrightarrow X$, $L_l : Y_l \longrightarrow X$. Suppose that

$$K_0 y = L_0 y = y, \ y \in Y_0, \tag{2}$$

and for all $0 \leq i \leq l \leq m$,

$$L_i y \in Y_{l-i}, \ y \in Y_m. \tag{3}$$

Moreover, assume that for $l = 1, 2, \ldots, m$,

$$\lim_{n\to\infty}\left\{n^l(P_ny-y)-\sum_{i=1}^{l-1}n^{l-i}K_iy\right\}=K_ly,\ y\in Y_l, \tag{4}$$

$$\lim_{n\to\infty}\left\{n^l(Q_ny-y)-\sum_{i=1}^{l-1}n^{l-i}L_iy\right\}=L_ly,\ y\in Y_l. \tag{5}$$

Theorem 1 *Under the above assumptions,*

$$\lim_{n\to\infty}\left\{n^m(P_nQ_ny-y)-\sum_{l=1}^{m-1}n^{m-l}\sum_{i=0}^{l}K_{l-i}L_iy\right\} \tag{6}$$

$$=\sum_{i=0}^{m}K_{m-i}L_iy,\ y\in Y_m.$$

Proof Let $y\in Y_m$. Then,

$$n^m(P_nQ_ny-y)-\sum_{l=1}^{m-1}n^{m-l}\sum_{i=0}^{l}K_{l-i}L_iy=P_nL_my+u_n+v_n+w_n, \tag{7}$$

where

$$u_n:=n^m(P_ny-y)-\sum_{i=1}^{m-1}n^{m-i}K_iy,$$

$$v_n:=\sum_{l=1}^{m-1}\left\{n^l\,(P_nL_{m-l}y-L_{m-l}y)-\sum_{i=1}^{l-1}n^{l-i}K_iL_{m-l}y\right\},$$

$$w_n:=P_n\left\{n^m\,(Q_ny-y)-\sum_{i=1}^{m-1}n^{m-i}L_iy-L_my\right\}.$$

According to (1),

$$\lim_{n\to\infty}P_nL_my=L_my. \tag{8}$$

Using (4) with $l=m$, we get

$$\lim_{n\to\infty}u_n=K_my. \tag{9}$$

Moreover, (4) yields

$$\lim_{n\to\infty} v_n = \sum_{l=1}^{m-1} K_l L_{m-l} y. \tag{10}$$

By using (1) and the Banach–Steinhaus theorem, we infer that the sequence $(\|P_n\|)_{n\geq 1}$ is bounded; i. e., there exists $M > 0$ such that $\|P_n\| \leq M$, $n \geq 1$. Therefore, we have

$$\|w_n\| \leq M \|n^m (Q_n y - y) - \sum_{i=1}^{m-1} n^{m-i} L_i y - L_m y\|.$$

From (5) with $l = m$, we infer that

$$\lim_{n\to\infty} w_n = 0. \tag{11}$$

Now (7), (8), (9), (10), and (11) show that

$$\lim_{n\to\infty} \left\{ n^m (P_n Q_n y - y) - \sum_{l=1}^{m-1} n^{m-l} \sum_{i=0}^{l} K_{l-i} L_i y \right\}$$

$$= L_m y + K_m y + \sum_{l=1}^{m-1} K_l L_{m-l} y$$

$$= \sum_{l=0}^{m} K_l L_{m-l} y,$$

and this concludes the proof.

Corollary 1 *Let $y \in Y_m$ such that $P_n Q_n y = y$. Then,*

$$\sum_{i=0}^{l} K_{l-i} L_i y = 0, \ l = 1, 2, \ldots, m. \tag{12}$$

Proof If $P_n Q_n y = y$, (6) yields

$$\lim_{n\to\infty} \sum_{l=1}^{m-1} n^{m-l} \sum_{i=0}^{l} K_{l-i} L_i y = - \sum_{i=0}^{m} K_{m-i} L_i y.$$

This entails (12).

Remark 1 For $m \in \{1, 2, 3\}$, Theorem 1 and Corollary 1 were proved in [4, Theorem 2.1]. Several examples and applications can be found in [3, 4].

3 The Operator B_n

Let $B_n : \mathcal{P}_m \longrightarrow \mathcal{P}_m$, $n \geq m$, be the classical Bernstein operator. It is known that

$$B_n p = p + \sum_{i=1}^{l} n^{-i} K_i p + o(n^{-l}), \quad l = 1, 2, \ldots, m-1, \tag{13}$$

see [1], where the operators K_i are described. On the other hand (see [2, (4.23)]),

$$B_n p = \sum_{j=0}^{m} \lambda_j^{(n)} p_j^{(n)} \mu_j^{(n)}(p), \quad p \in \mathcal{P}_m,$$

where $\lambda_j^{(n)}$ are the eigenvalues of B_n, $p_j^{(n)}$ the eigenpolynomials, and $\mu_j^{(n)}$ the dual functionals. Hence,

$$p = \sum_{j=0}^{m} p_j^{(n)} \mu_j^{(n)}(p), \tag{14}$$

and

$$\lambda_0^{(n)} = \lambda_1^{(n)} = 1, \quad \lambda_j^{(n)} = \frac{n!}{(n-j)!n^j} = 1 + \sum_{i=1}^{j-1} \frac{s(j, j-i)}{n^i}, \quad j \geq 2,$$

where $s(m, l)$ denote the Stirling numbers of first kind, defined by $x(x-1)\ldots(x-m+1) = \sum_{l=0}^{m} s(m, l)x^l$. Therefore,

$$B_n p - p = \sum_{j=2}^{m} p_j^{(n)} \mu_j^{(n)}(p) \sum_{i=1}^{j-1} \frac{s(j, j-i)}{n^i}$$

$$= \sum_{i=1}^{m-1} \frac{1}{n^i} \sum_{j=i+1}^{m} s(j, j-i) p_j^{(n)} \mu_j^{(n)}(p).$$

$$B_n p = p + \sum_{i=1}^{l} \frac{1}{n^i} \sum_{j=i+1}^{m} s(j, j-i) p_j^{(n)} \mu_j^{(n)}(p) \tag{15}$$

$$+ \sum_{i=l+1}^{m-1} \frac{1}{n^i} \sum_{j=i+1}^{m} s(j, j-i) p_j^{(n)} \mu_j^{(n)}(p).$$

Denote $K_{ni}p := \sum_{j=i+1}^{m} s(j, j-i)p_j^{(n)}\mu_j^{(n)}(p)$, and remark that $\sum_{i=l+1}^{m-1} \frac{1}{n^i} \sum_{j=i+1}^{m} s(j, j-i)p_j^{(n)}\mu_j^{(n)}(p) = o(n^{-l})$, since according to [2], the sequences $(p_j^{(n)})_{n\geq 0}$ and $(\mu_j^{(n)}(p))_{n\geq 0}$ are convergent to p_j^* and $\mu_j^*(p)$, respectively.

Thus, we have

Theorem 2 *For each $p \in \mathcal{P}_m$,*

$$B_n p = p + \sum_{i=1}^{l} \frac{1}{n^i} K_{ni}p + o(n^{-l}), \; l = 1, 2, \ldots, m-1, \tag{16}$$

where

$$K_{ni}p \longrightarrow \sum_{j=i+1}^{m} s(j, j-i)p_j^*\mu_j^*(p) =: \widetilde{K}_i p, \; i = 1, \ldots, l. \tag{17}$$

Moreover, $\widetilde{K}_i = K_i, i = 1, \ldots, l$.

Proof (16) and (17) are consequences of (15) and the above remarks. From (13) and (16), we infer that

$$\sum_{i=1}^{l} n^{-i} (K_{ni}p - K_i p) = o(n^{-l}), \; l = 1, \ldots, m-1.$$

This entails $\lim_{n\to\infty} K_{ni}p = K_i p$, i. e., $\widetilde{K}_i = K_i, i = 1, \ldots, l$.

4   The Operator B_n^{-1}

In the spirit of Sect. 3, consider $B_n^{-1} : \mathcal{P}_m \longrightarrow \mathcal{P}_m, n \geq m$.

From (14), we get

$$B_n^{-1}p = \sum_{j=0}^{m} p_j^{(n)}\mu_j^{(n)}(p) \frac{1}{\lambda_j^{(n)}}.$$

We have

$$\frac{1}{\lambda_0^{(n)}} = \frac{1}{\lambda_1^{(n)}} = 1, \; \frac{1}{\lambda_j^{(n)}} = 1 + \sum_{i=1}^{j-1} a_{ji} \frac{(n-i-1)!}{(n-1)!}, \; j \geq 2, \tag{18}$$

where the a_{ji} can be written in terms of a forward difference of order $j - i - 1$, i. e.,

$$a_{ji} = \frac{1}{(j-i-1)!} \Delta^{j-i-1} i^{j-1}. \tag{19}$$

To prove (19), we consider the Newton form of the interpolation polynomial of order $j-1$ for the monomial x^{j-1} with respect to the equidistant knots $j-1, j-2, \ldots, 1, 0$, evaluated at $x = n$. Thus,

$$n^{j-1} = \sum_{i=0}^{j-1} \left(\prod_{l=i+1}^{j-1} (n-l) \right) \cdot \frac{1}{(j-i-1)!} \Delta^{j-i-1} i^{j-1}.$$

Multiplying the equation by $\frac{(n-j)!}{(n-1)!}$ leads to

$$\frac{1}{\lambda_j^{(n)}} = \frac{(n-j)! n^j}{n!}$$

$$= \sum_{i=0}^{j-1} \frac{(n-i-1)!}{(n-1)!} \cdot \frac{1}{(j-i-1)!} \Delta^{j-i-1} i^{j-1}$$

$$= 1 + \sum_{i=1}^{j-1} \frac{(n-i-1)!}{(n-1)!} \cdot \frac{1}{(j-i-1)!} \Delta^{j-i-1} i^{j-1}.$$

This prove (19).

Consequently,

$$B_n^{-1} p - p = \sum_{j=2}^{m} p_j^{(n)} \mu_j^{(n)}(p) \sum_{i=1}^{j-1} a_{ji} \frac{(n-i-1)!}{(n-1)!}$$

$$= \sum_{i=1}^{m-1} \frac{(n-i-1)!}{(n-1)!} \sum_{j=i+1}^{m} a_{ji} p_j^{(n)} \mu_j^{(n)}(p).$$

$$B_n^{-1} p = p + \sum_{i=1}^{l} \frac{(n-i-1)!}{(n-1)!} \sum_{j=i+1}^{m} a_{ji} p_j^{(n)} \mu_j^{(n)}(p)$$

$$+ \sum_{i=l+1}^{m-1} \frac{(n-i-1)!}{(n-1)!} \sum_{j=i+1}^{m} a_{ji} p_j^{(n)} \mu_j^{(n)}(p).$$

Denote $L_{ni} p := \sum_{j=i+1}^{m} a_{ji} p_j^{(n)} \mu_j^{(n)}(p)$, and remark that $\sum_{i=l+1}^{m-1} \frac{(n-i-1)!}{(n-1)!} \sum_{j=i+1}^{m} a_{ji} p_j^{(n)} \mu_j^{(n)}(p) = o(n^{-l})$. Thus, we have

Theorem 3 *For each $p \in \mathcal{P}_m$,*

$$B_n^{-1}p = p + \sum_{i=1}^{l} \frac{(n-i-1)!}{(n-1)!} L_{ni}p + o(n^{-l}), \; l = 1, 2, \ldots, m-1, \qquad (20)$$

where

$$L_{ni}p \longrightarrow \sum_{j=i+1}^{m} a_{ji}p_j^*\mu_j^*(p) =: \widetilde{L}_i p \; i = 1, \ldots, l.$$

Problem 1 Taking into account (16) and (20), is there a relation connecting the operators $\widetilde{K}_i$ and $\widetilde{L}_i$, similar to (12)?

References

1. Abel, U., Ivan, M.: Asymptotic expansion of the multivariate Bernstein polynomials on a simplex. Approx. Theory Appl. **16**, 85–93 (2000)
2. Cooper, Sh., Waldron, Sh.: The Eigenstructure of the Bernstein operator. J. Approx. Theory **105**, 133–165 (2000)
3. Gonska, H., Heilmann, M., Lupaş, A., Raşa, I.: On the composition and decomposition of positive linear operators III: A non-trivial decomposition of the Bernstein operator, http://arxiv.org/abs/1204.2723 Aug 30, 2012, pp. 1–28
4. Nasaireh, F., Raşa, I.: Another look at Voronovskaja type formulas, J. Math. Inequal. **12**(1), 95–105 (2018)

Chapter 12
Mathematics and Machine Learning

Srinivas Pyda and Srinivas Kareenhalli

Abstract Machine learning is a branch of computer science that gives computers the ability to make predictions without explicitly being programmed. Machine learning enables computers to learn, as they process more and more data and make even more accurate predictions. Machine learning is becoming all pervasive in our daily lives, from speech recognition, medical diagnosis, customized content delivery, and product recommendations to advertisement placements to name a few. Knowingly or unknowingly, there is a very high chance that one would have encountered some form of machine learning several times in one's daily activities. In cloud data centers, machine learning presents an opportunity to make systems autonomous and thus transforming data centers into those that are less error prone, secure, self tuning, and highly available. Mathematics forms the bedrock of machine learning. This paper aims at highlighting the concepts in mathematics that are essential for building machine learning systems. Topics in mathematics like linear algebra, probability theory and statistics, multivariate calculus, partial derivatives, and algorithmic optimizations are quintessential to implementing efficient machine learning systems. This paper will delve into a few of the aforementioned areas to bring out core concepts necessary for machine learning. Topics like principal component analysis, matrix computation, gradient descent algorithms are a few of them covered in this paper. This paper attempts to give the reader a panoramic view of the mathematical landscape of machine learning.

Keywords Eigenvalues · Machine learning · Partial differential equations
Linear algebra

S. Pyda (✉)
Oracle America, Redwood Shores, CA 94065, USA
e-mail: Srinivas.Pyda@oracle.com

S. Kareenhalli
Oracle India, Bengaluru, India
e-mail: Srinivas.Kareenhalli@oracle.com

© Springer Nature Singapore Pte Ltd. 2018   135
D. Ghosh et al. (eds.), *Mathematics and Computing*, Springer Proceedings
in Mathematics & Statistics 253, https://doi.org/10.1007/978-981-13-2095-8_12

1 Introduction

We are seeing a huge explosion in the data that is being generated from online content and mobile data like messages, photos, videos and data generated by Internet of Things (IoT). The online logs generated from Web servers, users histories, operating system logs, and logs in databases are growing in size by leaps and bounds. This presents a great opportunity to be able to analyze the data from these sources to save costs and serve businesses and consumers better. In large enterprises, the data gathering used to be defensive in nature mainly for ensuring compliance for regulatory purposes. The role of data is changing with it becoming center of innovation.

According to a Gartner report [1], the number of connected "things" is estimated to be 20 billion by the year 2020. These include but are not limited to IoT from automotive systems, health monitoring devices, and smart meters to name a few.

In data centers, huge amounts of log data are being generated. This data can be analyzed to improve services, reduce costs, and make services more secure and available.

Processing this huge explosion of data using static analytics and algorithms will be both inefficient and impractical. The need of the hour is computers which can make predictions without being explicitly programmed. This is where machine learning comes into enable systems to automatically detect patterns in the data and make intelligent predictions. The machine learning algorithms should be capable of learning and enhance the accuracy of predictions as more and more data are processed. In cloud data centers, machine learning can be used to make systems autonomous which can predict usage patterns and allocate resources accordingly. Detect intrusions and take action to restrict damage or data loss. Machine learning can also be used to tune the systems to keep them performing optimally.

This paper will delve into the mathematics behind the machine learning and cover two areas in greater detail. The last section will cover the future of machine learning, especially in the cloud and database services.

2 Areas of Mathematics in Machine Learning

2.1 Recommender Systems (Supervised Learning)

Supervised learning is class of machine learning methods where there is a labeled dataset to learn from. The machine learning algorithms learn from the data and predict values for a new data item. Some examples of such systems are housing price predictor and stock values predictor.

Consider there are n different parameters in a dataset, representing the features of a house (like area of the house, area of the lot, location of the house, number of bedrooms). Let us denote them as x_i ($i = 1 \ldots n$). Let the outcome be price of the house as y_i ($i = 1 \ldots m$) taking on continuous values. Let there be m data items in the dataset. The machine learning system can learn from this dataset and predict price of a new house on the market based on the features of the house.

A popular machine learning method is the linear regression model. This technique tries to approximate a model by learning a set of weights for the features in the dataset and come up with a regression model. The weights for the features learnt using an iterative algorithm called as gradient ascent algorithm.

An initial set of weights are assigned to the features, and the error due to these weights is computed and compared to actual output values. The algorithm then tries to minimize the error by iterative changing the values of the weights, till the algorithm converges. The convergence is achieved when the error values do not change significantly. Let the initial set of weights be assigned as follows, and the hypotheses are represented as follows:

$$h(x) = \theta_0 + \theta_1 x_1 + \theta_2 x_2 + \cdots + \theta_n x_n$$

The error due to these weights is expressed as a cost function of the weights and is calculated as

$$C(\theta) = \frac{1}{2} \sum_{i=1}^{m} \left(h(x^{(i)}) - y^{(i)} \right)^2$$

The goal of the algorithm is to minimize $C(\theta)$ so that the error of prediction is minimal. This is achieved by adjusting the weights incrementally using the partial derivative of the cost function with respect to the weights. This is iteratively done till the cost function is minimized.

This machine learning technique is covered in greater details in Sect. 3 of this paper.

2.2 Classifier Systems (Unsupervised Learning)

Unlike in the previous section where the dataset had an output label associated with it, there are datasets which do not have any output label associated with them. Building machine learning algorithms to classify or discover underlying patterns on this kind of data is called unsupervised learning.

Consider a collection of articles (say Web sites or Web articles), how do we classify these articles based on "similarity." For instance, classifying the articles based on interests like "sports," "entertainment," "politics," "news," "art.".

One of the most popular machine learning algorithms for clustering is k-means clustering algorithm. This algorithm involves computing the similarity (distance) between documents and a cluster center and assigning the document to a cluster center based on its distance from the cluster center. To compute the distance or similarity between the documents, the collection of articles/documents first needs to be converted to a vector notation called Vector Space Model [2]. A common notation is the tf-idf. A document is first tokenized into set of terms, and tf-idf is defined as follows:

- TF: Term frequency is the frequency of the term in the document. Since the documents can be of different lengths, term frequency is divided by the number of terms in the documents as a way of normalization.

$$\text{tf}(t, d) = \frac{\text{frequency of term } t \text{ in the document}}{\text{total terms in the document}}$$

- IDF: Inverse document frequency is the ratio of total number of documents to the number of documents containing term t.

$$\text{idf}(t) = \ln\left(\frac{\text{Total number of documents}}{1 + \text{number of documents with term } t}\right)$$

$$\text{tf} - \text{idf}(t) = \text{tf}(t, d) * \text{idf}(t)$$

Each document can be represented as a vector of tf$-$idf(t) weights. The axes are terms, and the document can be modeled as a vector. By using length-normalized unit vector, the weights of the documents of varying lengths become comparable in weights.

Cluster centers are assigned in random, and data items are assigned to the closest cluster center using a distance measurement (Euclidean distance or cosine distance). The cluster centers are reevaluated and reassigned. This process is repeated iteratively till convergence.

Euclidean distance between two vectors is defined as

$$d(\vec{u}, \vec{v}) = ||\vec{u}, \vec{v}|| = \sqrt{(u_1 - v_1)^2 + (u_2 - v_2)^2 + \cdots + (u_n - v_n)^2}$$

Cosine similarity between two vectors is defined as

$$\text{cosine}(\vec{u}, \vec{v}) = \frac{\sum_{i=1}^{n} u_i v_i}{\sqrt{\sum_{i=1}^{n} u_i^2} \sqrt{\sum_{i=1}^{n} v_i^2}}$$

Randomly initialize k cluster centroids $\mu_1, \mu_2, \mu_3 \ldots \mu_k$

```
repeat till convergence {
    for i := 1 ..m
        c_i := index from 1 to k of cluster
              centroid closest to x_i
              ,closeness measure euclidean
    distance.
        for k := 1..K
            μ_k := mean of data points
                  assigned to the cluster k
}
```

Some more applications that can use k-means clustering algorithm:

- Clustering similar images, e.g., cluster by type of images (flower, ocean, sunset, clouds, cars, etc.)
- Use clustering to structure Web query results.
- Cluster product categories based on user buying patterns.
- Cluster similar neighborhood based on real estate or crime or other criteria for better forecasting.
- Customer or market segmentation based on geography, demography, price, lifestyle, etc.
- Cluster population based on medical condition.

2.3 Anomaly Detection

Anomaly detection is referred to the identification of items or events that are anomalous to other items present in the dataset. Anomalous detection techniques can be used to detect abnormal brain scans, cancerous cells from healthy cells, detecting frauds and intrusions, and detecting structural and manufacturing defects. Anomaly detection can be used to provide high availability of IT systems in data centers by recognizing anomalies and passing it downstream for resolution.

Anomaly detection is being used by databases (Oracle) [3] to detect anomalous intrusions and detect performance events to provide better availability and automated performance tuning of systems.

Consider the application where in a datacenter, anomalies are to be detected in systems that are behaving abnormally. Let us consider n features x_i $(1 \ldots n)$ like x_1 = cpu used by a system, x_2 = number of disk requests, x_3 = memory usage, x_4 = swap usage, x_5 = network usage, x_6 = device interrupts, etc. Let the dataset be m data items. Each feature can be treated as a Gaussian distribution as follows:

$$x_1 \sim \aleph\left(\mu_1, \sigma_1^2\right), x_2 \sim \aleph\left(\mu_2, \sigma_1^2\right), \ldots x_n \sim \aleph\left(\mu_n, \sigma_n^2\right)$$

where μ is the mean and σ is the variance. The probability distribution of $P(x_i)$ is given by the Gaussian distribution (normal distribution) as follows:

$$P\left(x_i; \mu_i, \sigma^2\right) = \frac{1}{\sqrt{2\pi}\sigma} e^{\frac{-(x_i-\mu_i)^2}{2\sigma_i^2}}$$

The curve of this probability distribution is a bell curve. The intuition is that probability of values under the curve being non-anomalous values is high and as we move along the axis away from the mean the probability of the value being non-anomalous becomes lower (i.e., they are probably anomalous). On the given dataset, we model the probability of the features as follows:

$$P(x) = p\left(x_1; \mu_1, \sigma_1^2\right) p\left(x_2; \mu_2, \sigma_2^2\right) \dots p\left(x_n; \mu_n, \sigma_n^2\right)$$
$$= \prod_{j=1}^{n} p\left(x_j; \mu_j, \sigma_j^2\right)$$

If $p(x)$ is less than a small threshold value ε, we flag that the data item x as anomalous. The anomaly detection algorithm [4] can be formalized as follows:

1. Formalize a set of n parameters
2. Estimate $\mu_1, \mu_2, \dots \mu_n, \sigma_1, \sigma_2 \dots, \sigma_n$ as

$$\mu_j = \frac{1}{m} \sum_{i=1}^{n} x_j^{(i)}$$

$$\sigma_j = \frac{1}{m} \sum_{i=1}^{n} \left(x_j^{(i)} - \mu_j\right)^2$$

3. Compute $P(x)$ for a new data x

$$P(x) = \prod_{j=1}^{n} p\left(x_j; \mu_j, \sigma_j^2\right) = \prod_{j=1}^{n} \frac{1}{\sqrt{2\pi}\sigma} e^{\frac{-(x_i-\mu_i)^2}{2\sigma_i^2}}$$

4. If $P(x) < \varepsilon$, then x is an anomaly.

If the features are not displaying a Gaussian distribution, a transformation function (like $\log(x)$, $x^{1/n}$) can be applied such that the transformed feature approximates to a Gaussian distribution.

If some of the features are correlated, then a multivariate Gaussian probability distribution can be applied, where the $P(x)$ is computed as

$$P(x; \mu, \Sigma) = \frac{1}{(2\pi)^{\frac{n}{2}} |\Sigma|^{\frac{1}{2}}} e^{\frac{-(x-\mu)^T \Sigma^{-1}(x-\mu)}{2}}$$

where μ is a vector of length n and Σ a matrix of dimension $n \times n$ and defined as follows for a dataset.

$$\mu = \frac{1}{m} \sum_{i=1}^{m} x^{(i)}$$

$$\Sigma = \frac{1}{m} \sum_{i=1}^{m} \left(x^{(i)} - \mu\right)\left(x^{(i)} - \mu\right)^{\mathrm{T}}$$

If $P(x) < \varepsilon$, then the dataset x is anomalous. Gaussian distribution is a special case of multivariate Gaussian distribution where the non-diagonal elements are 0.

Multivariate Gaussian distribution is a very useful tool in anomaly detection and widely used in machine learning. Anomaly detection is useful when there is a small number of anomalous compared to the total number of data items in the dataset.

2.4 Neural Networks

Neural network [5] is one of the most popular machine techniques used to model a host of machine learning problems like image recognition and speech recognition. Neural networks can efficiently model nonlinear decision boundaries compared to other linear models like logistic regression. Linear regression will have to use higher-order polynomials to model nonlinear decision boundaries adding to complexity.

Consider the dataset shown in Fig. 1, where the negative and positive examples are shown for a two-feature dataset. Clearly here the decision boundary is nonlinear and to model this using logistic regression would involve using higher-order polynomials. This is where neural networks come in handy to model such nonlinear decision boundaries.

Consider an example where we have a set of images and want to classify if the images are that of a street signage. The supervised labeled set consists of a set of images with positive classification (images that are those of street signs) and a set of images with negative classification (denoting images are not that of a street sign).

Fig. 1 An example of non-linear decision boundary

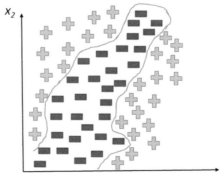

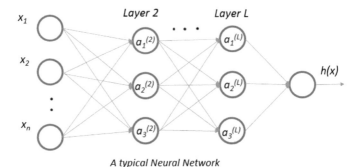

A typical Neural Network

Fig. 2 An example of a typical L layered neural network

Neural network algorithms try to mimic the behavior of the brain. Neural networks involve a series of layers through which the hypotheses are modeled. Figure 2 shows a typical neural network model. This neural network would output a 1 or 0 for images which it thinks are images of street signs or not respectively. This is a single-class classification problem.

A typical neural network consists of an input layer (layer 1) which takes in the input parameter and feeds into another layer called the intermediate or hidden layer. The output of the hidden layer can feed into more hidden layers. The final layer will output the hypothesis. The nodes in the hidden layers are called as hidden nodes or activation nodes. Each activation node implements an activation function (sigmoid function $g(z)$). For example, in the above two layered neural network the activation node 1 in layer 2 would take on values as shown below

$$a_1^{(2)} = g\left(w_{11}^{(1)}x_1 + w_{12}^{(1)}x_2 + \cdots + w_{1n}^{(1)}x_n\right)$$

The output of the last (output) layer would be:

$$h(x) = a_1^{(L+1)} = g\left(w_{11}^{(L)}a_1^{(L)} + w_{12}^{(L)}a_2^{(L)} + w_{13}^{(L)}a_3^{(L)}\right)$$

The weights learned are denoted by $w_{ij}^{(l)}$ where l is the *l*th layer, i is the input, and j represents the *j*th activation node. For example, $w_{13}^{(2)}$ represents the weight of 1st input to the 3rd activation node in layer 2. Sigmoid function is defined as

$$g(z) = \frac{1}{(1 + e^{-z})}$$

The nature of the sigmoid function is such that $g(z)$ tends to 1 as $z \to \infty$ and tends to -1 as $z \to -\infty$. The output of the sigmoid function is always bounded in the interval [0,1].

The intuition behind the neural network is that output of each layer itself acts as input to the subsequent layer resulting in modeling nonlinear decision boundary. It is not uncommon to have tens and hundreds of activation layers.

Neural network algorithm will be discussed in greater detail in the coming sections.

2.5   Principal Component Analysis

Principal component analysis (PCA) is a powerful tool in machine learning for determining the principal component features of a dataset. In lot of datasets, commonly there are features which are correlated. Using PCA, the number of highly correlated features can be prioritized into fewer uncorrelated features called principal components. This is also known as dimensionality reduction.

Apart from helping in visualizing the principal components, PCA helps in reducing the cost of machine learning algorithms. Computational cost of machine learning algorithms is dependent on the number of dimensions. Using PCA to reduce dimensions helps in reducing computational costs of machine learning algorithms. Reducing the dataset to its principal components also reduces the amount of dataset, with minimal compromise in the correlation between the data features.

In neural networks, using PCA, dimensions can be prioritized and low variance dimensions can be dropped and the algorithms converge faster.

2.5.1   PCA Algorithm

Assume a dataset X_j ($j = 1 \ldots m$) with n features and m data items.

The idea behind PCA is to project data points in a n-dimensional space onto a lower-dimensional space while preserving as much information as possible. Consider Fig. 3 shown below for a two-dimensional dataset.

PCA aims at orthogonally projecting the data onto the lower-dimensional linear space such that [6].

- Minimizes the distance between the points and the projections (sum of brown lines)
- Maximizes the variance of projected data (yellow line).

Fig. 3 Illustrating goals of PCA for a 2-d dataset

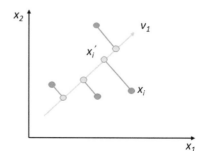

First principal component is along the direction of largest variance. Subsequent principal components are orthogonal to the previous principal component and points in the direction of the largest variance of remaining data space.

Compute the covariance matrix for the dataset.

$$\Sigma = \frac{1}{m} \sum_{i=1}^{m} (X_i - \bar{X})(X - \bar{X})^{\mathrm{T}}$$

where $\bar{X} = \frac{1}{m} \sum_{i=1}^{m} X_i$

The PCA basis vectors are the eigenvectors of the covariance matrix $\sum$. The eigenvalues will determine the importance of the eigenvectors. Larger the eigenvalues, more important the eigenvectors. By choosing a subset of the PCA vectors with largest eigenvalues, the dimensionality of the dataset can be reduced.

PCA can be used to discard dimensions of less significance, remove noise, and get compact description of the data.

Consider a dataset of m facial images, each of 256×256 pixels. Each data item has $N = 64$ K dimensions. Covariance matrix is of $N \times N$ dimensions. N eigenvectors and values can be computed in $O(N^3)$ complexity, and first p eigenvectors and eigenvalues can be computed in $O(pN^2)$ complexity. For $N = 64$ K, this is very computationally intensive. Invariably $m \ll N$.

using $L = X^{\mathrm{T}}X$ instead of $\Sigma = XX^{\mathrm{T}}$

Let v be the eigen vector of L

$Lv = \lambda v$

$X^{\mathrm{T}}Xv = \lambda v$

$X(X^{\mathrm{T}}Xv) = X(\lambda v)$

$(XX^{\mathrm{T}})Xv = \lambda(Xv)$

$\Sigma Xv = \lambda(Xv)$

So Xv is the eigenvector of Σ. Complexity of computation of eigenvector of L is much less computationally intensive compared to Σ [7].

3   Details of Two Areas

3.1   Linear/Logistic Regression

Let us consider the application of an intelligent smartphone review system. Users submit reviews on various smartphones based on the user experience. Let us look

Fig. 4 A simple linear classifier model with binary prediction

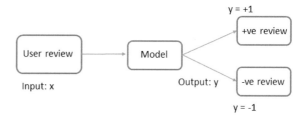

at a machine learning system that classifies these user reviews using a simple linear classifier.

Consider the following reviews:

Review	User experience
"Incredible phone. A great value for money"	Positive experience
"Sports a very good camera, however battery life is bad"	Mixed experience
"Best smartphone that I have ever bought"	Very positive experience
"Voice quality is poor, dropping calls is abysmal"	Negative experience

The classifier model, as shown in Fig. 4, would take user review as input and predict the rating for the product.

The dataset can be viewed as set of words x_i ($i = 1 \ldots n$), consider n different parameters in the review dataset (training dataset), commonly referred to as features. Let the outcome (recommendation) be represented as y_i ($i = 1 \ldots m$) taking on values -1 (do not recommend the item) or 1 (recommend the item).

The dataset can be used to train a set of parameter (coefficients) for each word as shown in Table 1 below. Some neutral words will get assigned coefficient value $= 0$, since they are not adding any sentiment to the review.

Table 1 Coefficients for words in the reviews

Word	Coefficient
Incredible	3.1
Great	1.7
Best	2.1
Good	1.0
Bad	−1.1
Poor	−1.6
Abysmal	−3.4
Phone, the, camera	0.0

Table 2 Table showing a simplistic model with coefficients for two words

Word	Coefficient
Incredible	2.0
Poor	−1.6

Fig. 5 An example of a linear decision boundary

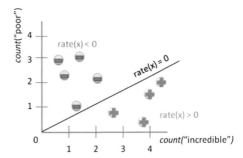

For each review, a rating rate(x) is assigned as weighted count of the words in it. If rate(x) > 0 y = +1 else y = −1.

Consider the table above depicting a simplistic model where the following words have nonzero coefficients (Table 2).

rate(x) = 2.0 * count("incredible") − 2.5 * count("poor")

The decision boundary can be plotted as shown in Fig. 5 above.

For linear classifiers, with three coefficients the decision boundary is a plane, and for more than three coefficients, it is a hyperplane. For other classifiers, the boundary will end up being complicated shapes. The features could have other functions like tf-idf weights instead of just count of terms. The rest of the section describes an algorithm to train a set of parameters based on gradient descent algorithm.

Let us start with a hypothesis h that can be used to approximate y as follows:

$$h(x) = \theta_0 + \theta_1 x_1 + \theta_2 x_2 + \cdots + \theta_n x_n$$

θ_i is called weight (or coefficient). By introducing an intercept term $x_0 = 1$, the hypothesis can be represented as

$$h(x) = \sum_{i=0}^{n} \theta_i x_i$$

In the above application, x_i is the count of the ith word in the review. In vectorized form, the hypothesis can be rewritten as

$$h(x) = [\theta_0 \theta_1 \ldots \theta_n] . \begin{bmatrix} x_0 \\ x_1 \\ . \\ x_n \end{bmatrix}$$

If the parameters are denoted by vector θ and features by vector x, the hypothesis can be simplified as

$$h(x) = \theta^{\mathrm{T}}.x$$

θ^{T} represents the transpose of vector θ. The goal is to learn the values of parameters θ such that the $h(x)$ is as close as possible to the outcome in the training set. To measure the accuracy of the hypothesis to the actual outcomes, a cost function can be defined as follows

$$C(\theta) = \frac{1}{2} \sum_{i=1}^{m} \left(h\left(x^{(i)}\right) - y^{(i)} \right)^2$$

The goal of supervised learning is to come up with a set of parameters θ such that it minimizes this cost function. Superscript i denotes the ith dataset item. This is the common **least squares regression model**.

3.1.1 Gradient Descent Algorithm

To achieve the goal of minimizing the cost function $C(\theta)$, we can start by an initial guess of values for θ and iteratively change the value of θ such that the cost function is smaller with every iteration till we converge to a set of parameters that minimizes $C(\theta)$. Consider the following update to the parameters

$$\theta_j := \theta_j - \alpha \frac{\partial}{\partial \theta_j} C(\theta) \tag{1}$$

α is referred to as the **step size** or **learning rate**. The algorithm involves updating all the parameters θ_j ($j = 1 \ldots n$) iteratively till convergence. In this algorithm, the parameter values step toward the steepest descent with each iteration.

Let us derive the partial derivative of the cost function with respect to θ_j.

$$\frac{\partial}{\partial \theta_j} C(\theta) = \frac{\partial}{\partial \theta_j} \left[\frac{1}{2} \sum_{i=1}^{m} \left(h\left(x^{(i)}\right) - y^{(i)} \right)^2 \right]$$

$$= \frac{1}{2} \sum_{i=1}^{m} \frac{\partial}{\partial \theta_j} (h(x^{(i)}) - y^{(i)})^2$$

$$= \frac{1}{2} \sum_{i=1}^{m} 2(h(x^{(i)}) - y^{(i)}) \frac{\partial}{\partial \theta_j} \left(h\left(x^{(i)}\right) - y^{(i)} \right)$$

$$= \frac{1}{2} (2) \sum_{i=1}^{m} (h(x^{(i)}) - y^{(i)}) x_j^{(i)}$$

$$= -\sum_{i=1}^{m}(y^{(i)} - h(x^{(i)}))x_j^{(i)}$$

Substituting this in (1) we get the step for θ_j

$$\theta_j = \theta_j - \alpha \left[-\sum_{i=1}^{m}(y^{(i)} - h(x^{(i)}))x_j^{(i)} \right]$$

$$= \theta_j + \alpha \sum_{i=1}^{m}(y^{(i)} - h(x^{(i)}))x_j^{(i)}$$

The gradient descent algorithm can be written as

```
while ( C(θ) > threshold ) {
```

$$\theta_j = \theta_j + \alpha \sum_{i=1}^{m}(y^{(i)} - h(x^{(i)}))x_j^{(i)}$$

```
    Compute C(θ)
}
```

The updates of all θj are performed simultaneously on all the values of the parameters. The above algorithm is called batch gradient descent. The whole dataset has to be scanned before the parameters are updated to make progress toward the global minimum. For large datasets, a variant called stochastic gradient descent [8] is used, where for each data encountered the parameters are updated. The stochastic gradient invariably converges quicker than batch gradient descent for large datasets.

3.1.2 Logistic Regression

For the application mentioned earlier like intelligent product review system, we need the outcome to be a binary value like "recommended" (value 1) or "not recommended" (value 0). For these classes of application, a discrete output of $h(x)$ as in linear regression is not intuitive since $y \in \{0,1\}$. Consider the following change to the hypothesis.

$$h(x) = g(\theta^T x) = \frac{1}{\left(1 + e^{-\theta^T x}\right)}$$

Function $g(z)$ is the **sigmoid function**. The graph of sigmoid function is shown in Fig. 6.

Fig. 6 Graph of Sigmoid function

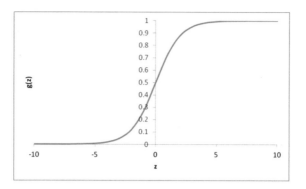

The nature of the sigmoid function is such that $g(z)$ tends to 1 as $z \to \infty$ and tends to -1 as $z \to -\infty$. The output of the sigmoid function is always bounded in the interval $[0, 1]$.

A salient feature of the sigmoid function is that its derivative can be expressed in terms of itself:

$$\frac{\partial}{\partial z}(g(z)) = g(z)(1 - g(z))$$

In the previous section, we have output a rating as $+1$ or -1 for a review based on the rate(x) function learned from the dataset. Defining the probability that an outcome is 1 for given set of features and parameters as

$$P(y = 1|x;\theta) = h(x)$$
$$P(y = 0|x;\theta) = 1 - h(x)$$

In the above example, this can be interpreted as the probability that a review is positive given a set of words in the review.

Combining the two into one expression, the probability can be rewritten as

$$P(y|x;\theta) = (h(x))^y (1 - h(x))^{1-y}$$

By defining a likelihood function $L(\theta)$, for m data items, assuming generated independently

$$L(\theta) = \prod_{i=1}^{m} p(y^{(i)}|x^{(i)};\theta)$$
$$= \prod_{i=1}^{m} \left(h\left(x^{(i)}\right)\right)^{y^{(i)}} \left(1 - h\left(x^{(i)}\right)\right)^{1-y^{(i)}}$$

The goal here is to maximize the probability for the set of parameters. For ease of derivation, the log of the likelihood function is maximized.

$$l(\theta) = \log(L(\theta))$$

$$= \sum_{i=1}^{m} \left[y^{(i)} \log(h(x^{(i)})) + (1 - y^{(i)}) \log(1 - h(x^{(i)})) \right]$$

Similar to linear regression, to maximize the likelihood function, we use a gradient ascent algorithm. If θ denotes vector of parameters,

$$\theta = \theta + \alpha \frac{\partial}{\partial \theta}(l(\theta))$$

Simplifying the partial derivative, we arrive at the following step

$$\theta_j := \theta_j + \alpha \left(y^{(i)} - h(x^{(i)}) \right) x_j^{(i)}$$

Now applying the gradient ascent, we can update the parameters simultaneously, iteratively till convergence.

On a new data item, we can predict the probability that the review is positive by using the parameters learnt using the training dataset

$$P(y = 1|x;\theta) = h(x) = g(\theta^T x)$$
$$\text{if } P(y = 1|x;\theta) > 0.5 \text{ outcome} = 1$$
$$< 0.5 \text{ outcome} = 0$$

The threshold of 0.5 can be set to a different value based on the application and the dataset. For instance in an application that detects cancerous cells, the threshold could be set conservatively.

3.2 Neural Networks

In Sect. 2.4, neural network was introduced as a method to model nonlinear decision boundary; in this section, we delve into neural networks in greater detail.

In earlier section, the neural network that was described was modeling data and classifying the output into single class. To model a multiclass classification, the same notion can be extended. Instead of the output y being a value [0,1], the output is a vector whose size is equal to the number of classes in the classification problem as shown in Fig. 7.

Consider a set of images where the images are labeled as that of sunrise, oceans, forests, or deserts. The output label for each image will be a vector of size 4. The

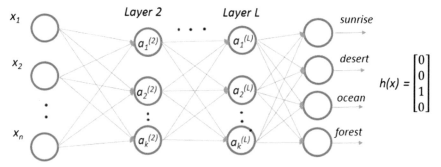

A Multiclass classification Neural Network

Fig. 7 Depiction of a typical multi-class neural network

values of the element in the vector (0 or 1) will denote which class the image belongs to. The neural network output would look something similar to what is shown below:

The output in the figure above classifies the image as that of an ocean.

Recall the cost function for the logistics regression (previous section) was

$$l(w) = \sum_{i=1}^{m} \left[y^{(i)} \log(h(x^{(i)})) + (1 - y^{(i)}) \log(1 - h(x^{(i)})) \right]$$

For a K-class classification neural network, the cost function can be generalized as follows:

$$l(w) = \sum_{i=1}^{m} \sum_{k=1}^{K} \left[y_k^{(i)} \log(h(x^{(i)})_k) + (1 - y_k^{(i)}) \log(1 - h(x^{(i)})_k) \right]$$

The second summation adds up the cost of logistic regression for each node in the output layer. As in the logistic regression, we minimize the cost function using an optimal set of weights. To achieve this, we need to compute the partial derivative of the cost function.

Let us define the error of the output as $^{(l)}$ as the error for layer l. For the output layer ($l = L$) for a L layered neural network

$$\delta^L = (a^L - y)$$

which is basically the error for each of the output nodes in the output layer. To get the delta values of the hidden layers, we follow a backpropagation algorithm and derive the delta values from the delta values of the subsequent layer. Values for $\delta^{(L-1)}$, $\delta^{(L-2)}, \ldots, \delta^{(2)}$ can be calculated as follows by the general formula for layer l.

$$\delta^l = \left(w^{(l)}\right)^{\mathrm{T}} \delta^{l+1} \cdot * a^{(l)} \cdot * \left(1 - a^{(l)}\right)$$

where $a^{(l)}$ is the activation function of layer l. Recall that for a sigmoid function $g(z)$

$$\frac{\partial}{\partial z}(g(z)) = g(z)(1 - g(z))$$

So the delta value $^{(l)}$ for layer l is

$$\delta^l = \left(w^{(l)}\right)^T \delta^{l+1} . * g'\left(z^{(l)}\right)$$

Propagating from right to left in a neural network, delta values for the units in all the layers can be derived for each of the layers in the network. The delta values are errors for each unit $a_j^{(l)}$ (activation unit j in layer l) and are derivative of the cost function.

$$\delta_j^l = \frac{\partial}{\partial z_j^{(l)}} cost(t)$$

The gradient for the hidden layer weights is simply the output error signal back-propagated to the hidden layer, then weighted by the input to the hidden layer. The gradient for the weights of layer l is

$$= \delta^{(l+1)}(a^l)^T$$

The backward propagation algorithm [9] can be formalized as follows:

- Initialize $\Delta_{ij}^l = 0$ for all values of l, i, j.
- Calculate the activations $a^{(l)}$ for layers $l = 2, 3, \dots L$
- Compute $^{(L)}$ as $a^l - y$ for layer L.
- Compute $\delta^{(L-1)}, \delta^{(L-2)}, \dots, {}^{(2)}$ as

$$- \delta^l = \left(w^{(l)}\right)^T \delta^{l+1} . * a^{(l)} . * \left(1 - a^{(l)}\right)$$

- Compute gradients for layer l as

$$- \Delta^{(l)} = \Delta^{(l)} + \delta^{(l+1)}(a^{(l)})^T$$

- $D_{ij}^l = \frac{1}{m}\Delta_{ij}^l$
- The delta matrix is the partial derivatives.

Update weights $w_{ij}^l = w_{ij}^l - D_{ij}^l$

In logistic regression, the initial weights can be assigned to 0 and the weights can be learnt the using gradient descent. In neural networks, assigning initial weights of 0 will cause all the nodes to update to the same values during backpropagation. Typically, the weights are assigned random weights in a neural network.

4 Future

The data centers and databases are headed toward self managing and autonomous systems that can self manage, self tune, detect and fix adversities. Oracle provides self-managing databases which are autonomous and self-driving.

The cost to acquire, store, and compute data will continue to fall. Amount of data will continue to grow. The machine learning building blocks are moving to cloud, where machine learning techniques are hosted in the cloud. With confluence of these will make every organization, a data company, and every application an intelligent application, moving to an algorithm economy. In the past, automation was limited to "blue-collar jobs," and we will see a future where automation by "white-collar machines" will be prevalent.

5 Conclusion

With the explosion of data, machine learning algorithms are need of the hour to dynamically analyze the data. The technological advancements in the processor speeds, distributed technology combined with explosion of data have seen resurgence in machine learning. Mathematics forms the bedrock of machine learning techniques. Machine learning is poised to take an even bigger role in our daily lives.

References

1. Gartner Research report, https://www.gartner.com/newsroom/id/3598917
2. Salton, G., Wong, A., Yang, C.S.: A vector space model for automatic indexing. ACM Commun. **18**(11), 613–620 (1975)
3. Oracle Corporation, "Oracle Autonomous database", https://www.oracle.com/database/autono mous-database/feature.html
4. Andrew, Ng., https://see.stanford.edu/Course/CS229
5. Lippmann, R., MIT Lincoln Lab. Lexington, MA, An introduction to computing with neural nets, http://ieeexplore.ieee.org/abstract/document/1165576
6. Shelns, J.: A tutorial on Principal Component Analysis. https://arxiv.org/pdf/1404.1100.pdf, Google Researc, Mountain View, CA
7. "PCA", Barnabas Pcozos and Aarti Singh, Machine Learning Department, Computer science Department, Carnegie Melon University, http://www.cs.cmu.edu/~aarti/Class/10701_Spring14/ slides/PCA.pdf
8. Bottu, L.: Large-Scale Machine Learning with Gradient Descent, http://leon.bottou.org/publica tions/pdf/compstat-2010.pdf, NEC Labs, Princeton, NJ
9. Jain, A.K., Mao, J., Mohiuddin, K.M.: Artificial Neural networks—A Tutorial, ieeex-plore.ieee.org/document/485891

Chapter 13
Numerical Study on the Influence of Diffused Soft Layer in pH Regulated Polyelectrolyte-Coated Nanopore

Subrata Bera, S. Bhattacharyya and H. Ohshima

Abstract Electroosmotic flow and its effect are numerically studied in the polyelectrolyte layer-coated cylindrical nanopore. The flow characteristic of the electrokinetic consists of the Nernst–Planck equation for species distribution, the Brinkman modified Navier–Stoke equation for fluid flow and the Poisson equation for induced electric potential. These nonlinear coupled governing equations for potential distribution, ionic species distribution and fluid flow are solved through a finite volume method in staggered grid system for cylindrical coordinate. This study established the importance of the bulk ionic concentration, electrolyte pH, the softness of the polyelectrolyte layer, the nanopore geometries and potential of the polyelectrolyte layer and nanopore wall. Three functional group as Succinoglycan, Glycine, and Proline functional group are considered in this study. The average electroosmotic flow rate increases with polyelectrolyte segment for a fixed pH value in the succinoglycan functional group. The axial velocity increases with the pH values for fixed polyelectrolyte segment. The increase of softness parameter decreases the average flow. The increase in pH values increases the average flow for different bulk ionic concentration. The increase of ionic current with the pH values are more prominent for the negatively charged surface than zero-charged potential. The electric body force increase with the pH values for both zero-charged nanopore and negatively charged nanopore.

Keywords Polyelectrolyte layer · Electroosmotic flow · Functional group
Nernst–Planck equation

S. Bera (✉)
Department of Mathematics, National Institute of Technology Silchar,
Silchar 788010, India
e-mail: subrata.br@gmail.com

S. Bhattacharyya
Department of Mathematics, Indian Institute of Technology Kharagpur,
Kharagpur 721302, India

H. Ohshima
Faculty of Pharmaceutical Sciences, Tokyo University of Science,
Noda, Chiba 2788510, Japan

© Springer Nature Singapore Pte Ltd. 2018
D. Ghosh et al. (eds.), *Mathematics and Computing*, Springer Proceedings
in Mathematics & Statistics 253, https://doi.org/10.1007/978-981-13-2095-8_13

1 Introduction

The grafting nanochannels in polyelectrolyte layer (PEL) has emerged as a novel technique for a large number of applications such as flow control, current rectification, ion sensing and manipulation, fabrication of nanofluidic diodes, liquid transport, and many more [1–3]. A thin layer of nonzero net charged density forms along the wall when the electrolyte comes in touch with the solid wall and is called the electric double layer (EDL). The EDL thickness is known as Debye length, and it is in nanometers order. Electroosmotic flow (EOF) occurs when the external electric field contacts with the net surplus charged ions in the EDL. Several interesting observations were seen experimentally when the characteristic length is an order of EDL thickness. When EOF is modeled with the thin EDL approximation using slip condition in velocity is called Helmholtz-Smoluchowski velocity [4]. Most of previous studies on EOF, the ion distribution is considered to obeys the equilibrium Boltzmann distribution and resulting Poisson–Boltzmann equation for the induced potential. But the Nernst–Planck equation for ions considered the convection, electromigration, and diffusion of ions. Several authors studied the various aspects of EOF in micro- and nanochannel in theoretically as well as experimentally. Conlisk and McFerran [5] developed a mathematical model for EOF and corresponding numerical solution in a rectangular nanochannel with overlapping EDL in the presence of the applied electric field. The combined effects of EOF and pressure-driven flow on species transport have studied by Bera and Bhattacharyya [6] by considering Nernst–Planck model for micro- and nanochannels.

There are many ways of modulation of electroosmotic flow. Polymer coatings are very useful to control the EOF rate. The electroosmotic flow in a semicircular cross section is studied by Wang et al. [7] under the Debye–Huckel approximation. The perturbation method was introduced by Chang et al. [8] to investigate the EOF of an incompressible, viscous, and electrically conducting Newtonian liquid through a microtube with slightly corrugated walls. Rojas et al. [9] theoretically studied the pulsatile electroosmotic flow (PEOF) within a circular microchannel. Liu et al. [10] established an analytical expression for the flow velocity and ionic current for the EOF in a charge-regulated circular channel, focusing on the effect of types of ions and their concentrations. The electroosmotic flow behavior of the nanopore can be influenced by its physicochemical properties, the applied external electric field, nature of liquid medium, and the potential of boundary surface and nearby PEL. Patwary et al. [11] established that the polyelectrolyte-grafted nanochannel which is highly efficient for electrochemomechanical energy conversion. Simple analytic expressions for the electrophoretic mobility of a soft particle were developed by Ohshima [12] within an ion-penetrable hard particle core surface of polyelectrolyte layer for low electric potential. Tessier and Slater [13] numerically investigated the EOF on coarse-grained molecular dynamics simulations. Cao and you [14] studied the coarse-grained molecular dynamics simulation method for mixed polymer brush-grafted nanochannels between two distinct species of polymers alternately grafted on the inner surface of nanochannels. The effect of PEL charged density on ions and fluid flow numerically investigated by Bera and Bhattacharyya [15] in a polyelectrolyte-coated nanopore. Ohshima [16] proposed a simple algorithms for the analytic solution of Poisson–Boltzmann equation in a charged narrow pore. They compared with the exact numerical solution for low-to-moderate values of the nanopore surface potential when the nanopore radius is less than the Debye length.

All these above studies considered a fixed charge density within the polyelectrolyte layer. But, many bacterial cell surfaces possess acidic and/or basic functional groups. Das [17] established the explicit relationships between surface potential of a charged soft interface and Donnan potential. Electroosmotic transport phenomena in a pH-dependent charge density were studied by Chen and Das [18] through polyelectrolyte-grafted nanochannel. Tseng et al. [19] theoretically investigated the influence of temperature distribution on the pH-regulated polyelectrolyte layer-coated particle.

The present study deals with the electroosmotic flow through PEL-coated nanopore in which the PEL charges is dependent on the ionization of corresponding functional group. The objective of the present study is to analyze the effects of bulk ionic concentration, pH value, softness of PEL, charged density of PEL and nanopore surface potential. Most of the authors studied linear EOF by considering Boltzmann distribution for ion. By taking the convection, diffusion, and electromigration effects, we have taken the Nernst–Plank equation. We have also considered the full Brinkman model in Navier–Stoke equations with body force term. The Poisson equations give the induced potential distribution in EDL. The characteristics of this electrokinetic flow are obtained by solving these nonlinear coupled equations through a finite volume method.

2   Mathematical Model

A canonical nanopore whose radius a and axial length z is considered in our study, as shown in Fig. 1. The nanopore wall bears negative potential ζ. A polyelectrolyte layer of thickness d is embedded in nanopore wall. We have assumed that the polyelectrolyte layer is homogeneously structured, ion-penetrable with fixed charge density ρ_{fix}. We consider that the polyelectrolyte segment distribution can be modeled as a soft step function and its distribution $h(r)$ is given by (as shown in Fig. 2)

$$h(r) = \begin{cases} 1 - \exp\left[-\frac{r-(a-d)}{\delta}\right], & a-d \leq r \leq a \\ 0 & 0 \leq a \leq a-d \end{cases} \tag{1}$$

Here δ is assumed to obey $\delta \ll d$, which measures the width of inhomogeneous distribution of PEL segments near front edge.

The polyelectrolyte layer (PEL) contains both the acidic functional groups/and basic functional groups namely AX and B, respectively. We have taken a volumetric charge density for uniform distribution of function group within the polyelectrolyte layer can be given as follows

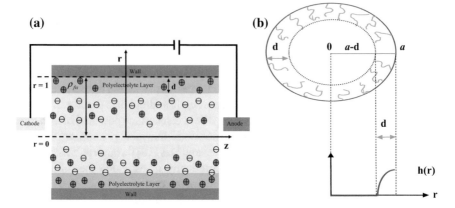

Fig. 1 **a** Schematic diagram of diffuse soft layer consisting of a rigid charged core in pH regulated the polyelectrolyte layer in a canonical nanopore and **b** soft function on the cross section of cylindrical nanopore

Fig. 2 Spatial distribution of polymer segments when $d = 0.4$ and arrow indicate increasing order of δ/d and varies from 0, 0.2, 0.4, 0.6, 0.8 and 1.0

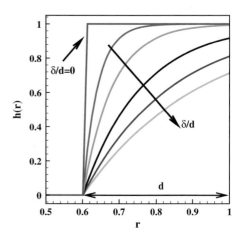

$$\rho_{fix}(r) = h(r)\rho(pH_0, pK_a, pK_b) \tag{2}$$

where charge density $\rho(pH_0, pK_a, pKb)$ comes due to the functional groups, which is given by Tseng et al. [19] as

$$\rho(pH_0, pK_a, pK_b) = \left[\frac{z_A N_A F}{1 + 10^{pK_a - pH_0}\exp(-e\phi/k_B T)} + \frac{z_B N_B F}{1 + 10^{pH_0 - pK_b}\exp(e\phi/k_B T)}\right] \tag{3}$$

We have taken a binary symmetric electrolyte with valance $z_A = -1$ and $z_B = 1$. Here, the total concentration for acidic functional groups is N_A and basic functional groups is N_B. Here, pH_0 is the bulk pH with $pK_a = -\log K_a$ and $pK_b = -\log K_b$. Here K_a is the ionization constant for acidic functional group and K_b is for the basic functional groups. We symbolically denote pH as the bulk pH value of the electrolyte in this discussion. Here the induced potential is $\phi(r)$, the Boltzmann constant is k_B, elementary electric charge is e, and absolute temperature is T.

The electric field $\mathbf{E}$ ($=\mathbf{E}_r, \mathbf{E}_z, \mathbf{E}_\theta$) has component along radial direction r, cross-radial direction z and axial direction z where a constant electric field E_0 is applied. The equation of total potential related to double layer potential (DLP) and polarization effects due to electric field. Therefore, the electric field connected to net charge density ρ_e plus charges density ρ_{fix} in PEL is given by the Poisson's equation

$$\nabla \cdot (\epsilon_e \mathbf{E}) = -\epsilon_e \nabla^2 \phi = \rho_e + \rho_{fix} \tag{4}$$

Here, the induced electric potential is Φ and permittivity $\epsilon_e = \epsilon_0 \epsilon_r$, where ϵ_0 and ϵ_r are the permittivity of vacuum and dielectric constant of the solution, respectively. Here net charge density $\rho_e = \sum_i z_i e n_i$; z_i and n_i are, respectively, the valance and ionic concentration. We have taken symmetric electrolyte of valance $z_i = \pm 1$ in the present study. We scaled the potential by ϕ_0 ($= k_B T/e$) and concentration by the bulk ionic concentration n_0. The bulk number density (n_0) and the bulk electrolyte concentration (C) are related by $FC = en_0$. The Poisson equation can be written in non-dimensional as

$$\left[\frac{\partial^2 \phi}{\partial z^2} + \frac{1}{r} \frac{\partial}{\partial r} \left(r \frac{\partial \phi}{\partial r} \right) \right] = -\frac{(\kappa a)^2}{2} (g - f) - h(r) Q_{fix} \tag{5}$$

We have scaled cylindrical coordinate r and z by nanopore radius a. The Debye layer thickness κ is the inverse of the EDL thickness (λ), where $\lambda = \sqrt{\epsilon_e k_B T / \sum_i (z_i e)^2 n_{i0}}$ and $\kappa a = a/\lambda$. The scale fixed charge density $Q_{fix}(r)$ within the diffused PEL is

$$Q_{fix}(r) = \left[\frac{z_A Q_A}{1 + 10^{pK_a - pH_0} \exp(-\phi)} + \frac{z_B Q_B}{1 + 10^{pH_0 - pK_B} \exp(\phi)} \right] \tag{6}$$

where the non-dimensional maximum charge density parameter are $Q_j = FN_j a^2 / \epsilon_e \phi_0$ ($j = A, B$) for acidic functional groups and basic functional groups.

The ion transport is described by the Nernst–Planck equation and is given as

$$\frac{\partial n_i}{\partial t} + \nabla \cdot \mathbf{N}_i = 0 \tag{7}$$

where $\mathbf{N}_i = -D_i \nabla n_i + n_i \omega_i z_i F \mathbf{E} + n_i \mathbf{q}$ is the net ionic flux of individual species. Here, Faraday's constant is F, D_i is the diffusivity and ω_i is the mobility of ith ionic species. Here velocity field $\mathbf{q} = (v, u)$ with the velocity components v and u along the radial r and axial z directions respectively. Here, velocity field $\mathbf{q}$ is nondimensionalized by the Helmholtz–Smoluchowski velocity U_{HS} ($=\epsilon_e E_0 \phi_0/\mu$) and time t is nondimensionalized by a/U_{HS}. The Reynolds number $Re = U_{HS} a/v$, Schmidt number $Sc = v/D_i$. Here, gas constant is R and the viscosity μ of the electrolyte is relate to $v = \mu/\rho$. Here, Peclet number Pe as $Pe = \frac{U_{HS} a}{D_i}$ Here, we denote g is the cationic concentration and f is anionic concentration in non-dimensional form. Hence, the non-dimensional equations of ion transport are expressed as follows

$$Pe\frac{\partial g}{\partial t} - \left[\frac{\partial^2 g}{\partial z^2} + \frac{1}{r}\frac{\partial}{\partial r}\left(r\frac{\partial g}{\partial r}\right)\right] + Pe\left[\frac{\partial(ug)}{\partial z} + \frac{1}{r}\frac{\partial(rvg)}{\partial r}\right] + \Lambda\frac{\partial g}{\partial z} - \left[\frac{\partial g}{\partial r}\frac{\partial \psi}{\partial r} + \frac{\partial g}{\partial z}\frac{\partial \psi}{\partial z}\right]$$

$$- \left[\frac{\partial g}{\partial r}\frac{\partial \phi}{\partial r} + \frac{\partial g}{\partial z}\frac{\partial \phi}{\partial z}\right] + \frac{(\kappa a)^2}{2}g\,(g-f) - hgQ_{fix} = 0 \tag{8}$$

$$Pe\frac{\partial f}{\partial t} - \left[\frac{\partial^2 f}{\partial z^2} + \frac{1}{r}\frac{\partial}{\partial r}\left(r\frac{\partial f}{\partial r}\right)\right] + Pe\left[\frac{\partial(uf)}{\partial z} + \frac{1}{r}\frac{\partial(rvf)}{\partial r}\right] + \Lambda\frac{\partial g}{\partial z} + \left[\frac{\partial f}{\partial r}\frac{\partial \psi}{\partial r} + \frac{\partial f}{\partial z}\frac{\partial \psi}{\partial z}\right]$$

$$+ \left[\frac{\partial f}{\partial r}\frac{\partial \phi}{\partial r} + \frac{\partial f}{\partial z}\frac{\partial \phi}{\partial z}\right] - \frac{(\kappa a)^2}{2}f\,(g-f) + hfQ_{fix} = 0 \tag{9}$$

The modified Navier–Stokes equation for electrokinetic flow is

$$\rho\left[\frac{\partial \mathbf{q}}{\partial t} + (\mathbf{q}\cdot\nabla)\mathbf{q}\right] = -\nabla p + \mu\nabla^2\mathbf{q} + \rho_e\mathbf{E} - \mu\lambda_s^2\mathbf{q} \tag{10}$$

$$\nabla\cdot\mathbf{q} = 0 \tag{11}$$

Here, fluid density and viscosity are given by ρ and μ, respectively, and $\lambda_s^2(r)$ is the position- dependent screening length. For diffuse polyelectrolyte layer and the softness parameter λ_s of the PEL can be expressed by Duval and Ohshima [20] as follows

$$\lambda_s = \lambda_0\,[h(r)]^{1/2} \tag{12}$$

where λ_0 is the softness degree of the homogeneous distribution of polyelectrolyte segments. Here, pressure is non-dimensionless by $\mu U_{HS}/a$.

The non-dimensional equations for fluid flow are given along axial direction z and radial direction r respectively as follows

$$Re\frac{\partial u}{\partial t} + Re\left(u\frac{\partial u}{\partial z} + v\frac{\partial u}{\partial r}\right) = -\frac{\partial p}{\partial z} - \frac{(\kappa a)^2}{2\Lambda}\left(-\Lambda + \frac{\partial \phi}{\partial z}\right)(g-f)$$

$$+ \left[\frac{\partial^2 u}{\partial z^2} + \frac{1}{r}\frac{\partial}{\partial r}\left(r\frac{\partial u}{\partial r}\right)\right] - \beta^2 hu \tag{13}$$

$$Re\frac{\partial v}{\partial t} + Re\left(u\frac{\partial v}{\partial z} + v\frac{\partial v}{\partial r}\right) = -\frac{\partial p}{\partial r} - \frac{(\kappa a)^2}{2\Lambda}\frac{\partial \phi}{\partial r}(g-f)$$

$$+ \left[\frac{\partial^2 v}{\partial z^2} + \frac{1}{r}\frac{\partial}{\partial r}\left(r\frac{\partial v}{\partial r}\right) - \frac{v}{r^2}\right] - \beta^2 hv \tag{14}$$

$$\frac{\partial u}{\partial z} + \frac{\partial v}{\partial r} + \frac{v}{r} = 0 \tag{15}$$

Here, the non-dimensional softness parameter is β. It can be expressed the softness degree of PEL (λ_0^{-1}) as $\beta = a/\lambda_0^{-1}$. The softness degree of PEL, $\lambda_0^{-1}(= \sqrt{\mu/\gamma})$ relates the hydrodynamic field inside the nanopore, while the conductance is not affected significantly by the flow field where γ is the hydrodynamics frictional coefficient. It (λ_0^{-1}) is the dimensional length and typically represents the characteristic penetration length of the fluid within soft structure. Here, we varied softness degree of PEL (λ_0^{-1}) so as to obtain the range of β between 1 and 20 [21, 22].

In the computational domain, we have used fully developed boundary condition in the upstream and downstream boundaries. We have also considered the no-slip condition along the channel walls. The rigid membrane surface is ion-impenetrable, i.e., $n \cdot N_i = 0$ with negative (ζ) potential on the walls. The axisymmetric condition is taken along the nanopore axis.

3　Numerical Schemes

The governing nonlinear coupled equations for potential, ion distribution, and fluid flow are solved by the finite volume method in staggered grid approach. The discretized form of these equations is obtained by integrating the governing equations over each control volumes. Different control volumes are used to integrate different equations. We considered the higher-order upwind scheme, QUICK (Quadratic Upwind Interpolation Convective Kinematics, [23] to discretize the convective and electromigration terms in both ion distribution and Navier–Stokes equations. These discretized governing equations are solved by the pressure correction-based iterative algorithm SIMPLE (Semi-Implicit Method for Pressure-Linked Equations, [24]).

We have taken non-uniform grid along radial direction r but the uniform grid along axial direction z. To verify the grid independency, we performed our computation for three different grid size when Grid 1: 400 × 250, Grid 2: 500 × 490, and Grid 3: 600 × 600 for EOF in cylindrical channel. We have also compared our result with the Ai et al. [25] and analytic solution. We have taken non-uniform grid size where δr is considered in range between 0.0025 to 0.01 with δz is either 0.0125 (for Grid 1) or $\delta z = 0.008$ (for Grid 3). In Grid 2, we have taken $\delta z = 0.01$ and $0.0025 \leq \delta r \leq 0.002$ and δt was taken as 0.0001. To validate of our numerical scheme, we have compared our computed solution for EOF in Fig. 3 with Ai et al. [25] and analytic solution for the axial velocity (u) of an electroosmotic flow (EOF) in a cylindrical channel. Here, 10 mM is the bulk electrolyte concentration without polyelectrolyte layer. The nanopore charge density $(\sigma) = -1$ mC/m^2 (i.e., $\zeta = -0.019$), and the applied electric field is -50 KV/m.

Fig. 3 Comparison of electroosmotic flow for the analytic solution and present solution with Ai et al. [25] of the axial velocity (u) in a cylindrical channel without PEL, i.e., $h(r) = 0$. The bulk concentration is 10 mM in KCl solution. The charge density of the nanopore (σ) = -1 mC/m^2 (i.e., $\zeta = -0.56$) and the applied imposed electric field is -50 KV/m

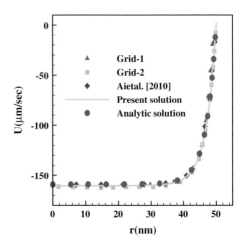

4 Results and Discussions

In this paper, we focus on the EOF effects on polyelectrolyte-coated cylindrical charged or uncharged nanopore. The nanopore geometry relates to experimental design which depend on nanofluidic devices, where the nanopore radius is 3–30 nm. It is proved that the continuum-based model is valid to capture their essential physics when the nanopore radius is larger than 3 nm. The thickness of the polyelectrolyte layer (d) is based on the biological lipids which typically ranges from 3 to 5 nm. We have considered the polyelectrolyte nanopore with radius $a = 10$ nm and the PEL thickness (d) is 4 nm. Here, we presented the results for various values of pH, bulk ionic concentration, wall potential, thickness of polyelectrolyte layer and surface charge density of polyelectrolyte layer.

We vary the bulk electrolyte concentration so that the Debye–Huckel parameter from $\kappa a \sim o(1)$ to $\kappa a \gg 1$. In our study, we have taken three functional groups as succinoglycan ($pK_a = 4.58$, $pK_b = 8.6$; proline ($pK_a = 1.99$, $pK_b = 10.96$; Wu et al. [3]), and glycine ($pK_a = 2.35$, $pK_b = 9.78$). We have consider $N_A = N_B = 10.23$ mM so that the scaled charge density becomes $Q_A = Q_B = 10$.

Figure 4a–c shows the non-dimensional distribution of ionic species g, f; induced potential ϕ and axial velocity u, respectively, for different values of PEL segment δ/d in succinoglycan functional group. Here, we have considered the softness parameter $\beta = 1$, bulk ionic concentration $C = 10$ mM in the surface potential $\zeta = -1$ with $Q_A = Q_B = 10$. It is clear for Fig. 4 that the increase of PEL segment increase the net charge density and so as axial electroosmotic velocity.

The distribution of axial velocity are shown in Fig. 5a for different vales of pH in succinoglycan functional group. We have considered different pH values such as 2, 4, 6, 8, 10, and 12 for acidic and basic groups. The pH values are taken lower and higher values close to the corresponding pK_a value in succinoglycan functional group (i.e., $pK_a = 4.58$). The axial velocity increase with pH values for fixed soft-

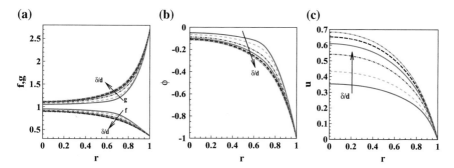

Fig. 4 Distribution of non-dimensional **a** ionic species distribution g, f; **b** induced potential ϕ and **c** axial velocity u for different δ/d in succinoglycan functional group in the PEL. Here, $pH = 2.0$, $\beta=1$, $C = 10$ mM, $\zeta = -1$ with $Q_A = Q_B = 10$. Arrows indicated the increasing order of δ/d as 0.1, 0.2, 0.4, 0.6, 0.8 and 1.0

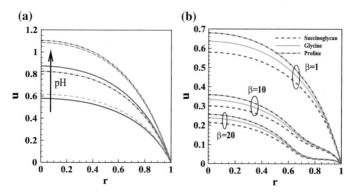

Fig. 5 Distribution of non-dimensional axial velocity u with **a** different pH when $\beta = 1$ in succinoglycan functional group and **b** different functional group for different softness parameter β when $pH = 2$. Here, $\delta/d = 0.5$, $C = 10$ mM, $\zeta = -1$ with $Q_A = Q_B = 10$. Arrows indicated the increasing order of pH as 2, 4, 6, 8, 10 and 12

ness parameter $\beta = 1$ and PEL segment $\delta/d = 0.5$. Figure 5b shows the distribution of axial velocity for different softness parameter β for three functional group such as succinoglycan, glycine, and proline . The axial velocity inversely varies with the softness parameter for those functional group.

Figure 6a shows the potential distribution for different pH in succinoglycan functional group for fixed ionic concentration, PEL segment, and softness parameter. We consider the pH values lower and higher of pK_a value of succinoglycan functional group. Fig. 6a that for lower values of pH with respect to pK_a, the potential distribution is positive. For higher values of pH, the factor $h(r)$ becomes to unity and the polyelectrolyte behaves like a layer with a constant charge density Q_{fix}, which is independent of the bulk pH and so the potential distribution is negative. When the pH values is lower close to pK_a, it observed a strong dependence of pH values

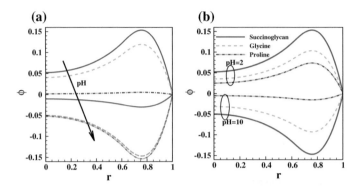

Fig. 6 Distribution of non-dimensional electric potential ϕ **a** different pH for succinoglycan functional group and **b** different functional groups of the PEL. Here, $\delta/d = 0.5$, $\beta = 1$, $C = 10\,\text{mM}$, $\zeta = 0$ with $Q_A = Q_B = 10$. Arrows indicated the increasing order of pH as 2, 4, 6, 8, 10, 12

on the PEL charge density. Figure 6b presents the non-dimensional distribution of potential for three functional group succinoglycan, glycine, and proline functional group for different values of softness parameter in fixed pH. It is evident from Fig. 6b that lower values of pH than pK_a, the potential distribution is always positive for three different functional group. Since the value of pK_a for succinoglycan functional group in higher than other functional group, the potential distribution is more high than others and reverse happens for high pH cases.

The non-dimensional average flow EOF velocity (u_m) in a cross section is defined by

$$u_m = \int_s \frac{u \cdot n\,ds}{\pi a^2} \tag{16}$$

Here, n is the unit vector in outward normal direction on the nanopore and s is the crosssectional area. The variation of dimensional average flow U_m with bulk pH is shown in Fig. 7a, b for different length of PEL segment δ/d in succinoglycan functional group when nanopore surface potential $\zeta = 0$ and $\zeta = -1$, respectively. For low pH value, average flow is negative, and it is increase with increase of pH when the nanopore surface potential $\zeta = 0$ as shown in Fig. 7a. But Fig. 7b shows that average flow is increase positively with the increase of pH when nanopore wall is negatively charged. For both cases $\zeta = 0$ and $\zeta = -1$, the average flow increases with the length of PEL segment δ/d for low pH values and reverse result happens for high pH values.

The ion concentration effects on the average flow are plotted in Fig. 8a, b for succinoglycan functional group when softness parameter and polyelectrolyte layer segment length are fixed. Figure 8a shows that the average flow increase from negative to positive with the increase of the bulk ionic concentration when nanopore surface potential $\zeta = 0$. But the average flow always increase positively with bulk ionic concentration for $\zeta = -1$.

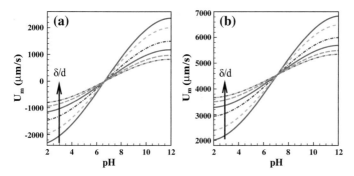

Fig. 7 Variation of dimensional average flow U_m with bulk pH for different polyelectrolyte segment δ/d in succinoglycan functional group. Here, $C = 10$ mM, $\beta = 1$, with $Q_A = Q_B = 10$. Arrows indicate the increasing order of δ/d as 0.1, 0.2, 0.4, 0.6, and 1.0. **a** $\zeta = 0$ and **b** $\zeta = -1$

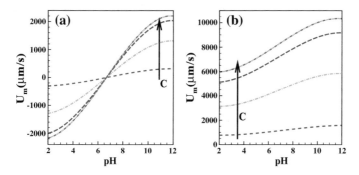

Fig. 8 Variation of dimensional average flow U_m with bulk pH for for different ionic concentration in succinoglycan functional group. Here, $\delta/d = 0.5$, $\beta = 1$ with $Q_A = Q_B = 10$. Arrows indicate the increasing order of C as 1 mm, 10 mM, 100 mM, and 1000 mM. **a** $\zeta = 0$ and **b** $\zeta = -1$

The variational of dimensional average flow with PEL segment are described in Fig. 9 for different values of softness parameter for succinoglycan, glycin, and proline functional group. Figure 9a, b indicates the average flow when nanopore surface potential $\zeta = 0$ and $\zeta = -1$ respectively. The increase of softness parameter decrease the average flow for all cases.

The current density is defined as follow

$$\mathbf{j} = e\Sigma z_i N_i = j_0 \left[-\Sigma z_i \nabla n_i - \Sigma z_i^2 n_i \nabla \Phi + Pe\mathbf{q}\Sigma z_i n_i \right] \tag{17}$$

where j_0 $(=D_i e n_0/a)$ is the scaled electric current density. We defined the average current density (I_z) along the z axial direction as

$$I_z = \int_s \mathbf{j} \cdot n ds \tag{18}$$

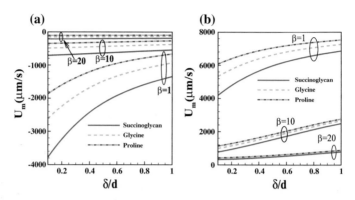

Fig. 9 Variation of dimensional average flow U_m with polyelectrolyte segment δ/d for different softness parameter β. Here, $pH = 2$, $C = 1000$ mM with $Q_A = Q_B = 10$. **a** $\zeta = 0$ and **b** $\zeta = -1$. solid, dash, and dash dot represent the succinoglycan, glycin, and proline functional group respectively

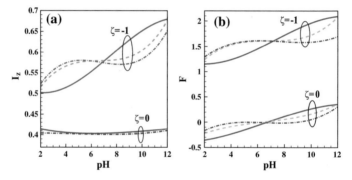

Fig. 10 Variation of scaled **a** average current density I_z and **b** total electric bodyforce F with bulk pH for different functional group. Here, $\delta/d = 0.5$, $C = 1$ mM and $\beta = 1$ with $Q_A = Q_B = 10$. **a** $\zeta = 0$ and **b** $\zeta = -1$

We also defined the total electrostatic body force (F) in the entire nanopore and is given by

$$F = \int_0^l h f_z \mathrm{d}z \tag{19}$$

where the average body force across the cylindrical cross section is defined as $f_z = \int_0^r \rho_e E \mathrm{d}r$. The scaled current density I_z and total electric bodyforce F for different functional group are shown in Fig. 10a, b when nanopore surface potential are $\zeta = 0$ and $\zeta = -1$, respectively.

5 Conclusions

We have focused the electroosmotic flow effects through polyelectrolyte layer-embedded nanopore. The governing equations for electrokinetic flow are considered as the Nernst–Planck equation for ion distribution, Brinkman-modified Navier–Stoke equation for flow and Poisson's equation for EDL potential. This coupled governing equations are solved by different algorithms in finite volume approach. The axial velocity increases with the polyelectrolyte segment for fixed pH value in the Succinoglycan functional group. The axial velocity increases with the pH values for fixed polyelectrolyte segment. The average flow rate is inversely proportional to the softness parameter. The increase in pH values increases the average flow for different bulk ionic concentration. The effect of ionic current is more prominent with the pH values for negatively charged nanopore than zero potential. The electric body force increase with the pH values for both zero charged nanopore and negatively charged nanopore.

Acknowledgements Authors (S. Bera) wish to thank the Sci. & Eng. Research Board in Dept. of Sci. and Tech., Govt. of India for supporting financial assistant in the project of File No: ECR/2016/000771.

References

1. Squires, A., Hersey, J.S., Grinstaff, M.W., Meller, A.: A nanopore-nanofiber mesh biosensor to control DNA translocation. J. Am. Chem. Soc. **135**, 16304–16307 (2013)
2. Bergen, W.G., Wu, G.: Intestinal nitrogen recycling and utilization in health and disease. J. Nutr. **139**, 821–825 (2009)
3. Wu, G., Bazer, F.W., Burghardt, R.C., Johnson, G.A., Kim, S.W., Knabe, D.A., Li, P., Li, X., McKnight, J.R., Satterfield, M.C., Spencer, T.E.: Proline and hydroxyprolinemetabolism: implications for animal and human nutrition. Amino Acids **40**, 1053–1063 (2011)
4. Probstein, R.F.: Physicochemical Hydrodynamics: An Introduction, 2nd edn. Wiley Interscience, New York (1994)
5. Conlisk, A.T., McFerran, J.: Mass transfer and flow in electrically charged micro-and nanochannels. Anal. Chem. **74**, 2139–2150 (2002)
6. Bera, S., Bhattacharyya, S.: On mixed electroosmotic-pressure driven flow and mass transport in microchannels. Int. J. Eng. Sci. **62**, 165–176 (2013)
7. Wang, C.-Y., Liu, Y.-H., Chang, C.C.: Analytical solution of electro-osmotic flow in a semicircular microchannel, ?Phys. Fluids **20**, 063105–063111 (2008)
8. Chang, L., Jian, Y., Buren, M., Liu, Q., Sunb, Y.: Electroosmotic flow through a microtube with sinusoidal roughness. J. Mol. Liq. **220**, 258–264 (2016)
9. Rojasa, G., Arcosa, J., Peraltaa, M., Méndezb, F., Bautistaa, O.: Pulsatile electroosmotic flow in a microcapillary with the slip boundary condition, Colloids and Surfaces A: Physicochem. Eng. Aspects **513**, 57–65 (2017)
10. Liu, B.-T., Tseng, S., Hsu, J.-P.: Analytical expressions for the electroosmotic flow in a charge-regulated circular channel. Electrochem. Commun. **54**, 1–5 (2015)
11. Patwary, J., Chen, G., Das, S.: Efficient electrochemomechanical energy conversion in nanochannels grafted with polyelectrolyte layers with pH-dependent charge density. Microfluid Nanofluid **20**, 37–51 (2016)

12. Ohshima, H.: Electrical phenomena of soft particles. A soft step function model. J. Phys. Chem. A. **116**, 6473–6480 (2012)
13. Tessier, F., Slater, G.W.: Modulation of electroosmotic flow strength with end-grafted polymer chains. Macromolecules **39**, 1250–1260 (2006)
14. Cao, Q., You, H.: Electroosmotic flow in mixed polymer brush-grafted nanochannels. Polymers **8**, 438–449 (2016)
15. Bera, S., Bhattacharyya, S.: Effect of charge density on electrokinetic ions and fluid flow through polyelectrolyte coated nanopore. In: ASME-Fluids Engineering Division Summer Meeting, V01BT10A008-V01BT10A008 (2017). https://doi.org/10.1115/FEDSM2017-69194.
16. Ohshima, H.: A simple algorithm for the calculation of the electric double layer potential distribution in a charged cylindrical narrow pore. Colloid Polym. Sci. **294**, 1871–1875 (2016)
17. Das, S.: Explicit interrelationship between Donnan and surface potentials and explicit quantification of capacitance of charged soft interfaces with pH-dependent charge. Colloids Surf. A: Physicochem. Eng. Aspects **462**, 6974 (2014)
18. Chen, G., Das, S.: Electroosmotic transport in polyelectrolyte-grafted nanochannels with pH-dependent charge density. J. Appl. Phys. **117**, 185304–185313 (2015)
19. Tseng, S., Lin, J.Y., Hsu, J.P.: Theoretical study of temperature influence on the electrophoresis of a pH-regulated polyelectrolyte. Anal. Chim. Acta. **847**, 80–89 (2014)
20. Duval, J.F.L., Ohshima, H.: Electrophoresis of diffuse soft particle. Langmuir **22**, 3533–3546 (2006)
21. van Dorp, S., Keyser, U.F., Dekker, N.H., Dekker, C., Lemay, S.G.: Origin of the electrophoretic force on DNA in solid-state nanopores. Nat. Phys. **5**, 347–351 (2009)
22. Yeh, L.-H., Zhang, M., Qian, S., Hsu, J.-P.: Regulating DNA translocation through functionalized soft nanopores. Nanoscale **4**, 2685–2693 (2012)
23. Leonard, B.P.: A stable and accurate convective modelling procedure based on quadratic upstream interpolation. Comput. Methods Appl. Mech. Eng. **19**, 59–98 (1979)
24. Fletcher, C.A.J.: Computational Techniques for Fluid Dynamics, vol-I & II Springer Ser. Comput. Phy. Springer, Heidelberg, New York (1991)
25. Ai, Y., Zhang, M., Joo, S.W.: Cheney. M.A., Qian. S.: Effects of electro osmotic flow on ionic current rectification in conical nanopores. J. Phys. Chem. C. **114**, 3883–3890 (2010)

Chapter 14
Quadruple Fixed Point Theorem for Partially Ordered Metric Space with Application to Integral Equations

Manjusha P. Gandhi and Anushri A. Aserkar

Abstract In this paper, two theorems have been established. The first theorem says the existences of a quadruple fixed point theorem in partially ordered metric space for nonlinear contraction mapping which is (α)-admissible and satisfies the mixed monotone property. The second result is proved for non-continuous mapping in addition to some other conditions. A suitable example of nonlinear contraction mapping validates the result. Moreover, an application to the integral equation is also presented.

Keywords Complete metric space · Partially ordered set · Quadruple fixed point Mixed monotone property · (α)-admissible

1 Introduction

The classical Banach's contraction principle has been improved and generalized by many researchers [1–8]. The existence of some new fixed point theorems for contraction mappings in partially ordered metric spaces was considered by Ran et al. [9], Bhaskar et al. [10], Nieto et al. [11, 12], and Agarwal et al. [13]. Bhashkar et al. [10] introduced the concept of a coupled fixed point and proved theorems in partially ordered complete metric spaces. Lakshmikantham et al. [5] proved coupled coincidence and coupled common fixed point theorems for nonlinear mappings in partially ordered complete metric spaces. Later, numerous results on coupled fixed point have been obtained [14–19]. Berinde et al. [20] came up with the idea of a tripled fixed point. Moreover, Samet et al. [21] proposed fixed point of order $N \geq 3$ for the first time. Karapnar [22] established quadruple fixed point theorems in partially ordered metric spaces. Several researchers [23–26] were motivated and

M. P. Gandhi (✉)
Department of Mathematics, Yeshwantrao Chavan College of Engineering, Nagpur, India
e-mail: manjusha_g2@rediffmail.com

A. A. Aserkar
Department of Mathematics, Rajiv Gandhi College of Engineering
and Research, Nagpur, India
e-mail: aserkar_aaa@rediffmail.com

© Springer Nature Singapore Pte Ltd. 2018
D. Ghosh et al. (eds.), *Mathematics and Computing*, Springer Proceedings
in Mathematics & Statistics 253, https://doi.org/10.1007/978-981-13-2095-8_14

proved theorems on quadruple fixed points under certain constraints. The present paper consists of three parts. In the first part, we prove two theorems. First theorem proves a quadruple fixed point theorem for a mapping satisfying the mixed monotone property as well as (α)-admissible condition. The second theorem proves for non-continuous mapping with some additional conditions. In addition, a suitable example validates the result. In the last section, the result is implicated for the existence of the solution of nonlinear integral equation. The theory of integral equations has many applications in the real world. For example, integral equations are often applicable in engineering, mathematical physics, economics, and biology.

2 Preliminaries

2.1 Quadruple Fixed Point: Let X be a nonempty set, and let $A : X \times X \times X \times X \to X$. An element (x, y, z, w) is called a quadruple fixed point of A if

$$A(x, y, z, w) = x, A(y, z, w, x) = y, A(z, w, x, y) = z, A(w, x, y, z) = w.$$

2.2 Mixed Monotone Property: Let $(X, \leq)$ be a partially ordered set, and let $A : X \times X \times X \times X \to X$ be a mapping. We say that A has the mixed monotone property if $A(x, y, z, w)$ is monotone non-decreasing in x and z and is monotone non-increasing in y and w, that is, for any $x, y, z, w \in X$.

$$x_1, x_2 \in X, x_1 \leq x_2 \Rightarrow A(x_1, y, z, w) \leq A(x_2, y, z, w)$$

$$y_1, y_2 \in X, y_1 \leq y_2 \Rightarrow A(x, y_2, z, w) \geq A(x, y_1, z, w)$$

$$z_1, z_2 \in X, z_1 \leq z_2 \Rightarrow A(x, y, z_1, w) \leq A(x, y, z_2, w)$$

$$w_1, w_2 \in X, w_1 \leq w_2 \Rightarrow A(x, y, z, w_2) \geq A(x, y, z, w_1)$$

2.3 Let ψ be the family of non-decreasing functions $\xi(t)$ such that $\sum_{n=1}^{\infty} \xi^n(t) < \infty$ for all $t > 0$, satisfying (i) $\xi(0) = 0$, (ii) $\xi(t) < t$ for all $t > 0$ (iii) $\lim_{r \to t^+} \xi(r) < t$ for all $t > 0$.

2.4 (α)-**admissible**: Let $A : X \times X \times X \times X \to X$ and $\alpha : X^4 \times X^4 \to [1, \infty)$ be two mappings. Then, A is said to be (α)-admissible if $\alpha((x, y, z, w), (p, q, r, s)) \geq 1$

$$\Rightarrow \alpha \left(\begin{array}{l} (A(x, y, z, w), A(y, z, w, x), A(z, w, x, y), A(w, x, y, z)), \\ (A(p, q, r, s), A(q, r, s, p), A(r, s, p, q), A(s, p, q, r)), \end{array} \right) \geq 1$$

for all $x, y, z, w, p, q, r, s \in X$.

3   Main Theorem

In this section, we establish two quadruple fixed point theorem with (α)-admissible mapping satisfying the mixed monotone property. In the second theorem, the continuity of the mapping is not considered.

Theorem 3.1 *Let* $(X, d, \leq)$ *be a partially ordered complete metric space.* $A : X \times X \times X \times X \to X$ *be a mapping having the mixed monotone property of* X. *Suppose that there exist* $\xi \in \psi$ *and* $\alpha : X^4 \times X^4 \to [1, \infty)$ *such that for* $x, y, u, v, p, q, r, s \in X$, *the following holds:*

$$\alpha((x, y, z, w), (p, q, r, s)) d(A(x, y, z, w), A(p, q, r, s))$$

$$\leq \xi \left(\frac{d(x, p) + d(y, q) + d(z, r) + d(w, s)}{4} \right) \tag{1}$$

for all $x \geq p, y \leq q, z \geq r, w \leq s$. *Also,*
(i) A is (α)-*admissible,*
(ii) There exist $(x_0, y_0, z_0, w_0) \in X$ *such that*
$\alpha\{(A(x_0, y_0, z_0, w_0), A(y_0, z_0, w_0, x_0), A(z_0, w_0, x_0, y_0), A(w_0, x_0, y_0, z_0)), (x_0, y_0, z_0, w_0)\} \geq 1$
(iii) A is continuous.
If there exists $x_0, y_0, z_0, w_0 \in X$ *such that*
$x_0 \leq A(x_0, y_0, z_0, w_0), y_0 \geq A(y_0, z_0, w_0, x_0), z_0 \leq A(z_0, w_0, x_0, y_0), w_0 \geq A(w_0, x_0, y_0, z_0)$, *then A has a quadruple fixed point.*

Proof Let $(x_0, y_0, z_0, w_0) \in X$ be such that

$$x_0 \leq A(x_0, y_0, z_0, w_0) = x_1, y_0 \geq A(y_0, z_0, w_0, x_0) = y_1,$$

$$z_0 \leq A(z_0, w_0, x_0, y_0) = z_1, w_0 \geq A(w_0, x_0, y_0, z_0) = w_1$$

Thus, $x_0 \leq x_1, y_0 \geq y_1, z_0 \leq z_1, w_0 \geq w_1$
    Again, $x_2 = A(x_1, y_1, z_1, w_1), y_2 = A(y_1, z_1, w_1, x_1), z_2 = A(z_1, w_1, x_1, y_1), w_2 = A(w_1, x_1, y_1, z_1)$
$\because$ A has the mixed monotone property

$$x_0 \leq x_1 \leq x_2 \, y_0 \geq y_1 \geq y_2, z_0 \leq z_1 \leq z_2, \; w_0 \geq w_1 \geq w_2$$

By continuing this process, we construct the sequence $\{x_n\}, \{y_n\}, \{z_n\}, \{w_n\}$ in X such that $x_{n+1} = A(x_n, y_n, z_n, w_n), y_{n+1} = A(y_n, z_n, w_n, x_n), z_{n+1} = A(z_n, w_n, x_n, y_n), w_{n+1} = A(w_n, x_n, y_n, z_n)$
    Since A has the mixed monotone property

$$x_n \leq x_{n+1}, y_{n+1} \geq y_n, z_{n+1} \leq z_n, w_{n+1} \geq w_n \tag{2}$$

Assume for some $n \in N$, $x_n = x_{n+1}$, $y_{n+1} = y_n$, $z_{n+1} = z_n$, $w_{n+1} = w_n$

Thus, (x_n, y_n, z_n, w_n) is a quadruple fixed point of A.

Thus, we assume $x_n \neq x_{n+1}, y_n \neq y_{n+1}, z_n \neq z_{n+1}, w_n \neq w_{n+1}$ for any $n \in N$,

$\because A$ is α-admissible

$$\therefore \alpha((x_0, y_0, z_0, w_0), (x_1, y_1, z_1, w_1)) \geq 1$$

$$\Rightarrow \alpha \left(\begin{array}{l} (A(x_0, y_0, z_0, w_0), A(y_0, z_0, w_0, x_0), A(z_0, w_0, x_0, y_0), A(w_0, x_0, y_0, z_0)), \\ A(x_1, y_1, z_1, w_1), A(y_1, z_1, w_1, x_1), A(z_1, w_1, x_1, y_1), A(w_1, x_1, y_1, z_1)), \end{array} \right) \geq 1$$

$$\therefore \alpha((x_1, y_1, z_1, w_1), (x_2, y_2, z_2, w_2)) \geq 1$$

Similarly, we may prove that

$$\alpha((y_1, z_1, w_1, x_1), (y_2, z_2, w_2, x_2)) \geq 1, \alpha((z_1, w_1, x_1, y_1), (z_2, w_2, x_2, y_2)) \geq 1$$

$$\alpha((w_1, x_1, y_1, z_1), (w_2, x_2, y_2, z_2)) \geq 1$$

Continuing and generalizing, we get

$$\alpha((x_n, y_n, z_n, w_n), (x_{n+1}, y_{n+1}, z_{n+1}, w_{n+1})) \geq 1$$
$$\alpha((y_n, z_n, w_n, x_n), (y_{n+1}, z_{n+1}, w_{n+1}, x_{n+1})) \geq 1$$
$$\alpha((z_n, w_n, x_n, y_n), (z_{n+1}, w_{n+1}, x_{n+1}, y_{n+1})) \geq 1$$
$$\alpha((w_n, x_n, y_n, z_n), (w_{n+1}, x_{n+1}, y_{n+1}, z_{n+1})) \geq 1 \quad (3)$$

Putting $(x, y, z, w) = (x_{n+1}, y_{n+1}, z_{n+1}, w_{n+1})$, $(p, q, r, s) = (x_n, y_n, z_n, w_n)$ in (1), we get

$$d(x_{n+1}, x_n) = d(A(x_n, y_n, z_n, w_n), A(x_{n-1}, y_{n-1}, z_{n-1}, w_{n-1}))$$
$$\leq \alpha((x_n, y_n, z_n, w_n), (x_{n-1}, y_{n-1}, z_{n-1}, w_{n-1}))d \left(\begin{array}{l} A(x_n, y_n, z_n, w_n), \\ A(x_{n-1}, y_{n-1}, z_{n-1}, w_{n-1}) \end{array} \right) \quad (4)$$
$$\leq \xi \left(\frac{d(x_n, x_{n-1}) + d(y_n, y_{n-1}) + d(z_n, z_{n-1}) + d(w_n, w_{n-1})}{4} \right)$$

Similarly, we may prove that

$$d(y_{n+1}, y_n) = d(A(y_n, z_n, w_n, x_n), A(y_{n-1}, z_{n-1}, w_{n-1}, x_{n-1}))$$
$$\leq \alpha((y_n, z_n, w_n, x_n), (y_{n-1}, z_{n-1}, w_{n-1}, x_{n-1}))d \left(\begin{array}{l} A(y_n, z_n, w_n, x_n), \\ A(y_{n-1}, z_{n-1}, w_{n-1}, x_{n-1}) \end{array} \right)$$
$$\leq \xi \left(\frac{d(y_n, y_{n-1}) + d(z_n, z_{n-1}) + d(w_n, w_{n-1}) + d(x_n, x_{n-1})}{4} \right)$$

$$(5)$$

$$d(z_{n+1}, z_n) = d(A(z_n, w_n, x_n, y_n), A(z_{n-1}, w_{n-1}, x_{n-1}, y_{n-1}))$$

$$\leq \alpha((z_n, w_n, x_n, y_n), (z_{n-1}, w_{n-1}, x_{n-1}, y_{n-1}))d \begin{pmatrix} A(z_n, w_n, x_n, y_n), \\ A(z_{n-1}, w_{n-1}, x_{n-1}, y_{n-1}) \end{pmatrix}$$

$$\leq \xi \left(\frac{d(z_n, z_{n-1}) + d(w_n, w_{n-1}) + d(x_n, x_{n-1}) + d(y_n, y_{n-1})}{4} \right)$$

(6)

$$d(w_{n+1}, w_n) = d(A(w_n, x_n, y_n, z_n), A(w_{n-1}, x_{n-1}, y_{n-1}, z_{n-1}))$$

$$\leq \alpha((w_n, x_n, y_n, z_n), (w_{n-1}, x_{n-1}, y_{n-1}, z_{n-1}))d \begin{pmatrix} A(w_n, x_n, y_n, z_n), \\ A(w_{n-1}, x_{n-1}, y_{n-1}, z_{n-1}) \end{pmatrix}$$

$$\leq \xi \left(\frac{d(w_n, w_{n-1}) + d(x_n, x_{n-1}) + d(y_n, y_{n-1}) + d(z_n, z_{n-1})}{4} \right)$$

(7)

$$\therefore max\{d(x_{n+1}, x_n), d(y_{n+1}, y_n), d(z_{n+1}, z_n), d(w_{n+1}, w_n)\}$$

$$\leq \xi \left(\frac{d(w_n, w_{n-1}) + d(x_n, x_{n-1}) + d(y_n, y_{n-1}) + d(z_n, z_{n-1})}{4} \right)$$

$$\therefore \frac{d(x_{n+1}, x_n) + d(y_{n+1}, y_n) + d(z_{n+1}, z_n) + d(w_{n+1}, w_n)}{4}$$

$$\leq \xi \left(\frac{d(w_n, w_{n-1}) + d(x_n, x_{n-1}) + d(y_n, y_{n-1}) + d(z_n, z_{n-1})}{4} \right)$$

(8)

Continuing with the same steps, we get

$$\therefore \frac{d(x_{n+1}, x_n) + d(y_{n+1}, y_n) + d(z_{n+1}, z_n) + d(w_{n+1}, w_n)}{4}$$

$$\leq \xi \left(\frac{d(x_1, x_0) + d(y_1, y_0) + d(z_1, z_0) + d(w_1, w_0)}{4} \right)$$

For $\epsilon > 0$, there exists $n \in N$ such that

$$\xi^n \left(\frac{d(x_1, x_0) + d(y_1, y_0) + d(z_1, z_0) + d(w_1, w_0)}{4} \right) \leq \frac{\epsilon}{4}$$

Let $m, n \in N$ be such that $m > n$

$$\therefore \frac{d(x_m, x_n) + d(y_m, y_n) + d(z_m, z_n) + d(w_m, w_n)}{4}$$

$$\leq \sum_{i=n}^{m-1} \frac{d(x_i, x_{i+1}) + d(y_i, y_{i+1}) + d(z_i, z_{i+1}) + d(w_i, w_{i+1})}{4}$$

$$\leq \sum_{i=n}^{m-1} \xi^i \left(\frac{d(x_1, x_0) + d(y_1, y_0) + d(z_1, z_0) + d(w_1, w_0)}{4} \right) < \frac{\epsilon}{4}$$

$$\therefore \frac{d(x_m, x_n) + d(y_m, y_n) + d(z_m, z_n) + d(w_m, w_n)}{4} < \frac{\epsilon}{4}$$

$$\therefore d(x_m, x_n) + d(y_m, y_n) + d(z_m, z_n) + d(w_m, w_n) < \epsilon$$

Now,

$$d(x_m, x_n) < d(x_m, x_n) + d(y_m, y_n) + d(z_m, z_n) + d(w_m, w_n) \leq \epsilon$$

Similarly,

$$d(y_m, y_n) < d(x_m, x_n) + d(y_m, y_n) + d(z_m, z_n) + d(w_m, w_n) \leq \epsilon$$

$$d(z_m, z_n) < d(x_m, x_n) + d(y_m, y_n) + d(z_m, z_n) + d(w_m, w_n) \leq \epsilon$$

$$d(w_m, w_n) < d(x_m, x_n) + d(y_m, y_n) + d(z_m, z_n) + d(w_m, w_n) \leq \epsilon$$

Hence, $\{x_n\}, \{y_n\}, \{z_n\}, \{w_n\}$ are Cauchy sequences in (X, d).
Since (X, d) is a complete metric space, $\{x_n\}, \{y_n\}, \{z_n\}, \{w_n\}$ must converge in it.
Let $x, y, z, w \in X$ such that

$$\lim_{n \to \infty} x_n = x, \ \lim_{n \to \infty} y_n = y, \ \lim_{n \to \infty} z_n = z, \ \lim_{n \to \infty} w_n = w$$

A is continuous and

$$x_{n+1} = A(x_n, y_n, z_n, w_n), \ y_{n+1} = A(y_n, z_n, w_n, x_n), \ z_{n+1}$$
$$= A(z_n, w_n, x_n, y_n), \ w_{n+1} = A(w_n, x_n, y_n, z_n)$$

Taking $\lim_{n \to \infty}$ to both sides, we get

$$\lim_{n \to \infty} x_{n+1} = \lim_{n \to \infty} A(x_n, y_n, z_n, w_n) \Rightarrow x = A(x, y, z, w)$$

$$\lim_{n \to \infty} y_{n+1} = \lim_{n \to \infty} A(y_n, z_n, w_n, x_n) \Rightarrow y = A(y, z, w, x)$$

$$\lim_{n \to \infty} z_{n+1} = \lim_{n \to \infty} A(z_n, w_n, x_n, y_n) \Rightarrow z = A(z, w, x, y)$$

$$\lim_{n \to \infty} w_{n+1} = \lim_{n \to \infty} A(w_n, x_n, y_n, z_n) \Rightarrow w = A(w, x, y, z)$$

Thus, A has a quadruple fixed point in (X, d).
In the next theorem, we omit the continuity of A.

Theorem 3.2 *Let $(X, d, \leq)$ be a partially ordered complete metric space. Let $A : X \times X \times X \times X \to X$ be a mapping having the mixed monotone property of X. Suppose that there exist $\xi \in \psi$ and $\alpha : X \times X \times X \times X \to [1, \infty)$ such that for $x, y, u, v \in X$, the following holds:*

$$\alpha((x, y, z, w), (p, q, r, s))d(A(x, y, z, w), A(p, q, r, s)) \leq \xi$$
$$\left(\frac{d(x, u) + d(y, q) + d(z, r) + d(w, s)}{4} \right)$$

for all $x \geq p, y \leq q, z \geq r, w \leq s$. Also,

(i) A is (α)-admissible,

(ii) There exist $(x_0, y_0, z_0, w_0) \in X$ such that

$$\alpha \{(x_0, y_0, z_0, w_0), (A(x_0, y_0, z_0, w_0), A(y_0, z_0, w_0, x_0), A(z_0, w_0, x_0, y_0),$$
$$A(w_0, x_0, y_0, z_0)) \} \geq 1$$

(iii) If $\{x_n\}, \{y_n\}, \{z_n\}, \{w_n\}$ are sequences in X, such that

$$\alpha((x_n, y_n, z_n, w_n), (x_{n+1}, y_{n+1}, z_{n+1}, w_{n+1})) \geq 1,$$
$$\alpha((y_n, z_n, w_n, x_n), (y_{n+1}, z_{n+1}, w_{n+1}, x_{n+1})) \geq 1$$

$$\alpha((z_n, w_n, x_n, y_n), (z_{n+1}, w_{n+1}, x_{n+1}, y_{n+1})) \geq 1,$$
$$\alpha((w_n, x_n, y_n, z_n), (w_{n+1}, x_{n+1}, y_{n+1}, z_{n+1})) \geq 1$$

If $\lim\limits_{n \to \infty} x_n = x, \lim\limits_{n \to \infty} y_n = y, \lim\limits_{n \to \infty} z_n = z, \lim\limits_{n \to \infty} w_n = w$, then

$$\alpha((x_n, y_n, z_n, w_n), (x, y, z, w)) \geq 1,$$

$$\alpha((y_n, z_n, w_n, x_n), (y, z, w, x)) \geq 1$$

$$\alpha((z_n, w_n, x_n, y_n), (z, w, x, y)) \geq 1$$

$$\alpha((w_n, x_n, y_n, z_n), (w, x, y, z)) \geq 1$$

If there exists $(x_0, y_0, z_0, w_0) \in X$ such that

$$x_0 \leq A(x_0, y_0, z_0, w_0), y_0 \geq A(y_0, z_0, w_0, x_0), z_0 \leq A(z_0, w_0, x_0, y_0),$$
$$w_0 \geq A(w_0, x_0, y_0, z_0)$$

then A has a quadruple fixed point.

Proof As already in Theorem 1 we have proved that $\{x_n\}, \{y_n\}, \{z_n\}, \{w_n\}$ are Cauchy sequences in X, therefore, there exists $x, y, z, w \in X$ such that

$$\lim\limits_{n \to \infty} x_n = x, \lim\limits_{n \to \infty} y_n = y, \lim\limits_{n \to \infty} z_n = z, \lim\limits_{n \to \infty} w_n = w$$

and hence

$$\alpha((x_n, y_n, z_n, w_n), (x, y, z, w)) \geq 1,$$

$$\alpha((y_n, z_n, w_n, x_n), (y, z, w, x)) \geq 1$$

$$\alpha((z_n, w_n, x_n, y_n), (z, w, x, y)) \geq 1$$

$$\alpha((w_n, x_n, y_n, z_n), (w, x, y, z)) \geq 1$$

Now,

$$
\begin{aligned}
d(A(x, y, z, w), x) &\leq d(A(x, y, z, w), A(x_n, y_n, z_n, w_n)) + d(x_{n+1}, x) \\
&\leq \alpha((x_n, y_n, z_n, w_n), (x, y, z, w)) d(A(x, y, z, w), A(x_n, y_n, z_n, w_n)) + d(x_{n+1}, x) \\
&\leq \xi \left(\frac{d(x_n, x) + d(y_n, y) + d(z_n, z) + d(w_n, w)}{4} \right) + d(x_{n+1}, x) \\
&\leq \left(\frac{d(x_n, x) + d(y_n, y) + d(z_n, z) + d(w_n, w)}{4} \right) + d(x_{n+1}, x)
\end{aligned}
$$

Similarly,

$$d(A(y, z, w, x), y) \leq \left(\frac{d(y_n, y) + d(z_n, z) + d(w_n, w) + d(x_n, x)}{4} \right) + d(y_{n+1}, y)$$

$$d(A(z, w, x, y), y) \leq \left(\frac{d(z_n, z) + d(w_n, w) + d(x_n, x) + d(y_n, y)}{4} \right) + d(z_{n+1}, z)$$

$$d(A(w, x, y, z), w) \leq \left(\frac{d(w_n, w) + d(x_n, x) + d(y_n, y) + d(z_n, z)}{4} \right) + d(w_{n+1}, w)$$

Taking $\lim_{n \to \infty}$ to both sides, we get

$$d(A(x, y, z, w), x) = 0 \Rightarrow A(x, y, z, w) = x$$

Similarly,

$$d(A(y, z, w, x), y) = 0 \Rightarrow A(y, z, w, x) = y$$

$$d(A(z, w, x, y), z) = 0 \Rightarrow A(z, w, x, y) = z$$

$$d(A(w, x, y, z), w) = 0 \Rightarrow A(w, x, y, z) = w \tag{9}$$

Example 3.3 Let $X = R$ and $d : X \times X \times X \times X \to R$ with $d = |x - y|$.
Let $A : X \times X \times X \times X \to R$ by $A(x, y, z, w) = \frac{1}{16} \ln((1 + |x|)(1 + |y|)(1 + |z|)(1 + |w|))$ for all $x, y, z, w \in X$.

Consider $\alpha : X^4 \times X^4 \to [1, \infty)$ be such that

$$\alpha((x, y, z, w), (p, q, r, s)) = \begin{cases} 2 & \text{if } x \geq p, y \leq q, z \geq r, w \leq s \\ 0 & \text{otherwise} \end{cases}$$

and $\xi(t) = \frac{1}{2} In(1 + |t|)$
Then, we get

$$d(A(x, y, z, w), A(p, q, r, s)) = \frac{1}{16} In((1 + |x|)(1 + |y|)(1 + |z|)(1 + |w|))$$

$$- \frac{1}{16} In((1 + |p|)(1 + |q|)(1 + |r|)(1 + |s|))$$

$$= \frac{1}{16} In(1 + |x|) + \frac{1}{16} In(1 + |y|) + \frac{1}{16} In(1 + |z|) + \frac{1}{16} In(1 + |w|)$$

$$- \left(\frac{1}{16} In(1 + |p|) + \frac{1}{16} In(1 + |q|) + \frac{1}{16} In(1 + |r|) + \frac{1}{16} In(1 + |s|) \right)$$

$$= \frac{1}{16} In \frac{1 + |x|}{1 + |p|} + \frac{1}{16} In \frac{1 + |y|}{1 + |q|} + \frac{1}{16} In \frac{1 + |z|}{1 + |r|} + \frac{1}{16} In \frac{1 + |w|}{1 + |s|}$$

$$\leq \frac{1}{4} \left(\begin{array}{c} \frac{1}{4} In(1 + |x - p|) + \frac{1}{4} In(1 + |y - q|) + \\ \frac{1}{4} In(1 + |z - r|) + \frac{1}{4} In(1 + |w - s|) \end{array} \right)$$

$$\leq \frac{1}{4} \left(In \left(\frac{4 + |x - p| + |y - q| + |z - r| + |w - s|}{4} \right) \right)$$

$$= \frac{1}{4} \left(In \left(1 + \frac{|x - p| + |y - q| + |z - r| + |w - s|}{4} \right) \right)$$

$$\therefore 2 \times (d(A(x, y, z, w), A(p, q, r, s)))$$
$$\leq \frac{1}{2} \left(In \left(1 + \frac{|x - p| + |y - q| + |z - r| + |w - s|}{4} \right) \right)$$

i.e.,

$$\alpha((x, y, z, w)(p, q, r, s)) d(A(x, y, z, w), A(p, q, r, s))$$
$$\leq \xi \left(\frac{d(x, u) + d(y, q) + d(z, r) + d(w, s)}{4} \right).$$

Thus, all the conditions of Theorem 1 are satisfied. Hence, $(0, 0, 0, 0)$ is a quadruple fixed point of A.

4 Application

In this section, we present an application of quadruple fixed point theorem for establishing the existence of solution of the following integral equation.

$$u(t) = \theta_1(s, t) \int_a^b \{F_1(s, u(s)) + F_2(S, u(s)) + F_3(s, u(s)) + F_4(s, u(s))\} \, ds + h(t)$$

$$(10)$$

where $t \in [a, b]$

Let $X = C([a, b], R)$ denote the class of R-valued continuous functions on the interval $[a, b]$ endowed with metric $d(u, v) = \max\limits_{t \in [a,b]} |u(t) - v(t)|$ for $u, v \in X$.

The partial order "$\leq$" on X by $x, y \in X$ $x \leq y \Rightarrow x(t) \leq y(t)$ for $t \in [a, b]$.

$(X, d, \leq)$ be partial ordered complete metric space.

We suppose that

(i) $F_1, F_2, F_3, F_4 : [a, b] \times R \to R$ is continuous.

(ii) $\theta_1(s, t) : [a, b] \times [a, b] \to R$ is continuous.

(iii) $h(t) : [a, b] \to R$ is continuous.

$$(iv) \quad 0 \leq F_1(s, x) - F_1(s, y) \leq \lambda \xi \left(\frac{(x - y)}{4} \right)$$

$$0 \leq F_2(s, y) - F_2(s, x) \leq \eta \xi \left(\frac{(x - y)}{4} \right)$$

$$0 \leq F_3(s, x) - F_3(s, y) \leq \delta \xi \left(\frac{(x - y)}{4} \right)$$

$$0 \leq F_4(s, y) - F_4(s, x) \leq \epsilon \xi \left(\frac{(x - y)}{4} \right)$$

for $\lambda, \eta, \delta, \epsilon > 0$ and $x, y \in R, x \geq y, \xi : [0, \infty) \to [0, \infty)$ is non-decreasing function such that $\xi(t) < t$ and $\lim\limits_{r \to t^+} \xi(r) < t$ for all $t > 0$.

(v) Let $sup(\lambda, \eta, \delta, \epsilon) = \beta$ and $4\gamma\beta \int_a^b (\theta_1(s, t)) \leq 1$ where $\gamma > 1$

(vi) Let there exists functions $x, y, z, w : [a, b] \to R(x, y, z, w)$ such that

$$x(t) \leq \int_a^b \theta_1(s, t) \{F_1(s, x(s)) + F_2(s, y(s)) + F_3(s, z(s)) + F_4(s, w(s))\} \, ds + h(t)$$

$$y(t) \geq \int_a^b \theta_1(s, t) \{F_1(s, y(s)) + F_2(s, z(s)) + F_3(s, w(s)) + F_4(s, x(s))\} \, ds + h(t)$$

$$z(t) \leq \int_a^b \theta_1(s, t) \{F_1(s, z(s)) + F_2(s, w(s)) + F_3(s, x(s)) + F_4(s, y(s))\} \, ds + h(t)$$

$$w(t) \geq \int_a^b \theta_1(s, t) \{F_1(s, w(s)) + F_2(s, x(s)) + F_3(s, y(s)) + F_4(s, z(s))\} \, ds + h(t)$$

for all $t \in [a, b]$

Theorem 4.1 *Consider the integral equation* (10) *and suppose that* $\theta_1, \theta_2, F_1, F_2,$ F_3, F_4 *satisfy all the conditions the assumptions, then equation* (10) *has a quadruple fixed point in* $C([a, b], R)$.

Proof Consider $A : X^4 \to X$ defined by

$$A(x_1, x_2, x_3, x_4)(t)$$

$$= \int_a^b \theta_1(s, t) \{F_1(s, x_1(s)) + F_2(s, x_2(s)) + F_3(s, x_3(s)) + F_4(s, x_4(s))\} \, ds + h(t)$$

$$(11)$$

for $x_1, x_2, x_3, x_4 \in X$
We will prove that it satisfies all the conditions of Theorem 1.
First, let us prove that it satisfies the mixed monotone property.
Let $(x_1, y_1) \in X$ with $x_1 \leq y_1$ and $t \in [a, b]$, then we have

$$A(y_1, y_2, y_3, y_4)(t) - A(x_1, x_2, x_3, x_4)(t)$$

$$= \int_a^b \theta_1(s, t) \{F_1(s, y_1(s)) - F_1(s, x_1(s))\} \, ds$$

$\because x_1(t) \leq y_1(t)$ and based on our assumption (iv)

$$\{F_1(s, y_1(s)) - F_1(s, x_1(s))\} \geq 0.$$

Thus,

$$A(y_1, y_2, y_3, y_4)(t) - A(x_1, x_2, x_3, x_4)(t) \geq 0$$

$$\Rightarrow A(x_1, x_2, x_3, x_4)(t) \leq A(y_1, y_2, y_3, y_4)(t)$$

Let $x_2, y_2 \in X$ with $x_2 \leq y_2$ and $t \in [a, b]$, then we have

$$A(x_2, x_3, x_4, x_1)(t) - A(y_2, y_3, y_4, y_1)(t)$$

$$= \int_a^b \theta_1(s, t) \{F_2(s, x_2(s)) - F_2(s, y_2(s))\} \, ds$$

$\because x_2(t) \leq y_2(t)$ and based on our assumption (iv)

$$\{F_2(s, x_2(s)) - F_1(s, y_2(s))\} \geq 0.$$

Thus,

$$A(x_2, x_3, x_4, x_1)(t) - A(y_2, y_3, y_4, y_1)(t) \geq 0$$

$$\Rightarrow A(y_2, y_3, y_4, y_1)(t) \leq A(x_2, x_3, x_4, x_1)(t)$$

Similarly, one proves the property for third and fourth component
  i.e.,

$$x_3(t) \leq y_3(t) \Rightarrow A(x_3, x_4, x_1, x_2)(t) \leq A(y_3, y_4, y_1, y_2)(t)$$

and

$$x_4(t) \leq y_4(t) \Rightarrow A(y_4, y_1, y_2, y_3)(t) \leq A(x_4, x_1, x_2, x_3)(t)$$

Let us proceed to find

$$d(A(x_1, x_2, x_3, x_4), A(y_1, y_2, y_3, y_4)) \text{ for } x_1 \leq y_1, x_2 \geq y_2, x_3 \leq y_3, x_4 \geq y_4$$

and with A having the mixed monotone property, we get $d(A(x_1, x_2, x_3, x_4), A(y_1, y_2, y_3, y_4))$

$$= \max_{t \in (a,b)} |A(x_1, x_2, x_3, x_4)(t) - A(y_1, y_2, y_3, y_4)(t)|$$

$$= \max_{t \in (a,b)} |A(y_1, y_2, y_3, y_4)(t) - A(x_1, x_2, x_3, x_4)(t)|$$

Now for $t \in [a, b]$ and equation (11)

$$d(A(y_1, y_2, y_3, y_4), A(x_1, x_2, x_3, x_4)) = \int_a^b \theta_1(s, t) \, ds$$

$$\{F_1(s, y_1(s)) - F_1(s, x_1(s)) + F_2(s, y_2(s)) - (F_2(s, x_2(s))) +$$

$$(F_3(s, y_3(s)) - F_3(s, x_3(s))) + (F_4(s, y_4(s)) - F_4(s, x_4(s)))\}$$

Using condition (v),

$$d(A(y_1, y_2, y_3, y_4), A(x_1, x_2, x_3, x_4))$$

$$\leq \int_a^b \theta_1(s, t) \left\{ \lambda \left(\xi \left(\frac{y_1 - x_1}{4} \right) \right) + \eta \left(\xi \left(\frac{x_2 - y_2}{4} \right) \right) + \delta \left(\xi \left(\frac{y_3 - x_3}{4} \right) \right) \right. $$
$$\left. + \epsilon \left(\xi \left(\frac{x_4 - y_4}{4} \right) \right) \right\} ds$$

$$d(A(y_1, y_2, y_3, y_4), A(x_1, x_2, x_3, x_4))$$

$$\int_a^b \theta_1(s, t) \left\{ \beta \left(\xi \left(\frac{y_1 - x_1}{4} \right) \right) + \beta \left(\xi \left(\frac{x_2 - y_2}{4} \right) \right) + \beta \left(\xi \left(\frac{y_3 - x_3}{4} \right) \right) \right. $$
$$\left. + \beta \left(\xi \left(\frac{x_4 - y_4}{4} \right) \right) \right\} ds$$

$$\leq \beta \int_a^b \theta_1(s, t) \left\{ \left(\xi \left(\frac{y_1 - x_1}{4} \right) \right) + \left(\xi \left(\frac{x_2 - y_2}{4} \right) \right) + \left(\xi \left(\frac{y_3 - x_3}{4} \right) \right) \right. $$
$$\left. + \left(\xi \left(\frac{x_4 - y_4}{4} \right) \right) \right\} ds$$

$$\because \xi \left(\frac{y_1 - x_1}{4} \right) \leq \xi \left(\left(\frac{y_1 - x_1}{4} \right) + \left(\frac{x_2 - y_2}{4} \right) + \left(\frac{y_3 - x_3}{4} \right) + \left(\frac{x_4 - y_4}{4} \right) \right)$$

$$\leq 4\beta \int_a^b (\theta_1(s, t)) \left\{ \xi \left(\left(\frac{y_1 - x_1}{4} \right) + \left(\frac{x_2 - y_2}{4} \right) + \left(\frac{y_3 - x_3}{4} \right) + \left(\frac{x_4 - y_4}{4} \right) \right) \right\} ds$$

$$\leq 4\beta \int_a^b (\theta_1(s, t)) \left\{ \xi \left(\frac{d(y_1, x_1)}{4} \right) + \left(\frac{d(x_2, y_2)}{4} \right) + \left(\frac{d(y_3, x_3)}{4} \right) + \left(\frac{d(x_4, y_4)}{4} \right) \right\} ds$$

$$\therefore \gamma d(A(y_1, y_2, y_3, y_4), A(x_1, x_2, x_3, x_4))$$

$$\leq 4\gamma \beta \int_a^b (\theta_1(s, t)) \left\{ \xi \left(\frac{d(y_1, x_1) + d(x_2, y_2) + d(y_3, x_3) + d(x_4, y_4)}{4} \right) \right\} ds$$

where

$$\alpha((y_1, y_2, y_3, y_4), (x_1, x_2, x_3, x_4)) = \gamma \geq 1$$

$$\therefore \gamma d(A(y_1, y_2, y_3, y_4), A(x_1, x_2, x_3, x_4))$$
$$\leq \left\{ \xi \left(\frac{d(y_1, x_1) + d(x_2, y_2) + d(y_3, x_3) + d(x_4, y_4)}{4} \right) \right\} \qquad (12)$$

Thus, all the conditions of the theorem are satisfied, so let (x, y, z, w) be a solution such that it satisfies condition (vi).

And so,

$$x \leq A(x, y, z, w), y \geq A(y, z, w, x), z \leq A(z, , x, y), w \geq A(w, x, y, z).$$

So, all the conditions of Theorem 1 are satisfied.

$\therefore$ We apply Theorem 1, and thus, we get a point

$$(\bar{x}, \bar{y}, \bar{z}, \bar{w}) \in C([a, b], R) \times C([a, b], R) \times C([a, b], R) \times C([a, b], R)$$

such that

$$(\bar{x}) = A(\bar{x}, \bar{y}, \bar{z}, \bar{w}), (\bar{y}) = A(\bar{y}, \bar{z}, \bar{w}, \bar{x}), (\bar{z}) = A(\bar{z}, \bar{w}, \bar{x}, \bar{y}), (\bar{w}) = A(\bar{w}, \bar{x}, \bar{y}, \bar{z})$$

Acknowledgements The authors are thankful to the affiliated college authorities for financial support given by them.

References

1. Arvanitakis, A.D.: A proof of the generalized Banach contraction conjecture. Proc. Am. Math. Soc. **131**(12), 36473656 (2003)
2. Choudhury, B.S., Das, K.P.: A new contraction principle in Menger spaces. Acta Math. Sin. **24**(8), 13791386 (2008)
3. Boyd, D.W., Wong, J.S.W.: On nonlinear contractions. Proc. Am. Math. Soc. **20**, 458464 (1969)
4. Aydi, H., Vetro, C., Sintunavarat, W., Kumam, P.: Coincidence and fixed points for contractions and cyclical contractions in partial metric spaces. Fixed Point Theory Appl. **2012**, 124 (2012)
5. Lakshmikantham, V., Ciric, LjB: Coupled fixed point theorems for nonlinear contractions in partially ordered metric spaces. Nonlinear Anal. **70**, 4341–4349 (2009)
6. Sintunavarat, W., Kumam, P.: Weak condition for generalized multi-valued (f,)-weak contraction mappings. Appl. Math. Lett. **24**, 460465 (2011)
7. Sintunavarat, W., Kumam, P.: Common fixed point theorem for cyclic generalized multi-valued contraction mappings. Appl. Math. Lett. **25**(11), 18491855 (2012)
8. Sintunavarat, W., Cho, Y.J., Kumam, P.: Common fixed point theorems for c-distance in ordered cone metric spaces. Comput. Math. Appl. **62**, 19691978 (2011)
9. Ran, A.C.M., Reurings, M.C.B.: A fixed point theorem in partially ordered sets and some applications to matrix equations. Proc. Am. Math. Soc. **132**, 14351443 (2004)

10. Bhaskar, T.G., Lakshmikantham, V.: Fixed point theory in partially ordered metric spaces and applications. Nonlinear Anal. **65**, 13791393 (2006)
11. Nieto, J.J., Rodriguez-Lopez, R.: Contractive mapping theorems in partially ordered sets and applications to ordinary differential equations. Order **22**, 223239 (2005)
12. Nieto, J.J., Lopez, R.R.: Existence and uniqueness of fixed point in partially ordered sets and applications to ordinary differential equations. Acta Math. Sin. Engl. Ser. **23**(12), 2205–2212 (2007)
13. Agarwal, R.P., El-Gebeily, M.A., ORegan D.: Generalized contractions in partially ordered metric spaces. Appl. Anal. **87**, 18 (2008)
14. Choudhury, B.S., Metiya, N., Kundu, A.: Coupled coincidence point theorems in ordered metric spaces. Ann. Univ. Ferrara. **57**, 116 (2011)
15. Karapnar, E.: Couple fixed point on cone metric spaces. Gazi Univ. J. Sci. **24**, 51–58 (2011)
16. Karapnar, E.: Coupled fixed point theorems for nonlinear contractions in cone metric spaces. Comput. Math. Appl. **59**, 36563668 (2010)
17. Aydi, H.: Some coupled fixed point results on partial metric spaces. Int. J. Math. Math. Sci. **2011**, Article ID 647091, 11 pages (2011)
18. Abbas, M., Khan, M.A., Radenovic, S.: Common coupled fixed point theorem in cone metric space for wcompatible mappings. Appl. Math. Comput. **217**, 195202 (2010)
19. Luong, N.V., Thuan, N.X.: Coupled fixed points in partially ordered metric spaces and application. Nonlinear Anal. **74**, 983992 (2011)
20. Berinde, V., Borcut, M.: Tripled fixed point theorems for contractive type mappings in partially ordered metric spaces. Nonlinear Anal. **74**, 48894897 (2011)
21. Samet, B., Vetro, C.: Coupled fixed point, f-invariant set and fixed point of N-order. Ann. Funct. Anal. **1**(2), 4656 (2010)
22. Karapnar, E.: Quartet fixed point for nonlinear contraction. http://arxiv.org/abs/1106.5472 (27 Jun 2011)
23. Karapnar, E.: A new quartet fixed point theorem for nonlinear contractions. J. Fixed Point Theory Appl. **6**(2), pp. 119–135 (2011)
24. Karapnar, E.: Quadruple fixed point theorems for weak ϕ-contractions. ISRN Math. Anal. **2011**, Article ID 989423, 15 pages (2011)
25. Karapnar, E., Luong, N.V.: Quadruple fixed point theorems for nonlinear contractions. Comput. Math. Appl. **64**, 18391848 (2012)
26. Karapnar, E., Berinde, V.: Quadruple fixed point theorems for nonlinear contractions in partially ordered metric spaces. Banach J. Math. Anal. **6**(1), 7489 (2012)

Chapter 15
Enhanced Prediction for Piezophilic Protein by Incorporating Reduced Set of Amino Acids Using Fuzzy-Rough Feature Selection Technique Followed by SMOTE

Anoop Kumar Tiwari, Shivam Shreevastava, Karthikeyan Subbiah
and Tanmoy Som

Abstract In this paper, the learning performance of different machine learning algorithms is investigated by applying fuzzy-rough feature selection (FRFS) technique on optimally balanced training and testing sets, consisting of the piezophilic and nonpiezophilic proteins. By experimenting using FRFS technique followed by Synthetic Minority Over-sampling Technique (SMOTE) at optimal balancing ratios, we obtain the best results by achieving sensitivity of 79.60%, specificity of 74.50%, average accuracy of 77.10%, AUC of 0.841, and MCC of 0.542 with random forest algorithm. The ranking of input features according to their differentiating ability of piezophilic and nonpiezophilic proteins is presented by using fuzzy-rough attribute evaluator. From the results, it is observed that the performance of classification algorithms can be improved by selecting the reduced optimally balanced training and testing sets. This can be obtained by selecting the relevant and non-redundant features from training sets using FRFS approach followed by suitably modifying the class distribution.

Keywords Feature selection · Imbalanced dataset · SMOTE · Fuzzy-rough set
Random forest · SVM

1 Introduction

Machine learning techniques are effectively implemented to solve a diversity of problems in pattern recognition, data mining, and bioinformatics [1, 28, 34]. Due

A. K. Tiwari · K. Subbiah
Department of Computer Science, Institute of Science (BHU), Varanasi, India

S. Shreevastava (✉) · T. Som
Department of Mathematical Sciences, Indian Institute of Technology (BHU),
Varanasi, India
e-mail: shivam.rs.apm@itbhu.ac.in

© Springer Nature Singapore Pte Ltd. 2018
D. Ghosh et al. (eds.), *Mathematics and Computing*, Springer Proceedings
in Mathematics & Statistics 253, https://doi.org/10.1007/978-981-13-2095-8_15

to the advancement of high-throughput assay systems in modern laboratories, large volume biological datasets are created every day. Data size is enlarging not only in the form of data instances (tuples) but also the dimensionality of data attributes (features). This may reduce the average accuracy and efficiency of most of the machine learning algorithms [5], especially in case of the existence of redundant or irrelevant features. Many researchers have outlined this issue in several bioinformatics problems [25]. High-dimensional bioinformatics datasets contain proteomics, genomics, clinical trial data, etc. Feature selection (FS) [14, 18] techniques focus on selecting subset of the original features while attaining the best for a predetermined goal, often the maximum accuracy (for test data). FS removes irrelevant and redundant features and acquires the best subsets of original features which most profitably differentiate(s) among classes. FS approaches are extensively explored as it is easier to interpret selected features than the extracted features. FS is required in numerous applications, such as object recognition, document classification, computer vision, and disease diagnosis. The class imbalance is another key issue, which directly affects the machine learning algorithms while solving many prediction problems in bioinformatics datasets. This class imbalance problem [9, 22] is almost ubiquitous in data mining, machine learning, and pattern recognition tasks [3]. This imbalance problem has been widely discussed in the literature. Many researchers have investigated that imbalanced data usually leads to performance loss [35, 36], and some kind of treatments, such as cost-sensitive learning, sampling, and ensemble learning, are capable to enhance prediction performance [15, 21, 23, 31–33]. Large difference among the overall instances (tuples) related to positive and negative classes causes imbalance data problem, which generates classifier biased problem. In this paper, we have presented a model to improve the prediction performance of piezophilic and nonpiezophilic groups in protein dataset by selecting relevant and non-redundant features from optimally balanced training sets using fuzzy-rough feature selection (FRFS) [10–13] technique. Now, the same features have been selected from testing sets followed by optimally balancing the testing sets using SMOTE [4, 19]. From the conducted experiments, it is observed that our model results in better performance than the reported results of Nath et al. [24]. Moreover, we have given a suitable schematic representation of our proposed methodology. Furthermore, we have given ranking of input features using fuzzy-rough attribute evaluator technique. Finally, we have given ROC curves [17] for four classifiers on different groups of testing set.

2 Materials and Methods

2.1 Dataset

We have taken the dataset of Nath et al. [24] to conduct our experiments. This was created as a two-dimensional habitat space-based dataset. It was created on the basis of pressure (nonpiezophiles and piezophiles) and on the basis of temperature (psychrophilic, mesophilic, and thermophilic). It consists of 2464 psychrophilic–piezophilic (PP)/2684 psychrophilic–nonpiezophilic (PNP), 2125 mesophilic–piezophilic (MP)/2566 mesophilic–nonpiezophilic (MNP), 1058 thermophilic –piezophilic (TP-I)/1025 thermophilic–nonpiezophilic (TNP-I), and 1099 thermo-philic–piezophilic (TP-II)/1249 thermophilic–nonpiezophilic (TNP-II). These datasets are imbalanced datasets as the ratio of positive (piezophilic) to negative (nonpiezophilic) class is different from ideal ratio (1:1), where the positive class (piezophilic) is the minority class in PP/PNP, MP/MNP, TP-II/TNP-II and is the majority class in case of TP-I/TNP-I.

2.2   Input Features

Nath et al. [24] have created separate training and testing datasets from the original dataset. Training sets are optimally balanced, and testing sets are imbalanced. The input feature vector consists of amino acid composition, which is the basic feature of any protein sequence and has adequate discriminating capability for classification of proteins. It can be calculated by applying the following expression:

$$P_{aa,k} = \frac{Z_{aa,k}}{Z_{res,k}} \times 100 \tag{1}$$

where

aa	denotes specific one of the twenty different amino acid residues,
$P_{aa,k}$	denotes the percentage frequency of the specific amino acid 'aa' in the kth sequence,
$Z_{aa,k}$	denotes the total count of the specific amino acid 'aa',
$Z_{res,k}$	denotes the total number of amino acid residues in the kth sequence.

where $P(k)$ denotes the percentage frequency of kth type residue (k changes from 1 to 20 indicating specific amino acids) and $Z(k)$ denotes the overall residues of kth type.

2.3   Classification Protocol

Our experiments are performed independently by using four different machine learning algorithms, which are widely used on biological datasets for classification and prediction tasks. From our experiments, it can be observed that random forest (RF) [2] and support vector machines with sequential minimization optimization (SMO) [27] are the better performing algorithms. A brief description of RF and SMO are given below.

RF: Random forest (proposed by Breiman [2]) is an ensemble learning approach comprising of many individual decision trees. The two factors determining the accuracy of random forest are the evaluation of correlation and strength between the individual tree classifiers. Feature randomization is characterized as an integral part of random forests. For individual tree, 2/3 of the training samples are adopted for tree construction and rest of the 1/3 samples are used for testing. This improves the performance of the tree and is defined as out of bag data [20].

SMO: Support vector machines (SVMs) work on the principle of structural risk minimization of statistical learning theory and are used to perform supervised learning task. SVM classifies input instances by mapping the Euclidean input instance (tuple) space into a greater dimensional space and the building of a hyperplane in the kernel feature space that is applied for dividing the two classes. We have conducted experiments using SMO algorithm [27] which is applied for training a SVM classifier in order to get faster optimization. SVMs are proven to be robust to noise and can cope with large feature space. SVMs have been successfully implemented in many biological domains and have presented promising results.

2.4 Optimal Balancing Protocol

When the real-world dataset is imbalance with the number of negative and positive class instances, then the evaluation parameters, such as overall accuracy with which most of the machine learning algorithms are optimized to perform, tend to be biased in favour of the majority class [8], which is not acceptable as it results in higher specificity and less sensitivity while predicting the minority class tuples (instances) [16]. In order to deal with this problem, we have balanced the reduced testing set in terms of an ideal balancing ratio of 1:1 by using Synthetic Minority Over-sampling Technique (SMOTE) [21, 31, 32]. A brief description of SMOTE is given below.

SMOTE: It is an over-sampling method that produces synthetic samples from the minority class. It is a nearest neighbor-based concept which advances by randomly picking a minority sample and its nearest neighbor samples. It then utilizes one of the nearest neighboring minority class instances to insert for generating an artificial minority class instance. The SMOTE samples are defined as the linear combinations of two similar samples related with minority class (p and p^k) and are defined by

$$s = p + i * (p^k - p) \tag{2}$$

where i varies from 0 to 1 and p^k is randomly selected among the five minority class nearest neighbors of p. In recent years, SMOTE has been successfully implemented to solve class imbalance problems. In WEKA [7], the default value of nearest neighbors for SMOTE is 5.

2.5   Feature Selection Protocol

The existence of identical and overlapped features in bioinformatics datasets makes the classification task difficult. Interclass feature overlaps, and the existence of similarities leads to vagueness and/or indiscernibility. Rough set concept [26] is invariably applicable for decision making in case of indiscernibility is present, and vague decision can be handled by fuzzy set theory [37]. These two theories (fuzzy set and rough set) can be combined to form fuzzy-rough set theory [6], which can cope with the uncertainty pertaining vagueness and indiscernibility for fuzzy and rough sets respectively, which is useful for addressing classification problems. In our proposed model, we have applied FRFS approach [10, 11] to select relevant and non-redundant features in order to enhance the prediction of piezophilic proteins. The FRFS algorithm is given as follows [12, 13]:

Fuzzy-Rough Quick Reduct Algorithm (C,D)
C, the set of all conditional attributes;
D, the set of decision attributes.
$R \leftarrow \{\}; \gamma'_{best} = 0; \gamma'_{prev} = 0$
do
$T \leftarrow R$
$\gamma'_{prev} = \gamma'_{best}$
for each $x \in (C - R)$
if $(\gamma'_{R \cup \{x\}})(D) > (\gamma'_T)(D)$
$T \leftarrow R \cup \{x\}$
$\gamma'_{best} = (\gamma'_T)(D)$
$R \leftarrow T$
until $\gamma'_{best} == \gamma'_{prev}$
return R

2.6   Performance Evaluation Metrics

The relative prediction performance of the four machine learning algorithms is calculated taking into account threshold-dependent and threshold-independent parameters. These parameters are determined from the values of the confusion matrix, namely true positives (TP) that is the number of correctly predicted piezophilic proteins, false negatives (FN) that is the number of incorrectly predicted piezophilic proteins, true negatives (TN) that is the number of correctly predicted nonpiezophilic proteins, and false positives (FP) that is the number of incorrectly predicted nonpiezophilic proteins.

Sensitivity: This parameter gives the percentage of correctly predicted piezophilic proteins and is given as follows:

$$Sensitivity = \frac{TP}{(TP + FN)} \times 100 \qquad (3)$$

Specificity: This parameter gives the percentage of correctly predicted non-piezophilic proteins and is calculated by:

$$Specificity = \frac{TN}{(TN + FP)} \times 100 \qquad (4)$$

Accuracy: This parameter calculates the percentage of correctly predicted piezophilic and nonpiezophilic proteins and is calculated as follows:

$$Accuracy = \frac{(TP + TN)}{(TP + FP + TN + FN)} \times 100 \qquad (5)$$

AUC: It represents the area under curve (AUC) of a receiver operating characteristics curve (ROC) [17]; the closer its value to 1, the better the piezophilic protein predictor; in the worst case, its value is 0, and in random ranking, its value is 0.5. It is one of the evaluation metrics which are robust to the imbalanced nature of the proteomics datasets.

Mathews correlation coefficient (MCC): It is calculated by using the following equation:

$$MCC = \frac{(TP \times TN - FP \times FN)}{\sqrt{(TP + FP)(TP + FN)(TN + FP)(TN + FN)}} \qquad (6)$$

It is extensively applied as a performance parameter for binary classification. The MCC value 1 is considered as the best for piezophilic protein predictor. In this study, the open source java-based machine learning platform WEKA [7] was used to conduct all the experiments.

3   Result and Discussion

In the current study, we experimented with four different machine learning algorithms, namely support vector machines with sequential minimization optimization (SMO) [27], multilayer perceptron (MLP) [30], rotation forest (ROF) [29], and random forest (RF) [2] on the reduced optimally balanced training and testing sets. We applied FRFS with rank search on training sets for selecting suitable features (as recorded in Table 1) and selected the same features from the corresponding testing sets. Reduced testing sets have been balanced by using varying degree of SMOTE. The values of different performance evaluation metrics for the four classifiers using tenfold cross validation are recorded in Table 2.

From experiments, it can be easily observed that the performance of SMO based on the values of different evaluation parameters is better than other classifiers on the training set and RF is the best performer on testing set for the differentiation of psychrophilic–piezophilic and psychrophilic–nonpiezophilic group. For the

Table 1 Different training sets dimensions and their reduct sizes based on FRFS

Dataset	Instances	Features	Reduct
PP/PNP	4000	20	14
MP/MNP	3000	20	14
TP/TNP-I	1400	20	15
TP/TNP-II	1400	20	12

Table 2 Evaluation metrics of different machine learning algorithms

Machine Learning Algorithms	Training Set					Testing Set				
	Sensitivity	Specificity	Accuracy	AUC	MCC	Sensitivity	Specificity	Accuracy	AUC	MCC
2(A) PP/PNP										
SMO	67.90	57.00	62.50	0.625	0.250	72.20	55.60	63.90	0.639	0.282
MLP	65.00	54.60	59.80	0.638	0.196	69.40	54.70	62.10	0.652	0.244
ROF	68.20	53.70	60.90	0.651	0.221	72.80	55.30	64.00	0.702	0.285
RF	61.50	55.50	58.50	0.617	0.169	72.40	62.30	67.30	0.745	0.348
2(B) MP/MNP										
SMO	66.50	68.20	67.40	0.674	0.347	66.90	67.80	67.30	0.673	0.347
MLP	67.80	62.80	65.30	0.720	0.306	67.70	63.50	65.60	0.716	0.312
ROF	62.20	68.70	65.50	0.715	0.310	68.00	71.10	69.50	0.762	0.391
RF	64.50	66.90	65.70	0.719	0.314	74.00	72.70	73.30	0.807	0.467
2(C) TP/TNP-I										
SMO	67.10	73.10	70.10	0.701	0.404	66.80	69.90	68.30	0.683	0.367
MLP	64.30	70.00	67.10	0.707	0.343	66.50	66.00	66.20	0.708	0.325
ROF	64.30	72.00	68.10	0.743	0.364	63.10	68.50	65.80	0.742	0.317
RF	63.90	67.10	65.50	0.715	0.310	65.90	67.70	66.80	0.751	0.336
2(D) TP/TNP-II										
SMO	75.60	69.30	72.40	0.724	0.449	78.40	72.90	75.60	0.756	0.512
MLP	72.60	65.60	69.10	0.746	0.382	71.80	70.30	71.10	0.766	0.421
ROF	73.40	65.40	69.40	0.759	0.390	77.60	72.10	74.90	0.821	0.498
RF	71.00	67.10	69.10	0.761	0.382	79.60	74.50	77.10	0.841	0.542

Table 3 Attribute ranking by fuzzy-rough feature selection algorithm

Rank / Dataset	1	2	3	4	5	6	7	8	9	10	11	12	13	14	15	16	17	18	19	20
PP/PNP	N	V	Q	I	S	K	T	C	H	E	L	A	R	G	P	M	F	W	D	Y
MP/MNP	T	S	G	M	R	A	L	K	Q	C	P	I	D	Y	W	V	N	E	H	F
TP/TNP-I	K	V	Q	I	R	L	P	N	A	Y	G	D	M	E	W	S	H	F	T	C
TP/TNP-II	I	K	C	L	R	T	N	V	S	G	Q	E	A	P	H	D	F	M	W	Y

discrimination of mesophilic–piezophilic and mesophilic–nonpiezophilic group, the values of evaluation metrics indicate that SMO is performing better than other machine learning algorithms in case of training set while RF is the best performer on testing set. SMO is the best performing classifier on both training and testing sets for discriminating thermophilic–piezophilic-I and thermophilic–nonpiezophilic-I group, while in case of discrimination of thermophilic–piezophilic-II and thermophilic–nonpiezophilic-II group, SMO is the best predictor on training set and RF gives the best performance result on testing set.

From the entire experiment, we can observe that RF is performing better and is closely followed by SMO on the basis of values of different evaluation metrics. The flow diagram of the proposed methodology is depicted in Fig. 1.

The existence of redundant features in a dataset affects the generalization ability of the model as well as the training time. We have used fuzzy-rough attribute evaluator technique to rank the 20 different amino acids based on their discerning ability, and the results are recorded in Table 3.

A suitable way to observe the overall performance of individual machine learning algorithms at different decision thresholds is the well-known receiver operating characteristic (ROC) curve, which allows a visual representation of the performance of different classifiers. The ROC curves for different machine learning algorithms on different reduced testing sets are given in Figs. 2, 3, 4, and 5, respectively. It can be observed that the performance of RF and SMO is better than other classifiers.

4 Conclusion

There are many aspects that can directly influence in attaining the real performance of the classifiers. The three key issues among these are selection of suitable input feature set, class imbalance, and selection of an appropriate learning algorithm. Redundant and irrelevant features available in biological datasets lead to accuracy loss and class

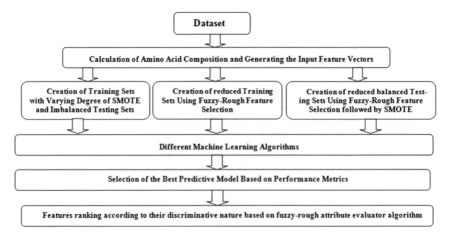

Fig. 1 Schematic representation of current study

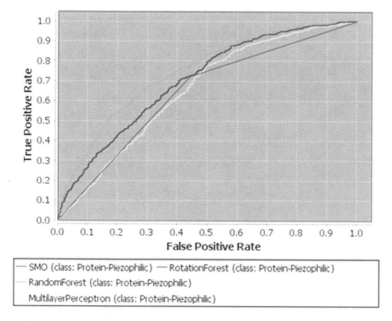

Fig. 2 AUC for four machine learning algorithms on reduced PP/PNP testing set

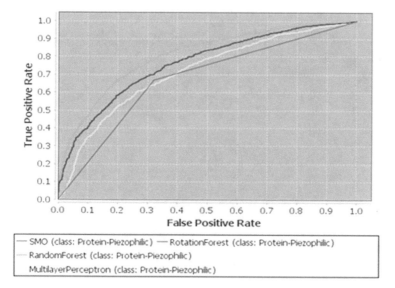

Fig. 3 AUC for four machine learning algorithms on reduced MP/MNP testing set

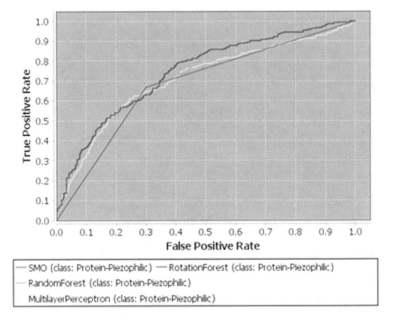

Fig. 4 AUC for four machine learning algorithms on reduced TP-I/TNP-I testing set

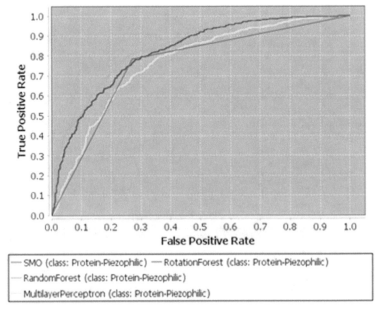

Fig. 5 AUC for four machine learning algorithms on reduced TP-II/TNP-II testing set

imbalance factor, which is usually observed in biological datasets and causes the classifier to be biased to majority class tuples (instances). Our experimental results validated the fact that selection of relevant and non-redundant features using FRFS technique followed by optimally balancing the ratios in both training and testing datasets results in higher sensitivity and higher accuracy through various machine learning algorithms. In our experiments, we explored that RF and SMO have more discriminating ability of piezophilic and nonpiezophilic proteins as we move up the temperature range from PP to TP, i.e., PP < MP < TP, and it is clearly visible from ROC curves of testing sets. Finally, the fuzzy-rough attribute evaluator ranking method is applied to rank all the input features according to their contribution toward discrimination of piezophilic and nonpiezophilic proteins.

In the future, we intend to apply our proposed model on some other bioinformatics datasets to enhance the prediction of positive and negative classes. Furthermore, we will apply our proposed model by using various search techniques for FRFS. Moreover, we can apply some more accurate feature selection techniques based on intuitionistic fuzzy-rough set models.

References

1. Baldi, P., Brunak, S.: Bioinformatics: The Machine Learning approach. MIT press (2001)
2. Breiman, L.: Random Forests. Mach. Learn. **45**(1), 5–32 (2001)
3. Chawla, N.V.: Data Mining for Imbalanced Datasets: An Overview. Data Mining and Knowledge Discovery Handbook, pp. 875–886. Springer (2009)
4. Chawla, N.V., Bowyer, K.W., Hall, L.O., Kegelmeyer, W.P.: SMOTE: synthetic minority oversampling technique. J. Artif. Intell. Res. **16**, 321–357 (2002)
5. Dash, M., Liu, H.: Feature selection for classification. Intell. Data Anal. **1**(1–4), 131–156 (1997)
6. Dubois, D., Prade, H.: Putting Rough Sets and Fuzzy Sets Together Intelligent Decision Support, pp. 203–232. Springer (1992)
7. Hall, M., Frank, E., Holmes, G., Pfahringer, B., Reutemann, P., Witten, I.H.: The WEKA data mining software: an update. ACM SIGKDD Explor. Newslett. **11**(1), 10–18 (2009)
8. He, H., Garcia, E.A.: Learning from imbalanced data. IEEE Trans. Knowl. Data Eng. **21**(9), 1263–1284 (2009)
9. Japkowicz, N., Stephen, S.: The class imbalance problem: a systematic study. Intell. Data Anal. **6**(5), 429–449 (2002)
10. Jensen, R., Shen, Q.: Fuzzy rough attribute reduction with application to web categorization. Fuzzy Sets Syst. **141**(3), 469–485 (2004a)
11. Jensen, R., Shen, Q.: Semantics-preserving dimensionality reduction: rough and fuzzy-rough-based approaches. IEEE Trans. Knowl. Data Eng. **16**(12), 1457–1471 (2004b)
12. Jensen, R., Shen, Q.: Fuzzy-rough sets assisted attribute selection. IEEE Trans. Fuzzy Syst. **15**(1), 73–89 (2007)
13. Jensen, R., Shen, Q.: Computational Intelligence and Feature Selection: Rough and Fuzzy Approaches, Vol. 8. Wiley (2008)
14. Langley, P.: Selection of relevant features in machine learning. Paper presented at the Proceedings of the AAAI Fall Symposium on Relevance
15. Lee, P.H.: Resampling methods improve the predictive power of modeling in class-imbalanced datasets. Int. J. Environ. Res. Public Health **11**(9), 9776–9789
16. Li, H., Pi, D., Wang, C.: The prediction of protein-protein interaction sites based on RBF classifier improved by SMOTE. Math. Prob, Eng (2014)

17. Ling, C., Huang, J., Zhang, H.: AUC: a better measure than accuracy in comparing learning algorithms. Adv. Artif. Intell. 991–991 (2003)
18. Liu, H., Motoda, H.: Feature Extraction, Construction and Selection: A Data Mining Perspective, vol. 453. Springer Science and Business Media (1998)
19. Lusa, L.: SMOTE for high-dimensional class-imbalanced data. BMC Bioinform. **14**(1), 106 (2013)
20. Nath, A., Chaube, R., Karthikeyan, S.: Discrimination of psychrophilic and mesophilic proteins using random forest algorithm. Paper presented at the 2012 International Conference on Biomedical Engineering and Biotechnology (iCBEB) (2012)
21. Nath, A., Karthikeyan, S.: Enhanced prediction and characterization of CDK inhibitors using optimal class distribution. Interdisc. Sci. Comput. Life Sci. **9**(2), 292–303 (2017)
22. Nath, A., Subbiah, K.: Inferring biological basis about psychrophilicity by interpreting the rules generated from the correctly classified input instances by a classifier. Comput. Biol. Chem. **53**, 198–203 (2014)
23. Nath, A., Subbiah, K.: Maximizing lipocalin prediction through balanced and diversified training set and decision fusion. Comput. Biol. Chem. **59**, 101–110 (2015)
24. Nath, A., Subbiah, K.: Insights into the molecular basis of piezophilic adaptation: extraction of piezophilic signatures. J. Theoret. Biol. **390**, 117–126 (2016)
25. Okun, O.: Feature Selection and Ensemble Methods for Bioinformatics: Algorithmic Classification and Implementations. Information Science Reference-Imprint of IGI Publishing (2011)
26. Pawlak, Z.: Rough sets. Int. J. Parallel. Program. **11**(5), 341–356 (1982)
27. Platt, J.: Sequential minimal optimization: a fast algorithm for training support vector machines (1998)
28. Prompramote, S., Chen, Y., Chen, Y.-P.P.: Machine learning in bioinformatics. In: Chen, Y.-P.P. (ed.) Bioinformatics Technologies, pp. 117–153. Springer, Berlin Heidelberg, Berlin, Heidelberg (2005)
29. Rodriguez, J.J., Kuncheva, L.I., Alonso, C.J.: Rotation forest: a new classifier ensemble method. IEEE Trans. Pattern Anal. Mach. Intell. **28**(10), 1619–1630 (2006)
30. Ruck, D.W., Rogers, S.K., Kabrisky, M., Oxley, M.E., Suter, B.W.: The multilayer perceptron as an approximation to a bayes optimal discriminant function. IEEE Trans. Neural Netw. **1**(4), 296–298 (1990)
31. Tiwari, A.K., Nath, A., Subbiah, K., Shukla, K.K.: Effect of varying degree of resampling on prediction accuracy for observed peptide count in protein mass spectrometry data. Paper presented at the 2015 11th International Conference on Natural Computation (ICNC) (2015)
32. Tiwari, A.K., Nath, A., Subbiah, K., Shukla, K.K.: Enhanced prediction for observed peptide count in protein mass spectrometry data by optimally balancing the training dataset. Int. J. Pattern Recogn. Artif. Intell. 1750040 (2017)
33. Vani, K.S., Bhavani, S.D.: SMOTE based protein fold prediction classification. In: Advances in Computing and Information Technology, pp. 541–550. Springer (2013)
34. Wang, L., Fu, X.: Data Mining with Computational Intelligence. Springer Science and Business Media (2006)
35. Weiss, G.M., Provost, F.: The effect of class distribution on classifier learning: an empirical study. Rutgers Univ (2001)
36. Weiss, G.M., Provost, F.: Learning when training data are costly: the effect of class distribution on tree induction. J. Artif. Intell. Res. **19**, 315–354 (2003)
37. Zadeh, L.A.: Fuzzy sets. Inf. Control **8**(3), 338–353 (1965)

Chapter 16
Effect of Upper and Lower Moving Wall on Mixed Convection of Cu-Water Nanofluid in a Square Enclosure with Non-uniform Heating

S. K. Pal and S. Bhattacharyya

Abstract Mixed convection of Cu-water nanofluid in a square enclosure with upper and lower moving lid has been investigated numerically. Non-uniform heating is imposed on the left wall, and the right wall is cooled at a constant temperature. Upper and lower walls are taken to be adiabatic. Finite volume-based SIMPLE algorithm has been used to solve the nonlinear equations. Results are presented graphically to describe the effect of nanoparticle volume fraction ($0.0 \leq \phi \leq 0.2$), Richardson number ($0.1 \leq Ri \leq 10.0$) and the moving walls (upper and lower) on flow field, thermal field and heat transfer rate at a fixed value of Reynolds number ($Re = 100$). Results show that heat transfer rate increases remarkably with the addition of nanoparticles. Non-uniform temperature distribution on the left wall affects the thermal field.

Keywords Mixed convection · Non-uniform heating · Heat transfer · Square enclosure

1 Introduction

Nanofluid is a colloid mixture of metallic and nonmetallic nano-sized particles with a base fluid. These nano-sized particles change the thermo-physical properties of the base fluid and exhibits a substantially larger thermal conductivity as compared to the conventional base fluid such as oil, water and ethylene glycol. Nanofluid has a wide range of application in those industries where heat transfer is a prime matter of concern. Choi et al. [1] investigated the potential benefits of copper nanometer-sized particles dispersed in ethylene glycol and concluded that significantly higher thermal conductivity can be achieved using the nanoparticles. Xuan and Li [2] experimentally investigated the heat transfer features of Cu-water nanofluid and concluded that suspended nanoparticles enhance the heat transfer process remarkably.

Over the years, the flow and heat transfer of nanofluid inside a closed enclosure has received a considerable attention because of its significant range of application

S. K. Pal (✉) · S. Bhattacharyya
Department of Mathematics, Indian Institute of Technology, Kharagpur 721302, India
e-mail: sanjibkumarpal1@gmail.com

© Springer Nature Singapore Pte Ltd. 2018
D. Ghosh et al. (eds.), *Mathematics and Computing*, Springer Proceedings
in Mathematics & Statistics 253, https://doi.org/10.1007/978-981-13-2095-8_16

in many industries such as cooling of electronic systems, room ventilation, nuclear reactors, gas production and lubrication. Tiwari and Das [3] numerically studied the behaviour of nanofluid inside a two-sided differentially heated lid-driven enclosure and found that nanoparticles increase heat transfer rate. Mahmoodi [4] numerically investigated the mixed convection of Al_2O_3-water nanofluid inside a rectangular enclosure and concluded that the average Nusselt number increases with the increase of nanoparticle volume fraction.

Along with nanoparticles, convection in enclosures with various wall temperature conditions also has been studied by many researchers because of its application in thermal engineering. Basak et al. [5] numerically studied the influence of linearly heated side walls on mixed convection in a square enclosure. They reported that multiple circulating cells were observed. A numerical investigation was carried out by Ramakrishna et al. [6] inside a square cavity for various thermal boundary conditions on bottom and side walls. Sivakumar and Sivasankaran [7] numerically investigated the mixed convection in an inclined square cavity with non-uniform temperature distribution on the both vertical side walls. Sivasankaran et al. [8] studied the effect of the upper moving wall direction on the mixed convection in an inclined square cavity with sinusoidal heating on the left wall. They used air as the working fluid and concluded that the moving wall's direction has significant impact on the flow and thermal field in the cavity.

To the best of our knowledge, there is no study to investigate the effect of upper and lower lid and non-uniform wall temperature on the mixed convection of Cu-water nanofluid in a square enclosure. Hence, the present study deals with the effects on flow and thermal fields of Cu-water nanofluid caused by the upper and lower wall movement and non-uniform sidewall temperature distribution.

2 Physical Model

A two-dimensional mixed convection flow of Cu-water nanofluid in a square enclosure of height H has been considered (Fig. 1a and b). The Cartesian coordinate system has its origin at the lower left corner of the square enclosure with lower the wall along the x^*-axis and left vertical wall along the y^* axis. The gravitational acceleration g is acting in the opposite direction of the y^* coordinate. The top and bottom walls of the cavity are kept insulated, and both walls are allowed to move with a velocity U_0 in the positive x-axis direction which induces shear in the cavity. A non-uniform temperature profile is applied on the left wall of the enclosure, and a constant temperature T_c is maintained at the right wall. The form of the non-uniform temperature profile is expressed as

$$T(y^*) = T_c + (T_{ref} - T_c)\left\{1.0 + 0.2\sin\left(\frac{10\pi y^*}{H}\right) + 0.2\sin\left(\frac{2\pi y^*}{H}\right)\right\} \quad (1)$$

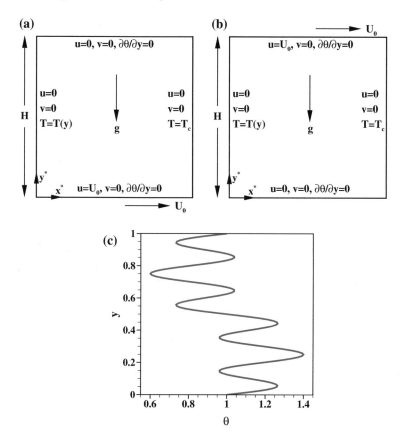

Fig. 1 Schematic diagram of the physical system with the boundary conditions for **a** lower moving lid and **b** upper moving lid. **c** Nondimensional temperature distribution on the left wall

3   Mathematical Formulation

Single phase model has been adopted for the study. The cavity is filled with Cu-water nanofluid, which is assumed to be Newtonian, incompressible and laminar. Nanoparticles are assumed to be of uniform size and shape and considered to be in thermal equilibrium with the base fluid. The thermo-physical properties of the nanofluid are assumed to be constant except the density which varies according to the Boussinesq approximation. Under the above assumptions continuity, momentum and energy equations for the buoyancy-driven flow inside the cavity employing the Boussinesq approximation can be expressed in nondimensional form as

$$\frac{\partial u}{\partial x} + \frac{\partial v}{\partial y} = 0 \tag{2}$$

$$\frac{\partial u}{\partial t} + u\frac{\partial u}{\partial x} + v\frac{\partial u}{\partial y} = -\frac{\partial p}{\partial x} + \frac{1}{Re}\frac{\rho_f}{\rho_{nf}}\frac{1}{(1-\phi)^{2.5}}\left(\frac{\partial^2 u}{\partial x^2} + \frac{\partial^2 u}{\partial y^2}\right) \qquad (3)$$

$$\frac{\partial v}{\partial t} + u\frac{\partial v}{\partial x} + v\frac{\partial v}{\partial y} = -\frac{\partial p}{\partial y} + \frac{1}{Re}\frac{\rho_f}{\rho_{nf}}\frac{1}{(1-\phi)^{2.5}}$$
$$\left(\frac{\partial^2 v}{\partial x^2} + \frac{\partial^2 v}{\partial y^2}\right) + Ri\frac{\rho_f}{\rho_{nf}}\left(1 - \phi + \phi\frac{\rho_p\beta_p}{\rho_f\beta_f}\right)\theta \qquad (4)$$

$$\frac{\partial\theta}{\partial t} + u\frac{\partial\theta}{\partial x} + v\frac{\partial\theta}{\partial y} = \frac{k_{nf}}{k_f}\frac{(\rho C_p)_f}{(\rho C_p)_{nf}}\frac{1}{Re\,Pr}\left(\frac{\partial^2\theta}{\partial x^2} + \frac{\partial^2\theta}{\partial y^2}\right) \qquad (5)$$

The dimensionless variables are defined by $x = x^*/H$, $y = y^*/H$, $t = t^*U_0/H$, $\theta = (T - T_c)/(T_{ref} - T_c)$, $u = u^*/U_0$, $v = v^*/U_0$, $p = \frac{p^*}{\rho_{nf}U_0^2}$. The dimensionless parameters are Reynolds number $Re = \frac{\rho_f U_0 L}{\mu_f}$, Prandtl number $Pr = \frac{\nu_f}{\alpha_f}$ and Richardson Number $Ri = \frac{Gr}{Re^2}$.

The effective density of nanofluid is given by $\rho_{nf} = (1 - \phi)\rho_f + \phi\rho_p$. Effective heat capacitance of nanofluid is given by $(\rho c_p)_{nf} = (1 - \phi)(\rho c_p)_f + \phi(\rho c_p)_p$, Xuan and Li [2]. The thermal diffusivity of nanofluid is expressed as $\alpha_{nf} = \frac{k_{nf}}{(\rho C_p)_{nf}}$, Chamkha and Abu-Nada [9]. There exists several modified models for the dynamic viscosity of nanofluids, but Brinkman model still gives reasonable result. So Brinkman model [10] has been adopted for effective viscosity of nanofluid, μ_{nf} and is given by $\mu_{nf} = \frac{\mu_f}{(1-\phi)^{2.5}}$, where ϕ is the nanoparticle volume fraction. The Maxwell–Garnett's model has been considered to determine the effective thermal conductivity of the nanofluid and is given by $\frac{k_{nf}}{k_f} = \frac{k_p + 2k_f - 2\phi(k_f - k_p)}{k_p + 2k_f + \phi(k_f - k_p)}$. The thermophysical properties for water and copper, at room temperature, used in this study, has been given in Table 1.

The boundary conditions are as follows:

$u = 1$ or $u = 0$, $v = 0$, $\frac{\partial\theta}{\partial y} = 0$ at top wall

$u = 0$ or $u = 1$, $v = 0$, $\frac{\partial\theta}{\partial y} = 0$ at the bottom wall

$u = 0$, $v = 0$, $\theta = 1.0 + (0.2\sin(10\pi y) + 0.2\sin(2\pi y))$ at the left wall

$u = 0$, $v = 0$, $\theta = 0$ at the right wall.

Table 1 Thermo-physical properties of water and copper

Parameter	Water	Copper
c_p (J/kgK)	4179	383
ρ (kg/m^3)	997.1	8954
k(W/mK)	0.6	400
β(K^{-1})	2.1×10^{-4}	1.67×10^{-5}

3.1 Nusselt Number

The heat transfer rate in terms of local Nusselt number (Nu) along the left nonhomogeneous hot wall is defined as $Nu = -\left(\frac{k_{nf}}{k_f}\right)\frac{\partial\theta}{\partial x}$.

Average Nusselt number at the left hot wall is calculated as $Nu_{av} = \int_0^1 Nu\, dy$.

4 Numerical Methods

Finite volume method is used to solve the nonlinear governing partial differential equations in its nondimensional form on a staggered grid system. In staggered grid system control volume are different for each computing variable. The velocity components are evaluated at the mid-point of the cell face which they are normal, and all the scalar quantities are stored at the centre of the cell. The equations are integrated over each control volume. QUICK algorithm is used to discretize the convective terms in the momentum and energy equation and a second-order central difference scheme is used to discretize the diffusive terms. Velocity-pressure coupling is done by SIMPLE algorithm. Uniform grid distribution is considered along both the axes and the resulting set of discretized equations are solved using block elimination method. The time step is chosen to be 10^{-4}, and at each iteration level, the pressure field is computed and updated by using SIMPLE algorithm. For any set of input parameters, the iteration process is repeated and until the convergence criterion $max_{ij} \mid \Phi_{ij}^{k+1} - \Phi_{ij}^{k} \mid \leq 10^{-6}$ is satisfied where subscripts i, j denote the cell index and superscripts k denotes the iteration index and Φ is the variable to compute.

Figure 2a represents the grid independence test on local Nusselt number on left hot wall for Richardson number $Ri = 1.0$, $Re = 100$ and nanoparticle volume fraction $\phi = 0.1$. Three sets of grid have been considered for the test, and it can be seen that 81×81 is optimal for this present study. More finer grid can give more accurate results, but it has been observed that the change in accuracy is less than 1%. Figure 2b shows the validation of the present code. In Fig. 2b, the average Nusselt number along the hot wall has been validated with the calculation of Abu-Nada and Chamkha [11] for inclination angle $0°$, $Ri = 1.0$, $Re = 10$ and $\phi = 0.1$. It can be seen that present result shows very good agreement with result given by Abu-Nada and Chamkha [11].

5 Results and Discussion

Mixed convection flow of Cu-water nanofluid in a square enclosure with non-uniform heat distribution on a sidewall has been investigated numerically for upper and lower lid movement when $0.0 \leq \phi \leq 0.2$ and $0.1 \leq Ri \leq 10.0$ at $Re = 100$. Throughout the study, the Reynolds number is kept fixed at 100 and Richardson number (Ri) has

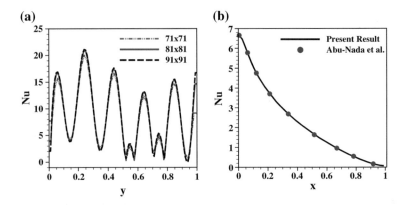

Fig. 2 a Grid-independence test for local Nusselt number along the left hot wall for $Ri = 1.0$, $Re = 100$ and $\phi = 0.1$, **b** comparison of the local Nusselt number along the hot wall with the calculation of Abu-Nada and Chamkha [11] for mixed convection of nanofluid in an square enclosure of inclination angle 0^0, $Ri = 1.0$, $Re = 10$ and $\phi = 0.1$

been varied by varying the Grashof number (Gr) between $10^3 \leq Gr \leq 10^5$. Water is considered as the base fluid with 6.2 as its Prandtl number.

5.1 Flow and Temperature Field

Figures 3 and 4 show the variation of streamlines and isotherms for mixed convection of Cu-water nanofluid for the movement of upper lid (first row of Figs. 3 and 4) and lower lid (second row of Figs. 3 and 4) respectively. To show the effect of nanoparticle volume fraction (ϕ), streamline and isotherms for pure water ($\phi = 0.0$) and nanofulid ($\phi = 0.2$) have been included in each figure where solid black lines represent pure fluid ($\phi = 0.0$) result and dotted red lines represent nano fluid ($\phi = 0.2$) result (Table 2).

Figure 3a–c shows the effect of upper lid movement on the streamline at $Re = 100$ for $Ri = 0.1$, 1.0, 10.0, respectively. Due to the combined effect of the buoyancy force and the temperature gradient between the hot and cold walls, hot fluid rises from bottom along the left vertical hot wall and comparatively heavier cold fluid occupies the bottom portion of the cavity along the right cold wall. Again, the motion of the upper lid in positive x-direction accelerates this movement and a primary vortex forms which move in clockwise direction. When $Ri = 0.1$, i.e. less than unity, then shear force dominates the buoyancy force and the nanofluid in the enclosure is primarily driven by the lid velocity. Due to strong shear force, the core of the primary eddy is near the moving lid. For $Ri = 1.0$, forced convection and natural convection have equal contribution on the flow field. Hence, the size of the primary eddy increases and the core region moves downwards due to natural convection effect.

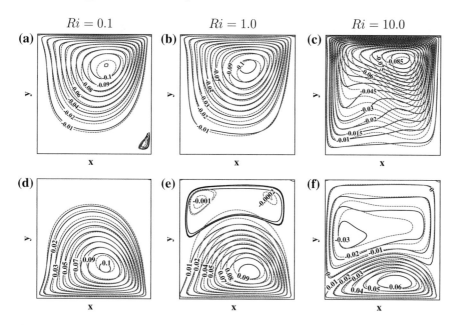

Fig. 3 Variation of streamline for different Richardson number (Ri), nanoparticle volume fraction (ϕ) at $Re = 100$ with **a–c** moving upper lid and **d–f** moving lower lid. Solid black lines are for pure fluid ($\phi = 0.0$), and dotted red lines are for nanofluid ($\phi = 0.2$)

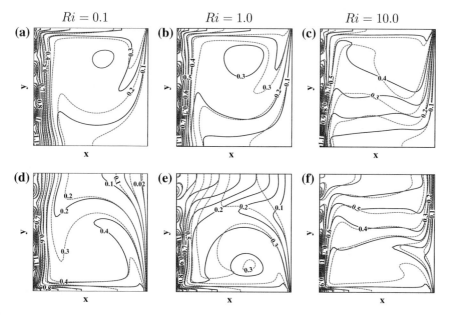

Fig. 4 Variation of isotherm for different Richardson number (Ri), nanoparticle volume fraction (ϕ) at $Re = 100$ with **a–c** moving lower lid and **d–f** moving upper lid. Solid black lines are for pure fluid ($\phi = 0.0$) and dotted red lines are for nanofluid ($\phi = 0.2$)

Table 2 Maximum absolute value of stream function ($|\psi_{max}|$) in the cavity for different lid movement, Ri and volume fraction (ϕ)

Lid	ϕ	$Ri = 0.1$ ($|\psi_{max}|$)	$Ri = 1.0$ ($|\psi_{max}|$)	$Ri = 10.0$ ($|\psi_{max}|$)
Upper	0.0	0.100152	0.102985	0.0775992
	0.1	0.104307	0.1056800	0.0831871
	0.2	0.104403	0.1061990	0.0888904
Lower	0.0	0.101411	0.0947755	0.0669556
	0.1	0.103499	0.0983553	0.0692968
	0.2	0.103998	0.1011970	0.0712635

At $Ri = 10.0$, natural convection plays dominant role on the flow field. Because of strong buoyancy force, the primary eddy is elongated horizontally and it occupies the whole cavity. At $Ri = 0.1$, a secondary vortex also has been formed at the right lower corner of the cavity which disappears for higher values of Ri because of the stronger buoyancy force. It also can be seen that streamline patterns for pure fluid ($\phi = 0.0$) and nanofluid ($\phi = 0.2$) case are almost similar for $Ri = 0.0$ and $Ri = 1.0$ while for $Ri = 10.0$ (Fig. 3c) streamline pattern differs significantly.

Figure 3d–f show the streamline pattern when lower lid is moving in positive x-direction at $Re = 100$ for $Ri = 0.1$, 1.0, 10.0, respectively. For $Ri = 0.1$ (forced convection dominated regime), the buoyancy force is overwhelmed by the shear force exerted by the lower moving lid and a single anticlockwise circulating cell has formed in the lower portion of the enclosure. The core of the circulation is displaced towards the lower right corner of the cavity due to shear effect. At $Ri = 1.0$ (mixed convection-dominated regime), the buoyancy force and shear force are relatively comparable in magnitude. Due to this combined effect, two circulations have formed in the enclosure circulating into opposite directions. Lower eddy circulates in anti-clockwise direction by the moving wall shear effect while the upper eddy circulates in clockwise direction due to buoyancy force and temperature gradient. Two centres have formed for the upper eddy at $\phi = 0.2$. This happens because of weak buoyancy force at $\phi = 0.2$, since as ϕ increases the contribution of buoyancy force decreases. But as Ri rises to 10 (i.e. natural convection-dominated regime), the two centres of the upper eddy submerged into a single clockwise circulating cell due to strong buoyancy force. At $Ri = 10$, lower cell shrinks and looses its strength while the upper cell becomes larger gaining strength.

Figure 4a–c shows the variation of isotherm for non-uniform wall temperature at different Richardson number (Ri) and nanoparticle volume fraction (ϕ) when upper lid of the square enclosure is moving in positive x-direction at $Re = 100$. It can be seen that isotherm lines are coiled on the hot wall due to non-uniform distribution of the temperature and the coiling is dense in the lower portion of the wall as compared to the upper portion of the wall. At $Ri = 0.1$ (forced convection-dominated regime), isotherms are clustered along the hot wall. This is due to the steep temperature gradients in the horizontal direction, and it indicates that the heat transfer near the wall is due to conduction. There is no significant heat distribution in the middle portion of the enclosure. When $Ri = 1.0$ (mixed convection regime), buoyancy force become stronger and the boundary layer vanishes. Also heat distribution increases in the middle of the enclosure due to mixed convection effect. At $Ri = 10.0$ (natural convection-dominated regime), the buoyancy force becomes dominant and the isotherms are distributed in the whole cavity. The thermal gradient near the wall is higher for nanofluid ($\phi = 0.2$) than the pure fluid ($\phi = 0.0$). This is because of the enhanced thermal conductivity at higher ϕ.

Figure 4d–f shows the variation of the isotherms when lower lid of the cavity is moving in the positive x-direction at different Richardson number (Ri) and nanoparticle volume fraction (ϕ) at $Re = 100$. At $Ri = 0.1$, isotherms are clustered along the left vertical wall and the lower corner of the left vertical wall. For $Ri = 1.0$ and 10.0, the isotherms are distributed all over the enclosure. Figures 4a–c and 4d–f illustrate the effect of the upper and lower moving lid on the temperature distribution as well as the effect of the non-uniform temperature distribution on the left vertical wall. The moving lid has significant impact on the temperature distribution as the moving lid drives the adjacent fluid in its direction which causes the temperature distribution in the same direction.

Figure 5a and b show the variation of the average Nusselt number (Nu_{av}) on the hot wall as a function of nanoparticle volume fraction (ϕ) for different Richardson Numbers (Ri) when upper and lower lid is moving in positive x-direction respectively. At a fixed Ri, Nu_{av} is a monotonic increasing function of nanoparticle volume fraction (ϕ). This is due to the fact that thermal conductivity of the nanofluid enhances with the increase of the nanoparticle volume fraction and hence a larger amount of heat gets absorbed and removed from the hot wall by the nanofluid. As a result Nu_{av} increases. Figure 5a shows that as Ri increases, Nu_{av} also increases. This is due to the fact that as Ri increases the buoyancy force also increases which reduces the thickness of the thermal boundary layer and heat transfer increases. But for the lower moving lid case (Fig. 5b), Nu_{av} decreases as Ri increases from 0.1 to 1.0. This phenomenon is largely illustrated in Fig. 5c. It can be seen that average Nusselt number (Nu_{av}) is a decreasing function of Richardson number (Ri) between 0.1 and 1.0 and increasing function between 1.0 and 10.0. At $Ri = 0.1$ the fluid flow is wholly dominant by shear force and so the heat transfer. But as Ri increases from 0.1 to 1.0, buoyancy force increases and at $Ri = 1.0$ shear force and buoyancy force become of same magnitude. These two forces have opposite effect on fluid flow when lower lid is moving in positive x-direction which can be seen from lower panel of Fig. 3. At $Ri = 1.0$, shear force moves lower half of the nanofluid into anticlockwise direction whereas the buoyancy force moves the upper half of the nanofluid into clockwise direction. Due to the combined effect of these two oppositely acting forces, heat transfer is minimum at $Ri = 1.0$. But further increment in Ri makes the buoyancy force stronger and hence Nu_{av} also increases.

Figure 6 shows the u-velocity and v-velocity profile for the movement of upper (Fig. 6a and b) and lower lid (Fig. 6c and d) at $x = 0.5$ and $Ri = 10.0$. Figure 6a shows that the u-velocity changes very little upto mid-height of the enclosure and after that it increases rapidly while the situation is almost opposite when lower lid

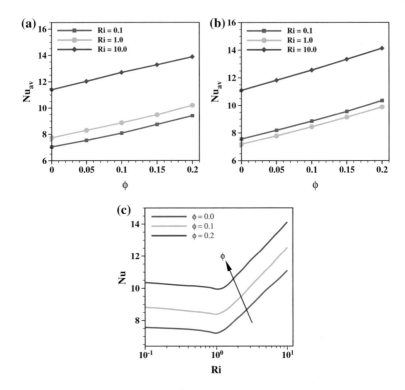

Fig. 5 Variation of average Nusselt number (Nu_{av}) at $Re = 100$ as a function of nanoparticle volume fraction (ϕ) at different Ri for **a** upper moving lid and **b** lower moving lid. **c** Variation of average Nusselt number (Nu_{av}) at $Re = 100$ as a function of Richardson number (Ri) at different nanoparticle volume fraction (ϕ) for lower moving lid

id moving. Figure 6c shows that the u-velocity increases in the lower portion of the enclosure. From Fig. 6b and d it can be concluded that v-velocity remains positive on the upper half and negative in the lower half of the enclosure for both the cases.

6 Conclusions

A numerical investigation of mixed convection of Cu-water nanofluid in a square enclosure with non-uniform temperature distribution on a side wall is made. Flow fields and thermal fields are illustrated by presenting the streamline and isotherm contour plots. The main findings of this study can be summarized as follows:

1. It is found that heat transfer rate is a strictly increasing function of Richardson number when upper lid is moving. But for moving lower wall, heat transfer rate

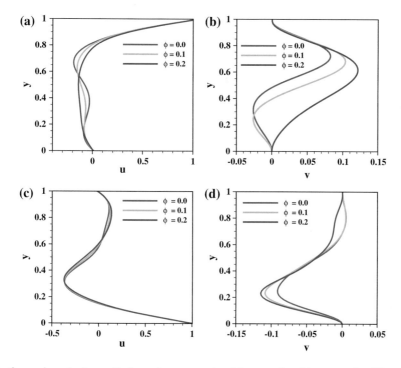

Fig. 6 u and v-velocity profile for **a–b** upper moving lid case and **c–d** lower moving lid case at $x = 0.5$ for $Ri = 10$, $Re = 100$ and $\phi = 0.0, 0.1, 0.2$

decreases in the interval $0.1 \leq Ri \leq 1.0$ with minimum value at $Ri = 1.0$ and it increases in the interval $1.0 \leq Ri \leq 10.0$.

2. Heat transfer rate is dependent on the choice of the moving wall. At natural convection regime, i.e. at $Ri = 10.0$, Nu_{av} has higher value when upper lid is moving. But at forced convection regime, i.e. at $Ri = 0.1$, Nu_{av} has higher value when lower lid is moving.
3. Non-uniform wall temperature effects the isotherm distribution on the hot wall. But it has negligible effect on the flow field.
4. Choice of moving lid has great impact on the flow field and temperature distribution.

References

1. Eastman, J.A., Choi, S.U.S., Li, S., Yu, W., Thompson, L.J.: Anomalously increased effective thermal conductivities of ethylene glycol-based nanofluids containing copper nanoparticles. Appl. Phys. Lett. **78**(6), 718–720 (2001)
2. Li, Q., Xuan, Y., Wang, J.: Investigation on convective heat transfer and flow features of nanofluids. J. Heat Transfer **125**(2003), 151–155 (2003). As references [2] and [9] are the

same, we have deleted the duplicate reference and renumbered accordingly. Please check and confirm.

3. Tiwari, R.K., Das, M.K.: Heat transfer augmentation in a two-sided lid-driven differentially heated square cavity utilizing nanofluids. Int. J. Heat Mass Transfer **50**(9), 2002–2018 (2007)

4. Mahmoodi, M.: Mixed convection inside nanofluid filled rectangular enclosures with moving bottom wall. Therm. Sci. **15**(3), 889–903 (2011)

5. Basak, T., Roy, S., Sharma, P.K., Pop, I.: Analysis of mixed convection flows within a square cavity with linearly heated side wall (s). Int. J. Heat Mass Transfer **52**(9), 2224–2242 (2009)

6. Ramakrishna, D., Basak, T., Roy, S., Pop, I.: A complete heatline analysis on mixed convection within a square cavity: effects of thermal boundary conditions via thermal aspect ratio. Int. J. Therm. Sci. **57**, 98–111 (2012)

7. Sivakumar, V., Sivasankaran, S.: Mixed convection in an inclined lid-driven cavity with non-uniform heating on both sidewalls. J. Appl. Mech. Tech. Phys. 55(4):634–649

8. Sivasankaran, S., Cheong, H.T., Bhuvaneswari, M., Ganesan, P.: Effect of moving wall direction on mixed convection in an inclined lid-driven square cavity with sinusoidal heating. Numer. Heat Transfer A **69**(6), 630–642

9. Chamkha, A.J., Abu-Nada, E.: Mixed convection flow in single-anddouble-lid driven square cavities filled with water Al_2O_3 nanofluid: effect of viscosity models. Eur. J. Mech. B Fluids **36**, 82–96 (2012)

10. Brinkman, H.C.: The viscosity of concentrated suspensions and solutions. J. Chem. Phys. **20**(4), 571–571 (1952)

11. Abu-Nada, E., Chamkha, A.J.: Mixed convection flow in a lid-driven inclined square enclosure filled with a nanofluid. Eur. J. Mech. B Fluids **29**(6), 472–482 (2010)

Chapter 17
On Love Wave Frequency Under the Influence of Linearly Varying Shear Moduli, Initial Stress, and Density of Orthotropic Half-Space

Sumit Kumar Vishwakarma, Tapas Ranjan Panigrahi and Rupinderjit Kaur

Abstract The present work studies Love wave propagation in an inhomogeneous anisotropic layer superimposed over an inhomogeneous orthotropic half-space under the influence of rigid boundary plane. The layer exhibits inhomogeneity which varies quadratically with depth, whereas the half-space has inhomogeneity in the shear moduli, density, and initial stress which varies linearly downward. The frequency equation is deduced in the closed form. It has been found that the dispersion equation is a function of phase velocity, wave number, inhomogeneity parameters, and initial stress. To analyze the result more profoundly, numerical simulation and graphical illustrations have been effectuated to depict the pronounced impact of the affecting parameters on the phase velocity of Love wave. As a special case, the procured dispersion relations have been found in well agreement with the standard Love wave equation.

Keywords Love wave · Inhomogeneous · Orthotropic · Anisotropic
Rigid plane

1 Introduction

It is very interesting to study Love wave propagation in an anisotropic media because the dispersion of seismic waves in anisotropic and orthotropic media is elementarily different from their dispersion in isotropic media. As the crustal layer of earth and mantle are not found to be homogeneous, it is very interesting to know the dispersion pattern of Love wave in an inhomogeneous medium as is studied sufficiently by Shearer [13]. It has been noticed that the propagation of Love wave is mostly affected by the elastic properties and the characteristic of the medium which it travels through. The earths' mantle (half-space) contains some hard and soft rocks or materials that may exhibit orthotropic property and porosity. In orthotropic medium, the thermal or mechanical properties being unique and independent in three mutually perpendicular

---
S. K. Vishwakarma (✉) · T. R. Panigrahi · R. Kaur
Department of Mathematics, BITS-Pilani, Hyderabad Campus, Hyderabad 500078, India
e-mail: sumitkumar@hyderabad.bits-pilani.ac.in; sumo.ism@gmail.com

© Springer Nature Singapore Pte Ltd. 2018
D. Ghosh et al. (eds.), *Mathematics and Computing*, Springer Proceedings
in Mathematics & Statistics 253, https://doi.org/10.1007/978-981-13-2095-8_17

directions make it an interesting medium. These facts motivated us to investigate further on Love wave propagation where the bearing of linear variation in the rigidity, density, and initial stresses can be studied. Destrade [5] studied in detail surface waves in orthotropic being incompressible in nature, whereas Kumar and Rajeev [11] analyzed the seismic wave motion to show the effect of voids at the boundary surface of orthotropic thermoelastic material. Ahmed and Dahab [1] demonstrated the remarkable effect of orthotropic granular layer on Love wave propagation, while a clear picture has been explained by Kumar and Choudhury [10] about the behavior and the response of orthotropic micropolar elastic medium via various sources.

Many problems in field of theoretical seismology are likely to be solved by demonstrating the earth as a layered medium with certain finite thickness and mechanical properties. An accurate and precise study on dispersion of elastic wave and its generation had been made by Chapman [4]. Propagation of surface seismic waves in the earths' crust due to its multiple applications in the field of geophysics, seismology, and applied mathematics has always been the subject of discussion along with various investigations. Vishwakarma et al. [14] demonstrated about the influence of the rigid boundary playing on the Love wave propagation in the elastic layer with void pores, while an interesting study made by Ke et al. [9] on Love wave dispersion under the effect of linearly varying properties of an inhomogeneous fluid saturated porous-layered half-space. In the theoretical study of seismic waves, mathematical expression provides the bridge between modeling results and field application. The propagation of elastic/seismic waves through the interior part of earth is governed by mathematical laws similar to the laws of light waves in optics.

The propagation of surface seismic wave such as Love waves in various inhomogeneous media has importance in multiple branches of engineering and applied science, like geophysics, seismology, earth science. Several studies have been carried out to understand the propagation technique of seismic waves in the inhomogeneous medium. Theories related to Love wave propagation in the anisotropic and inhomogeneous media have significant practical importance. It not only helps to investigate the internal structure of the earth and exploration of natural resources buried in the earths' surface but also about the composition of several layers under immense stress owing to different physical causes, i.e., presence of overlying layers, variation in temperature and gravitational field. This wave disperses when the solid medium near the surface has inhomogeneous elastic properties. Fortunately, Biot [2] developed the incremental deformation theory for pre-stressed medium. Adapting the same theory, earth being a spherical body with finite dimension, there exist remarkable influence of earths' crust on seismic surface waves. This phenomenon motivated us to investigate boundary waves or surface waves, i.e., waves that remain confined to certain surfaces during their dispersion. The formulations, solutions, and numerical simulations of many problems related to linear wave propagation for variety of geomedia may be found in the work of Gupta et al. [7, 8].

However, no attempt has been made to show the influence of inhomogeneous orthotropic half-space under initial stress on Love wave propagation. Therefore, in the present study, the half-space has been taken as inhomogeneous orthotropic medium followed by an inhomogeneous anisotropic layer resting over it. The inhomogeneity taken in the orthotropic mantle varies linearly along depth down toward the central core of the earth. This linear inhomogeneity has been taken in shear moduli, density, and initial stress of the half-space whereas the layer exhibits a quadratic variation in directional rigidities along horizontal and vertical direction and density. Suitable boundary condition under the assumption of rigid boundary plane has been considered and imposed on the displacement of the wave which have been found for individual layers. The frequency equation (dispersion equation) has been derived in closed form along with various particular cases. When all the inhomogeneities vanish, the frequency equation reduces to a classical equation of Love wave given by Love [12]. Numerical magnitude of the phase velocity has been calculated with the help of values of the material constants given by Biot [2] from experiments, and the effect of inhomogeneity parameter associated with directional rigidities, density, and initial stress is discussed and demonstrated using graphs.

2 Statement of the Problem

The geometry of the problem consists of an inhomogeneous anisotropic earth crust of finite thickness H resting over an inhomogeneous orthotropic half-space under the influence of linearly varying initial stress. Cartesian coordinate system has been employed with z-axis directed downwards and origin being at the interface where crustal layer and half-space meet as shown in the 3D diagram of Fig. 1. The upper boundary plane of the layer has been kept rigid where displacement of the wave vanishes. The inhomogeneities considered in the layer are as follows:

$$N = \overline{N} \left(1 + az\right)^2 , L = \overline{L} \left(1 + az\right)^2 , \rho = \overline{\rho} \left(1 + az\right)^2 \tag{1}$$

where $\overline{N}$ and $\overline{L}$ are the values of directional rigidities along x and z directions and $\overline{\rho}$ is the density at $z = 0$, a is called inhomogeneity parameter with dimension same as that of inverse of length.

The inhomogeneities taken in the anisotropic half-space are

$$Q_1 = \overline{Q_1} \left(1 + \alpha z\right), Q_3 = \overline{Q_3} \left(1 + \beta z\right), P = \overline{P} \left(1 + \gamma z\right), \rho_1 = \overline{\rho_1} \left(1 + \delta z\right) \tag{2}$$

where $\overline{Q_1}, \overline{Q_3}, \overline{P}$, and $\overline{\rho_1}$ are shear moduli, initial stress, and density of the medium at the interface $z = 0$ and α, β, γ, and δ are the inhomogeneity parameter associated with it having dimension equal to that of inverse length. Variation of rigidity, density, and initial stress along the depth inside the earth effects the propagation of seismic waves to a great extent. The inhomogeneity that exists is caused by variation in rigidity and density. The crust region of our planet is composed of various inhomogeneous layers with different geological parameters. As pointed out by Bullen [3], the density inside the earth varies at different rates with different layers within the earth. He approximated density law inside the earth as a quadratic polynomial in depth parameter for 413–984 km depth. For depth from 984 km to central core,

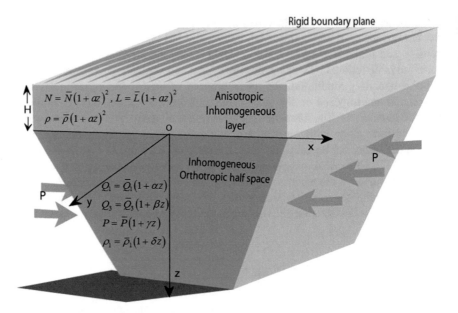

Fig. 1 Three-dimensional geometry of the problem

Bullen approximated the density as a linear function of depth parameter, and hence based on these theories, we have taken quadratic and linear variations.

3 Solution

3.1 Finding Displacement in Anisotropic Inhomogeneous Layer

Let u_1, v_1 and w_1 be the displacement components in the x, y, and z direction, respectively. Starting from the general equation of motion and using the conventional Love waves conditions, viz. $u = 0$, $w = 0$ and $v = v_1(x, z, t)$, the only y component. Then, the equation of motion in the absence of body force can be written as Biot [2]

$$N\frac{\partial^2 v_1}{\partial x^2} + \frac{\partial}{\partial z}\left(L\frac{\partial v_1}{\partial z}\right) = \rho\frac{\partial^2 v_1}{\partial t^2} \tag{3}$$

For a wave propagation along x-direction, we may assume

$$v_1 = V(z)e^{ik(x-ct)} \tag{4}$$

Using Eqs. (3) and (4) takes the form

$$\frac{d^2V}{dz^2} + \frac{1}{L}\frac{dL}{dZ}\frac{dV}{dz} + \frac{K^2}{L}\left(c^2\rho - N\right)V = 0 \tag{5}$$

After putting $V = \frac{V_1}{L}$ in equation, we get

$$\frac{d^2V_1}{dz^2} - \frac{1}{2L}\frac{d^2L}{dz^2}V_1 + \frac{1}{4L^2}\left(\frac{dL}{dz}\right)^2 V_1 + \frac{K^2}{L}\left(c^2\rho - N\right)V_1 = 0 \tag{6}$$

Using the inhomogeneity taken in Eqs. (1) and (6) changes to

$$\frac{d^2V_1}{dz^2} + m_1^2 V_1 = 0 \tag{7}$$

$$\text{where, } m_1^2 = \frac{K^2}{\overline{L}}\left(c^2\overline{\rho} - \overline{N}\right) \tag{8}$$

The solution of Eq. (7) may be assumed as

$$V_1 = A_1 e^{im_1 z} + B_1 e^{-im_1 z}$$

Thus, Eq. (4), the displacement in the inhomogeneous anisotropic layer may be taken as

$$v_1 = \frac{A_1 e^{im_1 z} + B_1 e^{-im_1 z}}{\sqrt{\overline{L}}\,(1 + az)} e^{iK(x-ct)} \tag{9}$$

3.2   Finding Displacement for Inhomogeneous Orthotropic Half-Space

The half-space taken in the problem is inhomogeneous orthotropic in nature under the influence of initial stress P along x direction as shown in Fig. 1. The system of equation pertaining to wave motion when there is no body forces is given by Biot [2]

$$\left.\begin{array}{l}\dfrac{\partial\sigma_{11}}{\partial x} + \dfrac{\partial\sigma_{12}}{\partial y} + \dfrac{\partial\sigma_{13}}{\partial z} - P\left(\dfrac{\partial w_z}{\partial y} - \dfrac{\partial w_y}{\partial z}\right) = \rho_1\dfrac{\partial^2 u_2}{\partial t^2} \\[3mm] \dfrac{\partial\sigma_{21}}{\partial x} + \dfrac{\partial\sigma_{22}}{\partial y} + \dfrac{\partial\sigma_{23}}{\partial z} - P\left(\dfrac{\partial w_z}{\partial x}\right) = \rho_1\dfrac{\partial^2 v_2}{\partial t^2}, \ \dfrac{\partial\sigma_{31}}{\partial x} + \dfrac{\partial\sigma_{32}}{\partial y} + \dfrac{\partial\sigma_{33}}{\partial z} - P\left(\dfrac{\partial w_y}{\partial x}\right) = \rho_1\dfrac{\partial^2 w_2}{\partial t^2}\end{array}\right\} \tag{10}$$

where u_2, v_2, and w_2 are the displacement components while w_x, w_y, and w_z are the rotational components along x, y, and z direction. Here, σ_{ij} are the incremental stress components and ρ_1 is the density of orthotropic medium. The relations between the strain and the incremental stress components are

$$\left.\begin{array}{l} \sigma_{11} = B_{11}e_{11} + B_{12}e_{22} + B_{13}e_{33}, \ \sigma_{12} = 2Q_3 e_{12}, \ \sigma_{22} = B_{21}e_{11} + B_{22}e_{22} + B_{23}e_{33} \\ \sigma_{23} = 2Q_1 e_{23}, \ \sigma_{33} = B_{31}e_{11} + B_{32}e_{22} + B_{33}e_{33}, \ \sigma_{31} = 2Q_2 e_{31} \end{array}\right\}$$

$$(11)$$

where B_{ij} and Q_i are the incremental normal elastic coefficients and shear moduli, respectively. Here e_{ij} are the strain components, which is defined by $e_{ij} = \frac{1}{2}\left(\frac{\partial u_i}{\partial x_j} + \frac{\partial u_j}{\partial x_i}\right)$, where $i, j = 1, 2, 3$.

Now, as per the characteristic of Love wave propagation, $u_2 = 0$, $w_2 = 0$, and $v_2 = v_2(x, z, t)$. Also, the inhomogeneity taken in Eq. (2) in Eq. (10) reduces to

$$\overline{Q}_3 (1 + \beta z) \frac{\partial^2 v_2}{\partial x^2} + \overline{Q}_1 \alpha \frac{\partial^2 v_2}{\partial z^2} + \overline{Q}_1 (1 + \alpha z) \frac{\partial^2 v_2}{\partial z^2} - \frac{\overline{P}}{2} (1 + \gamma z) \frac{\partial^2 v_2}{\partial x^2} = \overline{\rho_1} (1 + \delta z) \frac{\partial^2 v_2}{\partial t^2}$$

$$(12)$$

We may now use separation of variable, i.e., $v_2 = V_2(z) e^{ik(x - ct)}$, where k is the wave number and c is the phase velocity. Eq. (12) may now be written as

$$\frac{d^2 V_2}{dz^2} + \frac{\alpha}{(1 + \alpha z)} \frac{d V_2}{dz} + k^2 \left\{ A_1 \frac{(1 + \gamma z)}{(1 + \alpha z)} + A_2 \frac{(1 + \delta z)}{(1 + \alpha z)} - A_3 \frac{(1 + \beta z)}{(1 + \alpha z)} \right\} V_2 = 0$$

$$(13)$$

$$\text{where, } A_1 = \frac{\overline{P}}{2\overline{Q}_1}, \ A_2 = \frac{c^2}{c_1^2}, \ A_3 = \frac{\overline{Q}_3}{\overline{Q}_1}, \ c_1^2 = \frac{\overline{Q}_1}{\overline{\rho_1}} \tag{14}$$

Now, substituting $V_2 = \frac{\psi(z)}{(1+\alpha z)^{1/2}}$ in Eq. (13) to eliminate $\frac{dV_2}{dz}$, we get

$$\frac{d^2 \psi}{dz^2} + k^2 \left\{ A_1 \frac{(1 + \gamma z)}{(1 + \alpha z)} + A_2 \frac{(1 + \delta z)}{(1 + \alpha z)} - A_3 \frac{(1 + \beta z)}{(1 + \alpha z)} + \frac{1}{4}\left(\frac{a}{k}\right)^2 \frac{1}{(1 + \alpha z)^2} \right\} \psi = 0$$

$$(15)$$

Putting $n = 2(1 + \alpha z)\left(\frac{k}{\alpha}\right)\sqrt{A_3\left(\frac{\beta}{\alpha}\right) - A_1\left(\frac{\gamma}{\alpha}\right) - A_2\left(\frac{\delta}{\alpha}\right)}$, we will get

$$\frac{d^2 \psi}{d\eta^2} + \left(\frac{1}{4\eta^2} + \frac{R}{\eta} - \frac{1}{4}\right)\psi = 0 \tag{16}$$

where, $R = \frac{1}{2}\left(\frac{k}{\alpha}\right)\left\{A_3\left(\frac{\beta}{\alpha}\right) - A_1\left(\frac{\gamma}{\alpha}\right) + A_2\left(\frac{\delta}{\alpha}\right)\right\}^{\frac{1}{2}}\left\{A_1\left(\frac{\gamma}{\alpha} - 1\right) + A_2\left(\frac{\delta}{\alpha} - 1\right) + A_3\left(1 - \frac{\beta}{\alpha}\right)\right\}$

Equation (16) is a well-known Whittakers' equation, and the solution of which can be written as

$$\psi = A_2 W_{R,0}(\eta) + B_2 W_{-R,0}(-\eta)$$

where $W_{R,0}(\eta)$ is Whittakers' function and the general expansion of $W_{R,m}(\eta)$ may be written as Whittaker and Watson [15]

$$W_{R,m}(\eta) = e^{-\frac{\eta}{2}} . R^\eta \left[1 + \frac{\left\{m^2 - \left(R - \frac{1}{2}\right)^2\right\}}{1!z} + \frac{\left\{m^2 - \left(R - \frac{1}{2}\right)^2\right\}\left\{m^2 - \left(R - \frac{3}{2}\right)^2\right\}}{2!z^2} + \cdots \right] \tag{17}$$

Thus, the displacement in inhomogeneous orthotropic half-space becomes

$$v_2 = \frac{\left\{A_2 W_{R,0}(\eta) + B_2 W_{-R,0}(-\eta)\right\}}{(1+\alpha z)^{\frac{1}{2}}} e^{ik(x-ct)} \tag{18}$$

But, as we go down deep toward the center of earth, the displacement vanishes, i.e, as $z \to \infty$, $v_2 \to 0$, and therefore, the displacement in Eq. (18) reduces to

$$v_2 = \frac{A_2 W_{R,0}(\eta)}{(1+\alpha z)^{\frac{1}{2}}} e^{ik(x-ct)} \tag{19}$$

4   Boundary Conditions and Dispersion Equation

(1) Due to the presence of rigid boundary plane at $Z = -H$, the displacement vanishes

$$v_1 = 0 \; at \; z = -H \tag{20a}$$

(2) Displacement being continuous at the interface implies that

$$v_1 = v_2 \; at \; z = 0 \tag{20b}$$

(3) At the contact plane $z = 0$, the continuity of the stress requires that

$$L\frac{\partial v_1}{\partial z} = Q_1 \frac{\partial v_2}{\partial z} \; at \; z = 0 \tag{20c}$$

Using the above boundary conditions one by one, and eliminating the arbitrary constants A_1, B_1, and A_2 for nontrivial solution, we will have the following determinant.

$$\begin{vmatrix} e^{-im_1 H} & e^{im_1 H} & 0 \\ \dfrac{1}{\sqrt{L}} & \dfrac{1}{\sqrt{L}} & -W_{R,0}\,(\overline{\eta}) \\ \sqrt{L}\,(im_1 - a) & -\sqrt{L}\,(im_1 + a) & -\overline{Q}_1 \left[\dfrac{\partial W_{R,0}(\eta)}{\partial \eta} \cdot \dfrac{d\eta}{dz}\right]_{z=0} \end{vmatrix} = 0 \qquad (21)$$

Expanding the above determinant, we get the following:

$$\cot(m_1 H) = \left(\frac{a}{m_1}\right)\left(1 + 2\left(\frac{\overline{Q}_1}{\overline{L}}\right)\left(\frac{k}{a}\right)\left\{A_3\left(\frac{\beta}{\alpha}\right) - A_1\left(\frac{\gamma}{\alpha}\right) - A_2\left(\frac{\delta}{\alpha}\right)\right\}^{1/2}\right)$$
$$\left[-\frac{1}{2} + \frac{R}{\overline{\eta}} + \frac{(R-0.5)^2}{\overline{\eta}^2}\left\{1 - \frac{(R-0.5)^2}{\overline{\eta}}\right\}^{-1}\right]$$

Substituting the value of m_1 in the above expansion, it reduces to

$$\cot\left(kH\sqrt{\frac{c^2}{c_0^2} - \frac{\overline{N}}{\overline{L}}}\right) = \left(1 + 2\left(\frac{\overline{Q}_1}{\overline{L}}\right)\left(\frac{k}{a}\right)\left\{A_3\left(\frac{\beta}{\alpha}\right) - A_1\left(\frac{\gamma}{\alpha}\right) - A_2\left(\frac{\delta}{\alpha}\right)\right\}^{1/2}\right)$$
$$\left[-\frac{1}{2} + \frac{R}{\overline{\eta}} + \frac{(R-0.5)^2}{\overline{\eta}^2}\left\{1 - \frac{(R-0.5)^2}{\overline{\eta}}\right\}^{-1}\right]\left(\frac{a}{k}\left(\frac{c^2}{c_0^2} - \frac{\overline{N}}{\overline{L}}\right)^{-1/2}\right)$$
$$(22)$$

where $c_0^2 = \frac{\overline{L}}{\overline{\rho}}$.

Equation (22) is the required frequency equation of Love wave propagation in an inhomogeneous anisotropic layer resting over an inhomogeneous orthotropic medium with rigid boundary plane at the top. We find that Eq. (22) is a function of dimensionless phase velocity $\left(\frac{c^2}{c_0^2}\right)$, dimensionless wave number kH along with the inhomogeneity parameters m, α, γ, and δ associated with the rigidities, densities, and initial stress of the medium taken in to consideration.

Particular Case:
Case-I: When there is no inhomogeneity in the layer $a \to 0$, then Eq. (22) reduces to

$$\cot\left(kH\sqrt{\frac{c^2}{c_0^2} - \frac{\overline{N}}{\overline{L}}}\right) = 2\left(\frac{\overline{Q}_1}{\overline{L}}\right)\left\{A_3\left(\frac{\beta}{\alpha}\right) - A_1\left(\frac{\gamma}{\alpha}\right) - A_2\left(\frac{\delta}{\alpha}\right)\right\}^{1/2}$$
$$\left[-\frac{1}{2} + \frac{R}{\overline{\eta}} + \frac{(R-0.5)^2}{\overline{\eta}^2}\left\{1 - \frac{(R-0.5)^2}{\overline{\eta}}\right\}^{-1}\right]\left(\frac{c^2}{c_0^2} - \frac{\overline{N}}{\overline{L}}\right)^{-1/2}$$

which is the frequency equation of Love wave in a homogeneous anisotropic layer over inhomogeneous orthotropic half-space.

Case-II: When the half-space is stress-free, i.e., $\overline{P} \to 0$, then Eq. (22) becomes

$$\cot\left(kH\sqrt{\frac{c^2}{c_0^2} - \frac{\overline{N}}{\overline{L}}}\right) = \left(1 + 2\left(\frac{\overline{Q_1}}{\overline{L}}\right)\left(\frac{k}{a}\right)\left\{A_3\left(\frac{\beta}{\alpha}\right) - A_2\left(\frac{\delta}{\alpha}\right)\right\}^{1/2}\right)$$

$$\left[-\frac{1}{2} + \frac{R}{\overline{\eta}} + \frac{(R-0.5)^2}{\overline{\eta}^2}\left\{1 - \frac{(R-0.5)^2}{\overline{\eta}}\right\}^{-1}\right]\left(\frac{a}{k}\left(\frac{c^2}{c_0^2} - \frac{\overline{N}}{\overline{L}}\right)\right)^{-1/2},$$

which is the frequency equation of Love wave in an inhomogeneous anisotropic layer resting over inhomogeneous orthotropic half-space with no initial stress.

Case-III: When $\overline{N} = \overline{L}, \overline{Q_1} = \overline{Q_2}, a \to 0, \alpha \to 0, \beta \to 0, \delta \to 0$ and $P \to 0$, then the frequency Eq. (22) becomes

$$\cot\left(kH\sqrt{\frac{c^2}{c_0^2} - 1}\right) = \frac{\overline{Q_1}}{\overline{L}}\frac{\sqrt{\frac{c^2}{c_1^2} - 1}}{\sqrt{\frac{c^2}{c_0^2} - 1}}$$

which is the standard classical dispersion equation of Love wave given by Love [12] and therefore validated the solution of the problem discussed.

5 Numerical Computations, Graphs, and Discussion

In order to illustrate the theoretical results obtained in the preceding sections, the data have been fetched from Gubbins [6] to study graphically the impact of inhomogeneity, rigid boundary, and the various elastic constants on the propagation of Love wave using frequency equation as obtained in Eq. (22). We will use the asymptotic linear expansion of Whittakers' function as given in Eq. (17). In all the graphs, horizontal axis has been taken as dimensionless wave number kH while vertical axis has been taken as dimensionless phase velocity $\left(\frac{c}{c_0}\right)^2$. Numerical values taken are as follows:

1. Inhomogeneous anisotropic layer: $\overline{N} = 7.34 \times 10^{10}$N/m^2, $\overline{L} = 5.98 \times 10^{10}$ N/m^{2}N/m^2, $\overline{\rho} = 3195$ kg/m^3
2. Inhomogeneous orthotropic half-space: $Q_1 = 5.82 \times 10^{10}$N/m^2, $\overline{Q_3} = 3.99 \times 10^{10}$N/m^2, $\overline{\rho_1} = 4500$ kg/m^3

Figure 2 reflects the effect of inhomogeneity parameter $\left(\frac{a}{k}\right)$ associated with the directional rigidities and density in the anisotropic layer. The value of $\left(\frac{a}{k}\right)$ for curve no. 1, curve no. 2, curve no. 3, and curve no. 4 has been taken as 0.1, 0.3, 0.5, and 0.7, respectively, whereas the value of $\frac{\overline{P}}{2\overline{Q}}, \frac{\alpha}{k}, \frac{\beta}{k}, \frac{\gamma}{k}$ and $\frac{\delta}{k}$ are 0.2, 0.1, 0.2, 0.2, and 0.1, respectively. The following observations and effects are obtained under the above considered values.

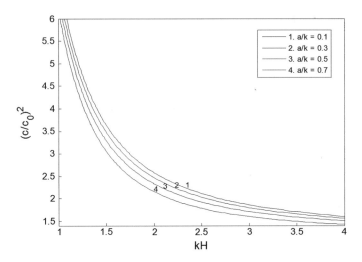

Fig. 2 Variation of dimensionless phase velocity against dimensionless wave number for different values of (m/k) when $\frac{\overline{P}}{2\overline{Q}_1} = 0.2$, $\frac{\alpha}{k} = 0.1$, $\frac{\beta}{k} = 0.2$, $\frac{\gamma}{k} = 0.2$, $\frac{\delta}{k} = 0.1$

2a. The phase velocity decreases as the wave number increases for all the values of $\left(\frac{a}{k}\right)$.

2b. While at a particular wave number as the value of $\left(\frac{a}{k}\right)$ increases from 0.1 to 0.7, the phase velocity decreases.

2c. Toward low wave number, the curves seem accumulating which reveals that the phase velocity remains unaffected as inhomogeneity changes.

2d. Toward higher wave number, the phase velocity decreases gradually, whereas it decreases rapidly for low wave number.

2e. Seeing the pattern of the curve, we can claim that the inhomogeneity present in the layer bears a remarkable effect on the phase velocity of Love wave.

Figure 3 has been drawn to analyze the bearing of dimensionless inhomogeneity parameter $\left(\frac{\alpha}{k}\right)$ on the phase velocity of Love wave. Curve no. 1 has been plotted for $\frac{\alpha}{k} = 0.2$, curve no. 2 for $\frac{\alpha}{k} = 0.4$, curve no. 3 for $\frac{\alpha}{k} = 0.6$ and curve no. 4 for $\frac{\alpha}{k} = 0.8$. The value of $\frac{\overline{P}}{2\overline{Q}}$, $\frac{a}{k}$, $\frac{\beta}{k}$, $\frac{\gamma}{k}$ and $\frac{\delta}{k}$ are 0.2, 0.1, 0.2, 0.2, and 0.1, respectively. The following results are obtained.

3a. The pattern of curves obtained here is quite similar to one obtained in Fig. 2.

3b. As the magnitude of $\left(\frac{\alpha}{k}\right)$ increases from 0.2 to 0.8, the phase velocity decreases at a fixed wave number.

3c. Curves being equally apart, a periodic effect of inhomogeneity parameter $\frac{\beta}{k}$ may be found throughout the figure.

Figure 4 describes the influence of inhomogeneity parameter $\left(\frac{\beta}{k}\right)$ for its increasing magnitude from 0.1 to 0.4 for curve no. 1–4. The following observations and effects are found.

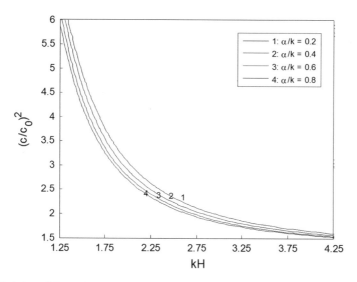

Fig. 3 Variation of dimensionless phase velocity against dimensionless wave number for different values of (α/k) when $\frac{\bar{P}}{2\bar{Q}_1} = 0.2$, $\frac{a}{k} = 0.1$, $\frac{\beta}{k} = 0.2$, $\frac{\gamma}{k} = 0.2$, $\frac{\delta}{k} = 0.1$

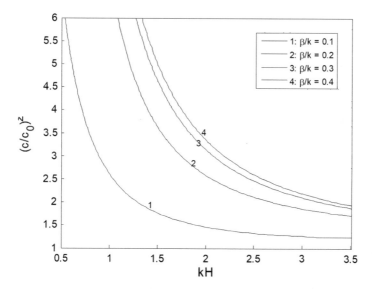

Fig. 4 Variation of dimensionless phase velocity against dimensionless wave number for different value of (β/k) when $\frac{\bar{P}}{2\bar{Q}_1} = 0.2$, $\frac{a}{k} = 0.1$, $\frac{\alpha}{k} = 0.1$, $\frac{\gamma}{k} = 0.2$, $\frac{\delta}{k} = 0.1$

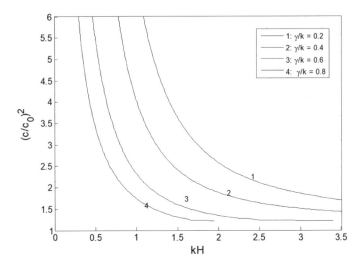

Fig. 5 Variation of dimensionless phase velocity against dimensionless wave number for different value of (γ/k) when $\frac{P}{2Q_1} = 0.2$, $\frac{a}{k} = 0.1$, $\frac{\alpha}{k} = 0.1$, $\frac{\gamma}{k} = 0.2$, $\frac{\delta}{k} = 0.1$

4a. Unlike Figs. 2 and 3, the phase velocity increases for the increases in the inhomogeneity parameter $\left(\frac{\beta}{k}\right)$ associated with shear Modulus Q_3.

4b. The curves are becoming closer as the magnitude of $\left(\frac{\beta}{k}\right)$ increases.

4c. The impact of the inhomogeneity is more pronounced for its least value.

4d. It can also be said that phase velocity may attain a constant magnitude as the inhomogeneity increases further.

Figure 5 illustrates a clear picture of the variation of phase velocity against wave number when initial stress in the half-space increases. Curves have been plotted for $\frac{\gamma}{k}$ equals to 0.2, 0.4, 0.6, and 0.8 for curve no. 1, curve no. 2, curve no. 3, and curve no. 4, respectively. The values of other parameter such as $\frac{P}{2Q}$, $\frac{a}{k}$, $\frac{\alpha}{k}$, $\frac{\beta}{k}$, $\frac{\gamma}{k}$, and $\frac{\delta}{k}$ have been taken as 0.2, 0.1, 0.1, 0.2, 0.1. We can enlist the following points about Fig. 5.

5a. The pattern is similar to some extent as that of one obtained in Fig. 4.

5b. Here the phase velocity diminishes as the magnitude of the inhomogeneity parameter linked with initial stress increases.

5c. The phase velocity for curve no. 3 and curve no. 4 is restricted upto to $kH = 3.5$ and $kH = 2$, respectively, thereby showing a significant effect of inhomogeneity in the half-space.

In Fig. 6, attempt has been made to show the influence of inhomogeneity parameter $\frac{\delta}{k}$ present in the density of the orthotropic half-space. We find that

6a. there is an decrement in the magnitude of phase velocity as the wave number diminishes for all the values of $\frac{\delta}{k}$.

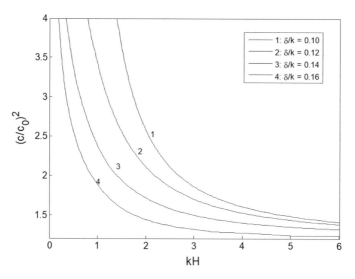

Fig. 6 Variation of dimensionless phase velocity against dimensionless wave number for different value of (δ/k) when $\frac{P}{2Q_1} = 0.2$, $\frac{a}{k} = 0.1$, $\frac{\alpha}{k} = 0.1$, $\frac{\beta}{k} = 0.2$, $\frac{\gamma}{k} = 0.2$

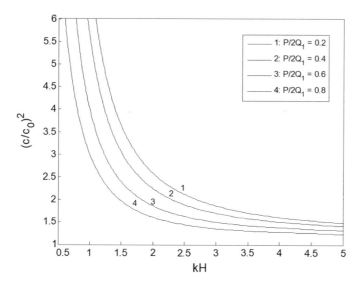

Fig. 7 Variation of dimensionless phase velocity against dimensionless wave number for different values of compressive initial stress $\left(\frac{P}{2Q_1} > 0\right)$ when $\frac{a}{k} = 0.1$, $\frac{\alpha}{k} = 0.1$, $\frac{\beta}{k} = 0.2$, $\frac{\gamma}{k} = 0.0$, $\frac{\delta}{k} = 0.1$

6b. At a particular wave number, the phase velocity also decreases for the increasing magnitude of inhomogeneity parameter in the density of orthotropic medium.

6c. When the phase velocity is least, the curves appearing closer to each other at high wave number showing a prominent effect of inhomogeneity parameter $\frac{\delta}{k}$.

Figure 7 depicts the impact of initial stress $\frac{\overline{P}}{2\overline{Q}_1}$ when $\frac{\gamma}{k} = 0$ shows the effect of compressive initial stress $\left(\frac{\overline{P}}{2\overline{Q}_1} > 0\right)$ on the phase velocity of Love wave propagating in an inhomogeneous anisotropic layer. It has been observed that as the magnitude of compressive initial stress becomes larger, the phase velocity decreases while it increases as the tensile stress increases.

6 Conclusion

Propagation of Love waves in an inhomogeneous anisotropic layer resting over an inhomogeneous orthotropic half-space with linearly varying inhomogeneity has been studied in details. Solutions in terms of displacement of the wave in the layer and half-space have been derived separately. We have used asymptotic linear expansion of Whittakers' function and obtained the dispersion relation (frequency equation) in compact form. Numerical investigations have been made on phase velocity against wave number and the effect of each one of the linearly varying inhomogeneity parameters associated with anisotropic layer and orthotropic half-space has been studied and discussed in detail. We observed that

I. Under the assumed condition, phase velocity $\left(\frac{c}{c_0}\right)^2$ increases with decrease in dimensionless wave number.

II. The phase velocity of Love wave decreases as the inhomogeneity parameter $\left(\frac{a}{k}\right)$ associated with directional rigidity and density of the layer increases.

III. The increasing magnitude of $\frac{\beta}{k}$ increases the phase velocity whereas $\frac{\alpha}{k}$, $\frac{\gamma}{k}$ and $\frac{\delta}{k}$ decreases the phase velocity as it increases.

IV. At a fixed wave number, the increasing value of compressive initial stress $\left(\frac{P}{2Q_1} > 0\right)$ decreases the velocity while increasing tensile stress $\left(\frac{P}{2Q_1} > 0\right)$ increases.

V. In the absence of all inhomogeneity and initial stress, the dispersion equation turns into the classical form of equation of Love wave and therefore revealing the validation of current work.

The consequences of the present study gives a theoretical framework for the adopted model, which may likely to be utilized to collect, investigate, and recognize the propagation pattern Love wave propagation in anisotropic layer over orthotropic half-space, which may further help in accessing the resources buried inside the earth such oils, gases, minerals, deposits, and other useful hydrocarbons. Apart from these, the outcomes of the present study may also be used widely in the design and development of heavy civil construction projects involving steel structures, disaster-resistant buildings, bridge, and towers, etc. Precisely the study may also be useful in the inter-departmental fields like rock mechanics, soil mechanics, geotechnical engineering, and applied science.

Acknowledgements Authors extend their sincere thanks to SERB-DST, New Delhi, for providing financial support under Early Career Research Award with Ref. no. ECR/2017/001185. Authors are also thankful to DST, New Delhi, for providing DST-FIST grant with Ref. no. 337 to Department of Mathematics, BITS-Pilani, Hyderabad campus. Authors also express their deep sense of respect and gratitude to honorable reviewers for their constructive suggestions to improve the quality of the manuscript.

References

1. Ahmed, S.M., Abd-Dahab, S.M.: Propagation of Love waves in an orthotropic Granular layer under initial stress overlying a semi-infinite Granular medium. J. Vib. Control **16**(12), 1845–1858 (2010)
2. Biot, M.A.: Mechanics Incremental Deformation. Wiley, New York (1965)
3. Bullen, K.E.: The problem of Earth's density variation. Bull. Seismological Soc. Am. **30**(3), 235–250 (1940)
4. Chapman, C.: Fundamentals of Seismic Wave Propagation. Cambridge University Press, Cambridge (2004)
5. Destrade, M.: Surface waves in orthotropic incompressible materials. J. Acoustical Soc. Am. **110**(2), 837–840 (2001)
6. Gubbins, D.: Seismology and Plate Tectonics. Cambridge University Press, Cambridge (1990)
7. Gupta, S., Majhi, D.K., Kundu, S., Vishwakarma, S.K.: Propagation of love waves in non-homogeneous substratum over initially stressed heterogeneous half-space. Appl. Mathe. Mech. **34**(2), 249–258 (2013)
8. Gupta, S., Vishwakarma, S.K., Majhi, D.K., Kundu, S.: Possibility of Love wave propagation in a porous layer under the effect of linearly varying directional rigidities. Appl. Mathe. Modell. **37**, 6652–6660 (2013)
9. Ke, L.L., Wang, Y.S., Zhang, Z.M.: Propagation of love waves in an inhomogeneous fluid saturated porous layered half-space with linearly varying properties. Soil Dyn. Earthquake Eng. **26**, 574–581 (2006)
10. Kumar, R., Choudhary, S.: Response of orthotropic micropolar elastic medium due to various sources. Meccanica **38**, 349–368 (2003)
11. Kumar, R., Rajeev, K.: Analysis of wave motion at the boundary surface of orthotropic thermoelastic material with voids and isotropic elastic half space. J. Eng. Phys. Thermophys. **84**(2), 463–478 (2003)
12. Love, A.E.H.: A Treatise on Mathematical Theory of Elasticity, 4th edn. Dover Publication, New York (1944)
13. Shearer, P.M.: Introduction to Seismology, 2nd edn. Cambridge University Press, Cambridge (2009)
14. Vishwakarma, S.K., Gupta, S., Majhi, D.K.: Influence of rigid boundary on the Love wave propagation in elastic layer with void pores. Acta Mechanica Solida Sinica **25**(5), 551–558 (2013)
15. Whittaker, E.T., Watson, G.N.A.: Course in Modern Analysis. Cambridge University Press, Cambridge (1990)

Chapter 18
The Problem of Oblique Scattering by a Thin Vertical Submerged Plate in Deep Water Revisited

B. C. Das, S. De and B. N. Mandal

Abstract The problem of oblique scattering by fixed thin vertical plate submerged in deep water is studied here, assuming linear theory, by employing single-term Galerkin approximation involving *constant as basis* multiplied by appropriate weight function after reducing it to solving a pair of first kind integral equations. Upper and lower bounds of reflection and transmission coefficients when evaluated numerically are seen to be very close so that their averages produce fairly accurate numerical estimates for these coefficients. Numerical estimates for the reflection coefficient are depicted graphically against the wave number for different values of various parameters. The numerical results obtained by the present method are found to be in an excellent agreement with the known results.

Keywords Submerged plate · Linearized theory · Galerkin technique · *Constant as basis* · Reflection coefficient

1 Introduction and Mathematical Formulation of the Problem

There is no explicit solutions for the problems of oblique scattering of water waves by a thin vertical barrier of various geometrical configurations present in deep water. However, there exists some approximate methods to solve these problems approximately in the sense that the reflection and transmission coefficients could be obtained

B. C. Das (✉) · S. De
Department of Applied Mathematics, Calcutta University,
92, A.P.C. Road, Kolkata 700009, India
e-mail: findbablu10@gmail.com

S. De
e-mail: soumenisi@gmail.com

B. N. Mandal
Physics and Applied Mathematics Unit, Indian Statistical Institute,
203 B.T Road, Kolkata 700108, India
e-mail: bnm2006@rediffmail.com

© Springer Nature Singapore Pte Ltd. 2018
D. Ghosh et al. (eds.), *Mathematics and Computing*, Springer Proceedings
in Mathematics & Statistics 253, https://doi.org/10.1007/978-981-13-2095-8_18

numerically. Oblique scattering problems involving a partially immersed or completely submerged thin vertical barrier was studied by [2–4, 6–8] by using various methods.

The problem of oblique scattering by a thin vertical plate submerged in deep water can be formulated mathematically as follows. Assuming linear theory and irrotational motion, let a train of surface water waves represented by the potential function $Re\{\phi_0(x, y, z)e^{i\nu z-i\sigma t}\}$ with

$$\phi_0(x, y, z) = e^{-Ky+i\mu x}, \tag{1.1}$$

where $\mu = K \cos \alpha$, $\nu = K \sin \alpha$, $K = \frac{\sigma^2}{g}$, g being the gravity, σ being the circular frequency, be incident obliquely at an angle α on a thin vertical plate represented by $x = 0, y \in L = (a, b)$ which is submerged in deep water. Here y−axis is taken vertically downwards and the (x, y)-plane denotes the mean free surface. Due to geometrical symmetry, the resulting motion in water can be described by the velocity potential $Re\{\phi(x, y, z)e^{i\nu z-i\sigma t}\}$ where $\phi(x, y)$ satisfies

$$(\nabla^2 - \nu^2)\phi = 0, y \ge 0, -\infty < x < \infty, \tag{1.2}$$

$$K\phi + \phi_y = 0 \text{ on } y = 0, \tag{1.3}$$

$$\phi_x = 0 \text{ on } x = 0, y \in L, \tag{1.4}$$

$$\nabla\phi \to 0 \text{ as } y \to \infty, \tag{1.5}$$

$$r^{\frac{1}{2}}\nabla\phi = O(1) \text{ as } r = (x^2 + (y - c)^2)^{\frac{1}{2}} \to 0, \text{ where } c = a, b \tag{1.6}$$

and

$$\phi(x, y) \to \begin{cases} Te^{-Ky+i\mu x} \text{ as } x \to \infty \\ e^{-Ky+i\mu x} + Re^{-Ky-i\mu x} \text{ as } x \to -\infty, \end{cases} \tag{1.7}$$

where T and R are the transmission and reflection coefficient, respectively, and are to be determined.

It may be noted that for the case of normal incidence ($\alpha = 0$), R and T can be obtained in closed forms. In fact, [1] solved the normal incidence problem using complex variable theory and obtained the corresponding reflection and transmission coefficients in closed forms involving some complicated (but computable) integrals. However, when $\alpha \neq 0$, closed form results cannot be obtained. For this case, [9] reduced it to the solution of an integral equation involving the unknown difference of potentials across the plate, the integral equation being solved by an expansion method similar to the expansion of its kernel involving different orders of $\sin \alpha$. Later, [10, 11] employed one-term Galerkin approximations involving the exact solutions for normally incident waves to solve the integral equations involving the unknown difference of potential across the plate and the unknown horizontal velocity across the gaps above and below the plate and obtained very accurate upper and lower bounds for the reflection and transmission coefficients for all angles of incidence and wave numbers. However, the aforesaid exact solutions of integral equations for normally incident waves are somewhat complicated, and as such, the upper and lower bounds for the reflection and transmission coefficients involve complicate integrals. In the present method, single-term Galerkin approximations in solving the integral equations are employed, but these approximations involve *constant* multiplied by appropriate weight functions. This process produces upper and lower bounds involving simple integrals which are quite easy to evaluate.

2   Method of Solution

A solution for the velocity potential $\phi(x, y)$ satisfying (1.2) and the conditions (1.7) is given by

$$\phi(x, y) = \begin{cases} T\phi_0(x, y) + \int_0^\infty A(k)S(k, y)e^{-k_1 x}dk, x > 0, \\ \phi_0(x, y) + R\phi_0(-x, y) + \int_0^\infty B(k)S(k, y)e^{k_1 x}dk, x < 0 \end{cases} \tag{2.1}$$

where $k_1 = \left(k^2 + \nu^2\right)^{\frac{1}{2}}$ with $k_1 = k$ when $\nu = 0$ and

$$S(k, y) = k \cos(ky) - K \sin(ky). \tag{2.2}$$

Let

$$f(y) = \frac{\partial \phi}{\partial x}(0, y), 0 < y < \infty, \tag{2.3}$$

and

$$g(y) = \phi(x + 0) - \phi(x - 0), 0 < y < \infty, \tag{2.4}$$

then

$$f(y) = 0 \text{ for } y \in L \tag{2.5}$$

and

$$g(y) = 0 \text{ for } y \in \overline{L} = (0, \infty) - L. \tag{2.6}$$

The unknown constants R, T and the unknown functions $A(k)$ and $B(k)$ are related to $f(y)$ and $g(y)$ as given by

$$T = 1 - R = -\frac{2iK}{\mu} \int_{\overline{L}} f(y)e^{-Ky}dy, \tag{2.7}$$

$$A(k) = -B(k) = -\frac{2}{\pi k_1(k^2 + K^2)} \int_{\overline{L}} f(y)S(k, y)dy, \tag{2.8}$$

$$R = -K \int_L g(y)e^{-Ky}dy, \tag{2.9}$$

$$A(k) = \frac{1}{\pi(k^2 + K^2)} \int_L g(y)S(k, y)dy. \tag{2.10}$$

In deriving relations (2.7) and (2.8), the condition (2.5) and in deriving the relations (2.9) and (2.10) the condition (2.6) have been utilized in the appropriate Havelock [5] inversion formula.

Use of the condition (1.4) in the form

$$\frac{\partial \phi}{\partial x}(\pm 0, y) = 0, y \in L$$

in the representation (2.1) for $\phi(x, y)$ produces an integral equation for $g(y)$ as given by

$$\int_L g(u)\mathcal{M}(y, u)du = \pi i \mu(1 - R)e^{-Ky}, y \in L \tag{2.11}$$

where

$$\mathcal{M}(y, u) = \lim_{\epsilon \to +0} \int_0^\infty \frac{k_1 S(k, y)S(k, u)}{k^2 + K^2} e^{-\epsilon k} dk, \tag{2.12}$$

the exponential term being introduced to ensure convergence of the integral. In the relation (2.12), we note that $\mathcal{M}(y, u)$ is a real and symmetric function of y and u.

Again, as $\phi(x, y)$ is continuous across the gap, use of the representation (2.1) along with the relation (2.8) produces an integral equation for $f(y)$ as given by

$$\int_{\overline{L}} f(u)\mathcal{N}(y, u)du = -\frac{\pi}{2} Re^{-Ky}, y \in \overline{L} \tag{2.13}$$

where

$$\mathcal{N}(y, u) = \int_0^\infty \frac{S(k, y)S(k, u)}{k_1(k^2 + K^2)} dk, \tag{2.14}$$

so that $\mathcal{N}(y, u)$ is also a real and symmetric function of y and u.

Let us write

$$F(y) = -\frac{2}{\pi R}f(y), y \in \overline{L}, \tag{2.15}$$

$$G(y) = \frac{1}{\pi i \mu(1-R)}g(y), y \in L, \tag{2.16}$$

then $G(y)$ and $F(y)$ satisfy the integral equations

$$\int_L G(u)\mathcal{M}(y, u)du = e^{-Ky}, y \in L, \tag{2.17}$$

$$\int_{\overline{L}} F(u)\mathcal{N}(y, u)du = e^{-Ky}, y \in \overline{L}. \tag{2.18}$$

It may be noted that functions $G(y)$ and $F(y)$ in (2.17) and (2.18), respectively, must be real.

The relations (2.7) and (2.9) are now recast as

$$\int_{\overline{L}} F(y)e^{-Ky}dy = C, \tag{2.19}$$

and

$$\int_L G(y)e^{-Ky}dy = \frac{1}{\pi^2 K^2 C}, \tag{2.20}$$

where

$$C = \frac{1-R}{i\pi R}\cos\alpha. \tag{2.21}$$

It is important to note that C is real.

3   Upper and Lower Bounds for C

Following [2], we define an inner product

$$<f, g> = \int_L f(y)g(y)dy. \tag{3.1}$$

Then, obviously $< f(y), g(y) >$ is symmetric and linear. Also, the operator $\mathcal{M}$ defined by

$$(\mathcal{M}g)(y) = < \mathcal{M}(y, u), g(u) > \tag{3.2}$$

is linear, self-adjoint, and positive semi-definite. For the solution of the integral equation (2.17), we choose a single-term Galerkin approximation as given by

$$G(y) \approx \lambda_1 g(y), y \in L \tag{3.3}$$

where λ_1 is an unknown constant and $g(y)$ is to be chosen suitably. Then,

$$\lambda_1 = \frac{< g(y), e^{-Ky} >}{< g(y), (\mathcal{M}g)(y) >}. \tag{3.4}$$

Hence, using the approximate solution (3.3) for $G(y)$ in the relation (2.20), we find

$$\frac{1}{\pi^2 K^2 C} = < G(y), e^{-Ky} > \geq < \lambda_1 g(y), e^{-Ky} >, \tag{3.5}$$

after using the same argument as in [2]. Thus, we find that

$$C \geq A \tag{3.6}$$

where A is an upper bound of C and is given by

$$A = \frac{1}{\pi^2 K^2} \frac{< g(y), (\mathcal{M}g)(y) >}{(< g(y), e^{-Ky} >)^2}. \tag{3.7}$$

Again, if we define another inner product by

$$\ll f, g \gg = \int_{\overline{L}} f(y) g(y) dy \tag{3.8}$$

and another operator $\mathcal{N}$ by

$$(\mathcal{N}f)(y) = \ll \mathcal{N}(y, u), f(u) \gg, \tag{3.9}$$

, then it is obvious that $\ll f, g \gg$ is linear, symmetric, and also the operator $\mathcal{N}$ is linear, self-adjoint, and positive semi-definite.

For solution of the integral Eq. (2.18), we choose single-term Galerkin approximation as

$$F(y) \approx \lambda_2 f(y), y \in \overline{L}, \tag{3.10}$$

where λ_2 is an unknown constant and $f(y)$ is to be chosen suitably, then

$$\lambda_2 = \frac{\ll f(y), e^{-Ky} \gg}{\ll f(y), (\mathcal{N}f)(y) \gg}. \tag{3.11}$$

Hence, using the approximate solution (3.10) for $F(y)$ in the relation (2.19), we find that

$$C = \ll F(y), e^{-Ky} \gg \geq \ll \lambda_2 f(y), e^{-Ky} \gg \tag{3.12}$$

after using the same argument as in [2]. Thus, we find that

$$C \geq B \tag{3.13}$$

where B is a lower bound of C and is given by

$$B = \frac{(\ll f(y), e^{-Ky} \gg)^2}{\ll f(y)(\mathcal{N}f)(y) \gg}. \tag{3.14}$$

Hence, for the unknown real constant C, we find

$$B \leq C \leq A \tag{3.15}$$

where A and B are given by (3.7) and (3.14), respectively. Thus, upper and lower bounds for $|R|$ and $|T|$ are obtained as

$$R_1 \leq |R| \leq R_2, T_1 \leq |T| \leq T_2 \tag{3.16}$$

where

$$R_1 = \frac{1}{\left(1 + \pi^2 A^2 \sec^2 \alpha\right)^{\frac{1}{2}}}, R_2 = \frac{1}{\left(1 + \pi^2 B^2 \sec^2 \alpha\right)^{\frac{1}{2}}}, \tag{3.17}$$

$$T_1 = \frac{\pi B \sec \alpha}{\left(1 + \pi^2 A^2 \sec^2 \alpha\right)^{\frac{1}{2}}}, T_2 = \frac{\pi A \sec \alpha}{\left(1 + \pi^2 B^2 \sec^2 \alpha\right)^{\frac{1}{2}}}. \tag{3.18}$$

Here $L = (a, b)$ so that $\overline{L} = (0, a) + (b, \infty)$. The function $g(y)$ in (3.3) is chosen as

$$g(y) = \frac{1}{b}\sqrt{(y - a)(b - y)}, a < y < b. \tag{3.19}$$

After substituting $g(y)$ in (3.7), A is obtained as

$$A = \frac{1}{\pi^2 K^2} \frac{\int_0^\infty \frac{k_1}{k^2+K^2} \left[kp(a,b,k) - Kq(a,b,k)\right]^2 dk}{(r(a,b,K))^2} \tag{3.20}$$

where

$$p(a,b,k) = \frac{1}{b} \int_a^b \sqrt{(y-a)(b-y)} \cos(ky) dy,$$

$$q(a,b,k) = \frac{1}{b} \int_a^b \sqrt{(y-a)(b-y)} \sin(ky) dy,$$

$$r(a,b,K) = \frac{1}{b} \int_a^b e^{-Ky} \sqrt{(y-a)(b-y)} dy.$$

Again, we choose $f(y)$ in (3.10) as

$$f(y) = \begin{cases} \sqrt{\frac{a}{a-y}}, & 0 < y < a, \\ e^{-Ky}\sqrt{\frac{b}{y-b}}, & b < y < \infty. \end{cases} \tag{3.21}$$

After substituting this $f(y)$ in the expressions (3.14), B is obtained as

$$B = \frac{[M(Ka) + N(Kb)]^2}{\int_0^\infty \frac{1}{k_1(k^2+K^2)} \left[kU(a,k) + KV(a,k) + kW(b,k,K) - KX(b,k,K)\right]^2 dk} \tag{3.22}$$

where

$$M(Ka) = \int_0^a e^{-Ky} \sqrt{\frac{a}{a-y}} dy,$$

$$N(Kb) = \int_b^\infty e^{-2Ky} \sqrt{\frac{b}{y-b}} dy,$$

$$U(a,k) = \int_0^a \sqrt{\frac{a}{a-y}} \cos(ky) dy,$$

$$V(a,k) = \int_0^a \sqrt{\frac{a}{a-y}} \sin(ky) dy,$$

$$W(b,k,K) = \int_b^\infty e^{-Ky} \sqrt{\frac{b}{y-b}} \cos(ky) dy,$$

$$X(b, k, K) = \int_b^\infty e^{-Ky} \sqrt{\frac{b}{y-b}} \sin(ky) dy.$$

The integrals appearing in (3.20) and (3.22) are simple to evaluate numerically.

4   Discussion of Numerical Results

Numerical estimation for the upper (R_2 or T_2) and lower (R_1 or T_1) bounds of the reflection and transmission coefficients $|R|$ and $|T|$ are obtained for different values of the various parameters. In Table 1, the lower and upper bounds of $|R|$ for different values of various parameters are presented. From this table, it is seen that the two bounds of $|R|$ coincide upto 3–4 decimal places. Similar results are found for the two bounds for $|T|$ which are not given here. The average of an upper and lower bound of $|R|(|T|)$ thus produces fairly good numerical estimate for $|R|(|T|)$. The numerical results obtained by the present method satisfy the energy identity $|R|^2 + |T|^2 = 1$. This provides a check on the correctness of the results obtained here. Because of the energy identity, we confine our attention on $|R|$ only. In Fig. 1, $|R|$ is depicted against the wave number $Kb(= \frac{\sigma^2 b}{g})$ for different values of $\mu(= \frac{a}{b})$ and for normal incidence($\alpha = 0$). From Fig. 1, it is seen that the curve of $|R|$ corresponding to normal incidence almost coincides with the curve of $|R|$ in Fig. 2 of [1]. This provides another check on the correctness of the results obtained using the present method. Geometrical significance of the limiting case $\mu = 0$ is that the plate intersects the free surface. This indicates that submerged plate behaves like partially immersed barrier in deep water. For each finite μ, $|R|$ first increases to maximum as Kb increases and then decreases to zero for further increases of Kb. Thus, for each finite value of μ, $|R| \to 0$ as $Kb \to \infty$.

In Figs. 2 and 3, $|R|$ is platted against the wavenumber Kb for different incident angles and for $\mu = 0.05$ and 0.1, respectively. From these figures, it is seen that the curve of $|R|$ almost coincides with the curve of $|R|$ in Figs. 2 and 3 of [11]. Further,

Table 1 Lower and upper bounds of the reflection coefficient $|R|$ for various values of the parameters Kb, α and $\mu(= \frac{a}{b}) = 0.5$

Kb	$\alpha = 15^0$		$\alpha = 45^0$		$\alpha = 75^0$		$\alpha = 85^0$	
	R_1	R_2	R_1	R_2	R_1	R_2	R_1	R_2
0.05	0.000442	0.000442	0.000346	0.000372	0.000118	0.000119	0.000043	0.000046
0.4	0.017022	0.017051	0.012437	0.012473	0.004553	0.004563	0.001508	0.001552
0.8	0.037450	0.037481	0.027287	0.027352	0.009962	0.009981	0.003352	0.003377
1.6	0.045591	0.045599	0.032631	0.032669	0.001170	0.011733	0.003920	0.003991
2.4	0.032430	0.032445	0.022473	0.022680	0.007881	0.00794	0.002663	0.002699
3.0	0.021992	0.021999	0.014773	0.014796	0.005074	0.005290	0.001603	0.001692

Fig. 1 Graph of $|R|$ versus
Kb for different values of
$\mu(= a/b)$

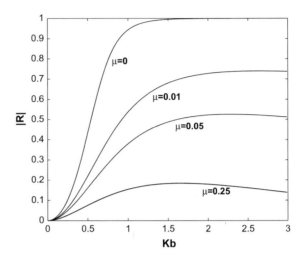

Fig. 2 Graph of $|R|$ versus
Kb for different values of α
and $\mu = 0.05$

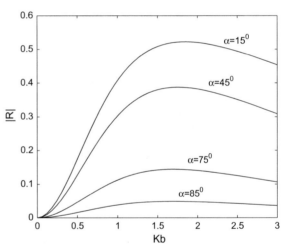

it is also observed that most of the cases Table 1 coincides upto 3 to 4 decimal places
with Table 1 of [11]. This provides another check on the correctness of the results
obtained using the present method. Reflection coefficient $|R|$ first increases as Kb
increases and then decreases for further increases of Kb in Figs. 2 and 3. Also, in
Fig. 4, the curve for $|R|$ depicted against α for different Kb and for $\mu = 0.5$. From
Fig. 4 and from Table 1, it is seen that for fixed μ, $|R|$ decreases as α increases from
$0°$ to $90°$. This is obvious since the incident wave then almost grazes along the plate.

Fig. 3 Graph of $|R|$ versus Kb for different values of α and $\mu = 0.1$

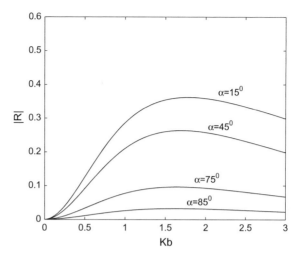

Fig. 4 Graph of $|R|$ versus α for different values of Kb and $\mu = 0.5$

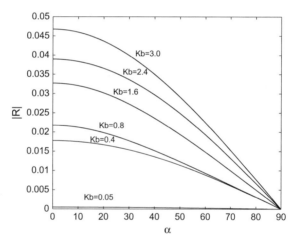

5 Conclusion

Here, we have used Havelock's expansion of water wave potential for the problem of water wave scattering by submerged plate to reduced the problem to the solution of pair of integral equations involving the difference of potentials and the horizontal component of velocity across the barriers. These integral equations are solved by using single-term Galerkin technique involving a *constant as basis*. Numerical evaluations of upper and lower bounds for the reflection coefficients are seen to be very close. Their averages give actual values of reflection coefficients for all practical purposes. The present method produces numerical results which are in good agreement with the earlier results obtained by [11].

Acknowledgements The first author acknowledges financial support from UGC, New Delhi. This work is also partially supported by SERB through the research project no. EMR/2016/005315.

References

1. Evans, D.V.: Diffraction of water waves by a submerged vertical plate. J. Fluid Mech. **40**, 433–451 (1970)
2. Evans, D.V., Morris, A.C.N.: The effect of a fixed vertical barrier on oblique incident surface waves in deep water. J.Inst. Maths. Applies, **9**, 198-204 (1972)
3. Faulkner, T.R.: The diffraction of an obliquely incident surface wave by a submerged plane barrier. ZAMP **17**, 699–707 (1965)
4. Faulkner, T.R.: The diffraction of an obliquely incident surface wave by a vertical barrier of finite depth. Proc. Camb. Phil. Soc. **62**, 829–38 (1966)
5. Havelock, T.H.: Forced surface waves on water. Phil. Mag. **8**, 569–576 (1929)
6. Jarvis, R.J., Taylor, B.S.: The scattering of surface waves by a vertical plane barrier. Proc. Camb. Phil. Soc. **66**, 417–22 (1969)
7. Mandal, B.N., Goswami, S.K.: A note on the scattering of surface wave obliquely incident on a submerged fixed vertical barrier. J. Phys. Soci. Jpn. **53**(9), 2980–2987 (1984a)
8. Mandal, B.N., Goswami, S.K.: A note on the diffraction of an obliquely incident surface wave by a partially immersed fixed vertical barrier. App. Sci. Res. **40**, 345–353 (1983)
9. Mandal, B.N., Goswami, S.K.: The scattering of an obliquely incident surface wave by a submerged fixed vertical plate. J. Math. Phys **25**, 1780–1783 (1984)
10. Mandal B.N., Dolai D.P.: Oblique water wave diffraction by thin vertical barrier in water of uniform finite depth. App. Ocean. Res. **16**, 195-203 (1994)
11. Mandal, B.N., Das, P.: Oblique diffraction of surface waves by a submerged vertical plate. J. Engng. Math. **30**, 459–470 (1996)

Chapter 19
A Note on Necessary Condition for L^p Multipliers with Power Weights

Rajib Haloi

Abstract In this article, we prove a necessary condition for L^p multipliers, $1 < p \leq 2$. The results are obtained by the use of Hausdorff–Young inequality that generalizes the result available for $p = 2$.

Keywords Fourier transform · Schwartz functions · A_p Weights
Hausdorff–Young inequality

AMS Subject Classification (2010): 42A38 · 26D15 · 42B10

1 Introduction

Let $L^p(\mathbb{R}, |x|^\alpha dx)$, $1 < p \leq 2$, $\alpha \geq 0$ denote the space of all measurable functions on $\mathbb{R}$ such that

$$\int_{\mathbb{R}} |f(x)|^p |x|^\alpha dx < \infty.$$

We prove a necessary condition for $L^p(\mathbb{R}, |x|^\alpha dx)$, $1 < p \leq 2$, $\alpha \geq 0$ multipliers. Let $S_{0,0}(\mathbb{R})$ be the space of all Schwartz functions whose Fourier transform has compact support not including the origin. We note that $S_{0,0}(\mathbb{R})$ is dense in $L^p(\mathbb{R}, |x|^\alpha dx)$ [see [3]]. For $f \in S_{0,0}(\mathbb{R})$, the multiplier operator is deined as

$$T_m(f) = (m\widehat{f})^\vee, \tag{1.1}$$

R. Haloi (✉)
Department of Mathematical Sciences, Tezpur University,
Napaam, Tezpur, Sonitpur 784028, Assam, India
e-mail: rajib.haloi@gmail.com

© Springer Nature Singapore Pte Ltd. 2018
D. Ghosh et al. (eds.), *Mathematics and Computing*, Springer Proceedings
in Mathematics & Statistics 253, https://doi.org/10.1007/978-981-13-2095-8_19

which has continuous extension to $L^p(\mathbb{R}, |x|^\alpha dx)$[2]. Here $\widehat{f}$ and $f^\vee$ denote the Fourier transform and the inverse Fourier transform of f, respectively. We begin with the definition of the multiplier space $M(s, \lambda)$ which is due to Strichartz [8].

Definition 1.1 Let $[\lambda]$ denote the greatest integer less than or equal to λ. For $\lambda > 0$ and $1 \leq s \leq \infty$, we define $M(s, \lambda)$ to be the set of all m with $[\lambda]$ weak derivatives on $\mathbb{R} \setminus \{0\}$ such that $B(m, s, \lambda) < \infty$, where

$$B(m, s, \lambda) = \|m\|_\infty + \sup_{r>0} r^{\lambda-1/s} \left(\int_{r \leq |x| \leq 2r} |m^{(\lambda)}(x)|^s dx \right)^{1/s}$$

if λ is an integer, and

$$B(m, s, \lambda) = \|m\|_\infty + \sup_{r>0} r^{\lambda-1/s} \left(\int_{r \leq |x| \leq 2r} \int_{r \leq |y| \leq 2r} \frac{|m^{(k)}(x) - m^{(k)}(y)|^s}{|x-y|^{1+p'(\lambda-k)}} dy dx \right)^{1/s}$$

if λ is not an integer with $k = [\lambda]$.

The characterization of the multiplier space for the multipliers defined on $L^2(\mathbb{R}, |x|^{2\alpha} dx)$ is done by Muckenhoupt et al. [3]. We state the following therorems due to Muckenhoupt et al. [3].

Theorem 1.1 [3] If $\alpha > \frac{1}{2}$ and 2α is not an odd integer and $m(x)$ is in $M(2, \alpha)$, then

$$\|(m\widehat{f})^\wedge\|_{2,2\alpha} \leq CB(m, 2, \alpha)\|f\|_{2,2\alpha}$$

for every $f \in S_{0,0}(\mathbb{R})$, where C depends only on α.

The sufficient part for $\alpha < -\frac{1}{2}$ is established by duality argument. Again for $0 < |\alpha| < \frac{1}{2}$, the characterization of the multipliers space is established in term of Reisz capacity [1]. Then, the characterization for $0 < |\alpha| < \frac{1}{2}$ is used to prove the remaining sufficiency part in [3].

Theorem 1.2 [3] If $\alpha \geq 0$, $m(x)$ is locally integrable on $\mathbb{R} \setminus \{0\}$ and

$$\|(m\widehat{f})^\wedge\|_{2,2\alpha} \leq A\|f\|_{2,2\alpha}$$

for all $f \in S_{0,0}(\mathbb{R})$, then m is in $M(2, \alpha)$ and there is a constant C, depending only on α, such that

$$B(m, 2, \alpha) \leq CA.$$

Further, Muckenhoupt et al. [5] proved the following sufficient condition that extends the results for the values of $p \neq 2$ in [3].

Theorem 1.3 [5] *Let* $1 < p \le 2$ *and* $\lambda > \frac{1}{p'}$. *If* $\alpha \in \mathbb{R}$ *such that* $-1 < \alpha < -1 + p(\lambda + \frac{1}{2})$, $(\alpha + 1)/p$ *is not an integer, then for* f *in* $S_{0,0}$ *and* $m \in M(p', \lambda)$, *we have*

$$\int_{-\infty}^{\infty} |(m\widehat{f})^{\vee}|^p |x|^\alpha dx \le CB(m, p', \lambda)^p \int_{-\infty}^{\infty} |f(x)|^p |x|^\alpha dx,$$

where C *is a constant independent of* m *and* f.

However, there is no known result for necessary conditions for multipliers space for $p \ne 2$ on $L^p(\mathbb{R}, |x|^\alpha dx)$. We prove the following necessary condition for the multiplier operator defined in (1.1) in terms of the space $M(p', \lambda)$.

Theorem 1.4 *If* $\lambda \ge 0$, $1 < p \le 2$, $m \in L^1_{loc}(\mathbb{R} - \{0\})$, *and*

$$\int_{-\infty}^{\infty} |(m\widehat{f})^{\vee}|^p |x|^{p\lambda} dx \le A^p \int_{-\infty}^{\infty} |f(x)|^p |x|^{p\lambda} dx \tag{1.2}$$

for all $f \in S_{0,0}$, *then* $m \in M(p', \lambda)$, *and there exists a constant* C *depending only on* p *and* λ *such that*

$$B(m, p', \lambda) \le CA.$$

2 Lemmas

In this section, we prove two important lemmas that are used to prove Theorem 1.4. The following lemma is analogous to a proposition by Stein [7, Proposition 4, page 139].

Lemma 2.1 *If* $f \in L^p(\mathbb{R})$ *for* $1 < p \le 2$, $\frac{1}{p} + \frac{1}{p'} = 1$, *and* $0 < \alpha < 1$, *then*

$$\left(\int_{\mathbb{R}} \int_{\mathbb{R}} \frac{|\widehat{f}(x) - \widehat{f}(y)|^{p'}}{|x - y|^{1+p'\alpha}} dy dx \right)^{1/p'} \le C \left(\int_{\mathbb{R}} |f(x)|^p |x|^{p\alpha} dx \right)^{1/p}$$

for some constant C *independent of* f.

Proof By a change of variable, we get

$$\left(\int_{\mathbb{R}} \int_{\mathbb{R}} \frac{|\widehat{f}(x) - \widehat{f}(y)|^{p'}}{|x - y|^{1+p'\alpha}} dy dx \right)^{1/p'} = \left(\int_{\mathbb{R}} |t|^{-(1+p'\alpha)} \left(\int_{\mathbb{R}} |\widehat{f}(y + t) - \widehat{f}(y)|^{p'} dy \right) dt \right)^{1/p'}.$$

Using the Hausdorff–Young inequality, we obtain

$$\left(\int_{\mathbb{R}} |t|^{-(1+p'\alpha)} \left(\int_{\mathbb{R}} |\widehat{f}(y+t) - \widehat{f}(y)|^{p'} dy\right) dt\right)^{1/p'}$$

$$\leq \left(\int_{\mathbb{R}} |t|^{-(1+p'\alpha)} \left(\int_{\mathbb{R}} |f(y)(e^{iyt}-1)|^p dy\right)^{p'/p} dt\right)^{1/p'}$$

$$= \left(\left(\int_{\mathbb{R}} |t|^{-(1+p'\alpha)} \left(\int_{\mathbb{R}} |f(y)(e^{iyt}-1)|^p dy\right)^{p'/p} dt\right)^{p/p'}\right)^{1/p}$$

$$= \left(\left(\int_{\mathbb{R}} \left(\int_{\mathbb{R}} \frac{|f(y)(e^{iyt}-1)|^p}{|t|^{(1+p'\alpha)p/p'}} dy\right)^{p'/p} dt\right)^{p/p'}\right)^{1/p}.$$

Again applying the Minkowski's integral inequality, we obtain

$$\left(\int_{\mathbb{R}} |t|^{-(1+p'\alpha)} \left(\int_{\mathbb{R}} |\widehat{f}(y+t) - \widehat{f}(y)|^{p'} dy\right) dt\right)^{1/p'}$$

$$\leq \left(\int_{\mathbb{R}} \left(\int_{\mathbb{R}} \frac{|f(y)(e^{iyt}-1)|^{p'}}{|t|^{(1+p'\alpha)}} dt\right)^{p/p'} dy\right)^{1/p}$$

$$= \left(\int_{\mathbb{R}} |f(y)|^p dy \left(\int_{\mathbb{R}} \frac{|(e^{iyt}-1)|^{p'}}{|t|^{1+p'\alpha}} dt\right)^{p/p'}\right)^{1/p}.$$

We note that

$$\left(\int_{\mathbb{R}} \frac{|(e^{iyt}-1)|^{p'}}{|t|^{1+p'\alpha}} dt\right)^{p/p'} = C|y|^{p\alpha}$$

for $0 < \alpha < 1$, $y \in \mathbb{R}$ and for some constant C [7, page 140]. Thus

$$\left(\int_{\mathbb{R}} \int_{\mathbb{R}} \frac{|\widehat{f}(x) - \widehat{f}(y)|^{p'}}{|x-y|^{1+p'\alpha}} dy dx\right)^{1/p'} \leq C \left(\int_{\mathbb{R}} |f(y)|^p |y|^{p\alpha} dy\right)^{1/p}.$$

$\square$

Lemma 2.2 *If $f \in L^p(\mathbb{R})$ for $1 < p \leq 2$, k is a nonnegative integer, $\widehat{f}$ has a weak derivative of order k on $\mathbb{R}$ and $\widehat{f}^{(k)} \in L^{p'}(\mathbb{R})$; and $k < \alpha < k+1$, then*

$$\left(\int_{\mathbb{R}}\int_{\mathbb{R}}\frac{|\widehat{f^{(k)}}(x)-\widehat{f^{(k)}}(y)|^{p'}}{|x-y|^{1+p'(\alpha-k)}}dydx\right)^{1/p'}\leq C\left(\int_{\mathbb{R}}|f(x)|^p|x|^{p\alpha}dx\right)^{1/p},$$

for some constants C, $\frac{1}{p}+\frac{1}{p'}=1$.

Proof Using the following property of the Fourier transform $\widehat{f^{(k)}}(\xi)=[\widehat{(-ix)^k f}](\xi)$ a.e. ξ, the proof can be obtained from Lemma 2.1. □

3 Proof of the Main Results

In this section, we complete the proof of the Theorem 1.4 by proving a sequence of Lemmas. The idea of the proof is based on Muckenhoupt et al. [3].

Lemma 3.1 *If we assume (1.2), then there exists a constant C depending only on p and λ such that $\|m\|_\infty \leq CA$.*

Proof We choose $\phi \in C_c^\infty(\mathbb{R})$ with $\phi(x) \geq 0$, $\int \phi(x)dx = 1$; and

$$\phi(x) = 1, \forall |x| \leq \frac{1}{4},$$

$$\phi(x) = 0, \forall |x| \geq \frac{1}{2}.$$

It is given that m is locally integrable on $\mathbb{R} \setminus \{0\}$. Thus, a.e. $x \neq 0 \in \mathbb{R}$ is a Lebesgue point for m. Let $y \neq 0$ be a Lebesgue point for m. Let r be fix number such that $0 < r < \frac{|y|}{2}$. Define

$$\hat{f}(t) = \frac{1}{r}\phi(\frac{t-y}{r}).$$

Then, $f(t) = e^{iyt}\check{\phi}(rt)$ and $f \in S_{0,0}$. Next, we claim for this f that

$$|(m\hat{f})^\vee(x)| \geq |m(y)| \text{ a.e. } y.$$

Now for a.e. y,

$$|(m\hat{f})^\vee(x)| \geq \left|\int_{\mathbb{R}} m(t)\hat{f}(t)e^{ixt}dt\right| - \left|\int_{\mathbb{R}} m(t)\hat{f}(t)(e^{iyt}-e^{ixt})dt\right|$$

$$\geq \left|\int_{\mathbb{R}} m(t)\hat{f}(t)dt\right| - \int_{\mathbb{R}} |m(t)||\hat{f}(t)||x||y-t|dt$$

$$\geq \left|\int_{\mathbb{R}} m(t)\hat{f}(t)dt\right| - 1/8\int_{\mathbb{R}} |m(t)||\hat{f}(t)|dt \qquad (3.1)$$

if t is such that $\hat{f}(t) \neq 0$, that is if $|t - y| < r/2$, and if $|x| \leq 1/4r$. Further, we choose $\phi \in C_c^\infty(\mathbb{R})$ such that

$$\left| \frac{1}{r} \int_{\mathbb{R}} m(t)\phi(\frac{t-y}{r})dt \right| \geq \frac{|m(y)|}{2}, \tag{3.2}$$

and

$$\frac{1}{r} \int_{\mathbb{R}} |m(t)||\phi(\frac{t-y}{r})dt| \leq 2|m(y)|. \tag{3.3}$$

Using (3.2) and (3.3) in (3.1), we get

$$|(m\hat{f})^\vee(x)| \geq |m(y)|/4 \tag{3.4}$$

a.e. s with $|x| \leq 1/4r$. Integrating for $|x| \leq 1/4r$, we get from (3.4) that

$$\int_{|x| \leq 1/4r} |m(y)/4|^p |x|^{p\lambda} dx \leq \int_{-\infty}^{\infty} |(m\hat{f})^\vee|^p |x|^{p\lambda} dx$$
$$\leq A \int_{-\infty}^{\infty} |f(x)|^p |x|^{p\lambda} dx$$
$$= A \int_{-\infty}^{\infty} |\check{\phi}(rx)|^p |x|^{p\lambda} dx$$
$$= ADr^{-1-p\lambda},$$

where $D^p = \int_{\mathbb{R}} |\check{\phi}(u)|^p |u|^{p\lambda} du < \infty$ as $\phi \in S$. Thus, we obtain

$$|m(y)| \leq CA$$

with C depending only on p and λ. Thus, the proof follows. $\qquad\square$

Lemma 3.2 *If we assume (1.2), then m has kth weak derivative and $m^{(k)} \in L^{p'}(I)$, where I is any compact interval in $\mathbb{R}$ not containing 0.*

Proof Let $k = [\lambda]$. For the existence of the weak derivative of m, we must show that there exist h with

$$\int m(t)\psi^{(k)}(t)dt = \int h(t)\psi(t)dt$$

for all $\psi \in C_c^\infty(\mathbb{R} - \{0\})$. Choose a sequence $\{f_n\}$ in $S_{0,0}$ such that $\hat{f}_n(\xi) = 1$ for $1/n \leq |\xi| \leq n$. Let $\psi \in C_c^\infty(\mathbb{R} - \{0\})$ and $n \in \mathbb{N}$ such that supp(ψ) is a subset of $\{1/n \leq |x| \leq n\}$. Now

$$\int m\psi^{(k)} = \int m\widehat{f_n}\psi^{(k)}$$

$$= \int (m\widehat{f_n})^\vee (\psi^{(k)})^\vee$$

$$= \int (m\widehat{f_n})^\vee (-it)^k \psi^\vee)$$

$$= \int [(m\widehat{f_n})^\vee (-it)^k]^\wedge \psi.$$

Define

$$h_n(x) = [(m\widehat{f_n})^\vee (-it)^k]^\wedge (x).$$

By the Hausdorff–Young inequality

$$\left(\int |h_n(t)|^{p'} dt \right)^{1/p'} = \left(\int |[(m\widehat{f_n})^\vee (-ix)^k]^\wedge (t)|^{p'} dt \right)^{1/p'}$$

$$\leq \left(\int |(m\widehat{f_n})^\vee (t)(-it)^k|^p dt \right)^{1/p}. \tag{3.5}$$

Now

$$\int |(m\widehat{f_n})^\vee (-it)^k|^p dt \leq \int_{|t| \geq 1} |(m\widehat{f_n})^\vee (-it)^\lambda|^p dt + \int_{|t| \leq 1} |(m\widehat{f_n})^\vee|^p dt.$$

The first term in the last inequality is finite by the hypothesis. The integrand in the second term is in S, so it follows from inequality (3.5) that $h_n \in L^{p'}$. For $m > n$, $\widehat{f_m}(x) = \widehat{f_n}(x)$, so we have

$$0 = \int_{\mathbb{R}} [m\widehat{f_m} - m\widehat{f_n}]\psi^{(k)}$$

$$= \int_{\mathbb{R}} [(m\widehat{f_m})^\vee - (m\widehat{f_n})^\vee](-ix)^k \psi^\vee$$

$$= \int_{\mathbb{R}} [h_m - h_n]\psi,$$

which is true for all $\psi \in C_c^\infty(\mathbb{R} - \{0\})$ with supp(ψ) is a subset of $\{1/n \leq |x| \leq n\}$. This implies that a.e. $x \in \{x : 1/n \leq |x| \leq n\}$,

$$h_m(x) = h_n(x),$$

and hence, $\{h_n\}$ is Cauchy sequence in $L^{p'}$. Thus, $\{h_n\}$ has convergent subsequence which converges a.e. We call the subsequence as $\{h_n\}$. So

$$h(x) = \lim h_n(x)$$

is defined a.e. $x \in \mathbb{R}$ and $h(x) = h_n(x)$ a.e. $x \in \{x : 1/n \le |x| \le n\}$.

Now we show that h is weak derivative of m of order k and $h \in L^{p'}(I)$ for compact interval I in $\mathbb{R}$ not containing 0. For $\psi \in C_c^\infty(\mathbb{R} - \{0\})$ with $\mathrm{supp}(\psi)$ is a subset of $\{1/n \le |x| \le n\}$, we have

$$\int_{\mathbb{R}} h\psi = \int_{\mathbb{R}} h_n \psi$$
$$= \int_{\mathbb{R}} [m\widehat{f_n}]^\vee (-it)^k]^\wedge \psi$$
$$= \int_{\mathbb{R}} m\psi^{(k)}.$$

Thus, $m^{(k)}(x) = (-1)^k h(x)$. Next, let I be any compact interval not containing 0. Choose $n \in \mathbb{N}$ such that $I \subseteq \{x : 1/n \le |x| \le n\}$. Then,

$$\int_I |h|^{p'} \le \int_{\{x:1/n \le |x| \le n\}} |h|^{p'}$$
$$\le \int_{\mathbb{R}} |h_n|^{p'} < \infty.$$

This shows that the kth order derivative of m is in $L^{p'}$ on any compact interval not containing the origin. $\square$

Lemma 3.3 *If we assume (1.2), then there is a constant C_1 depending only on p and λ such that*

$$B(m, p', \lambda) \le C_1 A.$$

Proof Because of Lemma 3.1, it is enough to show that there exists a constant A independent of r such that

$$r^{\lambda - 1/p'} \left(\int_{r \le |x| \le 2r} |m^{(\lambda)}(x)|^{p'} dx \right)^{1/p'} \le CA$$

for λ integer and

$$r^{\lambda-1/p'} \left(\int_{r\leq|x|\leq 2r} \int_{r\leq|y|\leq 2r} \frac{|m^{(k)}(x) - m^{(k)}(y)|^{p'}}{|x-y|^{1+p'(\lambda-k)}} dy dx \right)^{1/p'} \leq CA$$

for λ non-integer with $k = [\lambda]$. We choose $\eta \in C_c^\infty$ such that

$$\eta(x) = 1, \forall |x| \leq 3/2,$$
$$\eta(x) = 0, \forall |x| \geq 7/4.$$

For fix $r > 0$, define f such that

$$\widehat{f}(x) = \eta(\frac{x}{r} - 2) + \eta(\frac{x}{r} + 2)$$

and so

$$f(x) = (e^{2irx} + e^{-2irx})\widecheck{\eta}(rx).$$

Then, $\widehat{f}(x) = 1$ on $r/2 \leq |x| \leq 3r$ and $\widehat{f}(x) = 0$ on $|x| \leq r/4$ and $|x| \geq 4r$. So,

$$[m(x)\widehat{f}(x)]^{(k)} = m^{(k)}(x), \tag{3.6}$$

for a.e. $x \in \{x : r/2 \leq |x| \leq 3r\}$. We first estimate $B(m, p', \lambda)$ for integer λ. By the Hausdorff–Young inequality and the Assumption (1.2), we have

$$r^{\lambda-1/p'} \left(\int_{r\leq|x|\leq 2r} |m^{(\lambda)}(x)|^{p'} dx \right)^{1/p'} = r^{\lambda-1/p'} \left(\int_{r\leq|x|\leq 2r} |(m\widehat{f})^{(\lambda)}(x)|^{p'} dx \right)^{1/p'}$$

$$\leq r^{\lambda-1/p'} \left(\int_{\mathbb{R}} |(m\widehat{f})^\vee(x)|x|^\lambda|^p dx \right)^{1/p}$$

$$\leq Ar^{\lambda-1/p'} \left(\int_{\mathbb{R}} |f(x)|^p |x|^{p\lambda} dx \right)^{1/p}$$

$$\leq AD_p,$$

where $D_p = r^{\lambda-1/p'} \left(\int_{\mathbb{R}} |f(x)|^p |x|^{p\lambda} dx \right)^{1/p}$. For λ non-integer, we use (3.6) and Lemma 2.2 to obtain

$$r^{\lambda-1/p'}\left(\int_{r\le|x|\le 2r}\int_{r\le|y|\le 2r}\frac{|m^{(k)}(x)-m^{(k)}(y)|^{p'}}{|x-y|^{1+p'(\lambda-k)}}dydx\right)^{1/p'}$$

$$=r^{\lambda-1/p'}\left(\int_{r\le|x|\le 2r}\int_{r\le|y|\le 2r}\frac{|(m\hat{f})^{(k)}(x)-(m\hat{f})^{(k)}(y)|^{p'}}{|x-y|^{1+p'(\lambda-k)}}dydx\right)^{1/p'}$$

$$\le r^{\lambda-1/p'}\left(\int_{\mathbb{R}}\int_{\mathbb{R}}\frac{|(m\hat{f})^{(k)}(x)-(m\hat{f})^{(k)}(y)|^{p'}}{|x-y|^{1+p'(\lambda-k)}}dydx\right)^{1/p'}$$

$$\le Cr^{\lambda-1/p'}\left(\int_{\mathbb{R}}|(m\hat{f})^{\vee}|^p|x|^{p\lambda}dx\right)^{1/p}$$

$$\le ACr^{\lambda-1/p'}\left(\int_{\mathbb{R}}|f(x)|^p|x|^{p\lambda}dx\right)^{1/p}$$

$$=ACD_p.$$

Here, the second inequality follows from the Lemma 2.1 and third inequality follows from the hypothesis. $\qquad\square$

4 Remark

We note that the condition in Theorem 1.4 cannot be sufficient. We recall the following results for the sufficient condition established by Muckenhoupt et al. [5] for $L^p(\mathbb{R})$ multipliers with power weights.

Theorem 4.1 [5] *If* $1<p<\infty$, $1\le s\le\infty$, $\lambda>\max(\frac{1}{s},|\frac{1}{p}-\frac{1}{2}|)$ *or* $\lambda=s=1$, $m\in m(s,\lambda)$, $\max(-1,-p\lambda,-1+p(-\lambda+\frac{1}{2}))<\alpha<\min(p\lambda,-1+p(\lambda+\frac{1}{2})$, $-1+p(\lambda+1-\frac{1}{s}))$ *and* $(\alpha+1)/p$ *is not an integer, then for* $f\in S_{0,0}$, *we have*

$$\int_{-\infty}^{\infty}|(m\hat{f})^{\vee}|^p|x|^{\alpha}dx\le CB(m,s,\lambda)^p\int_{-\infty}^{\infty}|f(x)|^p|x|^{\alpha}dx,$$

where C is a constant independent of m and f.

Further Muckenhoupt [6] proved the following necessity conditions for L^p multipliers with power weights on α.

Theorem 4.2 [6] *If* $1<p<\infty$, $1\le s\le\infty$, $\lambda\ge\frac{1}{s}$ *and assume*

$$\int_{-\infty}^{\infty}|(m\hat{f})^{\vee}|^p|x|^{\alpha}dx\le C\int_{-\infty}^{\infty}|f(x)|^p|x|^{\alpha}dx,$$

for all $m \in M(s, \lambda)$ and $f \in S_{0,0}$, then

(1) $\alpha > -1$,
(2) $\max(-p\lambda, -1 + p(-\lambda + \frac{1}{2})) \leq \alpha \leq \min(p\lambda, -1 + p(\lambda + \frac{1}{2}), -1 + p(\lambda + 1 - \frac{1}{s}))$,
(3) $(\alpha + 1)/p$ is not an integer.

It is clear that for $1 < p \leq 2$ and $\lambda \geq 1/p'$, the result (3) of Theorem 4.2 is not satisfied. So only possibility is for $0 < \lambda < 1/p'$ which is the A_p range of the weight. As the Hilbert transform is bounded, so in this case there are non-constants multipliers [3, page 183]. Thus, we conclude that Theorem 1.4 cannot be sufficient.

Acknowledgements The author would like to thank Professor Parasar Mohanty and Professor Sobha Madan for technical discussion with them. The author acknowledges the financial support by AICTE-NEQIP, Tezpur University . The author would like to acknowledge the excellent facilities of Indian Institute of Technology Kanpur that are availed during the preparation of the article.

References

1. Dahlberg, B.J.: Regularity properties of Riesz potentials. Indiana Univ. Math. J. **28**(2), 257–268 (1979)
2. Grafakos, L.: Classical and Modern Fourier Analysis. Pearson Education, New Jersey (2004)
3. Muckenhoupt, B., Hunt, R., Wheeden, R.: L^2 multipliers with power weights. Adv. Math. **49**, 170–216 (1983)
4. Muckenhoupt, B., Young, W.S.: L^p multipliers with weights $|x|^{kp-1}$. Trans. Amer. Math. Soc. **275**(2), 623–639 (1983)
5. Muckenhoupt, B., Wheeden, R., Young, W.S.: Sufficiency conditions for L^p multipliers with power weights. Trans. Amer. Math. Soc. **300**(2), 433–461 (1987)
6. Muckenhoupt, B.: Necessity conditions for L^p multipliers with power weights. Trans. Amer. Math. Soc. **300**(2), 503–520 (1987)
7. Stein, E.M.: Singular Integrals and Differentiability Properties of Functions. Princeton University Press, Princeton (1970)
8. Strichartz, R.S.: Multipliers for sperical harmonic expansions. Trans. Amer. Math. Soc. **167**, 115–124 (1972)

Chapter 20
On $M/G^{(a,b)}/1/N$ Queue with Batch Size- and Queue Length-Dependent Service

G. K. Gupta and A. Banerjee

Abstract In this paper, we analyze finite buffer $M/G/1$ queue where service is offered in groups/batches according to 'general bulk service' rule by a single server. The service time distribution is considered to be generally distributed and allowed to change dynamically depending on the batch size under service and queue length just before the service initiation of the batch under consideration. Using the embedded Markov chain technique and supplementary variable technique, we obtain the joint distribution of the queue length and batch size at various epoch. At the end, we present several numerical results in the form of self-explanatory table and graphs to bring out some interesting features of the model.

Keywords Batch size-dependent queue · Finite buffer · Queue length-dependent queue · Supplementary variable technique · Embedded Markov chain technique General bulk service rule

1 Introduction

In asynchronous transfer mode (ATM) networks, based on continuous-bit-rate (CBR) traffic services, packetized voice or video samples are transmitted over the communication channel. An important issue, called 'congestion', may arise frequently in packet-switched network. Generally, a high-rate traffic flow causes congestion to the system and degrades the performance of the system significantly. The mechanisms of congestion control prevent congestion of the system either, before it happens, or remove congestion, after it has happened. A significant amount of literature on queuing study is found to be focused on congestion control mechanism to regulate service

G. K. Gupta (✉) · A. Banerjee
Department of Mathematical Sciences, Indian Institute
of Technology (BHU) Varanasi, Varanasi 221005, Uttar Pradesh, India
e-mail: gopalgupta.bhu90@gmail.com; gopal.rs.apm13@iitbhu.ac.in

A. Banerjee
e-mail: anuradha.mat@iitbhu.ac.in

© Springer Nature Singapore Pte Ltd. 2018
D. Ghosh et al. (eds.), *Mathematics and Computing*, Springer Proceedings
in Mathematics & Statistics 253, https://doi.org/10.1007/978-981-13-2095-8_20

(transmission) rates in communication networks; see [1, 2] and reference therein. In queueing literature, congestion control mechanism is achieved by controlling either arrival rates or service rates or both.

Queueing models are useful in different real-world practical management situations to optimize the total system cost by keeping the QoS. Bulk service queuing systems have wide applications such as in transportation system, manufacturing systems, computer networking systems, telecommunication systems. Over the past decades, the bulk service queueing system is focused by the researchers; see, e.g., [3–5]. It should be noted here that the service time of a batch, containing more customers, should be longer than the batches with lesser number of customers within it. Therefore, assuming the service time of the batch to be dependent on the batch size is more applicable in describing the congestion control mechanism of the bulk service queue. The batch size-dependent service queue has been recently studied in [6–11], etc. However, the queue length dependency along with the batch size-dependent service rate (time) may increase the system's productivity in terms of decreasing in blocking probability.

In view of the above discussion, we come to the conclusion that there are less number of literature found in bulk service queue where batch size as well as queue length-dependent service rate has been considered (see [12]). To support the raising interest in the study of batch size-dependent bulk service queueing models together with queue length-dependent service, this paper devotes our current work. We consider a finite buffer $M/G^{(a,b)}/1$ queue, where a single server serves a group of customers of varying batch size according to GBS rule, the server changes its service times (rates) only at the beginning of the service depending on the number of customers taken for service, i.e., batch size under service, as well as on the number of remaining customers left in the queue, i.e., queue length. We analyze our model using the supplementary variable technique and the embedded Markov chain technique. The former one is used to develop a relation between joint distribution of queue content and serving batch size at departure epoch and arbitrary epoch, and the latter one is used to obtain the joint distribution of queue content and serving batch size at departure epoch.

The outline of the rest of this paper is as follows: after giving the formal description of the model in Sect. 2, in Sect. 2.1 we obtain the joint distribution of queue content and serving batch size at departure epoch by using the embedded Markov chain technique. Using the supplementary variable technique, we obtain a relation between departure epoch and arbitrary epoch joint distributions of queue content and serving batch size in Sect. 2.2. Section 3 is assigned to present for the various performance measures. Several numerical examples are presented in Sect. 4. Some conclusions are drawn in Sect. 5 followed by the references.

2   Model Description and Steady-State Analysis

We consider a bulk service queue with single server and buffer size is finite. The customers' arrival follows the Poisson process with parameter λ and service is provided in batches according to the "GBS" rule. That is, when queue length is less than 'a' (≥ 1), server waits till the queue length reaches 'a' and then initiates service for that group of 'a' customers. However, if the queue length is greater than or equal to 'a' and less than or equal to 'b', the entire group of 'r' ($a \leq r \leq b$) customers are served at a time. However, when the queue length is greater than 'b', then server serves first 'b' customers for service and rest of them will have to wait in the queue. Further, it is assumed that the queue size is fixed to N ($> b$). The service time of the batches is considered to be dependent on the size of the batch taken for service as well as on the queue length (excluding the batch with the server) at the beginning of the service. Let $S_{n,r}(t)$ ($a \leq r \leq b$, $0 \leq n \leq N$) denote the service time distribution with probability density function (pdf) $s_{n,r}(.)$, LST $s_{n,r}^*(.)$ and mean service time $\tilde{s}_{n,r}$ (rate $\mu_{n,r} = 1/\tilde{s}_{n,r}$). Now, define the state space of the system under consideration at time t as follows.

– $N_q(t)$: the queue length at time t.
– $N_s(t)$: the server content at time t when server is busy.
– $U(t)$: the remaining service time of a batch of customers under service, if any.

The state probabilities, at time t, are defined as follows:

– $P_{n,0}(t) \equiv prob.\{N_q(t) = n,\ N_s(t) = 0\};\quad 0 \leq n \leq a - 1,$
– $P_{n,r}(x, t)dx \equiv prob.\{N_q(t) = n,\ N_s(t) = r,\ x \leq U(t) \leq x + dx\};\quad 0 \leq n \leq N,$
  $a \leq r \leq b, x \geq 0,$

The corresponding steady-state probabilities are defined as follows:

$$\lim_{t \to \infty} P_{n,0}(t) = P_{n,0};\ 0 \leq n \leq a - 1,$$
$$\lim_{t \to \infty} P_{n,r}(x, t) = P_{n,r}(x);\ 0 \leq n \leq N,\ a \leq r \leq b.$$

Our main objective is to obtain the joint distribution of the queue content as well as server content at departure epoch and arbitrary epoch. In the following sections, we will proceed to obtain the required distributions by using the embedded Markov chain technique and supplementary variable technique.

2.1   Joint Probability Distribution at Departure Epoch

This section devotes to obtain the steady-state joint probability distribution of the queue length and number of customers with the serving batch at departure epoch. Now by observing the state of the system at two consecutive batch departure epochs, we obtain a two-dimensional Markov chain with state space

$\{(n, r) : \ 0 \le n \le N, \ a \le r \le b\}$ (see [6]). The corresponding one-step transition probability matrix (TPM) $\mathscr{P} = [p_{i,j}]$ of dimension $(N + 1)(b - a + 1)$, where each $p_{i,j}$ is a matrix of dimension $(b - a + 1) \times (b - a + 1)$, is given by

$$\mathscr{P} = \begin{array}{c} \\ \\ 0 \\ \vdots \\ a-1 \\ a \\ a+1 \\ \vdots \\ b \\ b+1 \\ \vdots \\ N \end{array} \overset{\begin{array}{ccccccc} 0 & 1 & \dots & N-b-1 & N-b & \dots & N-1 & N \end{array}}{\begin{pmatrix} D_0^{(0,1)} & D_1^{(0,1)} & \dots & D_{N-b-1}^{(0,1)} & D_{N-b}^{(0,1)} & \dots & D_{N-1}^{(0,1)} & \bar{D}_N^{(0,1)} \\ \vdots & \vdots & \ddots & \vdots & \vdots & \ddots & \vdots & \vdots \\ D_0^{(0,1)} & D_1^{(0,1)} & \dots & D_{N-b-1}^{(0,1)} & D_{N-b}^{(0,1)} & \dots & D_{N-1}^{(0,1)} & \bar{D}_N^{(0,1)} \\ D_0^{(0,1)} & D_1^{(0,1)} & \dots & D_{N-b-1}^{(0,1)} & D_{N-b}^{(0,1)} & \dots & D_{N-1}^{(0,1)} & \bar{D}_N^{(0,1)} \\ D_0^{(0,2)} & D_1^{(0,2)} & \dots & D_{N-b-1}^{(0,1)} & D_{N-b}^{(0,2)} & \dots & D_{N-1}^{(0,1)} & \bar{D}_N^{(0,2)} \\ \vdots & \vdots & \ddots & \vdots & \vdots & \ddots & \vdots & \vdots \\ D_0^{(0,b-a+1)} & D_1^{(0,b-a+1)} & \dots & D_{N-b-1}^{(0,b-a+1)} & D_{N-b}^{(0,b-a+1)} & \dots & D_{N-1}^{(0,b-a+1)} & \bar{D}_N^{(0,b-a+1)} \\ 0 & D_0^{(1,b-a+1)} & \dots & D_{N-b-2}^{(1,b-a+1)} & D_{N-b-1}^{(1,b-a+1)} & \dots & D_{N-2}^{(1,b-a+1)} & \bar{D}_{N-1}^{(1,b-a+1)} \\ \vdots & \vdots & \ddots & \vdots & \vdots & \ddots & \vdots & \vdots \\ 0 & 0 & \dots & 0 & D_0^{(N-b,b-a+1)} & \dots & D_{b-1}^{(N-b,b-a+1)} & \bar{D}_b^{(N-b,b-a+1)} \end{pmatrix}}$$

here, each 0 and $D_j^{(n,i)}$ in $\mathscr{P}$ are the matrices of dimension $(b - a + 1)$ and are described as follows:

$D_j^{(0,i)} = e_i^T \otimes \kappa_j^{(0,i+a-1)}, \ 1 \le i \le b - a + 1, \ 0 \le j \le N - 1,$

$D_j^{(n,b-a+1)} = e_{b-a+1}^T \otimes \kappa_j^{(n,b)}, \ 1 \le n \le N - b, \ 0 \le j \le N - n - 1,$

$\bar{D}_N^{(0,i)} = e_i^T \otimes \bar{\kappa}_{N-1}^{(0,i+a-1)}, \ 1 \le i \le b - a,$

$\bar{D}_j^{(n,b-a+1)} = e_{b-a+1}^T \otimes \bar{\kappa}_{j-1}^{(n,b)}, \ b \le j \le N, \ 0 \le n \le N - b, \ n + j = N.$

where

- e_i is a column vector of dimension $(b - a - 1)$ with 1 at the ith position and 0 elsewhere.
- $\kappa_j^{(n,r)}$ is a column vector of dimension $(b - a - 1)$, consisting of $\xi_j^{(n,r)}$'s, and $\xi_j^{(n,r)}$'s are the probabilities of j arrival during the service period of a batch of size r $(a \le r \le b)$ servicing with service time distribution $S_{n,r}(.)$ and is obtained as

$$\xi_j^{(n,r)} = \begin{cases} \int_0^\infty \frac{e^{-\lambda t}(\lambda t)^j}{j!} dS_{0,r}(t), & a \le r \le b - 1, \\ \int_0^\infty \frac{e^{-\lambda t}(\lambda t)^j}{j!} dS_{n,b}(t), & r = b, \ 0 \le n \le N - b. \end{cases}$$

- $\bar{\kappa}_j^{(n,r)}$ is a column vector of dimension $(b - a - 1)$ consisting of $\left(1 - \sum_{i=0}^j \xi_i^{(n,r)}\right)$'s.

Let us now define the departure epoch joint probability as $p_{n,r}^+$ $(0 \le n \le N, a \le r \le b)$, which represents that there are n customers are left in the queue at departure epoch of a batch of size r. Then, $p_{n,r}^+$ can be determined by solving the system of equations $\pi \mathscr{P} = \pi$, where $\pi = (\pi_0^+, \pi_1^+, ..., \pi_N^+)$ and each π_n^+ $(0 \le n \le N)$ is a row vector of order $(b - a + 1)$, and is given by $\pi_n^+ = (p_{n,a}^+, p_{n,a+1}^+, ..., p_{n,b}^+)$. Once we obtain the joint probabilities $p_{n,r}^+$ $(0 \le n \le N, a \le r \le b)$, the marginal distribution of the queue length at departure epoch, represented by p_n^+, is obtained

as $p_n^+ = \sum_{r=a}^b p_{n,r}^+, 0 \le n \le N.$

2.2 Joint Probability Distribution at Arbitrary Epoch

This section devotes to obtain the joint distribution of the queue content and server content at an arbitrary epoch. Toward this end, we will first obtain the governing equations of the system in steady state.

Relating the state of the system at time t and $t + dt$, we find the Kolmogrov equations of the system, in steady state, as follows.

$$0 = -\lambda P_{0,0} + \sum_{r=a}^{b} P_{0,r}(0), \tag{1}$$

$$0 = -\lambda P_{n,0} + \lambda P_{n-1,0} + \sum_{r=a}^{b} P_{n,r}(0), \quad 1 \leq n \leq a - 1, \tag{2}$$

$$-\frac{\partial P_{0,a}(x)}{\partial u} = -\lambda P_{0,a}(x) + \lambda P_{a-1,0} s_{0,a}(x) + \sum_{r=a}^{b} P_{a,r}(0) s_{0,a}(x), \tag{3}$$

$$-\frac{\partial P_{0,r}(x)}{\partial u} = -\lambda P_{0,r}(x) + \sum_{k=a}^{b} P_{r,k}(0) s_{0,r}(x), \quad a + 1 \leq r \leq b, \tag{4}$$

$$-\frac{\partial P_{n,r}(x)}{\partial u} = -\lambda P_{n,r}(x) + \lambda P_{n-1,r}(x), \quad a \leq r \leq b - 1, \ 1 \leq n \leq N - 1, \tag{5}$$

$$-\frac{\partial P_{n,b}(x)}{\partial u} = -\lambda P_{n,b}(x) + \lambda P_{n-1,b}(x) + \sum_{r=a}^{b} P_{n+b,r}(0) s_{n,b}(x), \quad 1 \leq n \leq N - b, \tag{6}$$

$$-\frac{\partial P_{n,b}(x)}{\partial u} = -\lambda P_{n,b}(x) + \lambda P_{n-1,b}(x), \quad N - b + 1 \leq n \leq N - 1, \tag{7}$$

$$-\frac{\partial P_{N,r}(x)}{\partial u} = \lambda P_{N-1,r}(x), \quad a \leq r \leq b, \tag{8}$$

We may note here that the joint probabilities $p_{n,r}^{+}$ and $P_{n,r}(0)$ are proportional to each other and hence can be written as

$$p_{n,r}^{+} = \sigma P_{n,r}(0), \quad 0 \leq n \leq N, \ a \leq r \leq b, \tag{9}$$

where σ is the proportionality constant and its value is obtained in Lemma 1.

Lemma 1 *The value of the proportionality constant σ, as appeared in (9), is given by*

$$\sigma^{-1} = \sum_{n=0}^{N} \sum_{r=a}^{b} P_{n,r}(0) = g^{-1} \left(1 - \sum_{n=0}^{a-1} P_{n,0} \right), \tag{10}$$

where $g = \tilde{s}_{0,a} \sum_{n=0}^{a} p_n^+ + \sum_{n=a+1}^{b} p_n^+ \tilde{s}_{0,n} + \sum_{n=b+1}^{N} p_n^+ \tilde{s}_{n-b,b}.$

Proof Summing both sides of (9) over the range of r and n and using the result $\sum_{n=0}^{N} p_n^+ = 1$, we obtain $\sigma^{-1} = \sum_{n=0}^{N} \sum_{r=a}^{b} P_{n,r}(0).$

Now multiplying (3)–(8) by $e^{-\theta x}$ and integrating with respect to x from 0 to ∞, we obtain

$$(\lambda - \theta) P_{0,a}^*(\theta) = \lambda P_{a-1,0} s_{0,a}^*(\theta) + \sum_{r=a}^{b} P_{a,r}(0) s_{0,a}^*(\theta) - P_{0,a}(0), \tag{11}$$

$$(\lambda - \theta) P_{0,r}^*(\theta) = \sum_{k=a}^{b} P_{r,k}(0) s_{0,r}^*(\theta) - P_{0,r}(0); \quad a+1 \le r \le b, \tag{12}$$

$$(\lambda - \theta) P_{n,r}^*(\theta) = \lambda P_{n-1,r}^*(\theta) - P_{n,r}(0); \quad a \le r \le b-1, \ 1 \le n \le N-1, \tag{13}$$

$$(\lambda - \theta) P_{n,b}^*(\theta) = \lambda P_{n-1,b}^*(\theta) + \sum_{r=a}^{b} P_{n+b,r}(0) s_{n,b}^*(\theta) - P_{n,b}(0); \quad 1 \le n \le N-b,$$

$$\tag{14}$$

$$(\lambda - \theta) P_{n,b}^*(\theta) = \lambda P_{n-1,b}^*(\theta) - P_{n,b}(0); \quad N-b+1 \le n \le N-1, \tag{15}$$

$$-\theta P_{N,r}^*(\theta) = \lambda P_{N-1,r}^*(\theta) - P_{N,r}(0); \quad a \le r \le b, \tag{16}$$

where

$$\int_0^\infty e^{-\theta x} P_{n,r}(x) dx = P_{n,r}^*(\theta); \ 0 \le n \le N, \ a \le r \le b, \ \theta \ge 0,$$

$$\int_0^\infty e^{-\theta x} s_{n,r}(x) dx = s_{n,r}^*(\theta); \ 0 \le n \le N, \ a \le r \le b, \ \theta \ge 0,$$

$$P_{n,r} \equiv P_{n,r}^*(0) = \int_0^\infty P_{n,r}(x) dx.$$

From Eqs. (1) and (2), we obtain

$$\lambda P_{n,0} = \sum_{m=0}^{n} \sum_{r=a}^{b} P_{m,r}(0); \quad 0 \le n \le a-1, \tag{17}$$

Now using (17) in (11), we get

$$(\lambda - \theta) P_{0,a}^*(\theta) = \sum_{m=0}^{a} \sum_{r=a}^{b} P_{m,r}(0) s_{0,a}^*(\theta) - P_{0,a}(0), \tag{18}$$

Now summing Eqs. (12)–(16) and (18), we obtain

$$\sum_{n=0}^{N}\sum_{r=a}^{b} P_{n,r}^*(\theta) = \sum_{n=0}^{a}\sum_{r=a}^{b}\left(\frac{1-s_{0,a}^*(\theta)}{\theta}\right)P_{n,r}(0) + \sum_{n=a+1}^{b}\sum_{r=a}^{b}\left(\frac{1-s_{0,n}^*(\theta)}{\theta}\right)P_{n,r}(0)$$

$$+ \sum_{n=b+1}^{N}\sum_{r=a}^{b}\left(\frac{1-s_{n-b,b}^*(\theta)}{\theta}\right)P_{n,r}(0), \tag{19}$$

Taking limit as $\theta \to 0$ in above expression, and using L'Hospital's rule and the normalization condition $\sum_{n=0}^{a-1}P_{n,0} + \sum_{n=0}^{N}\sum_{r=a}^{b}P_{n,r} = 1$, we obtain

$$1 - \sum_{n=0}^{a-1}P_{n,0} = \tilde{s}_{0,a}\sum_{n=0}^{a}\sum_{r=a}^{b}P_{n,r}(0) + \sum_{n=a+1}^{b}\sum_{r=a}^{b}\tilde{s}_{0,n}P_{n,r}(0) + \sum_{n=b+1}^{N}\sum_{r=a}^{b}\tilde{s}_{n-b,b}P_{n,r}(0), \tag{20}$$

After algebraic manipulation from (20), we obtain the desired result (10).

The joint probability distribution of queue content and number of customers with server at an arbitrary epoch is obtained in Theorem 1.

Theorem 1 *The steady-state arbitrary epoch joint probabilities $\{P_{n,0}, P_{n,r}\}$ are related with the departure epoch joint probabilities $\{p_n^+, p_{n,r}^+\}$ as follows.*

$$P_{n,0} = E^{-1}\sum_{i=0}^{n} p_i^+, \quad 0 \le n \le a-1, \tag{21}$$

$$P_{n,a} = E^{-1}\left[\sum_{k=0}^{a} p_k^+ - \sum_{i=0}^{n} p_{i,a}^+\right], \quad 0 \le n \le N-1, \tag{22}$$

$$P_{n,r} = E^{-1}\left[p_r^+ - \sum_{i=0}^{n} p_{i,r}^+\right], \quad 0 \le n \le N-1, \ a+1 \le r \le b-1, \tag{23}$$

$$P_{n,b} = E^{-1}\left[\sum_{i=b}^{min(b+n,N)} p_i^+ - \sum_{i=0}^{n} p_{i,b}^+\right], \quad 0 \le n \le N-1, \tag{24}$$

where $E = \lambda g + \sum_{n=0}^{a-1}(a-n)p_n^+$ *and* $g = \tilde{s}_{0,a}\sum_{n=0}^{a}p_n^+ + \sum_{n=a+1}^{b}p_n^+\tilde{s}_{0,n} + \sum_{n=b+1}^{N}p_n^+\tilde{s}_{n-b,b}.$

Proof The desired results (22)–(24) are obtained by substituting $\theta = 0$ in (11)–(16), solving recursively, after some algebraic manipulations (as described in [6]).

Evaluation of $P_{N,r}$ $(a \le r \le b)$ By using the normalizing condition, the probabilities $P_{N,r}(a \le r \le b)$ cannot be determined. The procedure for determining those probabilities from Eq. (16) is described in this section.

Let us first differentiate (11)–(15) with respect to θ and set $\theta = 0$, and we obtain

$$\lambda P_{0,a}^{*(1)}(0) = P_{0,a} - \lambda P_{a-1,0}\tilde{s}_{0,a} - \tilde{s}_{0,a}\sum_{r=a}^{b}P_{a,r}(0), \tag{25}$$

$$\lambda P_{0,n}^{*(1)}(0) = P_{0,n} - \tilde{s}_{0,n}\sum_{k=a}^{b}P_{n,k}(0); \quad a+1 \le n \le b, \tag{26}$$

$$\lambda P_{n,r}^{*(1)}(0) = P_{n,r} + \lambda P_{n-1,r}^{*(1)}(0); \quad a \le r \le b-1, \ 1 \le n \le N-1, \tag{27}$$

$$\lambda P_{n,b}^{*(1)}(0) = P_{n,b} + \lambda P_{n-1,b}^{*(1)}(0) - \sum_{r=a}^{b}P_{n+b,r}(0)\tilde{s}_{n,b}; \quad 1 \le n \le N-b, \tag{28}$$

$$\lambda P_{n,b}^{*(1)}(0) = P_{n,b} + \lambda P_{n-1,b}^{*(1)}(0); \quad N-b+1 \le n \le N-1, \tag{29}$$

where $P_{n,r}^{*(1)}(0)$ is the derivative of $P_{n,r}^{*}(\theta)$ with respect to θ at $\theta = 0$. Solving Eqs. (25)–(29) recursively and Lemma 1, we obtain the values of $P_{n,r}^{*(1)}(0)$ ($a \le r \le b$, $0 \le n \le N-1$) in known terms.

Now to obtain $P_{N,r}$ ($a \le r \le b$), we differentiate (16) with respect to θ and set $\theta = 0$, and obtain

$$P_{N,r} = -\lambda P_{N-1,r}^{*(1)}(0); \quad a \le r \le b.$$

Henceforth, we have completely obtained all the joint probability distributions of queue length and server content. Now, we obtain several marginal distributions which will help us to compute the useful performance measures, as follows.

- the distribution of queue content, p_n^{queue} ($0 \le n \le N$), is given by $p_n^{queue} = $
$$\begin{cases} P_{n,0} + \sum_{r=a}^{b}P_{n,r}, & 0 \le n \le a-1, \\ \sum_{r=a}^{b}P_{n,r}, & a \le n \le N. \end{cases}$$
- the distribution of the system content, p_n^{sys} ($0 \le n \le N+b$), is given by $p_n^{sys} = $
$$\begin{cases} P_{n,0}, & 0 \le n \le a-1, \\ \sum_{r=a}^{min(b,n)}P_{n-r,r}, & a \le n \le N+a, \\ \sum_{r=n-N}^{b}P_{n-r,r}, & N+a+1 \le n \le N+b. \end{cases}$$

- the distribution of queue length when server is busy is given by, $p_n^{busy} = \sum_{r=a}^{b}P_{n,r}$, ($0 \le n \le N$),
- the distribution of the number of customer with the server, p_r^{ser} ($r = 0$ and $a \le r \le b$), is given by $p_r^{ser} = \begin{cases} \sum_{n=0}^{a-1}P_{n,0}, & r = 0, \\ \sum_{n=0}^{N}P_{n,r}, & a \le r \le b. \end{cases}$
- the probability that the server is in idle state, is given by $P_{idle} = \sum_{n=0}^{a-1}P_{n,0}$, and probability that the server is in busy state, is given by $P_{busy} = \sum_{r=a}^{b}p_r^{ser}$.

3 Performance Measure

As all the state probabilities are known, the significant performance measures of the present model c are evaluated as follows:

1. Expected queue length: $L_q = \sum_{n=0}^{N} n p_n^{queue}$.

2. Expected system length: $L = \sum_{n=0}^{N+b} n p_n^{sys}$.

3. Expected server content: $L_s = \sum_{r=a}^{b} r p_r^{ser}$.

4. Expected queue length when server is idle: $L_q^{idle} = \sum_{n=0}^{a-1} n P_{n,0}$.

5. Expected queue length when server is busy: $L_q^{busy} = \sum_{n=0}^{N} n . p_n^{busy}$.

6. Expected queue length when server is busy with r ($a \le r \le b$) customers: $L_{q,r}^{busy} = \sum_{n=0}^{N} n P_{n,r}$.

7. Probability of blocking: $P_{Block} = p_N^{busy}$.

8. Using Little's law, the expected waiting time of a customer in the queue is given by $W_q = L_q / \bar{\lambda}$ and the expected waiting time of a customer in the system is given by $W = L / \bar{\lambda}$, where $\bar{\lambda}$ is the effective arrival rate of the system and is given by $\bar{\lambda} = \lambda (1 - P_{Block})$.

4 Numerical Results

This section devotes to present several numerical examples in the form of tables and graphs to adjudicate the analytical results obtained in previous sections. For this purpose, we consider $M/G_{n,r}^{(4,9)}/1/25$ queue with state-dependent service rate as $\mu_{0,r} = (b - r + 1)\mu$ ($a \le r \le b - 1$) and $\mu_{n,b} = \mu + 0.5n$ ($0 \le n \le N - b$), where $\mu = 1.5$. Tables 1 and 2 present the departure epoch joint probabilities and arbitrary epoch joint probabilities for the above queuing system with E_4 service time distribution and arrival rate $\lambda = 29.0$. These results are presented here to show the numerical compatibility of our analytical results. The important performance measures of the queueing model under consideration are also presented at the bottom of Table 2. We have also presented here a comparative study in forms of graph to bring out the qualitative aspects of our current study. For this purpose, we have considered the following two cases.

Table 1 Departure epoch joint distribution for $M/E_4^{(4,9)}/1/25$ queue with $a = 4, b = 9, N = 25,$ $\lambda = 29.0$

n	$p_{n,4}^+$	$p_{n,5}^+$	$p_{n,6}^+$	$p_{n,7}^+$	$p_{n,8}^+$	$p_{n,9}^+$	p_n^+
0	0.0361923	0.0041659	0.0021518	0.0009450	0.0002947	0.0000335	0.0437831
1	0.0645893	0.0081906	0.0047096	0.0023322	0.0008337	0.0001954	0.0808509
2	0.0720419	0.0100647	0.0064424	0.0035976	0.0014743	0.0006609	0.0942818
3	0.0642835	0.0098941	0.0070502	0.0044396	0.0020856	0.0016717	0.0894248
4	0.0501906	0.0085106	0.0067509	0.0047938	0.0025816	0.0034760	0.0763035
5	0.0358284	0.0066931	0.0059102	0.0047326	0.0029216	0.0062348	0.0623207
6	0.0239774	0.0049347	0.0048509	0.0043802	0.0030997	0.0099328	0.0511758
7	0.0152823	0.0034651	0.0037918	0.0038610	0.0031321	0.0143932	0.0439255
8	0.0093751	0.0023419	0.0028528	0.0032757	0.0030462	0.0192659	0.0401575
9	0.0055770	0.0015348	0.0020813	0.0026949	0.0028728	0.0240327	0.0387934
10	0.0032347	0.0009807	0.0014804	0.0021616	0.0026416	0.0280985	0.0385975
11	0.0018367	0.0006135	0.0010310	0.0016975	0.0023780	0.0309460	0.0385028
12	0.0010243	0.0003769	0.0007052	0.0013093	0.0021025	0.0322733	0.0377915
13	0.0005625	0.0002280	0.0004749	0.0009943	0.0018303	0.0320510	0.0361409
14	0.0003047	0.0001361	0.0003155	0.0007449	0.0015720	0.0304890	0.0335623
15	0.0001631	0.0000803	0.0002072	0.0005516	0.0013343	0.0279464	0.0302829
16	0.0000864	0.0000469	0.0001346	0.0004041	0.0011207	0.0269328	0.0287256
17	0.0000454	0.0000271	0.0000867	0.0002934	0.0009326	0.0247909	0.0261760
18	0.0000236	0.0000155	0.0000553	0.0002112	0.0007696	0.0212840	0.0223593
19	0.0000122	0.0000088	0.0000350	0.0001509	0.0006303	0.0172570	0.0180943
20	0.0000063	0.0000050	0.0000221	0.0001071	0.0005127	0.0134749	0.0141279
21	0.0000032	0.0000028	0.0000138	0.0000755	0.0004144	0.0103188	0.0108286
22	0.0000016	0.0000016	0.0000086	0.0000529	0.0003331	0.0078661	0.0082639
23	0.0000008	0.0000009	0.0000053	0.0000369	0.0002664	0.0060357	0.0063459
24	0.0000004	0.0000005	0.0000033	0.0000256	0.0002119	0.0046952	0.0049370
25	0.0000004	0.0000006	0.0000051	0.0000559	0.0007647	0.0234197	0.0242465
Total	0.3846442	0.0623207	0.0511758	0.0439255	0.0401575	0.4177764	1.0000000

Case 1. Batch size as well as queue length-dependent service time, i.e., $\mu_{0,r} = (b - r + 1)\mu$ $(a \le r \le b - 1)$ and $\mu_{n,b} = \mu + 0.5n$ $(0 \le n \le N - b)$, where $\mu = 1.5$.

Case 2. Only batch size-dependent service time, i.e., $\mu_{0,r} = (b - r + 1)\mu$ $(a \le r \le b - 1)$ and $\mu_{n,b} = 4.150918$ $(0 \le n \le N - b)$ where $\mu = 1.5$.

The purpose behind choosing the value of $\mu_{n,b} = 4.150918$ $(0 \le n \le N - b)$ for Case 2 is that the average service time for both the cases; i.e., Cases 1 and 2 remain the same. It must be noticed here that when server serves a batch of size r $(a \le r \le b - 1)$ then for both the cases, service times remain unaffected by the queue length. This is because of the fact whenever server is starting a service of a batch of size r $(a \le r \le b - 1)$, then he finds that queue length is always zero.

Table 2 Arbitrary epoch joint distribution for $M/E_4^{(4,9)}/1/25$ queue with $a = 4$, $b = 9$, $N = 25$, $\lambda = 29.0$

n	$P_{n,0}$	$P_{n,4}$	$P_{n,5}$	$P_{n,6}$	$P_{n,7}$	$P_{n,8}$	$P_{n,9}$	p_n^{queue}
0	0.0065450	0.0520887	0.0086933	0.0073284	0.0064250	0.0059589	0.0057941	0.0928334
1	0.0186311	0.0424335	0.0074690	0.0066244	0.0060763	0.0058343	0.0115347	0.0986032
2	0.0327249	0.0316643	0.0059644	0.0056613	0.0055386	0.0056139	0.0171915	0.1043588
3	0.0460926	0.0220548	0.0044854	0.0046074	0.0048749	0.0053022	0.0225909	0.1100082
4		0.0145520	0.0032132	0.0035982	0.0041583	0.0049162	0.0274739	0.0579118
5		0.0091961	0.0022126	0.0027147	0.0034508	0.0044795	0.0315589	0.0536128
6		0.0056118	0.0014750	0.0019896	0.0027960	0.0040161	0.0346010	0.0504896
7		0.0033273	0.0009570	0.0014228	0.0022189	0.0035479	0.0367435	0.0482174
8		0.0019259	0.0006069	0.0009963	0.0017292	0.0030926	0.0377765	0.0461274
9		0.0010922	0.0003775	0.0006852	0.0013264	0.0026631	0.0375263	0.0436707
10		0.0006087	0.0002309	0.0004639	0.0010032	0.0022683	0.0360308	0.0406058
11		0.0003341	0.0001392	0.0003098	0.0007495	0.0019128	0.0335167	0.0369621
12		0.0001810	0.0000828	0.0002044	0.0005538	0.0015985	0.0303110	0.0329315
13		0.0000969	0.0000487	0.0001334	0.0004051	0.0013249	0.0267552	0.0287642
14		0.0000513	0.0000284	0.0000862	0.0002938	0.0010899	0.0231461	0.0246958
15		0.0000270	0.0000164	0.0000553	0.0002113	0.0008904	0.0197065	0.0209069
16		0.0000140	0.0000094	0.0000351	0.0001509	0.0007229	0.0193050	0.0202373
17		0.0000073	0.0000053	0.0000222	0.0001070	0.0005835	0.0155991	0.0163244
18		0.0000037	0.0000030	0.0000139	0.0000755	0.0004684	0.0124174	0.0129820
19		0.0000019	0.0000017	0.0000087	0.0000529	0.0003742	0.0098377	0.0102771
20		0.0000010	0.0000009	0.0000054	0.0000369	0.0002976	0.0078234	0.0081652
21		0.0000005	0.0000005	0.0000033	0.0000256	0.0002356	0.0062809	0.0065465
22		0.0000002	0.0000003	0.0000020	0.0000177	0.0001858	0.0051050	0.0053112
23		0.0000001	0.0000002	0.0000012	0.0000122	0.0001460	0.0042028	0.0043625
24		0.0000001	0.0000001	0.0000008	0.0000084	0.0001143	0.0035009	0.0036245
25		0.0000001	0.0000001	0.0000012	0.0000177	0.0003910	0.0210598	0.0214699
Total	0.1039936	0.1852744	0.0360222	0.0369753	0.0423158	0.0580289	0.5373898	1.0000000
(p_r^{ser})	(P_{idle})	(p_4^{ser})	(p_5^{ser})	(p_6^{ser})	(p_7^{ser})	(p_8^{ser})	(p_9^{ser})	

$L = 13.69263$, $W = 0.4825193$, $L_q = 6.952619$, $W_q = 0.2450057$, $P_{Block} = 0.02146987$,
$L_s = 6.740011$, $L_q^{idle} = 0.2223588$, $L_q^{busy} = 6.73026$,
$L_{q,4}^{busy} = 0.373122$, $L_{q,5}^{busy} = 0.08705351$, $L_{q,6}^{busy} = 0.1116945$, $L_{q,7}^{busy} = 0.1704021$, $L_{q,8}^{busy} = 0.3493061$, $L_{q,9}^{busy} = 5.638682$

Whereas when the server is starting service with a batch of size b, then he finds a queue of n ($0 \leq n \leq N - b$) customers. As a result in this situation the service rate, for Case 1, is decreasing depending on the queue length. However, for the Case 2, the service rate of a batch of size b is considered to be the reciprocal of the average service time of a batch of size b with n in the queue as considered in case 1, i.e., $(N - b + 1) / \sum_{n=0}^{N-b} (1/\mu_{n,b})$. This normalization of the service rate has been done for getting better result in comparing Cases 1 and 2. Therefore, the average

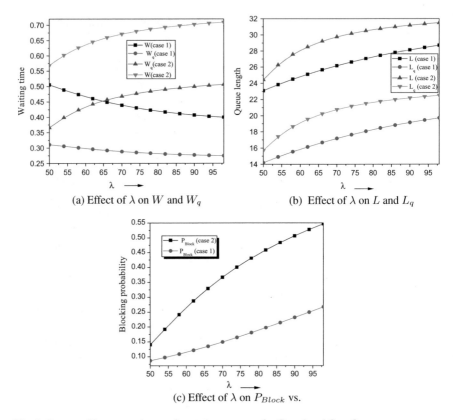

(a) Effect of λ on W and W_q

(b) Effect of λ on L and L_q

(c) Effect of λ on P_{Block} vs.

Fig. 1 Impact of λ on some key performance measures for Case 1 and Case 2

service time of a batch of size b is $E(S_b) = 0.240911$ and average service rate of a batch of size b is $\mu_{n,b} = 4.150918$ for both the cases, i.e., Case 1 and 2. Figure 1 depicts the impact of arrival rate λ on various performance measures for Case 1 and Case 2. In particular, Fig. 1a depicts that the value of W_s and W_q decreases with the increase in arrival rate for Case 1, while these are increasing for Case 2. This behavior well justifies the contribution of effect of queue length-dependent service together with batch size-dependent service. Again an important observation may be noted from Fig. 1b is that the expected system/queue length is much lower for Case 1 in comparison to Case 2. One of the most important performance measures for any queuing model is P_{Block}. Figure 1c reveals that the congestion control is achieved more significantly in terms of decrease in P_{Block} in our current study in comparison to the queuing model considered by Banerjee and Gupta [6].

Figure 2 is presented here for the purpose of revealing the behavior of some important performance measures w.r.t. λ and different service time distribution, viz. deterministic and E_4, for the Case 1. Figure 2a and b reveals that the idle probability of the server is decreasing while blocking probability is increasing with increase in

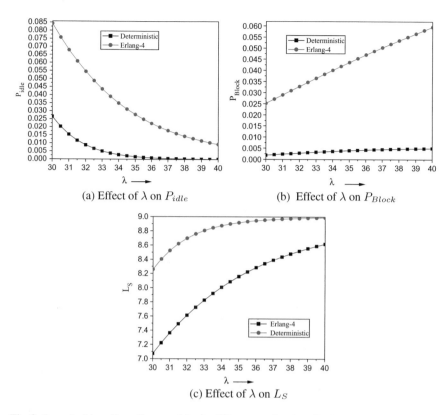

Fig. 2 Impact of λ on P_{idle}, P_{Block} and L_S for different service time distributions

the arrival rate, which is quite obvious for our current study. Figure 2c shows that expected server content increases with the increase in the value of λ as expected.

5 Conclusion

In this paper, we have analyzed a bulk service queue with finite buffer size. The service time, which depends on the size of the batches under service as well as workload which is measured as the queue length before service initiation, is considered to be generally distributed. We have presented here the procedure for obtaining the joint probabilities in steady state at various time epochs. Several numerical examples to compare the impact of our current study with the one presented by Banerjee and Gupta [6] are presented to explore the qualitative aspects of our considered model. The effect of arrival rate on some important performance measures reveals that the congestion control is achieved more significantly in our current study. The considered model can be extended to study the models with correlated arrival process.

Acknowledgements The authors are truly thankful to the anonymous reviewers and editors for their valuable comments and suggestions.

References

1. Jain, R.: Congestion control and traffic management in atm networks: Recent advances and a survey, Comput. Netw. Syst. **28**(13)
2. Leung, K.K.: Load-dependent service queues with application to congestion control in broadband networks. Perform. Eval. **50**(1), 27–40 (2002)
3. Neuts, M.F.: A general class of bulk queues with poisson input. Ann. Mathe. Statist. **38**, 759–770 (1967)
4. Powell, W.B., Humblet, P.: The bulk service queue with a general control strategy: theoretical analysis and a new computational procedure. Oper. Res. **34**(2), 267–275 (1986)
5. Neuts, M.F.: Transform-free equations for the stationary waiting time distributions in the queue with poisson arrivals and bulk services. Ann. Oper. Res. **8**(1), 1–26 (1987)
6. Banerjee, A., Gupta, U.C.: Reducing congestion in bulk-service finite-buffer queueing system using batch-size-dependent service. Perform. Eval. **69**(1), 53–70 (2012)
7. Banerjee, A., Gupta, U.C., Chakravarthy, S.R.: Analysis of a finite-buffer bulk-service queue under markovian arrival process with batch-size-dependent service. Comput. Oper. Res. **60**, 138–149 (2015)
8. Banerjee, A., Gupta, U.C., Sikdar, K.: Analysis of finite-buffer bulk-arrival bulk-service queue with variable service capacity and batch-size-dependent service: $M^X/G_r^Y/1/N$. Int. J. Mathe. Operational Res. **5**(3), 358–386 (2013)
9. Banerjee, A., Sikdar, K., Gupta, U.C.: Computing system length distribution of a finite-buffer bulk-arrival bulk-service queue with variable server capacity. Int. J. Operational Res. **12**(3), 294–317 (2011)
10. Bar-Lev, S.K., Blanc, H., Boxma, O., Janssen, G., Perry, D.: Tandem queues with impatient customers for blood screening procedures. Meth. Comput. Appl. Probab. **15**(2), 423–451 (2013)
11. Bar-Lev, S.K., Parlar, M., Perry, D., Stadje, W.: Applications of bulk queues to group testing models with incomplete identification. Eur. J. Operational Res. **183**, 226–237 (2007)
12. Germs, R., Van Foreest, N.: Loss probabilities for the $M^X/G^Y/1/(K+B)$ bulk queue. Probab. Eng. Inf. Sci. **24**(4), 457–471 (2010)

Chapter 21
A Fuzzy Random Continuous (Q, r, L) Inventory Model Involving Controllable Back-order Rate and Variable Lead-Time with Imprecise Chance Constraint

Debjani Chakraborty, Sushil Kumar Bhuiya and Debdas Ghosh

Abstract In this article, we analyze a fuzzy random continuous review inventory system with the mixture of back-orders and lost sales, where the annual demand is treated as a fuzzy random variable. The study under consideration assumes that the lead-time is a control variable and the lead-time crashing cost is being introduced as a negative exponential function of the lead-time. In a realistic situation, the back-order rate is dependent on the lead-time. Significantly large lead-times might lead to stock-out periods being longer. As a result, many customers may not be prepared to wait for back-orders. Instead of constant back-order rate, we introduce the back-order rate as a decision variable, which is a function of the lead-time throughout the amount of shortage. Moreover, a budgetary constraint is imposed on the model in the form of an imprecise chance constraint to capture the possible way of measuring the imprecisely defined uncertain information of the budget constraint. We develop a methodology to determine the optimum order quantity, reorder point, lead-time, and back-order rate such that the total cost is minimized in the fuzzy sense. Finally, a numerical example is presented to illustrate the proposed methodology.

Keywords Inventory · Imprecise chance constraint · Fuzzy random variable
Possibilistic mean value

D. Chakraborty (✉) · S. K. Bhuiya
Department of Mathematics, Indian Institute of Technology Kharagpur,
Kharagpur 721302, West Bengal, India
e-mail: debjani@maths.iitkgp.ernet.in

S. K. Bhuiya
e-mail: skbhuiya@maths.iitkgp.ernet.in; bhuiya.sushilkumar247@gmail.com

D. Ghosh
Department of Mathematical Sciences, Indian Institute
of Technology (BHU) Varanasi, Varanasi 221005, Uttar Pradesh, India
e-mail: debdas.mat@iitbhu.ac.in; debdas.email@gmail.com

© Springer Nature Singapore Pte Ltd. 2018
D. Ghosh et al. (eds.), *Mathematics and Computing*, Springer Proceedings
in Mathematics & Statistics 253, https://doi.org/10.1007/978-981-13-2095-8_21

263

1 Introduction

Inventory control plays a significant role in every production house. The continuous review inventory model is one of the most important and useful problems in industrial applications. In the continuous review inventory system, the occurrence of the shortage is a major concern. In most of the real-life situations, when such a condition arises, back-orders and lost sales happen simultaneously. Thus, the inventory model, which constitutes both back-order and lost sale cases, is more efficient than the ones based on the individual cases. Montgomery et al. [23] first introduced the inventory model with a mixture of back-orders and lost sales. After the pioneering work of [23], numerous related studies have been developed considerably in the problem of mixture of back-orders and lost sales (see, among others [16, 22, 31]).

In the earlier literature dealing with inventory systems, the lead-time is commonly considered as a prescribed constant or a stochastic variable. Hence, the lead-time becomes uncontrollable [26]. But, production management philosophies like just in time (JIT) show that there are advantages and benefits associated with the effort of control of the lead-time. By shortening lead-time [35], we can decrease the safety stock, minimize the loss due to stock-out, improve the service level to the customer, and increase the competitive capability in business. Liao and Shyu [21] first introduced the problem of lead-time reduction in a continuous review inventory model, where the order quantity was predetermined, and the lead-time was assumed to be a decision variable. Thereafter, several researchers (see, among others [2, 14, 20, 22, 25, 28–30]) have studied lead-time reduction in different types of inventory system.

In addition to lead-time, another key aspect of the inventory system is back-orders. Most of the earlier work in the field of inventory control, it is assumed that the back-order rate is constant. However, in a realistic situation, the back-order rate is dependent on the lead-time. Bigger lead-times might lead to stock-out periods being longer; and as a result, many customers may not be willing for back-orders. This phenomenon reveals that under the longer length of the lead-time, the period of shortage becomes longer. It signifies that the proportion of customers that can wait goes down; as a result, back-order rate decreases. The interdependence between the back-order rate and the lead-time has been proposed by Ouyang and Chuang [28]. They have considered the back-order rate to be dependent on the length of the lead-time through the amount of shortage and that the back-order rate is a control variable. After the work of [28], researchers have been attracted on the problem of controlling back-order rate, and they have extended the inventory control in various directions (see, among others [20, 22, 33]).

On the other hand, most of the real-life business situation, the decision maker has to work under limited budget. According to Hadley and Whitin [15], the most significant real-world constraint is the budgetary restriction on the amount of capital that can be contributed to procure the items of inventory. Keeping this in mind, many inventory models (see, among others [1, 18, 24]) have been developed under budgetary constraint in stochastic environment.

During the mid-1980s, researchers have noticed that the fuzziness is also an intrinsic property of key parameters of the inventory system, particularly when given or obtained data is undetectable, insufficient or partially ignorant. After that, fuzzy set theory has been extensively employed in the problem of inventory system for capturing the uncertainties in the non-probabilistic sense. Park [32] introduced the fuzzy mathematics in the inventory system by developing economic order quantity (EOQ) model in which trapezoidal fuzzy numbers were represented the ordering and holding costs. Gen et al. [13] developed a continuous review inventory model where the values of the parameters of inventory system are considered to be triangular fuzzy numbers. Ouyang and Yao [27] extended min-max distribution-free procedure in the fuzzy environment by developing a continuous review inventory model with variable lead-time in which the annual demand was assumed as the triangular fuzzy number. Tütüncü et al. [36] and Vijayan and Kumaran [37] studied the continuous review inventory model by fuzzifying the cost parameters into fuzzy numbers. Tütüncü et al. obtain the solution using a simulation-based analysis, while an iterative algorithm was used to derive the optimal solution by Vijayan and Kumaran. Recently, Shekarian et al. [34] presented a comprehensive review of the most relevant works of fuzzy inventory model.

It can be noticed that the models, primarily the ones as mentioned above, capture the uncertainty of the parameters of inventory system by characterizing the corresponding variable as either fuzzy or random variable. In a real-life inventory system, fuzziness and randomness often co-occur. Kwakernaak [19] first described the fuzziness and randomness of an event simultaneously. Dutta et al. [12] first incorporated the mixture of fuzziness and randomness into annual demand and developed a single periodic review inventory model. Chang et al. [5] and Dutta and Chakraborty [11] analyzed and extended the continuous review inventory model into fuzzy random circumstances. Chang et al. [5] treated the lead time as the fuzzy random variable and annual expected demand as the fuzzy number. On the other hand, Dutta and Chakraborty [11] considered both the lead-time and annual demand as discrete fuzzy random variables. Dey and Chakraborty [10] considered the annual demand as a fuzzy random variable for developing a periodic review inventory model. Dey and Chakraborty [9] proposed a methodology for constructing a fuzzy random data set from the partially known information. This method is applied on the fuzzy random periodic review model developed by Dey and Chakraborty [10]. Moreover, Dey and Chakraborty [8] also extended the model [10] by incorporating negative exponential crashing cost and lead-time as a variable. Kumar and Goswami [17] extended the min-max distribution-free approach in fuzzy random environments by developing a continuous review production–inventory system. Now, with increased complexity of inventory problem domain, it is hard to define budgetary constraint with proper certainty and precision. Chance-constrained programming [6] can be providing a procedure to construct the constraints in the presence of randomness. However, the imprecision and randomness may appear combined in the information of the restriction. Keeping the issue of vagueness in mind, Chakraborty [4] redefined the chance constraint as the imprecise chance constraint in which the probability of satisfying the imprecise constraint is considered to be vague in nature and to be imprecisely greater than or equal to a specified probability. Recently, Dey et al. [7] incorporated imprecise chance constraint into a fuzzy random continuous review inventory model with a mixture of back-orders and lost sales.

An analysis of the literature reveals that there are some studies of the continuous review inventory system that consider both the fuzziness and randomness simultaneously. But, existing research does not assemble the controllable lead-time and back-order rate in the mixed fuzzy random framework. Here, our intention is to address this research gap of the continuous review inventory model under fuzzy random environment.

Thus, in this paper, we consider a fuzzy random continuous review (Q, r, L) inventory model inclusive of back-orders and lost sales by including the annual demand as the fuzzy random variable. The lead-time is taken as a decision variable, and the crashing cost is being introduced by the negative exponential function of the lead-time. The back-order rate is also a decision variable, which is a function of the lead-time through an amount of shortages. A budgetary constraint has been considered on the model in the form of an imprecise chance constraint. A methodology has been developed to determine the optimal values of the decision-making variable such that the annual cost of the inventory model is minimized in the fuzzy sense. Finally, a numerical example is provided to illustrate the proposed methodology.

The rest of paper is organized as follows: Sect. 2 presents some basic concepts of fuzzy set theory. In Sect. 3, development of proposed methodology is discussed. We present a numerical example to illustrate the methodology in Sect. 4. Paper has been summarized in Sect. 5.

2 Preliminaries

In this section, we review some basic concepts of the fuzzy set theory in which will be used in this paper.

Definition 1 (*Triangular fuzzy number* [38]). A normalized triangular fuzzy number $\tilde{A} = (a, b, c)$ is a fuzzy subset of the real line $\mathbb{R}$, whose membership function $\mu_{\tilde{A}}(x)$ satisfies the following conditions:

  (i) $\mu_{\tilde{A}}(x)$ is a continuous function from $\mathbb{R}$ to the closed interval $[0, 1]$,
  (ii) $\mu_{\tilde{A}}(x) = \frac{x-a}{b-a}$ is strictly increasing function on $[a, b]$,
  (iii) $\mu_{\tilde{A}}(x) = 1$ for $x = b$,
  (iv) $\mu_{\tilde{A}}(x) = \frac{c-x}{c-b}$ is strictly decreasing function on $[b, c]$,
  (v) $\mu_{\tilde{A}}(x) = 0$ elsewhere.

Without any loss of generality, all fuzzy quantities are assumed as triangular fuzzy numbers throughout this paper.

Definition 2 (*α-cut of fuzzy set* [38]). Let $\tilde{A}$ be a fuzzy set. The α-cut of the fuzzy set $\tilde{A}$, denoted by $\tilde{A}_\alpha = [A_\alpha^-, A_\alpha^+]$, is defined as follows:

$$\tilde{A}_\alpha = \begin{cases} \{x \in \mathbb{R} : \mu_{\tilde{A}}(x) \geq \alpha\} & \text{if } \alpha \in (0, 1] \\ cl\{x \in \mathbb{R} : \mu_{\tilde{A}}(x) > 0\} & \text{if } \alpha = 0. \end{cases} \tag{1}$$

Definition 3 (*Fuzzy random variable* [19]). Let $(\Omega, \mathcal{B}, \mathcal{P})$ be a probability space and $F(\mathbb{R})$ be the set all all fuzzy numbers, then a mapping $\tilde{\chi} : \Omega \to F(\mathbb{R})$ is said to be a fuzzy random variable (or FRV for short) if for all $\alpha \in [0, 1]$, the two real-valued mappings $\chi_\alpha^- : \Omega \to \mathbb{R}$ and $\chi_\alpha^+ : \Omega \to \mathbb{R}$ are real-valued random variable.

Definition 4 (*Expectation of fuzzy random variable* [19]). If $\tilde{X}$ is a fuzzy random variable, then the fuzzy expectation of $\tilde{X}$ is a unique fuzzy number. It is defined by

$$E(\tilde{X}) = \int \tilde{X} dP = \left\{ \left(\int X_\alpha^- dP, \int X_\alpha^+ dP \right) : 0 \leq \alpha \leq 1 \right\}, \tag{2}$$

where the α-cut of fuzzy random variable is $[\tilde{X}]_\alpha = [X_\alpha^-, X_\alpha^+]$ for all $\alpha \in [0, 1]$.

Definition 5 (*Possibilistic mean value of a fuzzy number* [3]). Let $\tilde{A}$ be a fuzzy number with α-cut $\tilde{A}_\alpha = [A_\alpha^-, A_\alpha^+]$, and therefore, the possibilistic mean value of $\tilde{A}$ is denoted by $\overline{M}(\tilde{A})$ and defined as

$$\overline{M}(\tilde{A}) = \int_0^1 \alpha(A_\alpha^- + A_\alpha^+) d\alpha. \tag{3}$$

3 Methodology

3.1 Model and Assumptions

The inventory position is reviewed continuously in the (Q, r) continuous review inventory system. When the stock position falls to the reordering point r, an order quantity Q is placed to order. In inventory system, a state is said to be the stock-out state if inventory level falls to zero, at any point in time. Considering the simultaneous occurrence of back-orders and lost sales in real-world scenario, the effect of both are included in the model. The following notations have been used to develop the model:

Notations

P	fixed ordering cost per order,
h	holding cost per unit per year,
π	stock-out cost per unit stock-out,

π_0 marginal profit per unit,
Q order quantity,
r reorder point,
β fraction of demand back-ordered during the stock-out period, $(0 \leq \beta \leq 1)$,
L lead-time (in years),
$R(L)$ lead-time crashing cost,
$\tilde{D}(\omega)$ annual demand ($\omega \in \Omega$ where $(\Omega, \mathcal{B}, \mathcal{P})$ is a probability space),
$\tilde{D}_L(\omega)$ lead-time demand ($\omega \in \Omega$),
x^+ $\max\{0, x\}$.

In continuous review inventory system, the safety stock or buffer stock is defined as the difference between reorder point r and the expected lead-time demand. Now, for all practical purposes, none of the manufacturer wants to have a negative safety stock. Therefore, nonnegative safety stock criterion is imposed on the model. To maintain the safety stock at nonnegative level, $r \geq \overline{M}(\tilde{D}_L)$ has been considered, where $\overline{M}(\tilde{D}_L)$ denotes the expected lead-time demand in possibilistic sense and defined by

$$\overline{M}(\tilde{D}_L) = \int\limits_0^1 \alpha \left[D_{L,\alpha}^- + D_{L,\alpha}^+ \right] d\alpha. \tag{4}$$

In order to incorporate fuzziness and randomness simultaneously [11], the annual demand is assumed to be a discrete fuzzy random variable $\tilde{D}(\omega)$ ($\omega \in \Omega$ where $(\Omega, \mathcal{B}, \mathcal{P})$). Let us suppose that the annual customer demand $\tilde{D}(\omega)$ is of the form $\{(\tilde{D}_1, p_1), (\tilde{D}_2, p_2), \ldots, (\tilde{D}_n, p_n)\}$, where each of $\tilde{D}_i$'s are triangular fuzzy numbers of the form $(\underline{D}_i, D_i, \overline{D}_i)$ with corresponding probabilities p_i's, $i = 1, 2, \ldots, n$. Moreover, the lead-time demand is reflected by any fluctuation of the annual demand. Thus instead of independent parameter, the lead-time demand is assumed to be connected to the annual demand through the length of the lead-time in the following form [11]:

$$\tilde{D}_L(\omega) = \tilde{D}(\omega) \times L. \tag{5}$$

Since annual demand $\tilde{D}(\omega)$ is a fuzzy random variable of the form $\tilde{D}_i = (\underline{D}_i, D_i, \overline{D}_i)$ with associated probability p_i, $i = 1, 2, \ldots, n$, the lead-time demand is also fuzzy random variable. Thus, the lead-time demand is of the form $\tilde{D}_{L,i} = (\underline{D}_{L,i}, D_{L,i}, \overline{D}_{L,i})$ with associated probability p_i, $i = 1, 2, \ldots, n$. Hence, the expected lead-time demand can be expressed in triangular form. The triangular form of expected lead-time demand is given by $E(\tilde{D}_L(\omega)) = \tilde{D}_L = (\underline{D}_L, D_L, \overline{D}_L)$. The annual demand $\tilde{D}(\omega)$ and the lead-time demand $\tilde{D}_L(\omega)$ can be represented by its α-cut as $[\tilde{D}(\omega)]_\alpha = [D_\alpha^-(\omega), D_\alpha^+(\omega)]$ and $[\tilde{D}_L(\omega)]_\alpha = [D_{L,\alpha}^-(\omega), D_{L,\alpha}^+(\omega)]$ where $\alpha \in [0, 1]$. The α-cut representation of the expected lead-time demand is defined as follows:

$$D_{L,\alpha}^-(\omega) = D_{L,\alpha}^-(\omega) \times L \quad \text{and} \quad D_{L,\alpha}^+(\omega) = D_{L,\alpha}^+(\omega) \times L \tag{6}$$

$$\Rightarrow \begin{cases} E\left(D^-_{L,\alpha}(\omega)\right) = \sum_{i=1}^n D^-_{i,\alpha} p_i \times L \\ E\left(D^+_{L,\alpha}(\omega)\right) = \sum_{i=1}^n D^+_{i,\alpha} p_i \times L \end{cases} \tag{7}$$

In this study, we consider the lead-time is a decision variable and the lead-time crashing cost is assumed to be a negative exponential function [8] of the lead-time, which is given by

$$\text{Crashing cost } R(L) = \varepsilon e^{-\delta L} \tag{8}$$

where we can estimate the parameters ε, δ by some of known values of the lead-time crashing cost for a few values of L.

A function of fuzzy random variable is itself a fuzzy random variable; therefore, total cost function is also a fuzzy random variable. Thus, the fuzzy total cost function is given by

$$\tilde{C}(Q, r, L) = h\left[\frac{Q}{2} + r - \tilde{D}(\omega)L\right] + \left[h(1-\beta) + \{\pi + \pi_0(1-\beta)\}\frac{\tilde{D}(\omega)}{Q}\right]\overline{M}(\tilde{D}_L - r)^+$$
$$+ \frac{\tilde{D}(\omega)}{Q}(R(L) + P) \tag{9}$$

where $\overline{M}(\tilde{D}_L - r)^+$ denote the expected shortage at each cycle in possibilistic sense and defined by

$$\overline{M}\left(\tilde{D}_L - r\right)^+ = \int_0^1 \alpha\left[\left((\tilde{D}_L - r)^+\right)^-_\alpha + \left((\tilde{D}_L - r)^+\right)^+_\alpha\right]d\alpha. \tag{10}$$

As mentioned earlier, in a realistic situation, the back-order rate is dependent on the lead-time. Significantly large lead-times might lead to stock-out periods longer, and as a result, many customers may not be willing for back-orders. This phenomenon reveals that with the longer length of lead-time, the time of shortage gets longer and with the increase of shortage the proportion of customers that can wait goes down resulting in the overall decrease of back-order rate. Therefore, we consider the back-orders rate, β, which is a decision variable instead of constant. During the stock-out period, the back-order rate β is a function of the lead-time through the amount of shortage $\overline{M}(\tilde{D}_L - r)^+$. The larger expected shortage quantity implies, the smaller back-order rate. Thus, we consider β as $\beta = \frac{1}{1 + \alpha \overline{M}(\tilde{D}_L - r)^+}$, where α the back-order parameter $(0 \leq \alpha < \infty)$.

Hence, the fuzzy total cost function can be written as

$$\tilde{C}(Q, r, L) = \left[\left\{ h + \pi_0 \left(\frac{\tilde{D}(\omega)}{Q} \right) \right\} \frac{\alpha(\overline{M}(\tilde{D}_L - r)^+)^2}{1 + \alpha \overline{M}(\tilde{D}_L - r)^+} + \pi \left(\frac{\tilde{D}(\omega)}{Q} \right) \overline{M}(\tilde{D}_L - r)^+ \right]$$
$$+ h \left[\frac{Q}{2} + r - \tilde{D}(\omega)L \right] + \frac{\tilde{D}(\omega)}{Q} (P + \varepsilon e^{-\delta L}) \tag{11}$$

In real-life situation, decision maker has to work under limited budget. Let us consider that the cost of each item and the total available budget are c and C, respectively. Then since the order quantities are Q when an order is placed, the following inequality required to hold:

$$cQ \leq C \tag{12}$$

The information about the cost of each unit of the item and total budget available is estimated from past data. Let $\hat{c} \sim N(m^c, \sigma^c)$ and $\hat{C} \sim N(m^C, \sigma^C)$ be normally distributed and independent random variables of the cost of each unit of the item and the total available budget, respectively. Further, the fulfillment of the budget constraint is an individual, organizational decision. Again the decision maker allows some relaxation of the restriction; i.e., both sides of the constraint may be tied with the vague relationship '$\lesssim$' which is the fuzzified version of '$\leq$.' As explained earlier, the decision maker may be more confident to select the probability level in linguistic terms. Thus, instead of crisp probability, a fuzzy probability measure, say around $p \in [0, 1]$ will be attached such that the constraint is satisfied with no less than this imprecise probability level. Because of this, the budgetary constraint (12) may be written in the form of the imprecise chance constraint as [4]

$$\widetilde{\text{Prob}} \left(\hat{c} Q \lesssim \hat{C} \right) \gtrsim p. \tag{13}$$

The goal of the decision maker is to determine the optimal order quantity, reorder point, lead-time, and back-order rate in order to minimize the total cost in fuzzy sense. Since the total cost function is a fuzzy random variable thus the expectation of total cost function is a unique fuzzy number. Let $\overline{M}(\tilde{C}(Q, r, L)(\omega))$ or simply $\overline{M}(Q, r, L)$ be the defuzzified representation of the expectation of the total cost. So mathematically, the problem can be formulated in the following optimization form:

$$(\text{P}_3) \quad \begin{cases} \min_{Q,r,L} \overline{M}(Q, r, L) \\ \text{such that } \widetilde{\text{Prob}} \left(\hat{c} Q \lesssim \hat{C} \right) \gtrsim p \\ Q, r, L \geq 0; \end{cases}$$

where the value of $\overline{M}(Q, r, L)$ is need to be determined. Therefore, the following steps are required to find for obtaining the optimal solution of decision-making variables:

(i) The expected lead-time demand and the exact expression for expected shortage $\overline{M}(\tilde{D}_L - r)^+$ for a given $r \in [\underline{D}_L, \overline{D}_L]$ in possibilistic sense;

(ii) The expected value of the total cost function, which are a fuzzy random variable and the defuzzified representation of this fuzzy random variable;

(iii) The crisp equivalent form of the imprecise chance constraint;

(iv) The optimal values of order quantity Q^*, reorder point r^*, lead-time L^*, and back-order rate β^* in order to minimize the total cost.

3.2   Determination of the Expected Shortage

The expected lead-time demand is $\tilde{D}_L = (\underline{D}_L, D_L, \overline{D}_L)$. Now, in order to maintain the nonnegative safety or buffer stock, the lower bound of reorder point r is $\overline{M}(\tilde{D}_L)$. When the expected lead-time demand $\tilde{D}_L$ in each cycle is greater than r, then there is a shortage of amount $(\tilde{D}_L - r)$. Since the lead-time demand $\tilde{D}_L$ is a triangular fuzzy number, the upper bound of the reorder point r is $\overline{D}_L$ due to the nonnegative safety stock condition. Thus to determine the expected amount of shortage in each cycle, two situations will arise depending upon the position of $r \in [\underline{D}_L, \overline{D}_L]$ subject to condition that the safety or buffer stock is nonnegative.

Situation 1. For r lying between $\underline{D}_L$ and D_L, we have the α-level set of the lead-time demand as [11]

$$(\tilde{D}_L)_\alpha = \begin{cases} [r, D^+_{L,\alpha}], & \alpha \leq L(r) \\ [D^-_{L,\alpha}, D^+_{L,\alpha}], & \alpha > L(r) \end{cases}$$

which implies

$$\left((\tilde{D}_L - r)^+\right)_\alpha = \begin{cases} [0, D^+_{L,\alpha} - r], & \alpha \leq L(r) \\ [D^-_{L,\alpha} - r, D^+_{L,\alpha} - r], & \alpha > L(r) \end{cases} \qquad (14)$$

Therefore, the possibilistic mean is obtained as follows:

$$\overline{M}\left(\tilde{D}_L - r\right)^+ = \int_0^1 \alpha \left[\left((\tilde{D}_L - r)^+\right)^-_\alpha + \left((\tilde{D}_L - r)^+\right)^+_\alpha\right] d\alpha$$

$$= \int_0^1 \alpha D^+_{L,\alpha} d\alpha + \int_{L(r)}^1 \alpha D^-_{L,\alpha} d\alpha - (1 - 0.5L^2(r)) \qquad (15)$$

Situation 2. For r lying between D_L and $\overline{D}_L$, we have the α-level set of the lead-time demand as [11]

$$(\tilde{D}_L)_\alpha = \begin{cases} [r, D^+_{L,\alpha}], & \alpha \le R(r) \\ \phi, & \alpha > R(r) \end{cases}$$

which implies

$$\left((\tilde{D}_L - r)^+ \right)_\alpha = \begin{cases} [0, D^+_{L,\alpha} - r], & \alpha \le R(r) \\ \phi, & \alpha > R(r) \end{cases} \tag{16}$$

Therefore, the possibilistic mean is obtained as follows:

$$\overline{M}\left(\tilde{D}_L - r\right)^+ = \int_0^1 \alpha \left[\left((\tilde{D}_L - r)^+ \right)_\alpha^- + \left((\tilde{D}_L - r)^+ \right)_\alpha^+ \right] d\alpha$$

$$= \int_0^{R(r)} \alpha D^+_{L,\alpha} d\alpha - 0.5 r R^2(r) \tag{17}$$

3.3 Defuzzification of the Fuzzy Expected Total Cost Function Using Possibilistic Mean Value

We have obtained the total cost function in (11), which is given by

$$\tilde{C}(Q, r, L) = h\left[\frac{Q}{2} + r - \tilde{D}(\omega)L \right] + \left[\left\{ h + \pi_0 \left(\frac{\tilde{D}(\omega)}{Q} \right) \right\} \frac{\alpha (\overline{M}(\tilde{D}_L - r)^+)^2}{1 + \alpha \overline{M}(\tilde{D}_L - r)^+} \right.$$

$$\left. + \pi \left(\frac{\tilde{D}(\omega)}{Q} \right) \overline{M}(\tilde{D}_L - r)^+ \right] + \frac{\tilde{D}(\omega)}{Q} (P + \varepsilon e^{-\delta L}) \tag{18}$$

where $\overline{M}(\tilde{D}_L - r)^+$ is given by Eqs. (15) or (17) according to the position of the target inventory level r in the interval $[\underline{D}_L, \overline{D}_L]$. For computational purpose, we defuzzified the fuzzy expected total cost function using its possibilistic mean value. Let $E(\tilde{C}(\omega))$ be the fuzzy expectation of the total cost function. Then, the possibilistic mean value of the fuzzy expected total cost function is given by

$$\overline{M}(Q, r, L) = \int_0^1 \alpha \left(E(C^-_\alpha(\omega)) + E(C^+_\alpha(\omega)) \right) d\alpha \tag{19}$$

Now, the α-level set of $E(\tilde{C}(\omega))$ is then given by

$$EC_\alpha(\omega) = E(C_\alpha(\omega)) = [E(C_\alpha^-(\omega)), E(C_\alpha^+(\omega))], \ \alpha \in [0, 1], \ \omega \in (\Omega, \mathcal{B}, \mathcal{P}), \text{ where}$$

$$
\begin{aligned}
E(C_\alpha^-(\omega)) &= \sum_{i=1}^{n} \left[\frac{D_\alpha^-(\omega)}{Q} \left\{ P + \pi \overline{M}(\tilde{D}_L - r)^+ + \pi_0 \frac{\alpha(\overline{M}(\tilde{D}_L - r)^+)^2}{1 + \alpha \overline{M}(\tilde{D}_L - r)^+} + \varepsilon e^{-\delta L} \right\} \right. \\
&= \sum_{i=1}^{n} \left[\frac{\{\underline{D}_i + \alpha(D_i - \underline{D}_i)\}}{Q} \left\{ P + \pi \overline{M}(\tilde{D}_L - r)^+ + \pi_0 \frac{\alpha(\overline{M}(\tilde{D}_L - r)^+)^2}{1 + \alpha \overline{M}(\tilde{D}_L - r)^+} + \varepsilon e^{-\delta L} \right\} \right. \\
&\left. + h \left\{ \frac{Q}{2} + r - \{\overline{D}_i - \alpha(\overline{D}_i - D_i)\} L + \frac{\alpha(\overline{M}(\tilde{D}_L - r)^+)^2}{1 + \alpha \overline{M}(\tilde{D}_L - r)^+} \right\} \right] p_i
\end{aligned}
\tag{20}
$$

and

$$
\begin{aligned}
E(C_\alpha^+(\omega)) &= \sum_{i=1}^{n} \left[\frac{D_\alpha^+(\omega)}{Q} \left\{ P + \pi \overline{M}(\tilde{D}_L - r)^+ + \pi_0 \frac{\alpha(\overline{M}(\tilde{D}_L - r)^+)^2}{1 + \alpha \overline{M}(\tilde{D}_L - r)^+} + \varepsilon e^{-\delta L} \right\} \right. \\
&\left. + h \left\{ \frac{Q}{2} + r - D_\alpha^-(\omega) L + \frac{\alpha(\overline{M}(\tilde{D}_L - r)^+)^2}{1 + \alpha \overline{M}(\tilde{D}_L - r)^+} \right\} \right] p_i \\
&= \sum_{i=1}^{n} \left[\frac{\{\overline{D}_i - \alpha(\overline{D}_i - D_i)\}}{Q} \left\{ P + \pi \overline{M}(\tilde{D}_L - r)^+ + \pi_0 \frac{\alpha(\overline{M}(\tilde{D}_L - r)^+)^2}{1 + \alpha \overline{M}(\tilde{D}_L - r)^+} + \varepsilon e^{-\delta L} \right\} \right. \\
&\left. + h \left\{ \frac{Q}{2} + r - \{\underline{D}_i + \alpha(D_i - \underline{D}_i)\} L + \frac{\alpha(\overline{M}(\tilde{D}_L - r)^+)^2}{1 + \alpha \overline{M}(\tilde{D}_L - r)^+} \right\} \right] p_i
\end{aligned}
\tag{21}
$$

Substituting the values of Eqs. (20) and (21) in (19), we find the possibilistic mean value of the fuzzy expected total cost function $\overline{M}(Q, r, L)$, which is given by

$$
\begin{aligned}
\overline{M}(Q, r, L) &= \frac{1}{Q} \left[P + \pi \overline{M}(\tilde{D}_L - r)^+ + \pi_0 \frac{\alpha(\overline{M}(\tilde{D}_L - r)^+)^2}{1 + \alpha \overline{M}(\tilde{D}_L - r)^+} + \varepsilon e^{-\delta L} \right] \\
&\quad \left\{ \frac{1}{6} \sum_{i=1}^{n} (\underline{D}_i + \overline{D}_i) p_i + \frac{2}{3} \sum_{i=1}^{n} D_i p_i \right\} \\
&\quad + h \left[\frac{Q}{2} + r - \left\{ \frac{1}{6} \sum_{i=1}^{n} (\underline{D}_i + \overline{D}_i) p_i + \frac{2}{3} \sum_{i=1}^{n} D_i p_i \right\} L \right. \\
&\quad \left. + \frac{\alpha(\overline{M}(\tilde{D}_L - r)^+)^2}{1 + \alpha \overline{M}(\tilde{D}_L - r)^+} \right]
\end{aligned}
\tag{22}
$$

3.4 Crisp Equivalent Form of the Imprecise Chance Constraint

The imprecise chance constraint is as follows:

$\widetilde{\text{Prob}}\left(\hat{c}Q \lesssim \hat{C}\right) \gtrsim p$, where $\hat{c} \sim N(m^c, \sigma^c)$ and $\hat{C} \sim N(m^C, \sigma^C)$ are normally distributed and independent random variables of the cost of each unit of item and the total available budget, respectively. Since this constraint cannot be dealt with this form, hence the imprecise chance constraint is transformed to its crisp equivalent form using the concept which is mentioned in [4].

Suppose $\hat{Z} = \hat{c}Q - \hat{C}$. Then, $\hat{Z}$ follows the normal distribution with mean m^Z and

standard deviation σ^Z where $m^Z = m^c Q - m^C$ and $\sigma^Z = \left[(\sigma^c)^2 Q^2 + (\sigma^C)^2\right]^{\frac{1}{2}}$.

Resorting the fuzzy ordering between the left- and right-hand sides of '$\lesssim$' in the

parenthesis (), $\hat{Z}$ is then replaced by its standard normal variable $\left(\frac{\hat{Z}-m^Z}{\sigma^Z}\right)$ as follows

$$\widetilde{\text{Prob}}\left(\frac{\hat{Z}-m^Z}{\sigma^Z} \lesssim \frac{-m^Z}{\sigma^Z}\right) \gtrsim p. \tag{23}$$

Now, for a fuzzy event $(Z \lesssim z)$, the following proposition, as proved by [4], holds:

$$F(z) \leq \widetilde{\text{Prob}}(Z \lesssim z) \leq F(z + \Delta z) \tag{24}$$

where Δz is the extent of softness permitted and fixed by decision maker. Therefore using the result of (24) in (23), we have

$$\widetilde{\text{Prob}}\left(\frac{\hat{Z}-m^Z}{\sigma^Z} \lesssim \frac{-m^Z}{\sigma^Z}\right) \leq \phi\left(\frac{-m^{Z'}}{\sigma^{Z'}}\right) \tag{25}$$

where $\hat{Z}' = \hat{c}Q - (\hat{C} + \Delta C) \leq \hat{Z}$ and $\phi(.)$ is the distribution function of standard normal variable. Here, ΔC (non-random) is the range of tolerance allowed and fixed by the decision maker for the fuzzy events $\hat{c}Q \lesssim \hat{C}$. Hence, we get the following fuzzy relation

$$\phi\left(\frac{-m^{Z'}}{\sigma^{Z'}}\right) \gtrsim p. \tag{26}$$

Assuming the following linear membership function of the above fuzzy relation with Δp assumed to be range of tolerance permitted,

$$\mu_{\phi(\cdot)}(p) = \begin{cases} 1 & \text{if } \phi(\cdot) > p \\ \frac{\phi(\cdot) - (p - \Delta p)}{\Delta p} & \text{if } p - \Delta p \leq \phi(\cdot) \leq p \\ 0 & \text{otherwise} \end{cases} \tag{27}$$

Hence, the crisp equivalent form of the imprecise chance constraint is given as

$$m^{Z'} + \sigma^{Z'} \phi^{-1}(p - \Delta p) \leq 0. \tag{28}$$

3.5 Optimal Solution

Our main goal is to find the optimal solution. In order to find the optimal order quantity, reorder point, lead-time, and back-order rate for decision making, the following steps are required to execute.

Step (i): Input the values of $P, h, \pi, \pi_0, \hat{c}, \hat{C}, p, \alpha, \epsilon, \delta$.

Step (ii): Calculate the possibilistic mean value of the fuzzy expected shortage using either (15) or (17) with the condition $0 \leq L(r) \leq 1$ and $0 \leq R(r) \leq 1$, respectively.

Step (iii): Determine the safety stock criteria, i.e., $r - \overline{M}(D_L) \geq 0$.

Step (iv): Calculate the possibilistic mean value of the fuzzy expected total cost from (22).

Step (v): Find the crisp equivalent form of imprecise chance constraint using (28).

Step (vi): Use the Lingo, Lindo, or Mathematica to solve the following minimization problem

$$(P_3) \begin{cases} \min_{Q,r,L} \overline{M}(Q, r, L) \\ \text{such that } m^{Z'} + \sigma^{Z'} \phi^{-1}(p - \Delta p) \leq 0 \\ r \geq \overline{M}(\tilde{D}_L) \\ r \leq \overline{D}_L \\ Q, r, L \geq 0. \end{cases}$$

4 Numerical Example

A Leather Good's company in a city, say X Leather private limited, sells a particular type of handbags. The cost of placing an order is assumed to be Rs. 200. The holding cost is Rs. 20 per item per year. The fixed penalty cost for the shortage is Rs. 50, and the cost of lost sales including marginal profit is Rs. 100. Suppose it is estimated that the expense of each handbag is normally distributed with mean Rs. 375 and standard deviation Rs. 5. The total budget available to the private limited is also normally distributed with mean Rs. 30,000 and standard deviation Rs. 75. The lead-time reduction cost is a negative exponential function of the lead-time, i.e., $R(L) = \varepsilon e^{-\delta L}$ with $\varepsilon = 156$ and $\delta = 114$. Now, the manager of X private limited is quite

Table 1 Demand information

Demand	Probability
(825, 1130, 1270)	.25
(775, 977, 1275)	.22
(1120, 1325, 1450)	.27
(1240, 1352, 1560)	.26

satisfied if the budgetary constraint attains to the probability of 'around 0.8'. The information about annual demand is given in Table 1.

Thence, the problem is to determine the optimal order quantity Q^*, reorder point r^*, lead-time L^*, and back-order rate β^* in such a way that the expected annual inventory cost incurred is minimum. From the above problem, we have the order cost $P = 200$, the inventory holding cost $h = 20$, the fixed shortage cost $\pi = 50$, the marginal profit $\pi_0 = 100$, the lead-time reduction cost $R(L) = 156e^{-114L}$, the cost of each handbag $\hat{c} \sim N(375, 5)$, the total budget $\hat{C} \sim N(30000, 75)$ and the probability $p ='$ around $0.8'$. Thus, the expected lead-time demand and possibilistic mean value of lead-time are given by

$$\tilde{D}_L = (1001.55, 1206.71, 1373.1)L \text{ and,} \tag{29}$$

$$\overline{M}(\tilde{D}_L) = \left\{ \frac{1}{6} \sum_{i=1}^{n} (\underline{D}_i + \overline{D}_i)p_i + \frac{2}{3} \sum_{i=1}^{n} D_i p_i \right\} L = 1200.248L. \tag{30}$$

Then, the defuzzified fuzzy expected total cost function is obtained as

$$\overline{M}(Q, r, L) = 20 \left[\frac{Q}{2} + r - 1200.248L + \frac{\alpha(\overline{M}(\tilde{D}_L - r)^+)^2}{1 + \alpha \overline{M}(\tilde{D}_L - r)^+} \right]$$
$$+ \frac{1200.248}{Q} \left[200 + 50\overline{M}(\tilde{D}_L - r)^+ + 100 \frac{\alpha(\overline{M}(\tilde{D}_L - r)^+)^2}{1 + \alpha \overline{M}(\tilde{D}_L - r)^+} + 156e^{-114L} \right] \tag{31}$$

Thus, mathematically, we need to solve the following optimization problem for determining the optimal solutions:

$$(P_4) \begin{cases} \min_{Q,r,L} \overline{M}(Q, r, L) \\ \text{such that } 140607.29Q^2 - 22575000Q + 906006016 \geq 0 \\ \quad r \geq 1200.248L \\ \quad r \leq 1373.1L \\ \quad \dfrac{r - 1001.55L}{205.16L} \leq 1 \\ \quad Q, r, L \geq 0. \end{cases}$$

Table 2 Optimal solutions of optimization problem (P$_4$) for different values of α

α	Q^*	r^*	β^*	L^*(in yearr)	$R(L)$	Total cost
0.0	79.36030	26.52812	1.0000000	0.02198384	12.00000	4467.313
0.5	79.36030	20.88508	0.8064689	0.01730796	20.85363	4641.311
1.0	79.36030	18.99215	0.6961684	0.01573879	24.93852	4730.848
10	79.36030	15.20582	0.2225064	0.01260105	35.66305	5039.961
∞	79.36030	14.84971	0.0000000	0.01230594	36.88325	5158.792

where $\overline{M}(\tilde{D}_L - r)^+ = (1200.248L - r) + (r - 1001.55L)^2 \left\{ \frac{1.18791 \times 10^{-5}r}{L^2} - \frac{.01187542}{L} \right\}$
$- \frac{7.91942 \times 10^{-6}}{L^2}(r - 1001.55L)^3$, $\Delta p = .01$ and $\Delta C = 100$. For the different values of α, the optimal solutions are presented in Table 2. Through numerical solutions, we have seen that as the back-order parameter α increases, the back-order rate decreases, and with the decreases of back-order rates, the total cost increases. It is also observed that the lead-time crashing cost increases as the length of the lead-time declines.

5 Conclusions

In this paper, we have proposed a fuzzy random continuous review inventory system with a mixture of back-orders and lost sales. The model is developed under the consideration that the order quantity, reorder point, back-order rate, and lead-time are the decision variables. We have considered the negative exponential function of lead-time and introduced a function of lead-time through an amount of shortages for controlling the lead-time and back-order rate, respectively, in the fuzzy random framework. We have considered the annual demand as a fuzzy random variable to capture the fuzziness and randomness simultaneously. To quantify the imprecise information, a budgetary constraint has been imposed on the model in the form of an imprecise chance constraint. We developed a methodology for obtaining the optimum decision-making variables in such a way that the total annual cost is minimized in the fuzzy sense. A numerical example has illustrated the proposed methodology. In future research on this model, it would be interesting to deal with imprecise probabilities. On the other hand, a possible extension of this model can be achieved by inclusion of the service-level constraint.

References

1. Abdel-Malek, Layek L., Montanari, R.: An analysis of the multi-product newsboy problem with a budget constraint. Int. J. Prod. Econ. **97**(3), 296–307 (2005)
2. Ben-Daya, M., Raouf, A.: Inventory models involving lead time as decision variable. J. Oper. Res. Soc. **45**(5), 579–582 (1994)

3. Carlsson, C., Fuller, R.: On possibilistic mean value and variance of fuzzy numbers. Fuzzy Sets Syst. **122**(2), 315–326 (2001)
4. Chakraborty, D.: Redefining chance-constrained programming in fuzzy environment. Fuzzy Sets Syst. **125**(3), 327–333 (2002)
5. Chang, H.C., Yao, J.S., Quyang, L.Y.: Fuzzy mixture inventory model involving fuzzy random variable lead time demand and fuzzy total demand. Eur. J. Oper. Res. **169**(1), 65–80 (2006)
6. Charnes, A., Cooper, W.W.: Chance-constrained programming. Manag. Sci. **6**(1), 73–79 (1959)
7. Dey, O., Giri, B.C., Chakraborty, D.: A fuzzy random continuous review inventory model with a mixture of backorders and lost sales under imprecise chance constraint. Int. J. Oper. Res. **26**(1), 34–51 (2016)
8. Dey, O., Chakraborty, D.: A fuzzy random periodic review system with variable lead-time and negative exponential crashing cost. Appl. Math. Model. **36**(12), 6312–6322 (2012)
9. Dey, O., Chakraborty, D.: A fuzzy random periodic review system: a technique for real-life application. Int. J. Oper. Res. **13**(4), 395–405 (2012)
10. Dey, O., Chakraborty, D.: Fuzzy periodic review system with fuzzy random variable demand. Eur. J. Oper. Res. **198**(1), 9113–120 (2009)
11. Dutta, P., Chakraborty, D.: Continuous review inventory model in mixed fuzzy and stochastic environment. Appl. Math. Comput. **188**(1), 970–980 (2007)
12. Dutta, P., Chakrabortyand, D., Roy, A.R.: A single-period inventory model with fuzzy random variable demand. Math. Comput. Model. **41**(8–9), 915–922 (2005)
13. Gen, M., Tsujimura, Y., Zheng, D.: An application of fuzzy set theory to inventory control models. Comput. Ind. Eng. **33**(3), 553–556 (1997)
14. Glock, C.H.: Lead time reduction strategies in a single-vendor-single-buyer integrated inventory model with lot size-dependent lead times and stochastic demand. Int. J. Prod. Econ. **136**(1), 37–44 (2012)
15. Hadley, G., Whitin, T.M.: Analysis of Inventory Systems. Prentice-Hall, Englewood Cliffs, NJ (1963)
16. Kim, O.H., Park, K.S.: (Q, r) inventory model with a mixture of lost sales and weighted backorders. J. Oper. Res. Soc. **36**, 231–238 (1985)
17. Kumar, R.S., Goswami, A.: A continuous review production-inventory system in fuzzy random environment: minmax distribution free procedure. Comput. Ind. Eng. **79**(1), 65–75 (2015)
18. Kundu, A., Chakrabarti, T.: A multi-product continuous review inventory system in stochastic environment with budget constraint. Optim. Lett. **6**(2), 299–313 (2012)
19. Kwakernaak, H.: Fuzzy random variables—I. Definitions and theorems. Inform. Sci. **15**(1), 1–29 (1978)
20. Lee, W.C.: Inventory model involving controllable back-order rate and variable lead time demand with the mixture of distribution. Appl. Math. Comput. **160**(3), 701–717 (2005)
21. Liao, C.J., Shyu, C.H.: An analytical determination of lead time with normal demand. Int. J. Oper. Prod. Manage. **11**(9), 72–78 (1991)
22. Lin, H.J.: A stochastic periodic review inventory model with back-order discounts and ordering cost dependent on lead time for the mixtures of distributions. Top **23**(2), 386–400 (2015)
23. Montgomery, D.C., Bazaraa, M.S., Keswani, A.K.: Inventory models with a mixture of backorders and lost sales. Naval Res. Logistics Q. **20**(2), 255–263 (1973)
24. Moon, I., Silver, E.A.: The multi-item newsvendor problem with a budget constraint and fixed ordering costs. J. Oper. Res. Soc. **51**(5), 602–608 (2000)
25. Moon, I., Choi, S.: TECHNICAL NOTEA note on lead time and distributional assumptions in continuous review inventory models. Comput. Oper. Res. **25**(11), 1007–1012 (1998)
26. Naddor, E.: Inventory Systems. Wiley, New York (1966)
27. Ouyang, L.Y., Yao, J.S.: A minimax distribution free procedure for mixed inventory model involving variable lead time with fuzzy demand. Comput. Oper. Res. **29**(5), 471–487 (2002)
28. Ouyang, L.Y., Chuang, B.R.: Mixture inventory model involving variable lead time and controllable back-order rate. Comput. Ind. Eng. **40**(4), 339–348 (2001)
29. Ouyang, L.Y., Chang, H.C.: Impact of investing in quality improvement on (Q, r, L) model involving the imperfect production process. Prod. Plan. Control **11**(6), 598–607 (2000)

30. Ouyang, L.Y., Yeh, N.C., Wu, K.S.: Mixture inventory model with back-orders and lost sales for variable lead time. J. Oper. Res. Soc. **47**, 829–832 (1996)
31. Padmanabhan, G., Vrat, P.: Inventory models with a mixture of backorders and lost sales. Int. J. Syst. Sci. **21**(8), 1721–1726 (1990)
32. Park, K.S.: Fuzzy-set theoretic interpretation of economic order quantity. IEEE Trans. Syst. Man Cybern. **17**(6), 1082–1084 (1987)
33. Sarkar, B., Moon, I.: Improved quality, setup cost reduction, and variable backorder costs in an imperfect production process. Int. J. Prod. Econ. **155**, 204–213 (2014)
34. Shekarian, E., Kazemi, N., Rashid, S.H.A., Olugu, E.U.: Fuzzy inventory models: a comprehensive review. Appl. Soft Comput. **45**(2–3), 260–264 (2017)
35. Tersine, R.J.: Principles of Inventory and Materials Management. Prentice Hall, Englewood Cliffs, NJ (1994)
36. Tütüncü, G.Y., Aköz, O., Apaydın, A., Petrovic, D.: Continuous review inventory control in the presence of fuzzy costs. Int. J. Prod. Econ **113**(2), 775–784 (2008)
37. Vijayan, T., Kumaran, M.: Inventory models with mixture of back-orders and lost sales under fuzzy cost. Eur. J. Oper. Res. **189**(1), 105–119 (2008)
38. Zimmermann, H.J.: Fuzzy Set Theory and its Applications. Springer Science & Business Media (2011)

Chapter 22
Estimation of the Location Parameter of a General Half-Normal Distribution

Lakshmi Kanta Patra, Somesh Kumar and Nitin Gupta

Abstract In this paper, estimation of the location parameter of a half-normal distribution is considered. Some unbiased as well as biased estimators are derived. Admissibility and minimaxity of Pitman estimator are proved. A complete class of estimators among multiples of the maximum likelihood estimator is obtained. We develop a one-sided asymptotic confidence interval for the location parameter. Numerical comparisons of the percentage risk improvements over maximum likelihood estimator of various estimators are carried out.

Keywords Half-normal distribution · Generalized Bayes estimator · Pitman estimator · Admissible estimator · Minimax estimator

1 Introduction

If Z is a standard normal random variable, then $Y = |Z|$ follows a standard half-normal distribution. The half-normal distribution is a special case of the folded normal and truncated normal distributions ([12], pp. 156, 170). Also if W has a chi-square distribution on one degree of freedom, then $Y = \sqrt{W}$ follows a standard half-normal distribution.

Let $X = \eta Y + \xi$, then X follows a general half-normal distribution and the probability density function of X is given by

$$f_X(x|\xi, \eta) = \frac{1}{\eta}\sqrt{\frac{2}{\pi}} \exp\left\{-\frac{(x-\xi)^2}{2\eta^2}\right\}, \quad x > \xi, \quad -\infty < \xi < \infty, \quad \eta > 0. \quad (1)$$

The general half-normal distribution is a special case of the generalized gamma distribution and also of the two-parameter chi-square distribution. This distribution

L. K. Patra (✉)
Indian Institute of Information Technology Ranchi, Ranchi, India
e-mail: patralakshmi@gmail.com

S. Kumar · N. Gupta
Indian Institute of Technology Kharagpur, Kharagpur 721302, India

© Springer Nature Singapore Pte Ltd. 2018
D. Ghosh et al. (eds.), *Mathematics and Computing*, Springer Proceedings
in Mathematics & Statistics 253, https://doi.org/10.1007/978-981-13-2095-8_22

was first introduced by Daniel [6] with $\xi = 0$. He introduced half-normal plot for interpreting factorial two-level experiments. This distribution is also an important limiting distribution of Azzalini's three-parameter skew-normal class, introduced in [2]. The half-normal distribution has also been used for modeling truncated data. It has applications in several areas such as stochastic frontier modeling [1], sports science and physiology [15, 17]. Unbiased estimators for location and scale parameters of a general half-normal distribution are proposed by Nogales and Perez [14]. They have numerically shown that the proposed estimators perform better than other existing estimators in the literature. Using the proposed estimators, they derived large sample confidence intervals for the location and scale parameters.

The data set introduced in [5] consists of percentage body fat measurements made on 202 elite athletes who trained at the Australian Institute of Sports. The data on the measurements of male athletes is seen to follow a general half-normal distribution. Azzalini and Capitanio [3] and Pewsey [15] have shown that for highly skewed data, the maximum likelihood estimator of a fitted skew-normal distribution often corresponds to a general half-normal distribution. So, inferential results for a half-normal distribution have relevance in modeling of the skewed data.

First, we state some results which are already available in the literature. Let $\underline{X} = (X_1, \ldots, X_n)$ be a random sample from a general half-normal distribution with the density (1). Pewsey [16] considered a general half-normal distribution and showed that the maximum likelihood estimators (MLEs) of the location parameter ξ and the scale parameter η are $\hat{\xi}_{ML} = X_{(1)}$ and $\hat{\eta}_{ML} = \sqrt{\frac{1}{n} \sum (X_i - X_{(1)})^2}$, respectively, where $X_{(1)} = \min\{X_1, \ldots, X_n\}$. These estimators are biased. He also derived large sample confidence intervals for the parameters. The bias-corrected estimator of the parameters of a general half-normal distribution based on maximum likelihood estimators was considered by Pewsey [17]. He also constructed a bias-corrected confidence interval for the location parameter. The bias-corrected estimators of η and ξ are given as

$$\hat{\eta}_{BC} = \sqrt{\frac{n}{n-1}} \hat{\eta}_{ML} \quad \text{and} \quad \hat{\xi}_{BC} = \hat{\xi}_{ML} - \Phi^{-1}\left(\frac{1}{2} + \frac{1}{2n}\right) \hat{\eta}_{BC} \tag{2}$$

respectively, where $\Phi(.)$ is the cumulative distribution function of a standard normal distribution. We also denote by $\phi(.)$ the probability density function of a standard normal distribution. Bayes estimation of the parameters of a general half-normal distribution is studied by Farsipour and Rasouli [7]. Wiper et al. [19] derived Bayes estimators for the parameters of a general half-normal distribution with the location parameter ξ and scale parameter η. They considered a non-informative prior $f(\xi, \tau) \propto 1/\tau$, where $\xi \in \mathbb{R}$, $\tau > 0$ and $\tau = 1/\eta^2$. For this prior, the joint posterior distribution of ξ and τ is a right-truncated normal-gamma distribution. Wiper et al. [19] showed that the marginal distribution of ξ is a truncated t-distribution and the marginal distribution of τ is a Gaussian-modulated gamma distribution. Finally, they have numerically compared the bias and root-mean-squared errors of the proposed estimators using simulation.

To the best of our knowledge, the decision theoretic properties like admissibility and minimaxity have not been explored for estimating parameters of a half-normal distribution. In this paper, we first prove a complete class result for estimating the location parameter when the scale parameter is known. The admissibility and minimaxity properties of a generalized Bayes estimator are established. The generalized Bayes estimator is also shown to be a limit of Bayes rules and is also seen to perform very well in terms of the mean squared error.

The organization of the paper is as follows. In Sect. 2, estimation of the location parameter ξ is considered when η is known. Some biased and unbiased estimators of ξ are derived. A complete class result is established. In Sect. 3, we prove that the Pitman estimator is the same as a generalized Bayes estimator and that it is also a limit of Bayes rules. The minimaxity and admissibility of the Pitman estimator are established. A simulation study is carried out to numerically compare the performance of various estimators.

2 Unbiased Estimation and a Complete Class Result

We consider the estimation of the location parameter ξ when scale parameter η is known. Without loss of generality, we take $\eta = 1$. The probability density function and the cumulative distribution function of the random variable X following a half-normal distribution are given by

$$f_X(x|\xi) = \sqrt{\frac{2}{\pi}} \exp\left\{-\frac{(x-\xi)^2}{2}\right\}, \ x > \xi, \ -\infty < \xi < \infty \tag{3}$$

and

$$F_X(x|\xi) = \begin{cases} 2\Phi(x-\xi) - 1 \ , & \text{if } x > \xi \\ 0 & , & \text{if } x \le \xi \end{cases}$$

respectively.

Let $\underline{X} = (X_1, \ldots, X_n)$ be a random sample from this distribution. We consider the problem of estimating ξ with respect to the squared error loss function

$$L(\xi, \delta) = (\xi - \delta)^2. \tag{4}$$

Note that the method of moment estimator (MME) of ξ is $T_1 = \overline{X} - \sqrt{\frac{2}{\pi}}$. This is also unbiased for ξ. Further, the maximum likelihood estimator (MLE) of ξ is $\widehat{\xi}_{ML} = X_{(1)}$. The joint density function of $\underline{X}$ is

$$f_{\underline{x}}(\underline{x}|\xi) = \left(\sqrt{\frac{2}{\pi}}\right)^n \exp\left\{-\sum(x_i - \xi)^2/2\right\}, \ x_i > \xi, \ -\infty < \xi < \infty \quad (5)$$

$$= \left(\sqrt{\frac{2}{\pi}}\right)^n e^{-\sum x_i^2/2} e^{\{n\xi(\bar{x}-\frac{\xi}{2})\}} \prod_{i=2}^n I_{(x_{(1)},\infty)}(x_{(i)}) I_{(\xi,\infty)}(x_{(1)}),$$

where $\underline{x} = (x_1, x_2, \ldots, x_n)$ and $I(.)$ is the indicator function. By factorization theorem, a sufficient statistic for the above family of distributions is $T = (\overline{X}, X_{(1)})$.

Now, we show that T is not complete. For this, we find a function $g(t)$ such that $E_\xi(g(T)) = 0$ for all $\xi \in \mathbb{R}$, but $P_\xi(g(T) = 0) \neq 1$ for some $\xi \in \mathbb{R}$.

The density function of $X_{(1)}$ is

$$f_{X_{(1)}}(y|\xi) = \begin{cases} 2^n n \left(\Phi(\xi - y)\right)^{n-1} \phi(\xi - y) , & \text{if } y > \xi \\ 0 & , \text{if } y \leq \xi. \end{cases} \quad (6)$$

It is seen that $E(X_{(1)}) = \xi + Q_n$, where

$$Q_n = 2^n \int_{-\infty}^0 (\Phi(z))^n \, dz. \quad (7)$$

Consequently, $E_\xi(X_{(1)} - Q_n) = \xi$, and so $T_0 = X_{(1)} - Q_n$ is an unbiased estimator of ξ.

If we take

$$g(T) = X_{(1)} - Q_n - \overline{X} + \sqrt{\frac{2}{\pi}}, \quad (8)$$

then $E_\xi(g(T)) = 0$ for all $\xi \in \mathbb{R}$, but $g(t) \neq 0$ with probability 1. This proves that T is not complete.

Now, we define a new unbiased estimator of ξ as $T_\alpha = \alpha T_1 + (1 - \alpha)T_0$, where $\alpha \in \mathbb{R}$. Note that

$$V_\xi(T_\alpha) = E_\xi(T_\alpha - \xi)^2 = E_\xi (\alpha T_1 + (1 - \alpha)T_0 - \xi)^2.$$

The choice of α which minimizes $V_\xi(T_\alpha)$ is

$$\alpha(n) = \frac{E_\xi(T_0^2 - T_1 T_0)}{E_\xi(T_1 - T_0)^2}.$$

Note that $\alpha(n)$ does not depend on ξ. So, we get the following result

Lemma 1 *The estimator $T_{\alpha(n)}$ is the best estimator in the class of estimators $\{T_\alpha : \alpha \in \mathbb{R}\}$ for estimating ξ with respect to squared error loss function (4).*

Remark 1 The minimizing choice $\alpha(n)$ depends on the sample size n. In Table 1, we report values of $\alpha(n)$ for various choices of n. These values have been evaluated using

Table 1 Values of $\alpha(n)$

n	5	10	20	30	40	50	100	200	500
$\alpha(n)$	0.16670	0.10684	0.06358	0.04651	0.03565	0.02917	0.01581	0.00755	0.003167

simulation of half-normal random variables based on 50,000 replications. From the table, we note that $\alpha(n)$ decreases as n increases. It is also seen from the simulated values that $\alpha(n)$ always lies between 0 and 1.

Next, we consider a class of estimators of ξ of the form $\delta_c = X_{(1)} + c$, where c is a real number. The following lemma follows immediately.

Lemma 2 *The unbiased estimator T_0 is the best estimator of ξ in the class of estimators $\{\delta_c = X_{(1)} + c : c \in \mathbb{R}\}$ for estimating ξ with respect to squared error loss function (4).*

2.1 A Complete Class Result

Definition 1 For estimating $\theta \in \Theta$, a class of estimators D is said to be complete, if for any estimator δ_1 not in D, there exists an estimator $\delta_0 \in D$ such that

$$R(\theta, \delta_1) \leq R(\theta, \delta_0) \ \forall \ \theta \in \Theta \text{ and } R(\theta, \delta_1) < R(\theta, \delta_0) \text{ for at least one } \theta \in \Theta.$$

Here, $R(., .)$ is the risk function with respect to a given loss function.

Here, we prove a complete class result of the location parameter ξ when $\xi > 0$. Maximizing the likelihood function over the restricted parameter space $\xi > 0$, we find that the MLE of ξ is $X_{(1)}$. Consider estimators of the form

$$\delta_c(\underline{X}) = cX_{(1)} \tag{9}$$

with c is a positive constant. Now

$$E(\delta_c) = c(\xi + Q_n), \quad \text{and} \quad E(\delta_c^2) = c^2(\xi^2 + 2\xi Q_n - R_n),$$

where

$$R_n = 2^{n+1} \int_{-\infty}^{0} z \, (\Phi(z))^n \, dz,$$

and Q_n is given by (7). Thus, we get the risk function of δ_c with respect to the loss (4) as

$$R(\delta_c, \xi) = c^2(\xi^2 + 2\xi Q_n - R_n) - 2\xi c(\xi + Q_n) + \xi^2.$$

The choice of c which minimizes $R(\delta_c, \xi)$ is

$$\widehat{c}(\xi) = \frac{\xi(\xi + Q_n)}{\xi^2 + 2\xi Q_n - R_n}.$$

Note that

$$\inf_{0 \le \xi < \infty} \widehat{c}(\xi) = 0 \quad \text{and} \quad \sup_{0 \le \xi < \infty} \widehat{c}(\xi) = 1.$$

Since the risk function $R(\delta_c, \xi)$ is convex in c for every ξ, it can be seen that if $c < 0$, then the estimator δ_0 improves upon δ_c; and if $c > 1$, then the estimator δ_1 improves upon δ_c. Further, the estimator δ_c, $0 \le c \le 1$ cannot be improved by any δ_c. (This technique was first developed by [4].) This proves the following theorem.

Theorem 1 *The class of estimators $\{\delta_c : 0 < c \le 1\}$ forms a complete class for estimating ξ, among all estimator of the form (9) when the loss function is squared error loss function.*

3 Asymptotic Confidence Interval for ξ

In this section, we will construct an asymptotic confidence interval for ξ based on the unbiased estimator $T_\alpha = \alpha T_1 + (1 - \alpha)T_0$ given in Sect. 2. Let $\text{Exp}(\sigma)$ denote an exponential distribution with the scale parameter σ. Pewsey [16] has shown that

$$\frac{X_{(1)} - \xi}{\Phi^{-1}\left(\frac{1}{2} + \frac{1}{2n}\right)}$$

has asymptotically an $\text{Exp}(1)$ distribution. Therefore, the limiting distribution of

$$\frac{(1 - \alpha)\left(X_{(1)} - \xi\right)}{\Phi^{-1}\left(\frac{1}{2} + \frac{1}{2n}\right)}$$

is $\text{Exp}(1 - \alpha)$.

By weak law of large numbers, $\overline{X}$ converges in probability to $\left(\xi + \sqrt{\frac{2}{\pi}}\right)$. Hence, $\left(\overline{X} - \xi - \sqrt{\frac{2}{\pi}}\right)$ converges in probability to 0. Now, Q_n goes to 0 and $\Phi^{-1}\left(\frac{1}{2} + \frac{1}{2n}\right)$ goes to $\Phi^{-1}\left(\frac{1}{2}\right)$ as n goes to ∞. Hence by Slutsky's Lemma (see [10]), $\frac{\alpha\left(\overline{X} - \xi - \sqrt{\frac{2}{\pi}}\right)}{\Phi^{-1}\left(\frac{1}{2} + \frac{1}{2n}\right)}$ converges in probability to 0.

Now, we have

$$\frac{T_\alpha - \xi}{\Phi^{-1}\left(\frac{1}{2} + \frac{1}{2n}\right)} = \frac{\alpha T_1 + (1 - \alpha)T_0 - \xi}{\Phi^{-1}\left(\frac{1}{2} + \frac{1}{2n}\right)}.$$

This can be further written as

$$\frac{T_\alpha - \xi}{\Phi^{-1}\left(\frac{1}{2} + \frac{1}{2n}\right)} = \frac{\alpha\left(\overline{X} - \sqrt{\frac{2}{\pi}} - \xi\right)}{\Phi^{-1}\left(\frac{1}{2} + \frac{1}{2n}\right)} + \frac{(1 - \alpha)(X_{(1)} - \xi)}{\Phi^{-1}\left(\frac{1}{2} + \frac{1}{2n}\right)} - \frac{(1 - \alpha)Q_n}{\Phi^{-1}\left(\frac{1}{2} + \frac{1}{2n}\right)}. \quad (10)$$

Clearly, from (10) it follows that $\frac{T_\alpha - \xi}{\Phi^{-1}\left(\frac{1}{2} + \frac{1}{2n}\right)}$ converges in distribution to a random variable Y having $\mathrm{Exp}(1 - \alpha)$ distribution.

So, a one-tailed asymptotic $100(1 - \gamma)\%$ confidence interval for ξ is given by

$$(1 - \alpha)\log(\gamma)\,\Phi^{-1}\left(\frac{1}{2} + \frac{1}{2n}\right) + T_\alpha \leq \xi \leq X_{(1)}. \quad (11)$$

In Tables 2, 3, and 4, the estimated coverage probabilities and widths of the confidence interval (11) of ξ are evaluated based on simulations for 90%, 95%, and 99% confidence levels. Sample size n is taken ranging from 20 to 100. Reported coverage probabilities and widths of confidence intervals are obtained from 50,000 simulations of samples of size n from general half-normal distribution with the parameters $\xi = 0$ and $\eta = 1$. From the tables, we observe the following:

(i) Coverage probability and width of the proposed confidence interval decrease as α increases.

(ii) Coverage probability of the proposed confidence interval is larger than the coverage probability of the one-sided confidence interval proposed by [19] for all values of α.

Table 2 Estimated coverage probability and width based on T_α

γ	90%							
n	20		30		50		100	
α	Coverage	Width	Coverage	Width	Coverage	Width	Coverage	Width
0.00	0.9701	0.2043	0.9680	0.1367	0.9662	0.0823	0.9648	0.0413
0.01	0.9694	0.2028	0.9672	0.1357	0.9654	0.0817	0.9644	0.0409
0.05	0.9666	0.1974	0.9648	0.1319	0.9627	0.0792	0.9613	0.0398
0.1	0.9625	0.1898	0.9607	0.1271	0.9571	0.0756	0.9535	0.0384
0.2	0.9496	0.1754	0.9460	0.1175	0.9364	0.0707	0.9160	0.0356
0.3	0.9277	0.1610	0.9160	0.1079	0.8885	0.0649	0.8399	0.0327
$\alpha(n)$	0.9658	0.1951	0.9650	0.1322	0.9643	0.0806	0.9638	0.0408

Table 3 Estimated coverage probability and width based on T_α

γ	95%							
n	20		30		50		100	
α	Coverage	Width	Coverage	Width	Coverage	Width	Coverage	Width
0.00	0.9865	0.2477	0.9860	0.1657	0.9840	0.9968	0.9828	0.0500
0.01	0.9862	0.2458	0.9855	0.1644	0.9836	0.0989	0.9826	0.0495
0.05	0.9845	0.2383	0.9837	0.1594	0.9819	0.0959	0.9809	0.0481
0.1	0.9824	0.2289	0.9809	0.1531	0.9791	0.0922	0.9769	0.0462
0.2	0.9742	0.2102	0.9718	0.1407	0.9670	0.0846	0.9544	0.0425
0.2	0.9596	0.1914	0.9529	0.1282	0.9351	0.0771	0.8954	0.0387
$\alpha(n)$	0.9841	0.2358	0.9837	0.1598	0.9830	0.9749	0.09825	0.0493

Table 4 Estimated coverage probability and width based on T_α

γ	99%							
n	20		30		50		100	
α	Coverage	Width	coverage	width	Coverage	Width	Coverage	Width
0.00	0.9982	0.3486	0.9975	0.2329	0.9972	0.1400	0.9971	0.0701
0.01	0.9981	0.3457	0.9974	0.2310	0.9971	0.13887	0.9965	0.0696
0.05	0.9976	0.3342	0.9971	0.2233	0.9964	0.1343	0.9960	0.0673
0.1	0.9972	0.3198	0.9964	0.2137	0.9958	0.1284	0.9950	0.0643
0.2	0.9949	0.2909	0.9940	0.1944	0.9927	0.1169	0.9895	0.0586
0.3	0.9908	0.2621	0.9882	0.1752	0.9837	0.1054	0.9569	0.0529
$\alpha(n)$	0.9770	0.3302	0.9972	0.2239	0.9968	0.1366	0.9965	0.0692

(iii) The width of the proposed confidence interval is larger than that of the one-sided confidence interval proposed by [19].

(iv) We propose to use confidence interval with choice α as $\alpha(n)$ as it gives very good coverage probability and also has a smaller width as compared to $\alpha = 0$.

4 Bayes Estimation

We obtain a generalized Bayes estimator of ξ with respect to a non-informative prior which is uniform on $\mathbb{R}$.

The prior distribution of ξ is $g(\xi) = 1, \ -\infty < \xi < \infty$.

$$h(\xi|\underline{x}) = \frac{\sqrt{n}\exp(-n(\xi - \bar{x})^2/2)}{\sqrt{2\pi}\Phi(\sqrt{n}(x_{(1)} - \bar{x}))}, \ -\infty < \xi < x_{(1)}.$$

So, the generalized Bayes estimator with respect to squared error loss function is given by

$$\delta(\underline{X}) = \overline{X} - \frac{\exp\{-n(X_{(1)} - \overline{X})^2/2\}}{\sqrt{2n\pi}\,\Phi(\sqrt{n}(X_{(1)} - \overline{X}))},$$

which is same as the Pitman estimator. This estimator is biased for ξ.

Definition 2 For estimating the location parameter of a location families of distributions, the best location invariant estimator with respect to squared error loss function is called the Pitman estimator (see [8], p. 186).

Next, we prove that the Pitman estimator is a limit of Bayes rules. For this, we consider the prior distribution of ξ as a normal distribution with mean 0 and standard deviation τ. After some algebra, we observe that the posterior distribution of ξ given $\underline{x}$ is a truncated normal distribution and the Bayes estimator is the mean of this distribution. This is obtained as

$$\delta_B = \frac{n\overline{X}}{a^2} - \frac{\exp\left\{-\frac{1}{2}\left(aX_{(1)} - \frac{n\overline{X}}{a}\right)^2\right\}}{a\sqrt{2\pi}\,\Phi\left(aX_{(1)} - \frac{n\overline{X}}{a}\right)}.$$

Further, $a = \sqrt{\frac{n\tau^2+1}{\tau^2}} \to \sqrt{n}$ as $\tau \to \infty$, and so

$$\lim_{\tau \to \infty} \delta_B = \overline{X} - \frac{\exp\{-n(X_{(1)} - \overline{X})^2/2\}}{\sqrt{2n\pi}\,\Phi(\sqrt{n}(X_{(1)} - \overline{X}))}$$

which is the same as the Pitman estimator. This shows that the Pitman estimator is a limit of Bayes rules.

4.1 Minimaxity and Admissibility of the Pitman Estimator

In this section, we prove that the Pitman estimator is minimax and admissible.

Theorem 2 *The Pitman estimator $\delta(\underline{X})$ is minimax for estimating ξ with respect to squared loss function.*

Proof The estimator $\delta(\underline{X})$ is the best location equivariant estimator. So by Theorem 3.3 of [9], $\delta(\underline{X})$ is minimax.

Next, we state a theorem of given in [18]. Let $X_1, \ldots, X_n$ be independent and identical random variables with the density $f(x - \xi)$, where ξ is unknown but the function f is known. Then, the Pitman estimator of ξ is given as

$$\hat{\xi}(\underline{X}) = \frac{\int \xi \prod f(X_i - \xi)d\xi}{\int \prod f(X_i - \xi)d\xi}. \tag{12}$$

It is the best location equivariant estimator with respect to the loss function (4).

Theorem 3 *If*

$$\int \prod f(x_i) \left\{ \frac{\int \xi^2 \prod f(x_i - \xi)d\xi}{\int \prod f(x_i - \xi)d\xi} - \left(\frac{\int \xi \prod f(x_i - \xi)d\xi}{\int \prod f(x_i - \xi)d\xi} \right)^2 \right\}^{3/2} \prod dx_i < \infty$$

then $\hat{\xi}(\underline{X})$ is defined by (12) is admissible with respect to squared error loss function.

Proof Proof of this theorem is essentially given in [18].

We use the above theorem to prove admissibility of the Pitman estimator of our problem.

Theorem 4 *The Pitman estimator $\delta(\underline{X})$ given by (12) is admissible for estimating ξ with respect to squared error loss function.*

Proof The Pitman estimator $\delta(\underline{X})$ is given by

$$\delta(\underline{X}) = \overline{X} - \frac{\exp\{-n(X_{(1)} - \overline{X})^2/2\}}{\sqrt{2n\pi}\Phi(\sqrt{n}(X_{(1)} - \overline{X}))} = \overline{X} - \frac{1}{\sqrt{n}}\frac{\phi(b)}{\Phi(b)} = \overline{X} - \frac{1}{\sqrt{n}}\nu(b),$$

where $b = \sqrt{n}(X_{(1)} - \overline{X})$ and we denote $\nu(y) = \frac{\phi(y)}{\Phi(y)}$. We can express

$$h(\xi|\underline{x}) = \sqrt{n}\frac{\phi(\beta)}{\Phi(b)}, \quad -\infty < \xi < x_{(1)},$$

where $\beta = \sqrt{n}(\xi - \overline{x})$. Now

$$\int_{-\infty}^{x_{(1)}} \xi^2 h(\xi|\underline{x})d\xi = \int_{-\infty}^{b} \left(\frac{\beta}{\sqrt{n}} + \overline{x} \right)^2 \frac{\phi(\beta)}{\Phi(b)} d\beta$$

$$= \frac{1}{n\Phi(b)}\{\Phi(b) - b\phi(b)\} - \frac{2\overline{x}}{\sqrt{n}}\frac{\phi(b)}{\Phi(b)} + \overline{x}^2$$

$$= \frac{1}{n}\{1 - b\nu(b)\} - \frac{2\overline{x}}{\sqrt{n}}\nu(b) + \overline{x}^2. \tag{13}$$

Again,

$$\left(\int_{-\infty}^{x_{(1)}} \xi h(\xi|\underline{x})d\xi \right)^2 = (\delta(\underline{x}))^2 = \left(\overline{x} - \frac{1}{\sqrt{n}}\nu(b) \right)^2 = \overline{x}^2 - \frac{2\overline{x}}{\sqrt{n}}\nu(b) + \frac{1}{n}(\nu(b))^2.$$

Subtracting (14) from (13), we get

$$\frac{1 - b\nu(b)}{n} - \frac{(\nu(b))^2}{n} = \frac{1 - \nu(b)(b + \nu(b))}{n} < \frac{1}{n} - \frac{(\bar{x} - x_{(1)})(b + \nu(b))}{\sqrt{n}} \leq \frac{1}{n}$$

Since $(\bar{x} - x_{(1)}) > 0$ and by Lemma 1 of [13], $b + \nu(b) > 0$, it follows that

$$\left[\int_{-\infty}^{x_{(1)}} \xi^2 h(\xi | \underline{x}) d\xi - \left(\int_{-\infty}^{x_{(1)}} \xi h(\xi | \underline{x}) \right)^2 \right]^{3/2} < \left(\frac{1}{n} \right)^{3/2}.$$

So by Theorem 3, the Pitman estimator $\delta(\underline{X})$ is admissible.

5 Numerical Comparisons

In this section, we numerically compare the percentage risk improvement (PRI) of estimators T_0, $\hat{\xi}_{BC}$, T_1, $T_{\alpha(n)}$, and δ over the MLE $X_{(1)}$. For $\eta = 1$ the estimator $\hat{\xi}_{BC}$ is given as $\hat{\xi}_{BC} = X_{(1)} - \Phi^{-1}\left(\frac{1}{2} + \frac{1}{2n}\right)$. Note that risk function of these estimators does not depend on ξ. For the purpose of simulation study, we have generated 50,000 random samples of size n from a general half-normal distribution with parameters $\xi = 0$ and $\eta = 1$. For various values of n, we tabulate PRIs of all the estimators in Table 5. The PRI of an estimator T over the MLE is defined as

$$PRI(T) = \frac{Risk(MLE) - Risk(T)}{Risk(MLE)} \times 100. \tag{14}$$

Following observations can be made from the tabulated values.

 (i) The percentage risk improvement over MLE of the Pitman estimator δ is the highest among all estimators.
 (ii) The PRIs of $T_{\alpha(n)}$ and δ are approximately the same.

Table 5 Percentage risk improvement of various estimators

$(n, \alpha(n))$	$(5, 0.1667)$	$(10, 0.10684)$	$(20, 0.06358)$	$(30, 0.04651)$	$(50, 0.02971)$	$(100, 0.0158)$
T_0	55.88	53.85	52.24	51.61	50.82	50.44
$\hat{\xi}_{BC}$	54.09	53.43	52.14	51.56	50.81	50.44
T_1	12.00	−46.81	−164.23	−287.52	−505.16	−1088.79
$T_{\alpha(n)}$	57.47	55.26	53.23	52.37	51.37	50.70
δ	57.54	55.30	53.27	52.40	51.38	50.71

(iii) The performance of the unbiased estimators T_0 and $T_{\alpha(n)}$ is better than the bias-corrected estimator $\hat{\xi}_{BC}$.

Thus, we will recommend using δ or $T_{\alpha(n)}$ as estimator of ξ.

6 Conclusion

In the present article, we have considered the estimation of the location parameter of a general half-normal distribution with respect to squared error loss function. We have obtained some unbiased as well as biased estimators. It is proved that the Pitman estimator is a limit of Bayes rules and also shown that the Pitman estimator is minimax and admissible. Based on the MLE, we have derived a complete class of estimators. A one-sided asymptotic confidence interval is also obtained for the location parameter. Simulation study is carried out for implementation purpose.

References

1. Aigner, D., Lovell, C.K., Schmidt, P.: Formulation and estimation of stochastic frontier production function models. J. Econometrics **6**(1), 21–37 (1977)
2. Azzalini, A.: A class of distributions which includes the normal ones. Scand. J. Stat. **12**(2), 171–178 (1985)
3. Azzalini, A., Capitanio, A.: Statistical applications of the multivariate skew normal distribution. J. R. Stat. Soc. Ser. B (Stat. Methodol.) **61**(3), 579–602 (1999)
4. Brewster, J.F., Zidek, J.: Improving on equivariant estimators. Ann. Stat. **2**(1), 21–38 (1974)
5. Cook, R.D., Weisberg, S.: An Introduction to Regression Graphics, vol. 405. Wiley, New York (2009)
6. Daniel, C.: Use of half-normal plots in interpreting factorial two-level experiments. Technometrics **1**(4), 311–341 (1959)
7. Farsipour, N.S., Rasouli, A.: On the Bayes estimation of the general half-normal distribution. Calcutta Stat. Assoc. Bull. **58**(1–2), 37–52 (2006)
8. Ferguson, T.S.: Mathematical Statistics: A Decision Theoretic Approach. Academic Press, New York (2014)
9. Girshick, M., Savage, L., et al.: Bayes and minimax estimates for quadratic loss functions. In: Proceedings of the Second Berkeley Symposium on Mathematical Statistics and Probability, vol. 1, pp. 53–74 . University of California Press, Berkeley (1951)
10. Gut, A.: Probability: A Graduate Course. Springer Science, New York (2012)
11. Haberle, J.: Strength and failure mechanisms of unidirectional carbon fibre-reinforced plastics under axial compression. Ph.D. thesis, Imperial College London (1992)
12. Johnson, N.L., Kotz, S., Balakrishnan, N.: Continuous Univariate Distributions, vol. 1. Wiley, New York (1994)
13. Katz, M.W.: Admissible and minimax estimates of parameters in truncated spaces. Ann. Math. Stat. **32**(1), 136–142 (1961)
14. Nogales, A., Perez, P.: Unbiased estimation for the general half-normal distribution. Commun. Stat. Theory Methods **44**(17), 3658–3667 (2015)
15. Pewsey, A.: Problems of inference for Azzalini's skewnormal distribution. J. Appl. Stat. **27**(7), 859–870 (2000)

16. Pewsey, A.: Large-sample inference for the general half-normal distribution. Commun. Stat. Theory Methods **31**(7), 1045–1054 (2002)
17. Pewsey, A.: Improved likelihood based inference for the general half-normal distribution. Commun. Stat. Theory Methods **33**(2), 197–204 (2004)
18. Stein, C.: The admissibility of Pitman's estimator of a single location parameter. Ann. Math. Stat. **30**(4), 970–979 (1959)
19. Wiper, M., Girion, F., Pewsey, A.: Objective bayesian inference for the half-normal and half-t distributions. Commun. Stat. Theory Methods **37**(20), 3165–3185 (2008)

Chapter 23
Existence of Equilibrium Solution of the Coagulation–Fragmentation Equation with Linear Fragmentation Kernel

Debdulal Ghosh and Jitendra Kumar

Abstract The existence of equilibrium solution of a coagulation–fragmentation equation is shown in this article. We study the problem for a linear fragmentation kernel. A numerical example is provided to explore the given investigation.

Keywords Coagulation–fragmentation equation · Singular kernels · Equilibrium solution

1 Introduction

The aim of this work is to investigate the existence of equilibrium state of the solution to the continuous coagulation–fragmentation equation (C-F equation) where the reaction rate satisfies certain restriction. It is to mention here that the C-F process represents the dynamic system that describes the mechanisms by which clusters can coalesce to form larger particles or fragment into smaller pieces. Many scientific fields apply this C-F process and the pertaining equation; for instances, aerosol science [3], animal grouping in population dynamics [9], red blood cell aggregation in hematology [10], astrophysics [11], colloidal chemistry, and polymer science [12, 13].

The general form of the C-F equation is the following integro-partial differential equation:

$$\frac{\partial c(x,t)}{\partial t} = \frac{1}{2} \int_0^x K(x-y, y) \, c(x-y, t) \, c(y, t) \, dy - c(x, t) \int_0^\infty K(x, y) \, c(y, t) \, dy$$

$$- \frac{1}{2} c(x, t) \int_0^x F(x-y, y) \, dy + \int_0^\infty F(x, y) \, c(x+y, t) \, dy, \tag{1}$$

D. Ghosh (✉) · J. Kumar
Department of Mathematics, Indian Institute of Technology Kharagpur,
Kharagpur 721302, West Bengal, India
e-mail: debdulal.email@gmail.com

J. Kumar
e-mail: jkumar@maths.iitkgp.ernet.in

D. Ghosh et al. (eds.), *Mathematics and Computing*, Springer Proceedings
in Mathematics & Statistics 253, https://doi.org/10.1007/978-981-13-2095-8_23

with the initial data

$$c(x, 0) = c_0(x) \geq 0, \quad \text{a.e.} \tag{2}$$

Equation (1) describes the time evolution of particles $c(x, t) \geq 0$ of size $x \geq 0$ at time $t \geq 0$. The functions K and F represent the nonnegative coagulation and fragmentation rate that changes the mass of the system. The first two terms on the right-hand side of (1) represent the birth and death terms, respectively, due to coagulation. The last two terms are, respectively, the death and birth terms due to fragmentation. More details of this equation can be found in [15]. In the literature, Eq. (1) is also known as population balance equation.

1.1 Literature Survey

An equilibrium solution of the C-F equation arises when the birth and death terms in Eq. (1) are equal. Toward identifying an equilibrium solution of the C-F equation, [2] has proved the existence of equilibrium solution by Laplace transform. In the articles of [1, 2, 14], the equilibrium solutions are in the form of $\exp(-\lambda x)$. A general equilibrium solution is also given in [4]. For linear coagulation kernel and constant fragmentation kernel, [5] have proved the existence of equilibrium solution and its convergence.

In this research article, we attempt to prove the existence of equilibrium solution for linear fragmentation kernel.

1.2 Problem Statement

For the continuous C-F equation (1), the detailed balance condition leads to the following *separate cancelation* condition [5]:

$$\left. \begin{array}{l} \dfrac{1}{2} \displaystyle\int_0^x K(x - y, y)\, \bar{c}(x - y)\, \bar{c}(y)\, dy - \dfrac{1}{2}\bar{c}(x) \displaystyle\int_0^x F(x - y, y)\, dy = 0, \\[3mm] \displaystyle\int_0^\infty F(x, y)\, \bar{c}(x + y)\, dy - \bar{c}(x) \displaystyle\int_0^\infty K(x, y)\, \bar{c}(y)\, dy = 0. \end{array} \right\} \tag{3}$$

In the present study, the problem under consideration *does not assume such separate cancelation condition*. Thus, the existence of equilibrium solution of the problem is not trivially followed. *In this article, a proof of the existence of equilibrium solution is presented.*

The outline of the presented work is as follows. In Sect. 2, the result on the existence and uniqueness of an equilibrium solution for the problem is given. Section 3 provides a numerical illustration of the performed analysis. Finally, Sect. 4 concludes the work by mentioning a brief future direction.

2 Existence and Uniqueness of an Equilibrium Solution

In the present study, we consider the following forms of *coagulation kernel K* and *fragmentation kernel F* for Eq. (1):

$$K(x, y) = x^{-\sigma} y^{-\sigma} \tag{4}$$

where $\sigma \in \left[0, \frac{1}{2}\right]$ and

$$F(x, y) = b\left[1 + (x + y)\right], \quad \text{with} \quad b > 0. \tag{5}$$

Let $\bar{c}(x)$ be an equilibrium solution of Eq. (1). Then, from Eq. (1) we obtain

$$\frac{1}{2} \int_0^x K(x - y, y)\, \bar{c}(x - y)\, \bar{c}(y)\, dy - \bar{c}(x) \int_0^\infty K(x, y)\, \bar{c}(y)\, dy$$
$$- \frac{1}{2} \bar{c}(x) \int_0^x F(x - y, y)\, dy + \int_0^\infty F(x, y)\, \bar{c}(x + y)\, dy = 0. \tag{6}$$

Fitting the coagulation and fragmentation kernels under consideration into Eq. (6), the first term of Eq. (6) reduces to

$$\frac{1}{2} \int_0^x \left[(x - y)^{-\sigma} y^{-\sigma}\right] \bar{c}(x - y)\, \bar{c}(y)\, dy$$
$$= \frac{1}{2} \left[\phi * \phi\right](x),$$

where $\phi(x) := x^{-\sigma} \bar{c}(x)$, and $\zeta * \vartheta$ represents the following integral

$$\zeta * \vartheta(x) = \int_0^x \zeta(x - t)\, \vartheta(t)\, dt.$$

Denoting

$$N_{-\sigma} = \int_0^\infty \phi(x)\, dx,$$

the second term of Eq. (6) gives

$$\bar{c}(x) \int_0^\infty \left[x^{-\sigma} y^{-\sigma} \right] \bar{c}(y) \ dy = \phi(x) N_{-\sigma}.$$

The third integral of Eq. (6) yields

$$\frac{1}{2} \bar{c}(x) \int_0^x F(x - y, y) \ dy = \frac{b}{2} \bar{c}(x) \int_0^x (1 + x) \ dy = \frac{b}{2} \bar{c}(x) \left[x + \frac{x^2}{2} \right]$$

$$= \frac{1}{2} \bar{c}(x) \ b \ x + \frac{b}{2} \bar{c}(x) \frac{x^2}{2}.$$

Lastly, the fourth integral of Eq. (6) gives

$$\int_0^\infty F(x, y) \bar{c}(x + y) \ dy = \int_0^\infty b [1 + (x + y)] \bar{c}(x + y) \ dy$$

$$= b \int_x^\infty (1 + z) \bar{c}(z) \ dz$$

$$= b \left[\int_0^\infty (1 + z) \bar{c}(z) \ dz - \int_0^x (1 + z) \bar{c}(z) \ dz \right]$$

$$= b \ N + b M - b * \bar{c}(x) - b * \rho,$$

where $\rho(x) := x \bar{c}(x)$.

Therefore, from Eq. (6), we obtain

$$\frac{1}{2} [\phi * \phi] - \left[\phi(x) N_{-\sigma} \right] - \frac{1}{2} b \ x \ \bar{c}(x) - \frac{b}{2} \bar{c}(x) \frac{x^2}{2} + bN$$

$$+ bM - b * \bar{c}(x) - b * \rho = 0.$$

Hence,

$$\bar{c}(x) = \frac{\phi * \phi + 2bN + 2bM - 2b * \bar{c} - 2b * \rho}{2x^{-\sigma} N_{-\sigma} + b \left[x + \frac{x^2}{2} \right]}. \tag{7}$$

The function $\bar{c}$ is an equilibrium solution to (1).

Denoting the right-hand side of Eq. (7) by $\mathscr{A}(\bar{c})$, we note that

(i) $\mathscr{A}$ is an operator from the continuous functions space $\mathscr{C}(0, \ \alpha]$ into itself, α is a positive real number, and

(ii) Letting c_1 and c_2 satisfy (7), we have

$$\begin{aligned}
&|\mathscr{A}(c_1) - \mathscr{A}(c_2)| \\
&= \frac{x^\sigma \left[(\phi_1 * \phi_1 - \phi_2 * \phi_2) + 2b * (c_1 - c_2) + 2b * (\rho_1 - \rho_2)\right]}{2N_{-\sigma} + bx^{1-\sigma}\left[1 + \frac{x}{2}\right]}
\end{aligned}
\tag{8}$$

where

$$\begin{aligned}
\phi_1(x) &:= x^{-\sigma} c_1(x), & \phi_2(x) &:= x^{-\sigma} c_2(x), \\
\rho_1(x) &:= x c_1(x), & \rho_2 &:= x c_2(x).
\end{aligned}$$

We consider the first term of the numerator of Eq. (8). We see that

$$\begin{aligned}
|\phi_1 * \phi_1 - \phi_2 * \phi_2| &\leq |\phi_1 - \phi_2| * |\phi_1 + \phi_2| \\
&= \int_0^x y^{-\sigma}(x - y)^{-\sigma}|c_1 - c_2|(y)|c_1 + c_2|(x - y)\, dy.
\end{aligned}$$

Therefore, under the supremum norm, $\|f\| := \sup_{x \in (0,\alpha]} |f(x)|$, we get

$$\begin{aligned}
\|\phi_1 * \phi_1 - \phi_2 * \phi_2\| &= \|(\phi_1 - \phi_2) * (\phi_1 + \phi_2)\| \\
&\leq \|c_1 - c_2\| . \|c_1 + c_2\| \sup_{x \in (0,\alpha]} \left| \int_0^x y^{-\sigma}(x - y)^{-\sigma}\, dy \right| \\
&= \|c_1 - c_2\| . \|c_1 + c_2\| \beta(1 - \sigma, 1 - \sigma)\alpha^{1-2\sigma},
\end{aligned}$$

where $\beta(\cdot, \cdot)$ is the well-known beta function.

Let $\beta_0 := \beta(1 - \sigma, 1 - \sigma)$.
Thence, we have

$$\|\mathscr{A}c_1 - \mathscr{A}c_2\| \leq \frac{(\|c_1\| + \|c_2\|)\,\beta_0\alpha^{1-\sigma} + 2b\alpha^{1+\sigma}\left[1 + \frac{\alpha}{2}\right]}{2\left[N_{-\sigma}\right]}\|c_1 - c_2\|.$$

Thus, the operator $\mathscr{A}$ is contractive if

$$\frac{\|c\|\,\beta_0\alpha^{1-\sigma} + b\alpha^{\sigma+1}\left[1 + \frac{\alpha}{2}\right]}{N_{-\sigma}} \leq 1,$$

that is, if

$$\|c\| \leq \frac{N_{-\sigma} - b\alpha^{\sigma+1}\left[1 + \frac{\alpha}{2}\right]}{\beta_0\alpha^{1-\sigma}} =: R_\alpha, \quad \text{say.}
\tag{9}$$

It is to notice here that $R_\alpha > 0$ under certain restriction on α.

In order to use the contraction mapping theorem, we require to check if the ball $\mathfrak{B}(R_\alpha)$ is invariant.

We observe that

$$\|\mathscr{A}c\| \leq \frac{\beta_0 \alpha^{1-\sigma} \|c\|^2 + 2b\,(N+M) + 2b\alpha^{\sigma+1}\left(1+\frac{\alpha}{2}\right)\|c\|}{2N_{-\sigma}}.$$

Through the inequality $\|\mathscr{A}c\| \leq \|c\|$, it is easy to see that the ball $\mathfrak{B}(R_\alpha)$ remains invariant if

$$\frac{\beta_0 \alpha^{1-\sigma} \|c\|^2 + 2b\,(N+M) + 2b\alpha^{\sigma+1}\left(1+\frac{\alpha}{2}\right)\|c\|}{2N_{-\sigma}} \leq \|c\|,$$

that is, if

$$\beta_0 \alpha^{1-\sigma} \|c\|^2 + 2\|c\|\left[b\alpha^{\sigma+1}\left(1+\frac{\alpha}{2}\right) - N_{-\sigma}\right] + 2b(N+M) \leq 0.$$

We denote $a_1 := \left[N_{-\sigma} - b\alpha^{\sigma+1}\left(1+\frac{\alpha}{2}\right)\right]$. Therefore, the immediately above relation gives

$$\|c\| \leq \frac{a_1 + \sqrt{a_1^2 - 2b(N+M)\beta_0\alpha^{1-\sigma}}}{\beta_0\,\alpha^{1-\sigma}}. \tag{10}$$

The expression under the square root in inequality (10) and the quantity R_α in (9) is nonnegative for a range of values of α. We work on this range of α values.

We are now at a position to prove the following lemma.

Lemma 1 *Let α be such that $R_\alpha > 0$ and the expression under the square root in (10) is nonnegative. Then, there exists a unique continuous solution to (7) on the interval $(0, \alpha]$ which lies in the ball $\mathfrak{B}(R_\alpha)$.*

Proof Existence and uniqueness of a continuous solution $\bar{c}$ in the ball $\mathfrak{B}(R_\alpha)$ follow from the contraction mapping theorem [6].

We prove the uniqueness of all solutions to (7), not necessarily inside the ball $\mathfrak{B}(R_\alpha)$.

Suppose that there exists another solution $\bar{d}$ to (7). The continuity of $\bar{d}$ follows from its integrability and we remark that the operator $\mathscr{A}$ maps any integrable function to a continuous one.

Let us consider the restriction of $\bar{d}$ to an interval $(0, \alpha_1]$, $\alpha_1 < \alpha$. Choosing α_1 small enough, we find that the ball $\mathfrak{B}(R_{\alpha_1})$ contains two solutions $\bar{c}$ and $\bar{d}$. Actually, R_{α_1} tends to ∞ as α_1 tends to 0. This result contradicts the uniqueness of the solution of (7) in the ball $\mathfrak{B}(R_{\alpha_1})$.

Lemma 2 *For all $x > 0$, there exists a unique continuous solution to Eq. (7).*

Proof We consider the operator $\mathscr{A}$ as a mapping $\mathscr{A} : \mathscr{C}[\alpha, 2\alpha] \to \mathscr{C}[\alpha, 2\alpha]$ and it is a solution $\tilde{c}(x)$ of (7) on $[\alpha, 2\alpha]$. The function $\tilde{c}(x)$ evidently satisfies the equality

$$
\tilde{c}(x) = \left[\int_{\alpha}^{x} y^{-\sigma}(x-y)^{-\sigma} \tilde{c}(y)\bar{c}(x-y)\, dy + \frac{1}{2}\int_{x-\alpha}^{\alpha} y^{-\sigma}(x-y)^{-\sigma}\bar{c}(y)\bar{c}(x-y)\, dy \right.
$$
$$
\left. + b\left(N + M - \int_{0}^{\alpha}(1+x)\bar{c}(x)\, dx - \int_{\alpha}^{x}(1+y)\tilde{c}(y)\, dy \right) \right]
$$
$$
\times \left[\frac{1}{x^{-\sigma}(N_{-\sigma}) + \frac{bx}{2}\left(1+\frac{x}{2}\right)} \right]. \tag{11}
$$

Here, the function $\bar{c}$ is a solution to (7) on $(0, \alpha]$. Its existence and uniqueness were proved in Lemma 1. By the standard results on integral equations, the linear Volterra equation (11) has a unique continuous solution $\tilde{c}(x)$ on the interval $[\alpha, 2\alpha]$.

Put $\bar{c}(x) = \tilde{c}(x)$ if $\alpha \le x \le 2\alpha$. Obviously, $\bar{c}$ satisfies (7) for all $x \in (0, 2\alpha]$. Its continuity follows form the proof of Lemma 1. We can now analogously extend the solution obtained to the interval $[2\alpha, 4\alpha]$, and so on. Hence, the result follows.

3 Numerical Results

In this section, we shown that for some initial condition, the time-dependent solution achieved to equilibrium state. To explore the numerical result, we use the finite volume scheme introduced by [7, 8]. In this example, we consider computational domain in $[10^{-9}, 512]$ and it is discretized into 20 *non-uniform* subintervals $\Lambda_i :=$ $[x_{i-1/2}, x_{i+1/2}], i = 1, 2, \ldots, 20$. The end points of Λ_i satisfies the relation $x_{i+1/2} = r x_{i-1/2}$ where $r > 1$ is the geometric ratio. The mid-point of each Λ_i is considered to be the *cell representative* or the *pivot*. We have used adaptive Runge–Kutta 4(5) solver in MATLAB-R2015 software to solve the system of ODEs.

In order to prove the existence result, we have taken coagulation kernel in the form $K(x, y) = (1 + x^{\lambda} + y^{\lambda})(xy)^{-\sigma}$, where $0 \le \sigma \le 0.5$ and $0 \le \lambda - \sigma \le 1$ and constant fragmentation kernel $F(x, y) = 1$, with the initial data $c_0 = (1 + x)^{-2}$. To observe the equilibrium of the system, we plot numerical number density function along with the moments $M_2(t)$, $M_0(t)$ and $M_{-\sigma}(t)$. The zeroth moment $M_0(t)$ represents the total particle number in the system. Therefore, the constant value of $M_0(t)$, after a certain time lapse, indicates a equilibrium system and the constant moments of $M_2(t)$ and $M_{-\sigma}(t)$ also support the above result.

3.1 *Example 1*

In this example, we consider the problem (1) with the kernels

$$K(x, y) = (1 + x^{0.5} + y^{0.5})(xy)^{-0.5}, \qquad F(x, y) = 1$$

and is supported by the initial data $c_0 = (1 + x)^{-2}$. From Fig. 1, the particle number density $c(x, t)$ has no change at three different times $t = 1, 3, 5$, and from Fig. 2, we can see that all the moments are constant after $t = 2$. So, we can say that the system has reached to equilibrium after $t = 2$.

Fig. 1 Particle number density

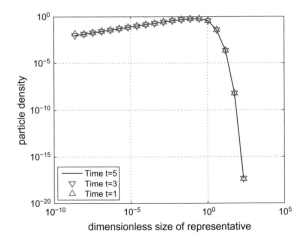

Fig. 2 Normalized moments

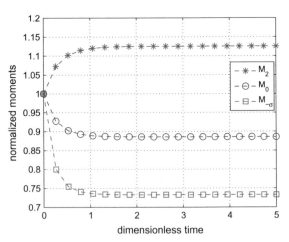

4 Conclusion

In this study, we have proved the existence of an equilibrium solution to for the C-F equation with a class of linear fragmentation kernel and singular coagulation kernel. One numerical example has been shown that explores the provided analysis. In order to prove the result, we have used that Banach contraction mapping theorem, a few inequalities related to improper integral and the properties of beta and gamma functions. As a future scope, one can attempt to extend the result for a larger class of fragmentation kernels.

References

1. Aizenman, M., Bak, T.A.: Convergence to equilibrium in a system of reacting polymers. Commun. Math. Phys. **65**(3), 203–230 (1979)
2. Barrow, J.D.: Coagulation with fragmentation. J. Phys. A Math. Gen. **14**(3), 729 (1981)
3. Drake, R.L.: A general mathematical survey of the coagulation equation. Top. Curr. Aerosol Res. (Part 2) **3**, 201–376 (1972)
4. Dubovskiĭ, P., Galkin, V.A., Stewart, I.W.: Exact solutions for the coagulation-fragmentation equation. J. Phys. A Math. Gen. **25**(18), 4737 (1992)
5. Dubovskiĭ, P., Stewart, I.W.: Trend to equilibrium for the coagulation-fragmentation equation. Math. Methods Appl. Sci. **19**(10), 761–772 (1996)
6. Edwards, R.: Functional analysis: theory and applications, Holt, Rinehart and Winston, New York, 1965. MR **36**, 4308 (1994)
7. Kumar, J., Kaur, G., Tsotsas, E.: An accurate and efficient discrete formulation of aggregation population balance equation. Kinet. Relat. Models **9**(2), 373–391 (2016)
8. Kumar, J., Saha, J., Tsotsas, E.: Development and convergence analysis of a finite volume scheme for solving breakage equation. SIAM J. Numer. Anal. **53**(4), 1672–1689 (2015)
9. Okubo, A.: Dynamical aspects of animal grouping: swarms, schools, flocks, and herds. Adv. Biophys. **22**, 1–94 (1986)
10. Perelson, A.S., Samsel, R.W.: Kinetics of red blood cell aggregation: an example of geometric polymerization. In: Kinetics of Aggregation and Gelation, pp. 137–144 (1984)
11. Safronov, V.S. Evolution of the protoplanetary cloud and formation of the earth and planets. In: Safronov, V.S. (ed.) Evolution of the Protoplanetary Cloud and Formation of the Earth and Planets, vol. 1, 212 p. Translated from Russian. Israel Program for Scientific Translations, Keter Publishing House, Jerusalem, Israel (1972)
12. Smoluchowski, M.: Drei vortrage uber diffusion. brownsche bewegung und koagulation von kolloidteilchen. Z. Phys. **17**, 557–585 (1916)
13. Smoluchowski, M.: Grundriß der koagulationskinetik kolloider lösungen. Colloid Polym. Sci. **21**(3), 98–104 (1917)
14. Stewart, I.W., Dubovskiĭ, P.: Approach to equilibrium for the coagulation-fragmentation equation via a Lyapunov functional. Math. Methods Appl. Sci. **19**(3), 171–185 (1996)
15. Stewart, I.W., Meister, E.: A global existence theorem for the general coagulation-fragmentation equation with unbounded kernels. Math. Methods Appl. Sci. **11**(5), 627–648 (1989)

Chapter 24
Explicit Criteria for Stability of Two-Dimensional Fractional Nabla Difference Systems

Jagan Mohan Jonnalagadda

Abstract In this article, we discuss a few stability properties of the Riemann–Liouville (or Caputo)-type linear two-dimensional fractional nabla difference system. For this purpose, we construct the equivalent Volterra difference system of convolution type and analyse its properties using the standard methods applied in the qualitative investigation of Volterra difference systems. Subsequently, we obtain sufficient conditions on stability of the considered fractional nabla difference system. We provide an example to illustrate the applicability of established results.

Keywords Fractional order · Nabla difference · Volterra system · Z-transform Stability

1 Introduction

Matignon [1] established the following well-known criteria for stability of the linear fractional differential system

$$\left(D^\alpha x\right)(t) = Ax(t), \quad t > 0, \tag{1}$$

of Riemann–Liouville (or Caputo) type:

Theorem 1 *Let $0 < \alpha < 1$ and $A \in \mathbb{R}^{k \times k}$. Then, (1) is asymptotically stable if and only if*

$$|arg \lambda| > \frac{\alpha \pi}{2} \tag{2}$$

for all the eigenvalues λ of A.

Later, many other stability results on systems of fractional differential equations have appeared [2]. On the other hand, stability theory of fractional nabla difference

J. M. Jonnalagadda (✉)
Department of Mathematics, Birla Institute of Technology and Science Pilani,
Hyderabad 500078, Telangana, India
e-mail: j.jaganmohan@hotmail.com

© Springer Nature Singapore Pte Ltd. 2018
D. Ghosh et al. (eds.), *Mathematics and Computing*, Springer Proceedings
in Mathematics & Statistics 253, https://doi.org/10.1007/978-981-13-2095-8_24

equations is less developed. Recently, [3, 4] obtained the nabla discrete analogue of Theorem 1 as follows:

Theorem 2 *Consider the fractional nabla difference system*

$$\left(\nabla^{\alpha}_{\rho(0)}u\right)(t) = Au(t), \quad t \in \mathbb{N}_1, \tag{3}$$

of Riemann–Liouville type. Let $0 < \alpha < 1$, $A \in \mathbb{R}^{k \times k}$ *and* $\det(I - A) \neq 0$. *If all the eigenvalues* λ *of A lie inside the region*

$$S_{\alpha} = \left\{ z \in \mathbb{C} : |arg\ z| > \frac{\alpha\pi}{2} \ or\ |z| > \left(2\cos\frac{arg\ z}{\alpha}\right)^{\alpha} \right\}, \tag{4}$$

then (3) is asymptotically stable.

But, in many applications, one needs explicit criteria on the entries of the matrix associated with the considered system. In this article, we wish to formulate explicit stability conditions for two-dimensional Riemann–Liouville type fractional nabla difference systems.

2 Preliminaries

Throughout this article, we use the following notations, definitions and known results of discrete calculus [5, 6]: denote the set of all real numbers and complex numbers by $\mathbb{R}$ and $\mathbb{C}$, respectively. For any $a \in \mathbb{R}$, define $\mathbb{N}_a = \{a, a+1, a+2, \ldots\}$. Assume that empty sums and products are taken to be 0 and 1, respectively.

2.1 Fractional Nabla Calculus

Definition 1 (*Gamma Function*) For any $t \in \mathbb{R} \setminus \{\ldots, -2, -1, 0\}$, the gamma function is defined by

$$\Gamma(t) = \int_0^{\infty} e^{-s}s^{t-1}ds, \quad t > 0,$$

$$\Gamma(t+1) = t\Gamma(t).$$

Definition 2 (*Rising Factorial Function*) For any $t \in \mathbb{R} \setminus \{\ldots, -2, -1, 0\}$ and $\alpha \in \mathbb{R}$ such that $(t + \alpha) \in \mathbb{R} \setminus \{\ldots, -2, -1, 0\}$, the rising factorial function is defined by

$$t^{\overline{\alpha}} = \frac{\Gamma(t+\alpha)}{\Gamma(t)}, \quad 0^{\overline{\alpha}} = 0.$$

We observe the following properties of rising factorial functions.

Theorem 3 *Assume that the following factorial functions are well defined.*

1. $t^{\overline{\alpha}}(t+\alpha)^{\overline{\beta}} = t^{\overline{\alpha+\beta}}$.
2. *If* $t \le r$, *then* $t^{\overline{\alpha}} \le r^{\overline{\alpha}}$.
3. *If* $\alpha < t \le r$ *then* $r^{\overline{-\alpha}} \le t^{\overline{-\alpha}}$.
4. $(t+1)^{\alpha-1} \le (t+1)^{\overline{\alpha-1}} \le t^{\alpha-1}$, $\quad 0 \le \alpha \le 1$.
5. $(t+b)^{\overline{a-b}} = t^{a-b}\left[1 + O\left(\frac{1}{t}\right)\right]$, $\quad |t| \to \infty$.

Definition 3 Let $u : \mathbb{N}_a \to \mathbb{R}$, $\alpha \in \mathbb{R}$ such that $0 < \alpha < 1$.

1. (Nabla Difference) The first-order backward (nabla) difference of u is defined by

$$\left(\nabla u\right)(t) = u(t) - u(t-1), \quad t \in \mathbb{N}_{a+1}.$$

2. (Fractional Nabla Sum) The αth-order nabla sum of u based at $\rho(a) = (a-1)$ is given by

$$\left(\nabla_{\rho(a)}^{-\alpha} u\right)(t) = \frac{1}{\Gamma(\alpha)} \sum_{s=a}^{t} (t - \rho(s))^{\overline{\alpha-1}} u(s), \quad t \in \mathbb{N}_a.$$

3. (R-L Fractional Nabla Difference) The Riemann–Liouville-type αth-order nabla difference of u based at $\rho(a) = (a-1)$ is given by

$$\left(\nabla_{\rho(a)}^{\alpha} u\right)(t) = \left(\nabla\left(\nabla_{\rho(a)}^{-(1-\alpha)} u\right)\right)(t)$$

$$= \frac{1}{\Gamma(-\alpha)} \sum_{s=a}^{t} (t - \rho(s))^{\overline{-\alpha-1}} u(s), \quad t \in \mathbb{N}_a.$$

4. (Caputo Fractional Nabla Difference) The Caputo-type αth-order nabla difference of u based at a is given by

$$\left(\nabla_{a*}^{\alpha} u\right)(t) = \left(\nabla_a^{-(1-\alpha)}\left(\nabla u\right)\right)(t)$$

$$= \left(\nabla_a^{\alpha} u\right)(t) - \frac{(t-a)^{\overline{-\alpha}}}{\Gamma(1-\alpha)} u(a), \quad t \in \mathbb{N}_{a+1}.$$

2.2 Volterra Difference Systems

Consider a linear Volterra difference system of convolution-type

$$u(t+1) = \sum_{j=0}^{t} B(t-j)u(j), \tag{5}$$

where $u(t) = \left(u_1(t), u_2(t), \ldots, u_k(t)\right)^T, u_i : \mathbb{N}_0 \to \mathbb{R}, 1 \le i \le k$ and $B(t) = \left[b_{ij}(t)\right],$
$b_{ij} : \mathbb{N}_0 \to \mathbb{R}, 1 \le i, j \le k,$ is a $k \times k$ matrix valued function defined on $\mathbb{N}_0$. We
assume that $B(t) \in l_1$, i.e.

$$\sum_{j=0}^{\infty} |B(j)| < \infty.$$

Now, we state the standard definitions of stability and asymptotic stability adapted
to the Volterra system (5).

Definition 4 Consider (5) along with the initial condition $u(0) = u_0$. Then, (5) is
said to be

1. stable, if for any real vector u_0 there exists $\varepsilon > 0$ such that the corresponding
   solution $u(t)$ of (5) satisfies $|u(t)| < \varepsilon$ for all $t \in \mathbb{N}_1$.
2. asymptotically stable, if $u(t) \to 0$ as $t \to \infty$ for any real vector u_0.
3. uniformly stable, if for any $\varepsilon > 0$, there exists a $\delta = \delta(\varepsilon) > 0$ such that if u_0 is
   any real vector with $|u_0| < \delta$ then the corresponding solution $u(t)$ of (5) satisfies
   $|u(t)| < \varepsilon$ for all $t \in \mathbb{N}_1$.
4. uniformly asymptotically stable, if it is uniformly stable and if there exists a $\eta > 0$
   such that for any $\varepsilon > 0$ there is $N = N(\varepsilon) \in \mathbb{N}_1$ such that if u_0 is any real vector
   with $|u_0| < \eta$ then the corresponding solution $u(t)$ of (5) satisfies $|u(t)| < \varepsilon$ for
   all $t \in \mathbb{N}_N$.

Definition 5 The Z-transform of a sequence of real numbers $\{v(t)\}_{t \in \mathbb{N}_0}$ is defined
by

$$\tilde{v}(z) = Z[v(t)] = \sum_{k=0}^{\infty} v(k)z^{-k},$$

where $z \in \mathbb{C}$ for which the series converges absolutely. The Z-transform of a sequence
of vectors $\{u(t)\}_{t \in \mathbb{N}_0}$ and a sequence of matrices $\{B(t)\}_{t \in \mathbb{N}_0}$ over $\mathbb{R}$ are given by
$\tilde{u}(z) = Z[u(t)] = \left(\tilde{u}_1(z), \tilde{u}_2(z), \ldots, \tilde{u}_k(z)\right)^T$ and $\tilde{B}(z) = Z[B(t)] = \left[\tilde{b}_{ij}(z)\right]$, where
$\tilde{u}_i(z) = Z[u_i(t)], 1 \le i \le k$ and $\tilde{b}_{ij}(z) = Z[b_{ij}(t)], 1 \le i, j \le k.$

Z-transform can be used to discuss the stability properties of (5) by analysing the
roots of the associated characteristic equation $\det\left(zI - \tilde{B}(z)\right)$, where I is the $k \times k$
identity matrix. In this connection, we recall a few important results which will be
used to establish the main results of this article.

Theorem 4 *A necessary and sufficient condition for uniform asymptotic stability of
(5) is* $\det\left(zI - \tilde{B}(z)\right) \ne 0$ *for all* $|z| \ge 1$.

An application of the preceding theorem will be introduced next. This will provide
explicit criteria for asymptotic stability. Let

$$\beta_{ij} = \sum_{t=0}^{\infty} |b_{ij}(t)|, \quad 1 \le i, j \le k.$$

Theorem 5 *The zero solution of (5) is uniformly asymptotically stable if either one of the following conditions holds:*

1. $\sum_{j=1}^{k} \beta_{ij} < 1$, *for each* $1 \leq i \leq k$.
2. $\sum_{i=1}^{k} \beta_{ij} < 1$, *for each* $1 \leq j \leq k$.

The following theorem provides criteria for uniform stability of (5).

Theorem 6 *The zero solution of (5) is uniformly stable if*

$$\sum_{i=1}^{k} \beta_{ij} \leq 1,$$

for each $1 \leq j \leq k$.

3 Main Results

In this section, we investigate a few stability properties of the two-dimensional Riemann–Liouville-type fractional nabla difference system

$$\left(\nabla_{\rho(0)}^{\alpha} U\right)(t) = A\, U(t), \quad 0 < \alpha < 1, \quad t \in \mathbb{N}_1, \tag{6}$$

where $U = \begin{pmatrix} u_1 \\ u_2 \end{pmatrix}$; $u_1, u_2 : \mathbb{N}_0 \to \mathbb{R}$, $A = \begin{pmatrix} a_{11} & a_{12} \\ a_{21} & a_{22} \end{pmatrix}$; $a_{11}, a_{12}, a_{21}, a_{22} \in \mathbb{R}$. Let $T = \text{Trace}(A)$ and $D = \det(A)$. We assume the following necessary and sufficient condition for the existence of unique solution of (6).

$$\det(I - A) \neq 0, \text{ i.e. } T - D \neq 1.$$

(I) First, we obtain the equivalent Volterra-type difference system of (6). Rewriting (6), for $t \in \mathbb{N}_1$, we have

$$\left(\nabla_{\rho(0)}^{\alpha} u_1\right)(t) = a_{11} u_1(t) + a_{12} u_2(t), \tag{7}$$

$$\left(\nabla_{\rho(0)}^{\alpha} u_2\right)(t) = a_{21} u_1(t) + a_{22} u_2(t). \tag{8}$$

Expanding the Riemann–Liouville operator in (7) and (8), for $t \in \mathbb{N}_1$, we get

$$\frac{1}{\Gamma(-\alpha)} \sum_{s=0}^{t} (t - \rho(s))^{-\alpha-1} u_1(s) = a_{11} u_1(t) + a_{12} u_2(t), \tag{9}$$

$$\frac{1}{\Gamma(-\alpha)} \sum_{s=0}^{t} (t - \rho(s))^{-\alpha-1} u_2(s) = a_{21} u_1(t) + a_{22} u_2(t). \tag{10}$$

Rearranging the terms in (9) and (10), we have

$$(1 - a_{11})u_1(t) = -\frac{1}{\Gamma(-\alpha)} \sum_{s=0}^{t-1} (t - \rho(s))^{\overline{-\alpha-1}} u_1(s) + a_{12}u_2(t), \quad t \in \mathbb{N}_1, \quad (11)$$

$$(1 - a_{22})u_2(t) = -\frac{1}{\Gamma(-\alpha)} \sum_{s=0}^{t-1} (t - \rho(s))^{\overline{-\alpha-1}} u_2(s) + a_{21}u_1(t), \quad t \in \mathbb{N}_1, \quad (12)$$

or

$$(1 - a_{11})u_1(t + 1) - a_{12}u_2(t + 1)$$
$$= -\frac{1}{\Gamma(-\alpha)} \sum_{s=0}^{t} (t + 2 - s)^{\overline{-\alpha-1}} u_1(s), \quad t \in \mathbb{N}_0, \quad (13)$$

$$-a_{21}u_1(t + 1) + (1 - a_{22})u_2(t + 1)$$
$$= -\frac{1}{\Gamma(-\alpha)} \sum_{s=0}^{t} (t + 2 - s)^{\overline{-\alpha-1}} u_2(s), \quad t \in \mathbb{N}_0. \quad (14)$$

The matrix form of (13) and (14) is given by

$$\begin{pmatrix} 1 - a_{11} & -a_{12} \\ -a_{21} & 1 - a_{22} \end{pmatrix} \begin{bmatrix} u_1(t + 1) \\ u_2(t + 1) \end{bmatrix} = -\sum_{s=0}^{t} \frac{(t + 2 - s)^{\overline{-\alpha-1}}}{\Gamma(-\alpha)} \begin{pmatrix} 1 & 0 \\ 0 & 1 \end{pmatrix} \begin{bmatrix} u_1(s) \\ u_2(s) \end{bmatrix}.$$

Thus,

$$U(t + 1) = \sum_{s=0}^{t} B(t - s)U(t), \quad t \in \mathbb{N}_0, \quad (15)$$

is the equivalent Volterra-type difference system of (6) with

$$B(t) = \begin{vmatrix} 1 - a_{11} & -a_{12} \\ -a_{21} & 1 - a_{22} \end{vmatrix}^{-1} \frac{-(t + 2)^{\overline{-\alpha-1}}}{\Gamma(-\alpha)} \begin{pmatrix} 1 - a_{22} & a_{12} \\ a_{21} & 1 - a_{11} \end{pmatrix}$$
$$= -\frac{1}{(1 - T + D)} \frac{(t + 2)^{\overline{-\alpha-1}}}{\Gamma(-\alpha)} \begin{pmatrix} 1 - a_{22} & a_{12} \\ a_{21} & 1 - a_{11} \end{pmatrix}. \quad (16)$$

(II) Next, we derive the characteristic equation of (15). Taking Z-transforms on both sides of (16), we get

$$\tilde{B}(z) = -\frac{z}{(1 - T + D)} \left[1 - \left(1 - \frac{1}{z}\right)^{\alpha} \right] \begin{pmatrix} 1 - a_{22} & a_{12} \\ a_{21} & 1 - a_{11} \end{pmatrix} \quad (17)$$

for all $z \in \mathbb{C}$ with $|z| \geq 1$. Let

$$S = -\frac{1}{(1 - T + D)}\left[1 - \left(1 - \frac{1}{z}\right)^{\alpha}\right]. \tag{18}$$

Consider $\det(zI - \tilde{B}(z))$

$$= \begin{vmatrix} z - zS + zSa_{22} & -zSa_{12} \\ -zSa_{21} & z - zS + zSa_{11} \end{vmatrix}$$

$$= z^2\left[(1 - S)^2 + S(1 - S)T + S^2 a_{11}a_{22}\right] - z^2 S^2 a_{11}a_{22}.$$

Thus, the characteristic equation of (15) becomes

$$z^2\left[(1 - S)^2 + S(1 - S)T\right] = 0. \tag{19}$$

(III) Finally, we formulate an explicit necessary and sufficient condition for asymptotic stability of the Volterra system (15). Applying Theorem 4, the system (15) is uniformly asymptotically stable if and only if

$$(1 - S)^2 + S(1 - S)T \neq 0, \tag{20}$$

for all $z \in \mathbb{C}$ with $|z| \geq 1$. Consider

$$(1 - S)^2 + S(1 - S)T = 0. \tag{21}$$

If $T = 1$, then $D \neq 0$ and the only root of (21) is $S = 1$ implies

$$\left(1 - \frac{1}{z}\right)^{\alpha} = 1 + D. \tag{22}$$

We analyse (22) with respect to D. If $1 + D < 0$, then (22) has no root z_r, and hence the condition (20) is satisfied trivially. If $1 + D \geq 0$, then the unique nonzero real root of the characteristic equation (19) is given by

$$z_r = \frac{1}{1 - (1 + D)^{\frac{1}{\alpha}}}.$$

To satisfy (20), we require $(1 + D) > 2^{\alpha}$.

Suppose $T \neq 1$. Then, the roots of (21) are

$$S = \frac{1}{1 - T} \text{ and } 1,$$

implies

$$\left(1 - \frac{1}{z}\right)^\alpha = 2 - T + D \tag{23}$$

and

$$\left(1 - \frac{1}{z}\right)^\alpha = 2 + \frac{D}{1 - T}, \tag{24}$$

respectively. We analyse (23) and (24) with respect to T and D.

1. If $2 - T + D < 0$, then (23) has no root z_r and hence the condition (20) is satisfied trivially. If $2 - T + D \geq 0$, then the unique nonzero real root of the characteristic equation (19) is given by

$$z_r = \frac{1}{1 - (2 - T + D)^{\frac{1}{\alpha}}}.$$

   To satisfy (20), we require $(2 - T + D) > 2^\alpha$.
2. If $2 + \frac{D}{1-T} < 0$, then (24) has no root z_r and hence the condition (20) is satisfied trivially. If $2 + \frac{D}{1-T} \geq 0$, then the unique nonzero real root of the characteristic equation (19) is given by

$$z_r = \frac{1}{1 - (2 + \frac{D}{1-T})^{\frac{1}{\alpha}}}.$$

   To satisfy (20), we require $2 + \frac{D}{1-T} > 2^\alpha$.

Compiling the above results, we provide a necessary and sufficient condition for the asymptotic stability of (15) in the following theorem.

Theorem 7 *The system (15) is uniformly asymptotically stable if and only if*

$$T = 1, D \in \mathbb{R} \setminus [-1, 2^\alpha - 1] \tag{25}$$

or

$$T \neq 1, (D - T) \quad and \quad \left(\frac{D}{1 - T}\right) \in \mathbb{R} \setminus [-2, 2^\alpha - 2]. \tag{26}$$

Now, we apply Theorems 5 and 6 to establish explicit criteria for asymptotic stability of (15). Consider

$$\beta_{11} = \sum_{t=0}^{\infty} |b_{11}(t)| = \left| \frac{1 - a_{22}}{1 - T + D} \right| \sum_{t=0}^{\infty} \frac{(t + 2)^{-\alpha - 1}}{\Gamma(-\alpha)}$$

$$= \left| \frac{1 - a_{22}}{1 - T + D} \right|.$$

Similarly, we get

$$\beta_{12} = \left| \frac{a_{12}}{1 - T + D} \right|,$$

$$\beta_{21} = \left| \frac{a_{21}}{1 - T + D} \right|,$$

$$\beta_{21} = \left| \frac{1 - a_{11}}{1 - T + D} \right|.$$

Theorem 8 *The zero solution of (15) is uniformly asymptotically stable if either one of the following conditions holds:*

$$|a_{12}| + |1 - a_{22}|, |1 - a_{11}| + |a_{21}| < |1 - T + D| \tag{27}$$

or

$$|a_{21}| + |1 - a_{22}|, |1 - a_{11}| + |a_{12}| < |1 - T + D|. \tag{28}$$

Theorem 9 *The zero solution of (15) is uniformly stable if*

$$|a_{21}| + |1 - a_{22}|, |1 - a_{11}| + |a_{12}| < |1 - T + D|. \tag{29}$$

Finally, we consider the following two-dimensional Caputo-type fractional nabla difference system

$$\left(\nabla_{0*}^{\alpha} U\right)(t) = A\, U(t), \quad 0 < \alpha < 1, \quad t \in \mathbb{N}_1. \tag{30}$$

Using Definition 3 in (30), we get

$$\left(\nabla_0^{\alpha} U\right)(t) = A\, U(t) + F(t), \quad t \in \mathbb{N}_1, \tag{31}$$

where

$$F(t) = \frac{t^{\overline{-\alpha}}}{\Gamma(1 - \alpha)} \begin{pmatrix} u_1(0) \\ u_2(0) \end{pmatrix}, \quad t \in \mathbb{N}_0. \tag{32}$$

Then,

$$U(t + 1) = \sum_{s=0}^{t} B(t - s)U(t) + G(t), \quad t \in \mathbb{N}_0, \tag{33}$$

is the equivalent Volterra-type difference system of (30) with

$$G(t) = \frac{1}{(1 - T + D)} \frac{t^{\overline{-\alpha}}}{\Gamma(1 - \alpha)} \begin{pmatrix} 1 - a_{22} & a_{12} \\ a_{21} & 1 - a_{11} \end{pmatrix} \begin{pmatrix} u_1(0) \\ u_2(0) \end{pmatrix}. \tag{34}$$

Consequently, the characteristic equation of (30) becomes $\det(zI - \tilde{B}(z))$, which is same as (19).

4 Conclusion

To summarise this article, we reformulate some of its results for the fractional nabla difference system (6) (or (30)). Theorems 7–9 imply the following assertions.

Corollary 1 *The system (6) (or (30)) is uniformly asymptotically stable if and only if either (25) or (26) holds.*

Corollary 2 *The zero solution of (6) (or (30)) is uniformly asymptotically stable if either (27) or (28) holds.*

Corollary 3 *The zero solution of (6) (or (30)) is uniformly stable if (29) holds.*

Example 1 Consider the fractional nabla difference system

$$\left(\nabla^{0.5}_{\rho(0)}u_1\right)(t) = -(0.75)u_1(t) - u_2(t), \quad t \in \mathbb{N}_1, \tag{35}$$

$$\left(\nabla^{0.5}_{\rho(0)}u_2\right)(t) = u_1(t) - u_2(t), \quad t \in \mathbb{N}_1. \tag{36}$$

Solution: Here $\alpha = 0.5$, $A = \begin{pmatrix} -0.75 & -1 \\ 1 & -1 \end{pmatrix}$ and $u_0 = \begin{pmatrix} 0.25 \\ 0.75 \end{pmatrix}$. Then, $T = -1.75$, $D = 1.75$, $T - D = -3.50$ and $\frac{D}{T-1} = -0.6363$. Clearly, condition (26) holds. Hence, the system (35)–(36) is uniformly asymptotically stable.

References

1. Matignon, D.: Stability results for fractional differential equations with applications to control processing. In: Computational Engineering in Systems and Application Multiconference, vol. 2, pp. 963–968. IMACS, IEEE-SMC, Lille, France (1996)
2. Li, C.P., Zhang, F.R.: A survey on the stability of fractional differential equations. Eur. Phys. J. Spec. Top. **193**(1), 27–47 (2011)
3. Čermák, J., Győri, I., Nechvátal, L.: Stability regions for linear fractional difference systems and their discretizations. Appl. Math. Comput. **219**, 7012–7022 (2013)
4. Čermák, J., Győri, I., Nechvátal, L.: On explicit stability conditions for a linear fractional difference system. Fractional Calc. Appl. Anal. **18**(3), 651–672 (2015)
5. Elaydi, S.: An Introduction to Difference Equations, 3rd edn. Springer, New York (2005)
6. Goodrich, C., Peterson, A.C.: Discrete Fractional Calculus. Springer International Publishing (2015). https://doi.org/10.1007/978-3-319-25562-0
7. Agarwal, R.P.: Difference Equations and Inequalities. Marcel Dekker, New York (1992)
8. Podlubny, I.: Fractional Differential Equations. Academic Press, San Diego (1999)
9. Kelly, W.G., Peterson, A.C.: Difference Equations: An Introduction with Applications, 2nd edn. Academic Press, San Diego (2001)

Chapter 25
Discrete Legendre Collocation Methods for Fredholm–Hammerstein Integral Equations with Weakly Singular Kernel

Bijaya Laxmi Panigrahi

Abstract In this paper, we discuss the discrete Legendre collocation methods for Fredholm–Hammerstein integral equations with the weakly singular kernel. Using sufficiently accurate quadrature rule, we obtain the convergence rates for the discrete Legendre collocation solutions to the actual solution in both L^2 and infinity norm. Numerical examples are presented to validate the theoretical estimates.

Keywords Hammerstein integral equations · Weakly singular kernels · Spectral methods · Collocation methods · Legendre polynomials

1 Introduction

We consider the following Fredholm–Hammerstein integral equation

$$u(s) - \int_{-1}^{1} k(s,t)\, \psi(t, u(t))\, \mathrm{d}t = f(s), \quad -1 \le s \le 1, \tag{1}$$

where k, f and ψ are known functions, u is the unknown function to be determined in a Banach space $\mathbb{X}$, and the kernel $k(.,.)$ is of weakly singular type of the form

$$k(s,t) = m(s,t)g_\alpha|s - t|,$$

$m(s,t) \in \mathcal{C}([-1,1] \times [-1,1])$ and

$$g_\alpha(x) = \begin{cases} x^{\alpha-1}, & \text{if } 1/2 < \alpha < 1, \\ \log x, & \text{if } \alpha = 1. \end{cases}$$

B. L. Panigrahi (✉)
Department of Mathematics, Sambalpur University,
Sambalpur 768019, Odisha, India
e-mail: blpanigrahi@suniv.ac.in; bijayalaxmi.panigrahi@gmail.com

© Springer Nature Singapore Pte Ltd. 2018   315
D. Ghosh et al. (eds.), *Mathematics and Computing*, Springer Proceedings
in Mathematics & Statistics 253, https://doi.org/10.1007/978-981-13-2095-8_25

This type of problem (1) arises as a reformulation of boundary value problems with certain nonlinear boundary conditions.

Many authors have studied numerical methods to solve nonlinear integral equations with the smooth kernel and also with weakly singular kernel [7–11, 13]. The Galerkin, collocation, Petrov–Galerkin degenerate kernel methods, and Nyström methods are commonly used projection methods for finding the numerical solution of Eq. (1). In all the projection methods, the infinite dimensional space $\mathbb{X}$ is approximated by the space of piecewise polynomials. However, to get better accuracy in piecewise polynomial-based projection methods, one has to solve a large system of nonlinear equations because of a large number of the partition. So, in the last some years, different spectral methods have been developed rapidly and the Legendre spectral methods have been applied to linear integral equations and nonlinear integral equations. The Legendre spectral projection methods for Fredholm–Hammerstein integral equations with smooth kernel have been studied in [4]. The important point is if $\mathcal{P}_n$ denotes either orthogonal or interpolatory projection from $\mathbb{X}$ into a subspace of global polynomials of degree $\leq n$, then $\|\mathcal{P}_n\|_\infty$ is unbounded. In [4], the similar convergence rates for the approximate solution of Fredholm–Hammerstein integral equations with smooth kernel have been obtained in both L^2 and infinity norm as in the case of piecewise polynomial bases.

However, the spectral projection methods lead to the algebraic nonlinear system, in which the coefficients are integrals appeared due to inner products and integral operator $\mathcal{K}$. Since these integrals are almost always evaluated numerically, in all the above methods the effect of error due to numerical integration has been ignored. So in the discrete methods, the integrals appeared in the nonlinear system of equations have been replaced by numerical quadrature rule. The discrete spectral methods for nonlinear integral equations have been discussed by [5]. However, in all these above methods, the nonlinear integral equations with smooth kernel have been considered. The integral equations with weakly singular kernels of the algebraic and logarithmic type cover many important applications, and this kind of problem arises from potential problems, Dirichlet problems, the description of the hydrodynamic interaction between elements of a polymer chain in solution, mathematical problems of radiative equilibrium, and transport problems.

In this paper, we apply the discrete Legendre spectral collocation methods to solve the Fredholm–Hammerstein integral equations with the weakly singular kernel. Our purpose in this paper is to obtain similar convergence rates as in using piecewise and global polynomial bases for smooth kernels.

The organization of this paper is as follows. In Sect. 2, we discuss the discrete Legendre collocation methods for Hammerstein integral equations with the weakly singular kernel. In Sect. 3, we discuss the convergence rates for both L^2 and infinity norm. In Sect. 4, we illustrate our result by the numerical example. Throughout this paper, we assume c is a generic constant.

2 Hammerstein Integral Equations

In this section, we will discuss on the collocation methods for solving Hammerstein integral equations with weakly singular kernels (1) using Legendre polynomial basis functions.

Let $\mathbb{X} = C[-1, 1]$ and $L^2[-1, 1]$ with norms $\|.\|_\infty$ and $\|.\|_{L^2}$, respectively. Throughout the paper, the following assumptions are made on f, $k(., .)$ and $\psi(., u(.))$:

(i) $f \in C[-1, 1]$.

(ii) For $m(s, t) \in C^r([-1, 1] \times [-1, 1])$, $r \geq 1$,

$$\|m\|_\infty = \sup_{s,t \in [-1,1]} |m(s, t)| \leq M < \infty,$$

$$\|m\|_{r,\infty} = \max_{0 \leq i, j \leq r, t, s \in [-1,1]} \left| \frac{\partial^{i+j}}{\partial s^i \partial t^j} m(s, t) \right|.$$

(iii) For $s, s' \in [-1, 1]$, $\| g_\alpha |s - t| - g_\alpha |s' - t| \|_{L^2} \to 0$ and $\| m_s(.) - m_{s'}(.) \|_{L^2} \to 0$ as $s \to s'$.

(iv) For $1/2 < \alpha < 1$, $\displaystyle \sup_{s \in [-1,1]} \int_{-1}^{1} |g_\alpha|s - t||^2 \, dt = M_2 < \infty$.

(v) The nonlinear function $\psi(t, u)$ is bounded and continuous over $[-1, 1] \times \mathbb{R}$. $\psi(t, u)$ is Lipschitz continuous in u, i.e., for any $u_1, u_2 \in \mathbb{R}$, $\exists c_1 > 0$ such that

$$|\psi(t, u_1) - \psi(t, u_2)| \leq c_1 |u_1 - u_2|, \ \forall t \in [-1, 1].$$

(vi) The partial derivative $\psi^{(0,1)}(t, u(t))$ of ψ with respect to the second variable exists and is Lipschitz continuous in u, i.e., for any $u_1, u_2 \in \mathbb{R}$, $\exists c_2 > 0$ such that

$$|\psi^{(0,1)}(t, u_1) - \psi^{(0,1)}(t, u_2)| \leq c_2 |u_1 - u_2|, \ \forall t \in [-1, 1].$$

This implies, $\psi^{(0,1)}(., .) \in C[-1, 1] \times \mathbb{R}$, $\|\psi^{(0,1)}\|_\infty \leq B$.

(vii) We assume that M, M_2, and c_1 satisfy the condition that $\sqrt{2M_2} M c_1 < 1$.

Define

$$z(t) = \psi(t, u(t)), \ t \in [-1, 1]. \tag{2}$$

It is easy to show by using chain rule for higher derivatives that $z \in C^r[-1, 1]$, because $\psi(., .) \in C^r([-1, 1] \times \mathbb{R})$ and $u \in C^r[-1, 1]$.

Then, the Hammerstein integral equation (1) can be written as an operator form

$$u = \mathcal{K}z + f, \tag{3}$$

where

$$\mathcal{K}z(s) = \int_{-1}^{1} k(s, t) z(t) \, dt. \tag{4}$$

For our convenience, we consider a nonlinear operator $\Psi : \mathbb{X} \to \mathbb{X}$ defined by

$$\Psi(u)(t) = \psi(t, u(t)).$$

Then, Eq. (2) becomes

$$z = \Psi(\mathcal{K}z + f). \tag{5}$$

Let $T(u) = \Psi(\mathcal{K}u + f)$, $u \in \mathbb{X}$, then the Eq. (5) can be written as

$$Tz = z. \tag{6}$$

Now, we will prove the existence and uniqueness of the solution of Eq. (6) in the next theorem.

Theorem 1 *Let $\mathbb{X} = C[-1, 1]$, $f \in \mathbb{X}$ and $g_\alpha |s - t|$ satisfy the assumption (iv) with $m(.,.) \in C[-1, 1] \times [-1, 1]$. Let $\psi(t, u(t)) \in C([-1, 1] \times \mathbb{R})$ satisfy the Lipschitz condition in the second variable and $\sqrt{2M_2} M c_1 < 1$. Then, the operator equation $Tz = z$ has a unique solution $z_0 \in \mathbb{X}$, i.e., $z_0 = Tz_0$.* □

Proof Using Cauchy–Schwarz inequality, we get

$$\|\mathcal{K}z\|_\infty = \sup_{s \in [-1,1]} |\mathcal{K}z(s)| \leq \sup_{t,s \in [-1,1]} |m(s,t)| \sup_{s \in [-1,1]} \int_{-1}^{1} |g_\alpha |s - t| z(t)| \, dt$$
$$\leq M\sqrt{M_2} \|z\|_{L^2}. \tag{7}$$

Since $f \in C[-1, 1]$, it follows that $u = \mathcal{K}z + f \in C[-1, 1]$. Let $z_1, z_2 \in C[-1, 1]$. Using the Lipschitz continuity of $\psi(., u(.))$ with Eq. (7), we get

$$\|Tz_1 - Tz_2\|_\infty = \|\Psi(\mathcal{K}z_1 + f) - \Psi(\mathcal{K}z_2 + f)\|_\infty$$
$$\leq c_1 \|\mathcal{K}(z_1 - z_2)\|_\infty$$
$$\leq c_1 M\sqrt{M_2} \|z_1 - z_2\|_{L^2} \leq \sqrt{2M_2} c_1 M \|z_1 - z_2\|_\infty. \tag{8}$$

By assumption (vii), $\sqrt{2M_2} M c_1 < 1$, hence T is a contraction mapping on $\mathbb{X}$. By using Banach contraction theorem, T has a unique fixed point in $\mathbb{X}$. Denote the unique solution as z_0. This completes the proof. □

To describe Legendre collocation methods for the solution of Hammerstein integral equation (1), we will first approximate the space $\mathbb{X}$ by a finite-dimensional space $\mathbb{X}_n$. Let $\mathbb{X}_n$ be the set of all polynomials of degree not more than n. Let $\{\tau_0, \tau_1, \ldots, \tau_n\}$ be the zeros of the Legendre polynomial of degree $n + 1$. For $z \in C[-1, 1]$, we define the Lagrange interpolation polynomial $\mathcal{Q}_n : \mathbb{X} \to \mathbb{X}_n$ by

$$\mathcal{Q}_n z(s) = \sum_{i=0}^{n} z(\tau_i) L_i(s), \quad s \in [-1, 1]$$

where

$$L_i(s) = \frac{\pi(s)}{(s - \tau_i)\pi'(\tau_i)}, \quad \pi(s) = (s - \tau_0)(s - \tau_1)\ldots(s - \tau_n).$$

Then, $\mathcal{Q}_n : \mathbb{X} \to \mathbb{X}_n$ satisfies

$$\mathcal{Q}_n u \in \mathbb{X}_n, \quad \mathcal{Q}_n u(\tau_i) = u(\tau_i), \quad i = 0, 1, \ldots, n, \quad u \in \mathbb{X}. \tag{9}$$

We quote the following lemma from [3, 6], which gives the properties of the interpolatory projection operator $\mathcal{Q}_n$.

Lemma 1 *Let $\mathcal{Q}_n : \mathbb{X} \to \mathbb{X}_n$ be the interpolatory projection operator defined by (9). Then, the following hold:*

(i) *$\{\mathcal{Q}_n : n \in \mathbb{N}\}$ is uniformly bounded in L^2 norm, that is, $\|\mathcal{Q}_n u\|_{L^2} \leq p\|u\|_\infty$, $u \in C[-1, 1]$, where p is a constant independent of n.*
(ii) *For any $u \in C^r[-1, 1]$, there exists a constant c independent of n such that*

$$\|\mathcal{Q}_n u - u\|_{L^2} \leq cn^{-r}\|u^{(r)}\|_{L^2}.$$

Then, the Legendre collocation method for Eq. (5) is seeking an approximate solution $z_n(s) = \sum_{i=0}^n \gamma_i L_i(s) \in \mathbb{X}_n$, which satisfies the following nonlinear system of equations

$$\sum_{i=0}^n \gamma_i L_i(\tau_j) = \Psi\left(\mathcal{K}\left(\sum_{i=0}^n \gamma_i L_i\right) + f\right)(\tau_j), \quad j = 0, 1, \ldots, n.$$

Using the interpolatory projection operator, the above system of nonlinear equations can be written in the following operator equation form.

$$z_n = \mathcal{Q}_n \Psi(\mathcal{K} z_n + f). \tag{10}$$

Corresponding approximate solution u_n of u is given by

$$u_n = \mathcal{K} z_n + f.$$

Using the projection operator $\mathcal{Q}_n$, we define $\mathcal{K}_n : \mathbb{X} \to \mathbb{X}$ by

$$\mathcal{K}_n(z)(s) = \int_{-1}^1 g_\alpha |s - t| \mathcal{Q}_n(m(s, t)z(t))\, dt, \tag{11}$$

which approximates the operator $\mathcal{K}$. For $z_n \in \mathbb{X}_n$, we have

$$\mathcal{K}_n(z_n)(s) = \sum_{i=0}^n w_i^\alpha(s) m(s, \tau_i) z_n(\tau_i),$$

where $w_i^\alpha(s) = \int_{-1}^{1} L_i(s)g_\alpha|s - t| dt$.

Denote $L_2^{(r)}[-1, 1] = \{u : D_s^i u \in L^2[-1, 1], i = 0, 1, \ldots, r\}$ with the norm

$$\|u\|_{L^2,r} = \sum_{i=0}^{r} \|D_s^i u\|_{L^2}.$$

Now in the following Lemma, we give the error bounds of the integral operator $\mathcal{K}$ with the approximate operator $\mathcal{K}_n$.

Theorem 2 *Let $m(s, t) \in \mathcal{C}^{(0,r)}([-1, 1] \times [-1, 1])$ and $z \in C^r[-1, 1]$. Then, there exists a positive constant c such that*

$$\|(\mathcal{K} - \mathcal{K}_n)z\|_\infty \leq cn^{-r}\|z\|_{L^2,r}. \tag{12}$$

Proof For fixed $s \in [-1, 1]$, denote $b_s(t) = m_s(t)z(t)$, where $m_s(t) = m(s, t)$. From Eqs. (11) and (4), we obtain

$$|(\mathcal{K} - \mathcal{K}_n)z(s)| = \left| \int_{-1}^{1} g_\alpha|s - t|(\mathcal{I} - \mathcal{Q}_n)(m(s, t)z(t))dt \right|.$$

Now by taking supremum over $s \in [-1, 1]$ and using Cauchy–Schwarz inequality with Lemma 1, we get

$$\|(\mathcal{K} - \mathcal{K}_n)z\|_\infty^2 \leq M_2 \sup_{s\in[-1,1]} \|(\mathcal{I} - \mathcal{Q}_n)b_s\|_{L^2}^2$$

$$= M_2 n^{-2r} \sup_{s\in[-1,1]} \left(\int_{-1}^{1} |[b_s(t)]^{(r)}|^2 dt \right). \tag{13}$$

Using Leibniz rule for differentiating the product of two terms and Cauchy–Schwarz inequality again, we get

$$\left([b_s(t)]^{(r)} \right)^2 = \left(\sum_{i=0}^{r} C_i^r D_t^{r-i} m(s, t) D_t^i z(t) \right)^2$$

$$\leq \|D_t^{r-i} m_s\|_\infty^2 \left(\sum_{i=0}^{r} (C_i^r)^2 \right) \left(\sum_{i=0}^{r} (D_t^i z)^2(t) \right) \tag{14}$$

Using Eq. (14) in Eq. (13), we obtain

$$\|(\mathcal{K} - \mathcal{K}_n)z\|_\infty^2 \leq M_2 n^{-2r} \|m\|_{r,\infty}^2 \left(\sum_{i=0}^{r} (C_i^r)^2 \right) \left(\int_{-1}^{1} \sum_{i=0}^{r} (D_t^i z)^2 (t) \mathrm{d}t \right)$$

$$\leq M_2 n^{-2r} \|m\|_{r,\infty}^2 \left(\sum_{i=0}^{r} (C_i^r)^2 \right) \left(\sum_{i=0}^{r} \|D_t^i z\|_{L^2}^2 \right)$$

$$\leq M_2 n^{-2r} \|m\|_{r,\infty}^2 \left(\sum_{i=0}^{r} (C_i^r)^2 \right) \|z\|_{L^2,r}^2.$$

Thus, we get

$$\|(\mathcal{K} - \mathcal{K}_n)z\|_\infty \leq \sqrt{M_2} n^{-r} \|m\|_{r,\infty} \left(\sum_{i=0}^{r} (C_i^r)^2 \right)^{1/2} \|z\|_{L^2,r} \leq cn^{-r} \|z\|_{L^2,r}.$$

This completes the proof.

Now by using the approximate discrete operator $\mathcal{K}_n$ instead of the integral operator $\mathcal{K}$, we obtain

$$\sum_{i=0}^{n} \xi_i L_i(\tau_j) = \Psi \left(\mathcal{K}_n \left(\sum_{i=0}^{n} \xi_i L_i \right) + f \right)(\tau_j), \quad j = 0, 1, \ldots, n. \tag{15}$$

Then, $\tilde{z}_n(t) = \sum_{j=0}^{n} \xi_j L_j(t)$ is the discrete Legendre collocation approximate solution of z of Eq. (5).

Using the interpolation operator $\mathcal{Q}_n$, the system of nonlinear equations (15) can be written in the following operator equation forms.

$$\tilde{z}_n = \mathcal{Q}_n \Psi (\mathcal{K}_n \tilde{z}_n + f). \tag{16}$$

Let $\widetilde{\mathcal{T}}_n(u) = \mathcal{Q}_n \Psi (\mathcal{K}_n u + f), u \in \mathbb{X}$, and Eq. (16) can be written as

$$\tilde{z}_n = \widetilde{\mathcal{T}}_n \tilde{z}_n. \tag{17}$$

The corresponding approximate solution $\tilde{u}_n$ of u is defined by $\tilde{u}_n = \mathcal{K}_n \tilde{z}_n + f$.

3 Convergence Rates

In this section, we will discuss convergence rates of approximated solutions with the exact solution of Fredholm–Hammerstein integral equations with weakly singular kernel, in both L^2 and infinity norm. To do this, we quote the following lemma.

Definition 1 [1] Let $\mathbb{X}$ be a Banach space and, $\mathcal{T}$ and $\mathcal{T}_n \in B(\mathbb{X})$. Then, $\{\mathcal{T}_n\}$ is said to be ν-convergent to $\mathcal{T}$ if $\|\mathcal{T}_n\| \leq c$, $\quad \|(\mathcal{T}_n - \mathcal{T})\mathcal{T}\| \to 0$, $\quad \|(\mathcal{T}_n - \mathcal{T})\mathcal{T}_n\| \to 0$ as $n \to \infty$.

Theorem 3 [2] *Let $\mathbb{X}$ be a Banach space and $\mathcal{T}, \mathcal{T}_n \in \mathbb{BL}(\mathbb{X})$. If $\mathcal{T}_n$ is norm convergent to $\mathcal{T}$ or $\mathcal{T}_n$ is ν-convergent to $\mathcal{T}$ and $(\mathcal{I} - \mathcal{T})^{-1}$ exists and bounded on $\mathbb{X}$, then $(\mathcal{I} - \mathcal{T}_n)^{-1}$ exists and uniformly bounded on $\mathbb{X}$ for sufficiently large n.*

Theorem 4 *Let $\mathcal{K}_n$ be the approximate integral operator defined by the Eq. (11), then the set of operators $\{\mathcal{K}_n : n = 1, 2, 3, \dots\}$ is collectively compact.*

Proof To prove $\{\mathcal{K}_n : n = 1, 2, 3, \dots\}$ is collectively compact, we need to show that the set $\bigcup_n \mathcal{K}_n(B)$ is a relatively compact set whenever $B \subset \mathbb{X}$ is bounded.

Let $S = \{\mathcal{K}_n(z) : z \in B\}$, and B is a closed unit ball in $\mathcal{C}[-1, 1] \subset L^2[-1, 1]$. To prove $\{\mathcal{K}_n(z)\}$ is a compact operator, we have to show that S is uniformly bounded and equicontinuous.
We have

$$\mathcal{K}_n(z)(s) = \int_{-1}^{1} g_\alpha |s - t| \mathcal{Q}_n(m(s, t)z(t)) \, dt,$$

Now by using Cauchy–Schwarz inequality and taking supremum over $s \in [-1, 1]$, we obtain

$$\|\mathcal{K}_n(z)\|_{L^2} \leq \sqrt{2}\|\mathcal{K}_n(z)\|_\infty \leq \sqrt{2M_2}\|\mathcal{Q}_n(m(s, t)z(t))\|_{L^2} \leq c\,pM\|z\|_{L^2}. \quad (18)$$

Thus, $\mathcal{K}_n$ is uniformly bounded in L^2 norm. Now to show the equicontinuity, for any $s, s' \in [-1, 1]$, we obtain

$$\mathcal{K}_n(z)(s) - \mathcal{K}_n(z)(s')$$
$$= \int_{-1}^{1} \left(g_\alpha |s - t| \mathcal{Q}_n(m(s, t)z(t)) - g_\alpha |s' - t| \mathcal{Q}_n(m(s', t)z(t)) \right) dt$$
$$\leq \int_{-1}^{1} \left(g_\alpha |s - t| - g_\alpha |s' - t| \right) \mathcal{Q}_n(m(s, t)z(t)) dt$$
$$+ \int_{-1}^{1} g_\alpha |s' - t| \mathcal{Q}_n \left(m(s, t)z(t) - m(s', t)z(t) \right) dt.$$

By using Cauchy–Schwarz inequality, we obtain

$$|\mathcal{K}_n(z)(s) - \mathcal{K}_n(z)(s')| \leq \left(\int_{-1}^{1} (g_\alpha |s - t| - g_\alpha |s' - t|)^2 dt \right)^{1/2} \|\mathcal{Q}_n(m(s, t)z(t))\|_{L^2}$$
$$+ M_2 p \|m(s, t) - m(s', t)\|_{L^2} \|z\|_\infty.$$

Using assumption (iii) in the above equation, we get $|\mathcal{K}_n(z)(s) - \mathcal{K}_n(z)(s')| \to 0$ as $s \to s'$ and $n \to \infty$. Thus, $\{\mathcal{K}_n(z)\}$ is equicontinuous on $[-1, 1]$. By using Arzela–Ascoli theorem, we conclude that $\{\mathcal{K}_n\}$ is collectively compact. This completes the proof.

We quote the following theorem which gives us the condition under which the solvability of one equation leads to the solvability of other equation.

Theorem 5 [13] *Let $\widehat{\mathcal{F}}$ and $\widetilde{\mathcal{F}}$ be continuous operators over an open set Ω in a Banach space $\mathbb{X}$. Let the equation $x = \widetilde{\mathcal{F}}x$ has an isolated solution $\tilde{x}_0 \in \Omega$, and let the following conditions be satisfied.*

(a) *The operator $\widehat{\mathcal{F}}$ is Frechet differentiable in some neighborhood of the point $\tilde{x}_0$, while the linear operator $\mathcal{I} - \widehat{\mathcal{F}}'(\tilde{x}_0)$ is continuously invertible.*

(b) *Suppose that for some $\delta > 0$ and $0 < q < 1$, the following inequalities are valid (the number δ is assumed to be so small that the sphere $\|x - \tilde{x}_0\| \leq \delta$ is contained within Ω).*

$$\sup_{\|x - \tilde{x}_0\| \leq \delta} \|(\mathcal{I} - \widehat{\mathcal{F}}'(\tilde{x}_0))^{-1}(\widehat{\mathcal{F}}'(x) - \widehat{\mathcal{F}}'(\tilde{x}_0))\| \leq q, \tag{19}$$

$$\alpha = \|(\mathcal{I} - \widehat{\mathcal{F}}'(\tilde{x}_0))^{-1}(\widehat{\mathcal{F}}(\tilde{x}_0) - \widetilde{\mathcal{F}}(\tilde{x}_0))\| \leq \delta(1 - q). \tag{20}$$

Then, the equation $x = \widehat{\mathcal{F}}x$ has a unique solution $\hat{x}_0$ in the sphere $\|x - \tilde{x}_0\| \leq \delta$. Moreover, the inequality

$$\frac{\alpha}{1 + q} \leq \|\hat{x}_0 - \tilde{x}_0\| \leq \frac{\alpha}{1 - q},$$

is valid.

Theorem 6 *The operators $\mathcal{T}$ and $\widetilde{\mathcal{T}}_n$ are Frechet differentiable on $\mathbb{X}$, and $\widetilde{\mathcal{T}}_n'(z_0)$ is ν-convergent to $\mathcal{T}'(z_0)$ in L^2-norm.*

Proof With the assumptions on the kernel and the nonlinear function ψ and by using the Lemma 4 of [11], we get that the operator $\mathcal{T}(z) = \Psi(\mathcal{K}z + f)$ is continuously Frechet differentiable on $\mathbb{X}$. Since $\mathcal{Q}_n$ is a linear operator, using [11, 12], it can be proved that $\widetilde{\mathcal{T}}_n(z) = \mathcal{Q}_n\Psi(\mathcal{K}_nz + f)$ is also Frechet differentiable on $\mathbb{X}$. Denote the Frechet derivatives of $\mathcal{T}(z)$ and $\widetilde{\mathcal{T}}_n(z)$ at the point z_0 as $\mathcal{T}'(z_0)$ and $\widetilde{\mathcal{T}}_n'(z_0)$, respectively. Then, $\mathcal{T}'(z_0) = \Psi'(\mathcal{K}z_0 + f)\mathcal{K}$, and $\widetilde{\mathcal{T}}_n'(z_0) = \mathcal{Q}_n\Psi'(\mathcal{K}_nz_0 + f)\mathcal{K}_n$.

Now, we need to show that $\widetilde{\mathcal{T}}_n'(z_0)$ is ν-convergent to $\mathcal{T}'(z_0)$ in L^2-norm. By using Lemma 1 and the estimate (18) with the assumptions, we obtain

$$
\begin{aligned}
\|\widetilde{\mathcal{T}}_n'(z_0)u\|_{L^2} &= \|\mathcal{Q}_n\Psi'(\mathcal{K}_nz_0 + f)\mathcal{K}_nu\|_{L^2} \\
&\leq p\|\Psi'(\mathcal{K}_nz_0 + f)\|_\infty\|\mathcal{K}_nu\|_\infty \\
&\leq p\Big(\|\Psi'(\mathcal{K}_nz_0 + f) - \Psi'(\mathcal{K}z_0 + f)\|_\infty + \|\Psi'(\mathcal{K}z_0 + f)\|_\infty\Big)\|u\|_{L^2} \\
&\leq c(\|(\mathcal{K}_n - \mathcal{K})z_0\|_\infty + B)\|u\|_{L^2} \leq c(n^{-r}\|z_0\|_{L^2,r} + B)\|u\|_{L^2}.
\end{aligned}
$$

This shows that $\|\widetilde{T}_n'(z_0)\|_{L^2}$ is uniformly bounded. Next, we consider

$$
\begin{aligned}
\left\|\left(\widetilde{T}_n'(z_0) - T'(z_0)\right)u\right\|_{L^2} &= \left\|\left(Q_n\Psi'(K_n z_0 + f)K_n - \Psi'(K z_0 + f)K\right)u\right\|_{L^2} \\
&\leq \left\|\left(Q_n\Psi'(K_n z_0 + f) - Q_n\Psi'(K z_0 + f)\right)K_n u\right\|_{L^2} \\
&\quad + \left\|\left(Q_n\Psi'(K z_0 + f)K_n - Q_n\Psi'(K z_0 + f)K\right)u\right\|_{L^2} \\
&\quad + \left\|\left(Q_n\Psi'(K z_0 + f)K - \Psi'(K z_0 + f)K\right)u\right\|_{L^2} \\
&\leq 2pc_2\|(K_n - K)z_0\|_\infty\|K_n u\|_\infty + \sqrt{2}pB\|(K_n - K)u\|_\infty \\
&\quad + \|(Q_n - I)\Psi'(K z_0 + f)K u\|_{L^2}.
\end{aligned}
$$

By using Theorem 2, the first two terms of the right hand side of the above equation $\to$ 0 as $n \to \infty$. Since $\Psi'(K z_0 + f)$ is bounded and K is a compact operator, $\Psi'(K z_0 + f)K$ is also a compact operator. Since Q_n converges pointwise to the identity operator I from Lemma 1 and $\Psi'(K z_0 + f)K$ is a compact operator, it follows that $\|(Q_n - I)\Psi'(K z_0 + f)K u\|_{L^2} \to 0$ as $n \to \infty$. Thus,

$$
\left\|\left(\widetilde{T}_n'(z_0) - T'(z_0)\right)u\right\|_{L^2} \to 0, \text{ as } n \to \infty.
$$

Let B be a closed unit ball in $C[-1, 1]$. Since $T'(z_0) = \Psi'(K z_0 + f)K$ is a compact operator, $S = \{T'(z_0)x : x \in B\}$ is a relatively compact set in $C[-1, 1]$. Then, it follows that

$$
\begin{aligned}
\left\|\left(\widetilde{T}_n'(z_0) - T'(z_0)\right)T'(z_0)\right\|_{L^2} &= \sup\{\|\left(\widetilde{T}_n'(z_0) - T'(z_0)\right)T'(z_0)u\|_{L^2} : u \in B\} \\
&= \sup\{\|\left(\widetilde{T}_n'(z_0) - T'(z_0)\right)u\|_{L^2} : u \in S\} \to 0, \text{ as } n \to \infty.
\end{aligned}
$$

Since Q_n is uniformly bounded in L^2 norm, $\Psi'(K_n z_0 + f)$ is also bounded and K_n is a compact operator, and then $\widetilde{T}_n'(z_0) = Q_n\Psi'(K_n z_0 + f)K_n$ is a compact operator. Proceeding in the similar way as in before, it can be easy to show that

$$
\left\|\left(\widetilde{T}_n'(z_0) - T'(z_0)\right)\widetilde{T}_n'(z_0)u\right\|_{L^2} \to 0 \text{ as } n \to \infty.
$$

This shows that $\widetilde{T}_n'(z_0)$ is v-convergent to $T'(z_0)$ in L^2-norm. This completes the proof.

Theorem 7 *Let $z_0 \in C^r[-1, 1]$ be an isolated solution of the Eq. (6). Assume that one is not an eigenvalue of the linear operator $T'(z_0)$. Then for sufficiently large n, the operators $(I - \widetilde{T}_n'(z_0))$ are invertible on $\mathbb{X}$ and there exist constants $A_1 > 0$ independent of n such that $\|(I - \widetilde{T}_n'(z_0))^{-1}\|_{L^2} \leq A_1$.*

Proof The proof completes by combining the Theorems 3 and 6.

Theorem 8 *Let $\mathcal{Q}_n : \mathbb{X} \to \mathbb{X}_n$ be the interpolatory projection operator defined by (9). Then Eq. (17) has an unique solution $\tilde{z}_n \in B(z_0, \delta) = \{z : \|z - z_0\|_{L^2} < \delta\}$ for some $\delta > 0$ and for sufficiently large n. Moreover, there exists a constant $0 < q < 1$, independent of n such that*

$$\frac{\beta_n}{1+q} \le \|\tilde{z}_n - z_0\|_{L^2} \le \frac{\beta_n}{1-q},$$

where $\beta_n = \|(\mathcal{I} - \tilde{T}_n'(z_0))^{-1}(\tilde{T}_n(z_0) - T(z_0))\|_{L^2}$.

Proof From Theorem 7, we have $(\mathcal{I} - \tilde{T}_n'(z_0))^{-1}$ that exists and it is uniformly bounded in L^2 norm; i.e., there exists $A_1 > 0$ such that $\|(\mathcal{I} - \tilde{T}_n'(z_0))^{-1}\|_{L^2} \le A_1$. Using Theorem 4 with the assumption (v), for any $z \in B(z_0, \delta)$ and $u \in \mathcal{C}[-1, 1]$, we get

$$\begin{aligned}
\|(\tilde{T}_n'(z) - \tilde{T}_n'(z_0))u\|_{L^2} &= \|[\mathcal{Q}_n\Psi'(\mathcal{K}_n z_0 + f)\mathcal{K}_n - \mathcal{Q}_n\Psi'(\mathcal{K}_n z + f)\mathcal{K}_n]u\|_{L^2} \\
&= \|\mathcal{Q}_n(\Psi'(\mathcal{K}_n z_0 + f)\mathcal{K}_n - \Psi'(\mathcal{K}_n z + f)\mathcal{K}_n)u\|_{L^2} \\
&\le p\|(\Psi'(\mathcal{K}_n z_0 + f) - \Psi'(\mathcal{K}_n z + f))\mathcal{K}_n u\|_\infty \\
&\le c\|\mathcal{K}_n(z_0 - z)\|_\infty \|\mathcal{K}_n u\|_\infty \le c\|z - z_0\|_{L^2}\|u\|_{L^2}.
\end{aligned}$$

Thus, $\|(\tilde{T}_n'(z) - \tilde{T}_n'(z_0))\|_{L^2} \le c\delta$. Hence, we obtain

$$\sup_{\|z - z_0\|_{L^2} \le \delta} \|(\mathcal{I} - \tilde{T}_n'(z_0))^{-1}(\tilde{T}_n'(z_0) - \tilde{T}_n'(z))\|_{L^2} \le A_1 c\delta \le q,$$

where $0 < q < 1$. This proves Eq. (19) of Theorem 5. Now by using Theorem 2 with Lemma 1, we obtain

$$\begin{aligned}
\|\tilde{T}_n(z_0) - T(z_0)\|_{L^2} &= \|\mathcal{Q}_n\Psi(\mathcal{K}_n z_0 + f) - \Psi(\mathcal{K} z_0 + f)\|_{L^2} \\
&\le \|\mathcal{Q}_n[\Psi(\mathcal{K}_n z_0 + f) - \Psi(\mathcal{K} z_0 + f)]\|_{L^2} \\
&\quad + \|(\mathcal{Q}_n - \mathcal{I})\Psi(\mathcal{K} z_0 + f)\|_{L^2} \\
&\le c\|(\mathcal{K}_n - \mathcal{K})z_0\|_\infty + \|(\mathcal{Q}_n - \mathcal{I})z_0\|_{L^2} \\
&\le cn^{-r}\|z_0\|_{L^2,r} + n^{-r}\|z_0\|_{L^2,r} \to 0, \quad \text{as } n \to \infty. \quad (21)
\end{aligned}$$

Hence,

$$\beta_n = \|(\mathcal{I} - \tilde{T}_n'(z_0))^{-1}(\tilde{T}_n(z_0) - T(z_0))\|_{L^2} \le A_1\|\tilde{T}_n(z_0) - T(z_0)\|_{L^2} \to 0,$$

as $n \to \infty$. Choose n large enough such that $\beta_n \le \delta(1 - q)$. Then, Eq. (20) of Theorem 5 is satisfied. Thus, by applying Theorem 5, we obtain

$$\frac{\beta_n}{1+q} \le \|z_0 - \tilde{z}_n\|_{L^2} \le \frac{\beta_n}{1-q}, \quad (22)$$

where $\beta_n = \|(\mathcal{I} - \widetilde{\mathcal{T}}_n(z_0))^{-1}(\widetilde{\mathcal{T}}_n(z_0) - \mathcal{T}(z_0))\|_{L^2}$. Using Eq. (21) with Eq. (22), we obtain

$$\|z_0 - \tilde{z}_n\|_{L^2} \le \beta_n \le A_1\|\widetilde{\mathcal{T}}_n(z_0) - \mathcal{T}(z_0)\|_{L^2} \le cn^{-r}\|z_0\|_{L^2,r} + n^{-r}\|z_0\|_{L^2,r}. \quad (23)$$

This completes the proof.

Theorem 9 *Let z_0 be the isolated solution of Eq. (6) and u_0 be the isolated solution of (3) such that $u_0 = \mathcal{K}z_0 + f$. Let $\tilde{u}_n = \mathcal{K}_n\tilde{z}_n + f$ be the discrete Legendre collocation approximation of u_0. Then, the following hold.*

$$\|u_0 - \tilde{u}_n\|_{L^2} = \mathcal{O}(n^{-r}), \quad \|u_0 - \tilde{u}_n\|_\infty = \mathcal{O}(n^{-r}).$$

Proof Using Theorems 4 and 2, we obtain

$$
\begin{aligned}
\|u_0 - \tilde{u}_n\|_{L^2} &= \|\mathcal{K}z_0 + f - (\mathcal{K}_n\tilde{z}_n + f)\|_{L^2} \\
&\le \|\mathcal{K}_n(z_0 - \tilde{z}_n)\|_{L^2} + \|(\mathcal{K}_n - \mathcal{K})z_0\|_{L^2} \\
&\le \sqrt{2}\|\mathcal{K}_n(z_0 - \tilde{z}_n)\|_\infty + \sqrt{2}\|(\mathcal{K}_n - \mathcal{K})z_0\|_\infty \\
&\le \sqrt{2}c\|z_0 - \tilde{z}_n\|_{L^2} + \sqrt{2}n^{-r}\|z_0\|_{L^2,r}.
\end{aligned}
$$

Using the estimate (23), we obtain

$$\|u_0 - \tilde{u}_n\|_{L^2} = \mathcal{O}(n^{-r}).$$

Now for the second estimate, using Theorem 2 with the estimate (23), we obtain

$$
\begin{aligned}
\|u_0 - \tilde{u}_n\|_\infty &\le \|\mathcal{K}_n(z_0 - \tilde{z}_n)\|_\infty + \|(\mathcal{K}_n - \mathcal{K})z_0\|_\infty \\
&\le c\|z_0 - \tilde{z}_n\|_{L^2} + cn^{-r}\|z_0\|_{L^2,r} \le cn^{-r}.
\end{aligned}
$$

This completes the proof.

Remark 1 From Theorem 9, we observe that the Legendre collocation solution converges to the exact solution with the order $\mathcal{O}(n^{-r})$ in both L^2 and infinity norm. We obtained the similar convergence rates for Legendre collocation methods for Fredholm–Hammerstein integral equations with weakly singular kernel using piecewise polynomial-based collocation methods.

4 Numerical Examples

In this section, we present an example to validate the errors of the approximation solutions by using Legendre collocation methods both in L^2 and infinity norm. To solve the problem by using Legendre collocation methods, we first choose Legendre polynomials as the basis functions of $\mathbb{X}_n$ evaluated from the recurrence relation,

Table 1 Discrete Legendre collocation method

n	$\|u_0 - \tilde{u}_n\|_{L^2}$	$\|u_0 - \tilde{u}_n\|_\infty$
2	2.457691e−02	6.874354e−03
3	9.347281e−03	3.576579e−03
4	3.566732e−03	9.348632e−04
5	1.008456e−03	3.569632e−04
6	7.869632e−04	1.068532e−05

$$\phi_0(x) = 1, \quad \phi_1(x) = x, \quad x \in [-1, 1],$$

and for $i = 1, 2, \cdots, n - 1$,

$$(i + 1)\phi_{i+1}(x) = (2i + 1)x\phi_i(x) - i\phi_{i-1}(x), \quad x \in [-1, 1].$$

Example 1 We consider the following integral equation

$$x(t) - \frac{1}{\sqrt{2}} \int_{-1}^{1} \frac{1}{\sqrt{|s - t|}} \cos\left(\frac{s + 1}{2} + x(s)\right)ds = f(t), \quad t \in [-1, 1],$$

where $f(t)$ is selected so that $x(t) = \cos\left(\dfrac{t + 1}{2}\right)$ is the solution.

For different values of n, we compute $\tilde{u}_n$ and compare the results with exact solution u_0. The computed errors in L^2 and infinity norm are presented in Table 1.

References

1. Ahues, M., Largillier, A., Limaye, B.V.: Spectral Computations for Bounded Operators. Chapman and Hall/CRC, New York (2001)
2. Atkinson, K.E.: The Numerical Solution of Integral Equations of the Second Kind. Cambridge University Press, Cambridge, UK (1997)
3. Canuto, C., Hussaini, M.Y., Quarteroni, A., Zang, T.A.: Spectral Methods: Fundamentals in Single Domains. Springer, Berlin (2006)
4. Das, P., Sahani, M.M., Nelakanti, G., Long, G.: Legendre spectral projection methods for Fredholm-Hammerstein integral equations. J. Sci. Comput. **68**, 213–230 (2016)
5. Das, P., Nelakanti, G., Long, G.: Discrete Legendre spectral projection methods for Fredholm-Hammerstein integral equations. J. Comp. Appl. Math. **278**, 293–305 (2015)
6. Guo, B.: Spectral Methods and their Applications. World Scientific, Singapore (1998)
7. Kaneko, H., Noren, R.D., Padilla, P.A.: Superconvergence of the iterated collocation methods for Hammerstein equations. J. Comput. Appl. Math. **80**(2), 335–349 (1997)
8. Kaneko, H., Xu, Y.: Superconvergence of the iterated Galerkin methods for Hammerstein equations. SIAM J. Numer. Anal. **33**(3), 1048–1064 (1996)

9. Kaneko, H., Noren, R.D., Xu, Y.: Numerical solutions for weakly singular Hammerstein equations and their superconvergence. J. Integral Equ. Appl. **4**(3), 391–407 (1992)
10. Kumar, S.: The numerical solution of Hammerstein equations by a method based on polynomial collocation. J. Aust. Math. Soc. Ser. B **31**(3), 319–329 (1990)
11. Kumar, S.: Superconvergence of a collocation-type method for Hammerstein equations. IMA J. Numer. Anal. **7**(3), 313–325 (1987)
12. Suhubi, E.S.: Functional Analysis. Kluwer Academic Publishers, Dordrecht (2003)
13. Vainikko, G.M.: A perturbed Galerkin method and the general theory of approximate methods for non-linear equations. USSR Comput. Math. Phys. **7**(4), 1–41 (1967)

Chapter 26
Norm Inequalities Involving Upper Bounds for Operators in Orlicz-Taylor Sequence Spaces

Atanu Manna

Abstract An Orlicz extension of the results obtained by Talebi (Indag Math (NS) 28(3):629–636, 2017 [1]) is given. Indeed, the upper bounds for the operator norm $\|A\|_{l_\varphi, t_\varphi^\alpha}$ are evaluated, where A is either generalized Hausdorff or Nörlund matrix, l_φ and t_φ^α, respectively, denote the Orlicz and Orlicz-Taylor sequence spaces.

Keywords Hausdorff matrix · Nörlund matrix · Taylor matrix · Orlicz function
Luxemburg norm

Mathematics Subject Classification (2010) Primary 26D15, 40G05, 47A30;
Secondary 46A45

1 Introduction

An *Orlicz function* is a map $\varphi : (0, \infty) \to (0, \infty)$ which is convex and satisfies $\varphi(0+) = 0$. Such a function is strictly increasing and continuous, so it has a unique inverse $\varphi^{-1} : (0, \infty) \to (0, \infty)$. Usually in the theory of Orlicz spaces, the domain of Orlicz function is extended to the real line by $\varphi(x) = \varphi(|x|)$ and $\varphi(0) = 0$ (see [2] for details).

A *supermultiplicative function* $\varphi : (0, \infty) \to (0, \infty)$ is such that for all positive u and v, the following holds

$$\varphi(uv) \geq \varphi(u)\varphi(v).$$

An immediate example of *supermultiplicative function* is $\varphi(t) = t^p$, $p \geq 1$. We would like to recall another example from [3]. Let a, b, p be fixed real numbers such that $a < 0$, $b > 0$, and $p > 1$. Choose $M_{a,b,p}$ be a function defined on the interval $[0, \frac{1}{b})$ with $M_{a,b,p}(0) = 0$ and

A. Manna (✉)
Faculty of Mathematics, Indian Institute of Carpet Technology,
Chauri Road, Bhadohi 221401, Uttar Pradesh, India
e-mail: atanu.manna@iict.ac.in

© Springer Nature Singapore Pte Ltd. 2018   329
D. Ghosh et al. (eds.), *Mathematics and Computing*, Springer Proceedings
in Mathematics & Statistics 253, https://doi.org/10.1007/978-981-13-2095-8_26

$$M_{a,b,p}(x) = x^p |\log(bx)|^a \text{ for } x \neq 0.$$

Define an Orlicz function φ by the formula

$$\varphi(x) = \frac{M_{a,b,p}(\delta x)}{M_{a,b,p}(\delta)} \text{ for } x \geq 0, \delta \in (0, \frac{1}{b}),$$

then φ is equivalent to $M_{a,b,p}$ at 0 with $\varphi(1) = 1$ and *supermultiplicative* on $[0, 1]$. Throughout our study, we consider those *supermultiplicative* Orlicz function φ such that $\varphi(1) = 1$ holds.

Let l^0 be the space of all real sequences and $x = (x_n) \in l^0$ (here and after $x = (x_n)_{n=0}^\infty$ will be replaced by $x = (x_n)$ in order to avoid ambiguity). Further, it is assumed that the Orlicz function φ is fixed. The sequence Orlicz spaces are denoted by l_φ and defined as follows:

$$l_\varphi = \left\{ x \in l^0 : \sum_{n=0}^\infty \varphi(r x_n) < \infty \text{ for some } r > 0 \right\}.$$

The space l_φ is a Banach space equipped with the Orlicz-Luxemburg norm $\| \cdot \|_\varphi$ defined as below:

$$\|x\|_\varphi = \inf \left\{ r > 0 : \sum_{n=0}^\infty \varphi\left(\frac{x_n}{r}\right) \leq 1 \right\}. \tag{1}$$

It is easy to prove that if $|x_n| \leq |y_n|$ for all $n = 0, 1, 2, \ldots$ then $\|x\|_\varphi \leq \|y\|_\varphi$. Thus, the norm $\| \cdot \|_\varphi$ is monotonic. Further, if $0 < \|x\|_\varphi < \infty$ then $\sum_{n=0}^\infty \varphi\left(\frac{x_n}{\|x\|_\varphi}\right)$ ≤ 1 holds (see [4] or [5], Lemma 1).

In particular, if $\varphi(t) = |t|^p$, $p \geq 1$ then we obtain p-summable sequence spaces l_p for $p \geq 1$ and Eq. (1) reduces to the l_p-norm $\| \cdot \|_p$ given below:

$$\|x\|_p = \left(\sum_{n=0}^\infty |x_n|^p \right)^{\frac{1}{p}}.$$

Let $A = (a_{n,k})$, $n, k = 0, 1, 2, \ldots$ be an infinite matrix with real entries and X, Y be two normed sequence spaces. Then A defines a matrix transformation from X to Y, is denoted by $A : X \to Y$ if for every sequence $x = (x_n) \in X$, the sequence $Ax = ((Ax)_n) \equiv (A_n(x))$, A-transform of x is in Y, where

$$A_n(x) = \sum_{k=0}^\infty a_{n,k} x_k, n = 0, 1, 2, \ldots.$$

Throughout the text, we are mainly concerned about the finding of upper bounds U (not depends on x) attached with the following inequality:

$$\|Ax\|_Y \le U \|x\|_X,$$

where $X = l_\varphi$ and $Y = t_\varphi^\alpha$. The notation $\| \cdot \|_X$ (or $\| \cdot \|_Y$) stands for norm on X (or on Y). We shall obtain the general value of U such that $\|A\|_{X,Y} \le U$. Some investigations and latest developments on bounds of operator norms, Refs. [6–10] are referred to the reader.

This paper consist of three sections besides this section. In the first one, that is, in Sect. 2, we introduce Orlicz-Taylor sequence spaces as an Orlicz extension of the Taylor sequence spaces introduced in [1] and obtain some inclusion relations. Section 3 is devoted to the study of obtaining upper bounds of operators norm $\|A\|_{l_\varphi,t_\varphi^\alpha}$, where A is generalized Hausdorff or Nörlund matrix. Finally, Sect. 4 gives the conclusion of this work.

2 Orlicz-Taylor Sequence Spaces

At the beginning, the definition of Taylor matrix $T(\alpha) = (t_{n,k}^\alpha)_{n,k\ge 0}$ of order α ($0 < \alpha < 1$) is recalled and given below:

$$t_{n,k}^\alpha = \begin{cases} \binom{k}{n}(1-\alpha)^{n+1}\alpha^{k-n} & \text{if } k \ge n, \\ 0 & \text{if } 0 \le k < n. \end{cases}$$

Let φ be an Orlicz function. Then we define the Orlicz-Taylor sequence spaces t_φ^α as the set of all sequences x whose $T(\alpha)$-transform belongs to l_φ, that is

$$t_\varphi^\alpha = \left\{ x : T(\alpha)x \in l_\varphi \right\}$$

$$= \left\{ x : \sum_{n=0}^\infty \varphi\left(r \sum_{k=n}^\infty \binom{k}{n}(1-\alpha)^{n+1}\alpha^{k-n}x_k \right) < \infty \text{ for some } r > 0 \right\}.$$

The space t_φ^α is a normed linear space endowed with the norm $\|x\|_\varphi^\alpha = \|T(\alpha)x\|_\varphi$. Denote $t_n^\alpha(x)$ as the $T(\alpha)$-transform of the sequence x, that is

$$t_n^\alpha(x) = \sum_{k=n}^\infty \binom{k}{n}(1-\alpha)^{n+1}\alpha^{k-n}x_k. \tag{2}$$

The inverse Taylor transform can be obtain easily from Eq. (2) and is given in the following expression:

$$x_n = \sum_{k=n}^\infty \binom{k}{n}(1-\alpha)^{-(k+1)}(-\alpha)^{k-n}t_k^\alpha(x). \tag{3}$$

For the sake of completeness, first we begin with a short proof of the following result, which states the completeness of t_φ^α.

Theorem 1 *The sequence space t_φ^α is a Banach space equipped with the norm $\| \cdot \|_\varphi^\alpha$.*

Proof Let (x^p) be a Cauchy sequence in t_φ^α. Then for any $\varepsilon > 0$, there exists a $p_0 \in \mathbb{N}$ such that

$$\|x^p - x^q\|_\varphi^\alpha < \varepsilon \text{ for each } p, q \geq p_0.$$

Choose a $r_\varepsilon > 0$ with $r_\varepsilon < \varepsilon$ such that for each $n \geq 0$

$$\sum_{n=0}^{\infty} \varphi\left(\frac{1}{r_\varepsilon} \sum_{k=n}^{\infty} \binom{k}{n}(1-\alpha)^{n+1}\alpha^{k-n}(x_k^p - x_k^q)\right) \leq 1 \text{ holds for each } p, q \geq p_0. \tag{4}$$

Using the assumption $\varphi(1) = 1$, one obtains

$$\frac{1}{r_\varepsilon} \sum_{k=n}^{\infty} \binom{k}{n}(1-\alpha)^{n+1}\alpha^{k-n}(x_k^p - x_k^q) \leq 1 \text{ for each } p, q \geq p_0 \text{ and } n \geq 0.$$

Then one can easily deduced that the sequence (x_k^p) is a Cauchy sequence of real numbers for each $k \geq 0$ and hence converges, that is, $x_k^p \to x_k$ for each $k \geq 0$ as $p \to \infty$. Therefore, using the continuity of φ, from inequality (4) one gets

$$\sum_{n=0}^{\infty} \varphi\left(\frac{1}{r_\varepsilon} \sum_{k=n}^{\infty} \binom{k}{n}(1-\alpha)^{n+1}\alpha^{k-n}(x_k^p - x_k)\right) \leq 1 \quad \text{for each } p \geq p_0.$$

Thus, $x \in t_\varphi^\alpha$ and $\|x^p - x\|_\varphi^\alpha \leq r_\varepsilon < \varepsilon$ for $p \geq p_0$. So $(t_\varphi^\alpha, \| \cdot \|_\varphi^\alpha)$ is a Banach space.

Now we would like to establish a inclusion between l_φ and t_φ^α. The following lemma plays a very important role to prove our further results.

Lemma 1 *Let φ be an Orlicz and supermultiplicative function, φ^{-1} be its inverse. Then $l_\varphi \subseteq t_\varphi^\alpha$ holds.*

Proof Let $x = (x_n) \in l_\varphi$ such that $x \neq 0$. Then applying the Jensen's inequality, one obtains

$$\sum_{n=0}^{\infty} \varphi\left(\frac{t_n^\alpha(x)}{r}\right) \le \sum_{n=0}^{\infty} \sum_{k=n}^{\infty} \binom{k}{n}(1-\alpha)^{n+1}\alpha^{k-n}\varphi\left(\frac{x_k}{r}\right) \quad \text{(by Jensen's inequality)}$$

$$= \sum_{n=0}^{\infty} \varphi\left(\frac{x_n}{r}\right) \sum_{k=0}^{n} \binom{n}{k}(1-\alpha)^{k+1}\alpha^{n-k}$$

$$= (1-\alpha)\sum_{n=0}^{\infty} \varphi\left(\frac{x_n}{r}\right)$$

$$= \sum_{n=0}^{\infty} \varphi\left(\frac{x_n}{r}\right)\varphi\left(\varphi^{-1}(1-\alpha)\right)$$

$$\le \sum_{n=0}^{\infty} \varphi\left(\frac{x_n}{r}\varphi^{-1}(1-\alpha)\right) \quad \text{(since } \varphi \text{ is supermultiplicative).}$$

Now put $r = \|x\|_\varphi \varphi^{-1}(1-\alpha)$. Then above inequality implies that

$$\sum_{n=0}^{\infty} \varphi\left(\frac{t_n^\alpha(x)}{r}\right) \le \sum_{n=0}^{\infty} \varphi\left(\frac{x_n}{r}\varphi^{-1}(1-\alpha)\right)$$

$$= \sum_{n=0}^{\infty} \varphi\left(\frac{x_n}{\|x\|_\varphi}\right) \le 1.$$

This gives

$$\|x\|_\varphi^\alpha \le r = \varphi^{-1}(1-\alpha)\|x\|_\varphi,$$

which yields the inclusion $l_\varphi \subseteq t_\varphi^\alpha$.

Lemma 2 *Let φ be an Orlicz and supermultiplicative function, φ^{-1} be its inverse. Then for $0 < \beta \le \alpha < 1$, the following inequality holds:*

$$\|x\|_\varphi^\alpha \le \varphi^{-1}\left(\frac{1-\alpha}{1-\beta}\right)\|x\|_\varphi^\beta. \tag{5}$$

Proof Let $x \in l^0$ be a sequence such that $x \ne 0$. Applying Eq. (3) in Eq. (2), the following is obtained:

$$t_n^\alpha(x) = \sum_{k=n}^{\infty} \binom{k}{n}(1-\alpha)^{n+1}\alpha^{k-n}x_k$$

$$= \sum_{k=n}^{\infty} \binom{k}{n}(1-\alpha)^{n+1}\alpha^{k-n} \sum_{j=k}^{\infty} \binom{j}{k}(1-\beta)^{-(j+1)}(-\beta)^{j-k}t_j^\beta(x)$$

$$= (1-\alpha)^{n+1} \sum_{j=n}^{\infty} \binom{j}{n} (1-\beta)^{-(j+1)} t_j^{\beta}(x) \sum_{k=n}^{j} \binom{j-n}{j-k} \alpha^{k-n}(-\beta)^{j-k}$$

$$= (1-\alpha)^{n+1} \sum_{j=n}^{\infty} \binom{j}{n} (1-\beta)^{-(j+1)} (\alpha-\beta)^{j-n} t_j^{\beta}(x)$$

$$= \sum_{j=n}^{\infty} \binom{j}{n} q^{j-n} (1-q)^{n+1} t_j^{\beta}(x) \quad \left(\text{setting } q = \frac{\alpha-\beta}{1-\beta} \right). \tag{6}$$

Hence by the Jensen's inequality applied to Eq. (6), one gets

$$\varphi\left(\frac{t_n^{\alpha}(x)}{r}\right) \le \sum_{j=n}^{\infty} \binom{j}{n} q^{j-n} (1-q)^{n+1} \varphi\left(\frac{t_j^{\beta}(x)}{r}\right) \quad \text{for some } r > 0. \tag{7}$$

Now summing both sides of inequality (7) from $n = 0$ to $n = \infty$, one obtains

$$\sum_{n=0}^{\infty} \varphi\left(\frac{t_n^{\alpha}(x)}{r}\right) \le \sum_{n=0}^{\infty} \sum_{j=n}^{\infty} \binom{j}{n} q^{j-n} (1-q)^{n+1} \varphi\left(\frac{t_j^{\beta}(x)}{r}\right)$$

$$\le \sum_{n=0}^{\infty} (1-q)^{n+1} \varphi\left(\frac{t_n^{\beta}(x)}{r}\right) \sum_{k=0}^{n} \binom{n}{k} q^{n-k}$$

$$\le (1-q) \sum_{n=0}^{\infty} \varphi\left(\frac{t_n^{\beta}(x)}{r}\right) = \frac{1-\alpha}{1-\beta} \sum_{n=0}^{\infty} \varphi\left(\frac{t_n^{\beta}(x)}{r}\right).$$

Proceeds in a parallel way as in Lemma 1, the Orlicz-Luxemburg norm implies that

$$\|x\|_{\varphi}^{\alpha} \le \varphi^{-1}\left(\frac{1-\alpha}{1-\beta}\right) \|x\|_{\varphi}^{\beta} \quad \text{as needed.}$$

Corollary 1 If $0 < \beta \le \alpha < 1$, then $t_{\varphi}^{\beta} \subseteq t_{\varphi}^{\alpha}$.

3   Matrix Operators on Orlicz-Taylor Sequence Spaces

In the following, two consecutive subsections, upper bounds of the generalized Hausdorff matrix operator and Nörlund matrix operator norms in Orlicz-Taylor sequence spaces are obtained.

3.1 Generalized Hausdorff Matrix Operator

In this portion, it is aimed to establish a Hardy type formula as an upper estimate for $\|H\|_{l_\varphi, t_\varphi^\alpha}$, where $H : l_\varphi \to t_\varphi^\alpha$. Suppose that $a > -1$ and $c > 0$. The definition of generalized Hausdorff matrix (see [4, 11]) will be recalled first. The generalized Hausdorff matrix is denoted by $H = (h_{n,k})$, $n, k = 0, 1, 2, \ldots$ and given by

$$h_{n,k} = \begin{cases} 0 & \text{if } k > n, \\ \binom{n+a}{n-k} \Delta^{n-k} \mu_k & \text{if } 0 \leq k \leq n, \end{cases}$$

where Δ is the difference operator defined by $\Delta \mu_k = \mu_k - \mu_{k+1}$ and $\mu = (\mu_0, \mu_1, \ldots)$ is a sequence of real numbers, normalized so that $\mu_0 = 1$ and

$$\mu_k = \int_0^1 \theta^{c(k+a)} d\mu(\theta),$$

where $d\mu(\theta)$ is a Borel probability measure on $[0, 1]$. Therefore, the equivalent expression of the matrix $H = (h_{n,k})$ is given by

$$h_{n,k} = \begin{cases} 0 & \text{if } k > n, \\ \int_0^1 \binom{n+a}{n-k} \theta^{c(k+a)} (1 - \theta^c)^{n-k} d\mu(\theta) & \text{if } 0 \leq k \leq n. \end{cases}$$

The case when $a = 0$ and $c = 1$, one obtains the ordinary Hausdorff matrix (see [6], p. 32), which include four famous classes of matrices as given below if we choose Lebesgue measure $d\theta$:

(a) Put $d\mu(\theta) = \beta(1 - \theta)^{\beta-1} d\theta$, then H leads to (C, β), the Cesàro matrix of order β;
(b) Put $d\mu(\theta) = \frac{|\log \theta|^{\beta-1}}{\Gamma(\beta)} d\theta$, then H reduces to (H, β), the Hölder matrix of order β;
(c) Put $d\mu(\theta) = $ point evaluation at $\theta = \beta$, then H reduces to (E, β), the Euler matrix of order β;
(d) Put $d\mu(\theta) = \beta\theta^{\beta-1} d\theta$, then H becomes (Γ, β), the Gamma matrix of order β.

Now the following hypothesis related to Orlicz function and Hausdorff matrix is considered:

'**Hypothesis OH**': Let φ be an Orlicz and supermultiplicative function, φ^{-1} be its inverse, and $\| \cdot \|_\varphi$ be the Orlicz-Luxemburg norm. Denote $(x)_q = \frac{\Gamma(x+q)}{\Gamma(x)}$ for $x \geq 0$ and $H = (h_{n,k})$, $h_{n,k} \geq 0$. Further, let $a > -1$, $c > 0$, $q > -a - 1$ and $\frac{1}{(n+a+1)_q}$ be non-increasing for $n \geq 0$.

Then the following ingenious result is due to Love (see [4], Theorem 2) and it is most important to prove our further results:

Lemma 3 ([4], Theorem 2) *Suppose that the 'Hypothesis OH' holds. Then for any nonnegative sequence $x = (x_k)$ and $\mu = (\mu_k)$ of real numbers normalized so that $\mu_0 = 1$, the following inequality holds:*

$$\|Hx\|_\varphi \le \widetilde{C}\|x\|_\varphi, \tag{8}$$

where

$$\widetilde{C} = \int_0^1 \varphi^{-1}(\theta^{-(q+1)c})d\mu(\theta). \tag{9}$$

Theorem 2 *Suppose that the 'Hypothesis OH' holds. Then the Hausdorff matrix H maps l_φ into t_φ^α and*

$$\|H\|_{l_\varphi, t_\varphi^\alpha} \le \widetilde{C}\varphi^{-1}(1-\alpha), \text{ where } \widetilde{C} \text{ will be evaluated from Eq. (9).}$$

Proof Let $x \in l_\varphi$ be any nonnegative and nonzero sequence of real numbers. Let $r > 0$ be a real number and applying the Jensen's inequality, one obtains

$$\sum_{n=0}^\infty \varphi\left(\frac{1}{r}\sum_{k=n}^\infty \binom{k}{n}(1-\alpha)^{n+1}\alpha^{k-n}\sum_{i=0}^k h_{k,i}x_i\right)$$

$$\le \sum_{n=0}^\infty \sum_{k=n}^\infty \binom{k}{n}(1-\alpha)^{n+1}\alpha^{k-n}\varphi\left(\frac{1}{r}\sum_{i=0}^k h_{k,i}x_i\right) \text{ (by Jensen's inequality)}$$

$$= \sum_{n=0}^\infty \varphi\left(\frac{1}{r}\sum_{i=0}^n h_{n,i}x_i\right)\sum_{k=0}^n \binom{n}{k}\alpha^{n-k}$$

$$\le (1-\alpha)\sum_{n=0}^\infty \varphi\left(\frac{1}{r}\sum_{i=0}^n h_{n,i}x_i\right).$$

Hence applying similar techniques as applied in Lemma 1 and the definition of Orlicz-Luxemburg norm, inequality (8) implies that

$$\|Hx\|_\varphi^\alpha \le \varphi^{-1}(1-\alpha)\|Hx\|_\varphi \le \widetilde{C}\varphi^{-1}(1-\alpha)\|x\|_\varphi,$$

which in turn implies that

$$\|H\|_{l_\varphi, t_\varphi^\alpha} \le \widetilde{C}\varphi^{-1}(1-\alpha).$$

This proves the theorem. ∎

Corollary 2 *Choose $c = 1$, $a = 0$. Then $\widetilde{C} = \int_0^1 \varphi^{-1}(\theta^{-(q+1)})d\mu(\theta)$ and Cesàro, Hölder, Euler, and Gamma operators map l_φ into t_φ^α. Further, one obtains the following result:*

(a) $\|(C, \beta)\|_{l_\varphi, t_\varphi^\alpha} \le \beta\varphi^{-1}(1-\alpha)\int_0^1 \varphi^{-1}(\theta^{-(q+1)})(1-\theta)^{\beta-1}d\theta$, $\beta > 0$;

(b) $\|(H, \beta)\|_{l_\varphi, t_\varphi^\alpha} \le \frac{1}{\Gamma(\beta)} \varphi^{-1}(1 - \alpha) \int_0^1 \varphi^{-1}(\theta^{-(q+1)}) |\log \theta|^{\beta-1} d\theta, \quad \beta > 0;$

(c) $\|(E, \beta)\|_{l_\varphi, t_\varphi^\alpha} \le \varphi^{-1}(1 - \alpha) \varphi^{-1}(\beta^{-(q+1)}), \quad 0 < \beta < 1;$

(d) $\|(\Gamma, \beta)\|_{l_\varphi, t_\varphi^\alpha} \le \beta \varphi^{-1}(1 - \alpha) \int_0^1 \varphi^{-1}(\theta^{-(q+1)}) \theta^{\beta-1} d\theta.$

Corollary 3 *Choose $c = 1$, $a = 0$ and denote $p^* = \frac{p}{p-1}$. Choose $\varphi(t) = t^p$, $p \ge 1$,*

which gives $\varphi^{-1}(t) = t^{\frac{1}{p}}$. Then $\tilde{C} = \int_0^1 \theta^{-(q+1)/p} d\mu(\theta)$ and Cesàro, Hölder, Euler, and Gamma operators map l_p into t_p^α. Further, one gets the following result from Corollary 2:

(a) $\|(C, \beta)\|_{l_p, t_p^\alpha} \le \beta(1 - \alpha)^{1/p} \frac{\Gamma(\beta+1)\Gamma(\frac{1}{p^*} - \frac{q}{p})}{\Gamma(\beta + \frac{1}{p^*} - \frac{q}{p})}, \beta > 0;$

(b) $\|(H, \beta)\|_{l_p, t_p^\alpha} \le (1 - \alpha)^{1/p} \frac{1}{\Gamma(\beta)} \int_0^1 \theta^{-(q+1)/p} |\log \theta|^{\beta-1} d\theta, \beta > 0;$

(c) $\|(E, \beta)\|_{l_p, t_p^\alpha} \le (1 - \alpha)^{1/p} \beta^{-(q+1)/p}, 0 < \beta < 1;$

(d) $\|(\Gamma, \beta)\|_{l_p, t_p^\alpha} \le (1 - \alpha)^{1/p} \frac{p\beta}{p\beta - q - 1}, p\beta > q + 1.$

Corollary 4 *Consider the similar assumptions as of Corollary 3 and additionally put $q = 0$, then one obtains the Corollary 3.2 established by Talebi in [1].*

3.2   Nörlund Matrix Operator

Now, the Nörlund matrix operator N maps, l_φ into t_φ^α is considered. Before proceeds further, the notion of Nörlund matrix is recalled. Let $p = (p_n)$ be a sequence of nonnegative numbers such that $p_0 > 0$ and denote $P_n = \sum_{k=0}^n p_k$ for $n \ge 0$. Then the Nörlund matrix $N \equiv N(p_n) = (a_{n,k})_{n,k \ge 0}$ associated with the sequence (p_n) is defined by

$$a_{n,k} = \begin{cases} 0 & \text{if } k > n, \\ \frac{p_{n-k}}{P_n} & \text{if } 0 \le k \le n. \end{cases}$$

Note that one can assume that $p_0 = 1$ because $N(p_n) = N(cp_n)$ holds for any $c > 0$. Bounds for the operator norms of Nörlund matrix operator are studied by Johnson et al. in [12]. Our interest in the following study is to estimate a general upper bound for the Nörlund matrix operator norm $\|N\|_{l_\varphi, t_\varphi^\alpha}$. The statement of the theorem is given below:

Theorem 3 *Suppose $p = (p_n)$ is a sequence of nonnegative numbers such that $p_0 = 1$. Then the Nörlund matrix N maps l_φ into t_φ^α and the following inequality holds:*

$$\|N\|_{l_\varphi, t_\varphi^\alpha} \leq \varphi^{-1}\left((1-\alpha)\sum_{k=0}^{\infty}\frac{p_k}{P_k}\right). \tag{10}$$

Proof Let $x \in l_\varphi$ be any nonnegative and nonzero sequence of real numbers and $r > 0$. Applying the Jensen's inequality, the following is obtained:

$$\sum_{n=0}^{\infty}\varphi\left(\frac{1}{r}\sum_{k=n}^{\infty}\binom{k}{n}(1-\alpha)^{n+1}\alpha^{k-n}\sum_{i=0}^{k}\frac{p_{k-i}}{P_k}x_i\right)$$

$$\leq \sum_{n=0}^{\infty}\sum_{k=n}^{\infty}\binom{k}{n}(1-\alpha)^{n+1}\alpha^{k-n}\varphi\left(\frac{1}{r}\sum_{i=0}^{k}\frac{p_{k-i}}{P_k}x_i\right)$$

$$\left(\text{by Jensen's inequality as }\sum_{k=n}^{\infty}\binom{k}{n}(1-\alpha)^{n+1}\alpha^{k-n}=1\right)$$

$$=\sum_{n=0}^{\infty}\varphi\left(\frac{1}{r}\sum_{i=0}^{n}\frac{p_{n-i}}{P_n}x_i\right)\sum_{k=0}^{n}\binom{n}{k}(1-\alpha)^{k+1}\alpha^{n-k}$$

$$\leq (1-\alpha)\sum_{n=0}^{\infty}\sum_{i=0}^{n}\frac{p_{n-i}}{P_n}\varphi\left(\frac{x_i}{r}\right)\quad\left(\text{by Jensen's inequality as }\sum_{i=0}^{n}\frac{p_{n-i}}{P_n}=1\right)$$

$$\leq (1-\alpha)\sum_{i=0}^{\infty}\sum_{n=i}^{\infty}\frac{p_{n-i}}{P_n}\varphi\left(\frac{x_i}{r}\right)$$

$$= (1-\alpha)\sum_{i=0}^{\infty}\sum_{k=0}^{\infty}\frac{p_k}{P_{k+i}}\varphi\left(\frac{x_i}{r}\right)$$

$$\leq (1-\alpha)\sum_{k=0}^{\infty}\frac{p_k}{P_k}\sum_{i=0}^{\infty}\varphi\left(\frac{x_i}{r}\right).$$

Using similar techniques as developed in the Lemma 1, the notion of Orlicz-Luxemburg norm implies that $\|Nx\|_\varphi^\alpha \leq \varphi^{-1}\left((1-\alpha)\sum_{k=0}^{\infty}\frac{p_k}{P_k}\right)\|x\|_\varphi$, which gives

$$\|N\|_{l_\varphi, t_\varphi^\alpha} \leq \varphi^{-1}\left((1-\alpha)\sum_{k=0}^{\infty}\frac{p_k}{P_k}\right),$$

and this completes the proof.

Corollary 5 *Choose a sequence* $p = (p_k)$ *such that* $\frac{p_k}{P_k} = \frac{1}{(k+1)^2}$, *then the Nörlund matrix* N, *maps* l_φ *into* t_φ^α *with*

$$\|N\|_{l_\varphi, t_\varphi^\alpha} \leq \varphi^{-1}\left(\frac{\pi^2}{6}(1-\alpha)\right).$$

Corollary 6 *Let $\varphi(t) = t^p$ for $p \geq 1$ in Corollary 5. Then the Nörlund matrix N maps l_p into t_p^α and one gets*

$$\|N\|_{l_p, t_p^\alpha} \leq \left(\frac{\pi^2}{6}(1-\alpha)\right)^{1/p},$$

which is Corollary 3.4 obtained by Talebi in [1].

4 Conclusion

Upper bounds of operator norms for generalized Hausdorff and Nörlund matrix operators in Orlicz-Taylor sequence spaces are obtained. This work strengthens the latest work presented by Talebi in [1] and shows a new direction of research. It is only the Jensen's inequality applied to prove all the results.

References

1. Talebi, G.: On the Taylor sequence spaces and upper boundedness of Hausdorff matrices and Nörlund matrices. Indag. Math. (N.S.) **28**(3), 629–636 (2017)
2. Musielak, J.: Orlicz Spaces and Modular Spaces, Springer Lecture Notes in Math., vol. 1034. Springer, Berlin (1983)
3. González, M., Sari, B., Wójtowicz, M.: Semi-homogeneous bases in Orlicz sequence spaces. Contemp. Math. **435**, 171–181 (2007)
4. Love, E.R.: Hardy's inequality in Orlicz-type sequence spaces for operators related to generalized Hausdorff matrices. Math. Z. **193**, 481–490 (1986)
5. Love, E.R.: Hardy's inequality for Orlicz-Luxemburg norms. Acta Math. Hung. **56**, 247–253 (1990)
6. Bennett, G.: Factorizing the classical inequalities. Mem. Am. Math. Soc. **120**(576), 1–130 (1996)
7. Mohapatra, R.N., Salzmann, F., Ross, D.: Norm inequalities which yield inclusion for Euler sequence spaces. Comput. Math. Appl. **30**(3–6), 383–387 (1995)
8. Talebi, G., Dehghan, M.A.: Approximation of upper bounds for matrix operators on Fibonacci weighted sequence spaces. Linear Multilinear Algebra **64**(2), 196–207 (2016)
9. Talebi, G., Dehghan, M.A.: Upper bounds for the operator norms of Hausdorff matrices and Nörlund matrices on the Euler-weighted sequence spaces. Linear Multilinear Algebra **62**(10), 1275–1284 (2014)
10. Lashkaripour, R., Foroutannia, D.: Inequalities involving upper bounds for certain matrix operators. Proc. Indian Acad. Sci. (Math. Sci.) **116**(3), 325–336 (2006)
11. Jakimovski, A., Rhoades, B.E., Tzimbalario, J.: Hausdorff matrices as bounded operators over l_p. Math. Z. **138**, 173–181 (1974)
12. Johnson Jr., P.D., Mohapatra, R.N., Ross, D.: Bounds for the operator norms of some Nörlund matrices. Proc. Am. Math. Soc. **124**(2), 543–547 (1996)

Chapter 27
A Study on Fuzzy Triangle and Fuzzy Trigonometric Properties

Debdas Ghosh and Debjani Chakraborty

Abstract This paper investigates fuzzy triangle, fuzzy triangular properties, and fuzzy trigonometry. A fuzzy triangle on the plane is constructed by three fuzzy points as its vertices. Using the proposed fuzzy triangle, basic fuzzy trigonometric functions are investigated. The extension principle and the concepts of same and inverse points in fuzzy geometry are used to define all the proposed ideas. It is shown that some well-known trigonometric identities for crisp angles may not hold with proper equality for fuzzy angles.

Keywords Fuzzy number · Fuzzy point · Same points · Inverse points · Fuzzy angle · Fuzzy triangle · Extension principle

AMS Subject Classification 03E72

1 Introduction

In the literature on fuzzy trigonometry, definition of a fuzzy triangle in a plane is given in four different ways: first, a fuzzy triangle is the intersection of three intersecting fuzzy half planes [12]; second, a fuzzy triangle is the union of three fuzzy line segments that are obtained by joining three fuzzy points (three vertices) [2]; third, a fuzzy triangle is a blurred image obtained by blurring the sides of a crisp triangle [9] and last, a fuzzy triangle is an approximate crisp triangle [7, 8]. Chaudhuri [5] has defined a fuzzy triangle as a fuzzy sets whose α-cuts are similar triangles.

D. Ghosh (✉)
Department of Mathematical Sciences, Indian Institute of Technology (BHU),
Varanasi 221005, Uttar Pradesh, India
e-mail: debdas.email@gmail.com

D. Chakraborty
Department of Mathematics, Indian Institute of Technology Kharagpur,
Kharagpur 721302, West Bengal, India
e-mail: debjani@maths.iitkgp.ernet.in

© Springer Nature Singapore Pte Ltd. 2018
D. Ghosh et al. (eds.), *Mathematics and Computing*, Springer Proceedings
in Mathematics & Statistics 253, https://doi.org/10.1007/978-981-13-2095-8_27

Fuzzy triangle defined in [5] cannot be a fuzzy triangle, and it is a fuzzy point [2] whose support is a triangular region. Membership value of a point in the fuzzy half plane defined in [12] depends on the perpendicular distance between the point and the boundary of the fuzzy half plane; membership value of the points increases for the increasing value of this perpendicular distance. At the core, the considered fuzzy half plane in [12] is not similar to the definition of crisp half plane. Over and above, core of a fuzzy half plane must be a crisp half plane, which also does not follow from the definition of fuzzy half plane, and hence, definition of fuzzy triangle therein may be questionable. In [12], boundary of α-cuts of a fuzzy triangle are equivalent triangles with identical vertex angles, and hence, it is shown that the well-known sine law for crisp triangles holds for fuzzy triangles also. The fuzzy trigonometric functions and angles in [12] are crisp for a fuzzy triangle. These should be fuzzy and cannot be crisp in general, because the considered environment is itself not precisely defined.

Buckley and Eslami [2] defined fuzzy triangle by three fuzzy points as its vertices. To form a fuzzy triangle, three intersecting fuzzy line segments are being adjoined. This definition may be acceptable in fuzzy environment. Fuzzy trigonometric functions have also been defined [2] for a right-angled fuzzy triangle using ratio of fuzzy distances; e.g., fuzzy sine function is the ratio of the perpendicular and hypotenuse. However, if fuzzy trigonometric functions are tried to be generalized for any fuzzy angle, then fuzzy distance of a vertex to the opposite side of a fuzzy triangle has to be measured. But how to measure is not given. By the same authors, in [3] the same has been defined through extension principle.

Liu and Coghill [10] have defined fuzzy trigonometric functions using fuzzy unit circle, which has been named as fuzzy qualitative circle. Boundary of the crisp circle has been partitioned fuzzily, and fuzzy qualitative angles are defined as 4-tuple trapezoidal fuzzy number. But it is very difficult to obtain the value of the trigonometric functions for arbitrary fuzzy angle in general.

Imran and Beg [7, 8] studied fuzzy triangle or f-triangle as an approximate triangle. It is reported that instead of drawing a triangle by ruler, any triangle drawn by free hand is a fuzzy triangle. Subsequently, similarity of fuzzy triangles is also studied. But we note that core of this fuzzy triangle is not a crisp triangle.

In [9], fuzzy triangle is defined by blurring boundary of a crisp triangle using smooth unit step function and implicit functions. But in the obtained shape, its 1-level sets contain all the points which lie outside the considered crisp triangle instead of the points on the boundary.

Recently, [15] has mentioned that the counterpart of a crisp triangle, C, in Euclidian geometry, is a fuzzy triangle. Fuzzy triangle is referred to as an fuzzy transform of C, with C playing the role of the prototype of fuzzy triangle. It is helpful to visualize a fuzzy transform of C as the result of execution of the instruction—draw C by hand with an unprecisiated spray pen. Here, the fuzzy transformation is an one-to-many function.

An overview on fuzzy geometrical concepts prior to the work of Buckley and Eslami is reported in [13]. Some simple construction on fuzzy geometrical concepts can also be obtained in [11].

In this paper, new concepts about fuzzy triangle, fuzzy triangular properties, and some basics of fuzzy trigonometry are proposed. After defining a fuzzy triangle, its side lengths, vertex angles, area and perimeter have been studied. In [1, 13, 14], some concepts about perimeter and area of fuzzy sets are given, but those measurements are crisp numbers. However, the proposed concept yields fuzzy numbers as measurement of side lengths and vertex angles. In the studied concepts of fuzzy triangle, above observation about fuzzy triangle reported by Zadeh [15] is followed. All the proposed concepts introduced here depend on the newly defined concepts of same and inverse points [4, 6]. The following is the outline of this paper.

Section 2 is covered by basic definitions and terminologies used in this paper. Construction of fuzzy triangle is proposed in Sect. 3. In Sect. 4, fuzzy trigonometric functions are introduced. A brief discussion about the work presented here and its future scope are added in the Sect. 5.

2 Preliminaries

The basic definitions adopted here are taken from [2, 6] with slight alteration. Small or capital letters with over tilde bar, i.e., $\widetilde{A}$, $\widetilde{B}$, $\widetilde{C}$, ...and $\widetilde{a}$, $\widetilde{b}$, $\widetilde{c}$, ...represent fuzzy subsets of $\mathbb{R}^n$, $n = 1, 2$. Membership function of a fuzzy set $\widetilde{A}$ of $\mathbb{R}^n$ is represented by $\mu(x|\widetilde{A})$, $x \in \mathbb{R}^n$ with $\mu(\mathbb{R}^n) \subseteq [0, 1]$, $n = 1, 2$.

Definition 1 (*α-cut of a fuzzy set* [6]) For a fuzzy set $\widetilde{A}$ of $\mathbb{R}^n$, $n = 1, 2$, an α-cut of $\widetilde{A}$ is denoted by $\widetilde{A}(\alpha)$ and is defined by:

$$
\widetilde{A}(\alpha) = \begin{cases} \{x : \mu(x|\widetilde{A}) \geq \alpha\} & \text{if } 0 < \alpha \leq 1 \\ closure\{x : \mu(x|\widetilde{A}) > 0\} & \text{if } \alpha = 0. \end{cases}
$$

The set $\{x : \mu(x|\widetilde{A}) > 0\}$ is called support of the fuzzy set $\widetilde{A}$.

To represent the construction of membership function of a fuzzy set $\widetilde{A}$, the notation $\bigvee\{x : x \in \widetilde{A}(0)\}$ is frequently used, which means $\mu(x|\widetilde{A}) = \sup\{\alpha : x \in \widetilde{A}(\alpha)\}$.

Definition 2 (*Fuzzy numbers* [2]) A fuzzy set $\widetilde{A}$ of $\mathbb{R}$ is called a fuzzy number if its membership function μ has the following properties:

(i) $\mu(x|\widetilde{A})$ is upper semi-continuous,
(ii) $\mu(x|\widetilde{A}) = 0$ outside some interval $[a, d]$, and
(iii) there exist real numbers b and c so that $a \leq b \leq c \leq d$ and $\mu(x|\widetilde{A})$ is increasing on $[a, b]$, decreasing on $[c, d]$ and $\mu(x|\widetilde{A}) = 1$ for each x in $[b, c]$.

Since $\mu(x|\widetilde{A})$ is upper semi-continuous for a fuzzy number $\widetilde{A}$, the set $\{x : \mu(x|\widetilde{A}) \geq \alpha\}$ is closed for all α in $\mathbb{R}$. So, α-cut of a fuzzy number $\widetilde{A}$, i.e., the set $\widetilde{A}(\alpha)$ is a closed and bounded interval of $\mathbb{R}$ for all α in $[0, 1]$.

For $b = c$, letting $f(x) = \mu(x|\widetilde{A})\forall x \in [a, b]$ and $g(x) = \mu(x|\widetilde{A})\forall x \in [c, d]$, the notation $(a, c, d)_{fg}$ is used in this paper to represent the above-defined fuzzy number. In particular, if $f(x)$ and $g(x)$ are linear functions, then fuzzy number is called a triangular fuzzy number and it is denoted by $(a/b/c)$.

Definition 3 (*Fuzzy number along a line* [6]) In defining a fuzzy number, conventionally a real line ($\mathbb{R}$) is taken as the universal set. Instead of a real line as the universal set, consider any line on the plane $\mathbb{R}^2$ where the x-axis represents real line, and let $\widetilde{p}$ be a fuzzy number. On the x-axis, the membership function of $\widetilde{p}$ may be written as $\mu((x, 0)|\widetilde{p}) = \mu(x|\widetilde{p})\forall x \in \mathbb{R}$. More explicitly:

$$\mu((x, y)|\widetilde{p}) = \begin{cases} \mu(x|\widetilde{p}) & \text{if } y = 0 \\ 0 & \text{elsewhere.} \end{cases}$$

Let $T : \mathbb{R}^2 \to \mathbb{R}^2$ be a transformation that includes rotation of the axes by angle θ and translation of the origin to $(\frac{ac}{a^2+b^2}, \frac{bc}{a^2+b^2})$, which is the point of intersection for $ax + by = c$ and its perpendicular line through origin. T can be expressed by $T(x, y) = (x\cos\theta - y\sin\theta + \frac{ac}{a^2+b^2}, x\sin\theta + y\cos\theta + \frac{bc}{a^2+b^2})$. T is a bijective transformation that transforms the x-axis to $ax + by = c$. Now, $\widetilde{p}$ may be considered as a fuzzy number on the line $ax + by = c$ and may be defined in the following way:

$$\mu((u, v)|\widetilde{p}) = \begin{cases} \mu((x, 0)|\widetilde{p}) & \text{if } (u, v) = T(x, 0), au + bv = c \\ 0 & \text{elsewhere.} \end{cases}$$

Definition 4 (*Fuzzy points* [2]) A fuzzy point at (a, b) in $\mathbb{R}^2$, written as $\widetilde{P}(a, b)$, is defined by its membership function:

 (i) $\mu((x, y)|\widetilde{P}(a, b))$ is upper semi-continuous,
 (ii) $\mu((x, y)|\widetilde{P}(a, b)) = 1$ if and only if $(x, y) = (a, b)$, and
 (iii) $\widetilde{P}(a, b)(\alpha)$ is a compact, convex subset of $\mathbb{R}^2$, for all α in $[0, 1]$.

Often the notations $\widetilde{P}_1, \widetilde{P}_2, \widetilde{P}_3, \ldots$ are used to represent fuzzy points.

Definition 5 (*Same points* [6]) Let (x_1, y_1) and (x_2, y_2) be two points on support of two continuous fuzzy points $\widetilde{P}(a, b)$ and $\widetilde{P}(c, d)$, respectively. Let L_1 be a line joining (x_1, y_1) and (a, b). As $\widetilde{P}(a, b)$ is a fuzzy point, along L_1 there exists a fuzzy number, $\widetilde{r}_1$ say, on the support of $\widetilde{P}(a, b)$. Membership function of this fuzzy number $\widetilde{r}_1$ can be written as $\mu((x, y)|\widetilde{r}_1) = \mu((x, y)|\widetilde{P}(a, b))$ for (x, y) on L_1, and 0 otherwise.

Similarly, along a line, L_2 say, joining (x_2, y_2) and (c, d), there exists a fuzzy number, $\widetilde{r}_2$ say, on the support of $\widetilde{P}(c, d)$. The points (x_1, y_1), (x_2, y_2) are said to be same points with respect to $\widetilde{P}(a, b)$ and $\widetilde{P}(c, d)$ if:

 (i) (x_1, y_1) and (x_2, y_2) are same points with respect to $\widetilde{r}_1$ and $\widetilde{r}_2$ and
 (ii) L_1, L_2 have made the same angle with the line joining (a, b) and (c, d).

Definition 6 (*Inverse points* [6]) Let (x_1, y_1) and (x_2, y_2) be two points in the support of two continuous fuzzy points $\widetilde{P}(a, b)$ and $\widetilde{P}(c, d)$, respectively. The points (x_1, y_1), (x_2, y_2) are said to be inverse points with respect to $\widetilde{P}(a, b)$ and $\widetilde{P}(c, d)$ if (x_1, y_1), $(-x_1, -y_1)$ are same point w.r.t. $\widetilde{P}(a, b)$ and $-\widetilde{P}(c, d)$.

Definition 7 (*Fuzzy distance* [6]) Fuzzy distance $(\widetilde{D} = \widetilde{D}(\widetilde{P}_1, \widetilde{P}_2))$ between two fuzzy points $\widetilde{P}_1$ and $\widetilde{P}_2$ is defined by its membership function: $\mu(d|\widetilde{D}) = \sup\{\alpha : d = d(u, v),$ where $u \in \widetilde{P}_1(0),\ v \in \widetilde{P}_2(0)$ are inverse points; $\mu(u|\widetilde{P}_1) = \mu(v|\widetilde{P}_2) = \alpha\}$. Here $d(,)$ is the Euclidean distance metric.

Definition 8 (*Fuzzy line segment* [6]) Fuzzy line segment $\widetilde{L}_{P_1P_2}$ joining two fuzzy points $\widetilde{P}_1$, $\widetilde{P}_2$ is defined by its membership function as $\mu((x, y)|\widetilde{L}_{P_1P_2}) = \sup\{\alpha :$ (x, y) lies on the line joining same points $(x_1, y_1) \in \widetilde{P}_1(0)$, $(x_2, y_2) \in \widetilde{P}_2(0)$ and $\mu((x_1, y_1)|\widetilde{P}_1) = \mu((x_2, y_2)|\widetilde{P}_2) = \alpha\}$.

Definition 9 (*Angle between two fuzzy line segments* [6]) Let $\widetilde{P}_1$, $\widetilde{P}_2$, $\widetilde{P}_3$ be three continuous fuzzy points. The angle between $\widetilde{L}_{P_1P_2}$, $\widetilde{L}_{P_2P_3}$ is denoted by $\widetilde{\Theta}$ and is defined by: $\mu(\theta|\widetilde{\Theta}) = \sup\{\alpha : \theta$ is angle between two line segments $\overline{L}_{uv}$ and $\overline{L}_{vw}$, where u, v and v, w are same points of membership value α; $u \in \widetilde{P}_1(0),\ v \in \widetilde{P}_2(0),$ $w \in \widetilde{P}_3(0)\}$.

In the next section, first we introduce the formation of fuzzy triangle and then the measurements of its side lengths, vertex angles, area, and perimeter.

3 Fuzzy Triangle

Let us suppose that three distinct fuzzy points $\widetilde{P}_1$, $\widetilde{P}_2$, and $\widetilde{P}_3$ are given and a fuzzy triangle $(\widetilde{\Delta}P_1P_2P_3)$ is to form. A construction procedure may be designed as follows. Considering three same points u, v, and w in the support of $\widetilde{P}_1$, $\widetilde{P}_2$, and $\widetilde{P}_3$, respectively, let us construct a triangle Δ having vertices as u, v and w. If $\mu(u|\widetilde{P}_1) = \alpha$, then obviously $\mu(v|\widetilde{P}_2) = \alpha$, $\mu(w|\widetilde{P}_3) = \alpha$. We may put membership value of Δ in $\widetilde{\Delta}P_1P_2P_3$ is also α. Now $\widetilde{\Delta}P_1P_2P_3$ can be considered as union of all of these Δ's—crisp triangles with different membership grades. Thus, a formal definition of a fuzzy triangle may be given by its membership function as
$\mu(x|\widetilde{\Delta}P_1P_2P_3) = \sup\{\alpha : x \in \Delta$, where Δ is constructed by the same points $u \in \widetilde{P}_1(0),\ v \in \widetilde{P}_2(0)$, and $w \in \widetilde{P}_3(0)$ as vertices; $\mu(u|\widetilde{P}_1) = \mu(v|\widetilde{P}_2) = \mu(w|\widetilde{P}_1) = \alpha\}$.

Remark 1 Fuzzy triangle defined in the above definition is exactly equal to $\widetilde{L}_{P_1P_2} \cup \widetilde{L}_{P_2P_3} \cup \widetilde{L}_{P_3P_1}$.

Example 1 Let us consider the fuzzy triangle, $\widetilde{\Delta}P_1P_2P_3$, whose vertices are three fuzzy points $\widetilde{P}_1(1, 2)$, $\widetilde{P}_2(5, 7)$ and $\widetilde{P}_3(6, 1)$. Let the membership functions are right elliptical/circular cone with supports

$$\widetilde{P}_1(1,2)(0) = \{(x, y) : \frac{(x-1)^2}{4} + (y-2)^2 \le 1\},$$

$$\widetilde{P}_2(5,7)(0) = \{(x, y) : (x-5)^2 + (y-7)^2 \le 4\} \text{ and}$$

$$\widetilde{P}_3(6,1)(0) = \{(x, y) : (x-6)^2 + (y-1)^2 \le 1\}.$$

Let us now evaluate membership value of $(2, 4)$ in the fuzzy triangle $\widetilde{\Delta}P_1P_2P_3$.

The same points with membership value $\alpha \in [0, 1]$ on $\widetilde{P}_1(1, 2)$, $\widetilde{P}_2(5, 7)$, and $\widetilde{P}_3(6, 1)$ are (see [6]):

$$A_{\alpha,\theta} : \left(1 + \frac{2(1-\alpha)\cos\theta}{\sqrt{4\sin^2\theta + \cos^2\theta}}, 2 + \frac{2(1-\alpha)\sin\theta}{\sqrt{4\sin^2\theta + \cos^2\theta}}\right),$$

$$B_{\alpha,\theta} : (5 + 2(1-\alpha)\cos\theta, 7 + 2(1-\alpha)\sin\theta) \text{ and}$$

$$C_{\alpha,\theta} : (6 + (1-\alpha)\cos\theta, 1 + (1-\alpha)\sin\theta),$$

respectively, where $\theta \in [0, 2\pi]$.

Apparently, there is a possibility that $(2, 4)$ may lie on the line segment $\widetilde{L}_{P_1P_2}$, but $(2, 4)$ cannot lie on the line segments $\widetilde{L}_{P_2P_3}$ and $\widetilde{L}_{P_1P_3}$. Now the condition that $(2, 4)$ lies $\widetilde{L}_{P_1P_2}$ or on the line segment $\overline{A_{\alpha,\theta}B_{\alpha,\theta}}$ (for some $\theta \in [0, 2\pi]$ and $\alpha \in [0, 1]$) is:

$$\frac{4 - (7 + 2(1-\alpha)\sin\theta)}{2 - (5 + 2(1-\alpha)\cos\theta)} = \frac{2 + 2k(1-\alpha)\sin\theta - (7 + 2(1-\alpha)\sin\theta)}{1 + 2k(1-\alpha)\cos\theta - (5 + 2(1-\alpha)\cos\theta)}, \text{ where } k = \frac{1}{\sqrt{4\sin^2\theta + \cos^2\theta}}$$

$$\Rightarrow \alpha = 1 - \frac{3}{(8 + 6k)\sin\theta - (10 + 6k)\cos\theta} = f(\theta), \text{ say.}$$

Here $f(\theta)$ must lie in $[0, 1]$, and hence admissible domain of $f(\theta)$ is $D_f = [63°, 222.66°]$. Maximum value of $f(\theta)$ over D_f occurred at $157.32°$ and the value is 0.8352, the possibility of containment of $(2,4)$ on $\widetilde{L}_{P_1P_2}$.

Thus, the point $(2, 4)$ lies on the triangle $\triangle A_{\alpha,\theta}B_{\alpha,\theta}C_{\alpha,\theta}$ for $\alpha = 0.8352$ and $\theta = 157.32°$, i.e, $A \equiv (0.7726, 2.1193)$, $B = (4.7081, 7.1531)$ and $C \equiv (5.8541, 1.0766)$. Hence, $\mu((2, 4)|\widetilde{\Delta}P_1P_2P_3) = 0.8352$.

In Fig. 1, α-cut of a fuzzy triangle is shown. The shaded regions represent $\widetilde{P}_1(\alpha)$, $\widetilde{P}_2(\alpha)$, and $\widetilde{P}_3(\alpha)$. The lines BM and CN are inclined at the same angle with the line joining P_2 and P_3. The pairs of points B, C and M, N are same points with membership value α in $\widetilde{P}_2$, $\widetilde{P}_3$. Likewise, A, B and B, C are pairs of same points with respect to $\widetilde{P}_1$, $\widetilde{P}_2$ and $\widetilde{P}_2$, $\widetilde{P}_3$, respectively. According to the proposed definition, fuzzy triangle with vertex $\widetilde{P}_1$, $\widetilde{P}_2$, and $\widetilde{P}_3$ is the union of all triangles like $\triangle ABC$.

Note 1 It is to observe that if support of any two vertices of a fuzzy triangle has non-empty intersection, there may be several same points which are coincident. Corresponding to those same points, the crisp triangle in the support of the fuzzy triangle reduces to a crisp line segment. Another case may happen that though the supports of vertices of a fuzzy triangle $\widetilde{\Delta}P_1P_2P_3$ are different but two or more of their core points are identical. In this case, since the core of the fuzzy triangle is a crisp line segment, it can form a fuzzy triangle. These are all degenerate cases of fuzzy triangle. So to get a fuzzy triangle, we need to have three fuzzy points having distinct core points.

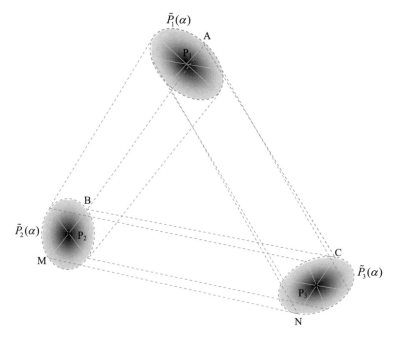

Fig. 1 Construction of a fuzzy triangle

In the following theorem, the α-cut of a fuzzy triangle is found.

Theorem 1 *Let $\widetilde{\Delta}P_1P_2P_3$ be a fuzzy triangle. Its α-cut is the set $\{x : x \in \Delta$ where Δ is a crisp triangle whose vertices are three same points $u \in \widetilde{P}_1(\alpha)$, $v \in \widetilde{P}_2(\alpha)$ and $w \in \widetilde{P}_3(\alpha)\}$.*

Proof The theorem is followed from the observation that $\widetilde{\Delta}P_1P_2P_3 = \bigvee_{\alpha \in [0,1]}\{\Delta_\alpha :$ Δ_α is a triangle with vertices as same points of $\widetilde{P}_1$, $\widetilde{P}_2$ and $\widetilde{P}_3$ with membership value $\alpha\}$.

Note 2 The result of the Theorem 1 directly shows that fuzzy triangle joining three fuzzy points having three distinct core points is unique, since once vertices of fuzzy triangle are changed, several crisp triangles in the support of the fuzzy triangle which eventually construct the fuzzy triangles are going to change, and hence, fuzzy triangle will have different membership function.

Now let us try to find side lengths of a fuzzy triangle. Length of the sides of the fuzzy triangle $\widetilde{\Delta}P_1P_2P_3$ may be defined by fuzzy distances (Definition 7) between the vertices, i.e., $\widetilde{D}(\widetilde{P}_1, \widetilde{P}_2)$, $\widetilde{D}(\widetilde{P}_2, \widetilde{P}_3)$ and $\widetilde{D}(\widetilde{P}_3, \widetilde{P}_1)$. Let us denote them as $\widetilde{p}_3, \widetilde{p}_1$, and $\widetilde{p}_2$, respectively. The vertex angles of $\widetilde{\Delta}P_1P_2P_3$ may be defined as $\widetilde{\angle}(\widetilde{L}_{P_1P_2}, \widetilde{L}_{P_2P_3})$, $\widetilde{\angle}(\widetilde{L}_{P_2P_3}, \widetilde{L}_{P_3P_1})$, and $\widetilde{\angle}(\widetilde{L}_{P_3P_1}, \widetilde{L}_{P_1P_2})$; the notations $\widetilde{\angle}P_2$, $\widetilde{\angle}P_3$ and $\widetilde{\angle}P_1$, respectively, may be used to represent. It is to note that vertex angle $\widetilde{\angle}P_i$ is situated opposite to the side with length $\widetilde{p}_i$, $i = 1, 2, 3$.

Example 2 Let $\widetilde{\Delta}P_1P_2P_3$ be a fuzzy triangle whose vertices $\widetilde{P}_1(1, 0)$, $\widetilde{P}_2(2, 0)$ and $\widetilde{P}_3(1.5, 2)$ are as follows.

The shape of $\widetilde{P}_1$ is a right circular cone with base $\widetilde{P}_1(0) = \{(x, y) : (x - 1)^2 + y^2 \leq \frac{1}{4}\}$ and vertex $(1, 0)$.

The shape of $\widetilde{P}_2$ is a right circular cone with base $\widetilde{P}_2(0) = \{(x, y) : (x - 2)^2 + y^2 \leq \frac{1}{4}\}$ and vertex $(2, 0)$.

The shape of $\widetilde{P}_3$ is a right elliptical cone with base $\widetilde{P}_3(0) = \{(x, y) : (x - 1.5)^2 + (y - 2)^2 \leq 1\}$ and vertex $(1.5, 2)$.

The same points with membership value $\alpha \in [0, 1]$ on $\widetilde{P}_1(1, 0)$, $\widetilde{P}_2(2, 0)$, and $\widetilde{P}_3(1.5, 2)$ are

$A_{\alpha,\theta} : \left(1 + \frac{(1-\alpha)}{2} \cos\theta, \frac{(1-\alpha)}{2} \sin\theta\right)$,

$B_{\alpha,\theta} : (2 + \frac{(1-\alpha)}{2} \cos\theta, \frac{(1-\alpha)}{2} \sin\theta)$ and

$C_{\alpha,\theta} : (1.5 + (1 - \alpha)\cos\theta, 2 + (1 - \alpha)\sin\theta)$, respectively, where $\theta \in [0, 2\pi]$.

To calculate length of the side $\widetilde{L}_{P_1 P_2}$, i.e., $\widetilde{p}_3$, first let us obtain the pair of inverse points in $\widetilde{P}_1$ and $\widetilde{P}_2$. The inverse points with membership value on $\widetilde{P}_1$ and $\widetilde{P}_2$ are $A_{\alpha,\theta}$ and $B_{\alpha,\pi+\theta}$. Here, $\min\limits_{0\leq\theta\leq 2\pi} d(A_{\alpha,\theta}, B_{\alpha,\pi+\theta}) = \alpha$ and $\max\limits_{0\leq\theta\leq 2\pi} d(A_{\alpha,\theta}, B_{\alpha,\pi+\theta}) = 2 - \alpha \forall \alpha \in [0, 1]$. Thus, $\widetilde{p}_3(\alpha) = [\alpha, 2 - \alpha] \forall \alpha \in [0, 1]$. Hence, membership value of $\widetilde{p}_3$ will be obtained as

$$\mu(d|\widetilde{p}_3) = \begin{cases} d & \text{if } 0 \leq d \leq 1 \\ 2 - d & \text{if } 1 \leq d \leq 2 \\ 0 & \text{elsewhere.} \end{cases}$$

Similarly, length of the sides $\widetilde{L}_{P_2 P_3}$ and $\widetilde{L}_{P_3 P_1}$, i.e., $\widetilde{p}_1$ and $\widetilde{p}_2$, respectively, will be obtained as

$$\mu(d|\widetilde{p}_1) = \mu(d|\widetilde{p}_2) = \begin{cases} \frac{2}{3}d - 0.3744 & \text{if } 0.5615 \leq d \leq 2.0616 \\ 2.3744 - \frac{2}{3}d & \text{if } 2.0616 \leq d \leq 3.5616 \\ 0 & \text{elsewhere.} \end{cases}$$

To evaluate the vertex angle $\angle P_2$ (Definition 9) of the fuzzy triangle $\widetilde{\Delta}P_1P_2P_3$, let us first calculate the angle between the line segments $\overline{A_{\alpha,\theta}B_{\alpha,\theta}}$ and $\overline{B_{\alpha,\theta}C_{\alpha,\theta}}$ joining same points of the vertices. Here, $\angle(\overline{A_{\alpha,\theta}B_{\alpha,\theta}}, \overline{B_{\alpha,\theta}C_{\alpha,\theta}}) = \tan^{-1} \frac{2+\frac{(1-\alpha)}{2}\sin\theta}{-0.5+\frac{(1-\alpha)}{2}\cos\theta}$. We obtain that the core of the angle $\angle P_2$ is $75.9627°$ and support is $\angle P_2(0) = \bigvee\limits_{0\leq\alpha\leq 1, 0\leq\theta\leq 2\pi} \left| \tan^{-1} \frac{2+\frac{(1-\alpha)}{2}\sin\theta}{-0.5+\frac{(1-\alpha)}{2}\cos\theta} \right| = [51.7839°, 90°]$.

Similarly, core and support of the angle $\angle P_1$ are $28.0749°$ and $[22.0228°, 157.6549°]$, respectively. For the angle $\angle P_2$, its core and support are $75.9627°$ and $[51.7839°, 90°]$, respectively.

We note that $\widetilde{\Delta}P_1P_2P_3 = \widetilde{L}_{P_1 P_2} \cup \widetilde{L}_{P_2 P_3} \cup \widetilde{L}_{P_3 P_1}$, since fuzzy line segments are also defined by collection of crisp line segments adjoining same points of the extreme fuzzy points. Side lengths of $\widetilde{\Delta}P_1P_2P_3$ are defined as length of the fuzzy line segments $\widetilde{L}_{P_1P_2}, \widetilde{L}_{P_2P_3}$, and $\widetilde{L}_{P_3P_1}$. Obviously, side lengths of a fuzzy triangle are fuzzy numbers, because distance between two fuzzy points, measured by inverse points, is a fuzzy number [6]. Vertex angles of $\widetilde{\Delta}P_1P_2P_3$ are the angles between fuzzy line segments $\widetilde{L}_{P_1P_2}, \widetilde{L}_{P_2P_3}$, and $\widetilde{L}_{P_3P_1}$. It is worthy to mention that vertex angles of a fuzzy triangle having vertices as three continuous fuzzy points are fuzzy numbers, since angle between two fuzzy line segments joining two fuzzy points can be easily shown as fuzzy number. But we observe that support of the fuzzy number obtained by addition of vertex angles of a fuzzy triangle may contain angle more than 180°. The following example gives one example supporting this observation.

Example 3 Let us consider the fuzzy triangle, $\widetilde{\Delta}P_1P_2P_3$, whose vertices are three fuzzy points $\widetilde{P}_1(3, 2), \widetilde{P}_2(1, 1)$, and $\widetilde{P}_3(1, 0)$. Membership functions of them are right circular/elliptical cone with supports as

$$\widetilde{P}_1(3, 2)(0) = \{(x, y) : \frac{(x-3)^2}{3^2} + \frac{(y-2)^2}{0.5^2} \leq 1\},$$

$$\widetilde{P}_2(1, 1)(0) = \{(x, y) : (x-1)^2 + (y-1)^2 \leq \frac{1}{4}\} \text{ and}$$

$$\widetilde{P}_3(1, 0)(0) = \{(x, y) : (x-1)^2 + y^2 \leq \frac{1}{4}\}.$$

Let us suppose vertex angle $\angle \widetilde{P}_2 = \widetilde{\angle}(\widetilde{L}_{P_1P_2}, \widetilde{L}_{P_2P_3})$ of the fuzzy triangle $\widetilde{\Delta}P_1P_2P_3$ is to evaluate.

The same points with membership value $\alpha \in [0, 1]$ on $\widetilde{P}_1(3, 2)$, $\widetilde{P}_2(1, 0)$, and $\widetilde{P}_3(1, 1)$ are

$$A_{\alpha,\theta} : (3 + 3(1-\alpha)\cos\theta, 2 + \frac{1-\alpha}{2}\sin\theta),$$

$$B_{\alpha,\theta} : (1 + \frac{1-\alpha}{2}\cos\theta, 1 + \frac{1-\alpha}{2}\sin\theta) \text{ and}$$

$$C_{\alpha,\theta} : (1 + \frac{1-\alpha}{2}\cos\theta, \frac{1-\alpha}{2}\sin\theta)$$

respectively, where $\theta \in [0, 2\pi]$.

Thus, $\angle P_2(0) = \bigvee_{\theta \in [0,2\pi], \alpha \in [0,1]} |\angle(\overline{A_{\alpha,\theta}B_{\alpha,\theta}}, \quad \overline{B_{\alpha,\theta}C_{\alpha,\theta}})| = [102.5306°, 194.0362°]$.

A geometric visualization of the scenario is given in Fig. 2.

Remark 2 As vertex angles of the fuzzy triangle are defined as angles between the side line segments of the fuzzy triangle, there is a possibility that vertex angle may contain more than 180° angle on its support. It is shown in the Fig. 2 that $\angle(\overline{A_{0,\pi}B_{0,\pi}}, \overline{B_{0,\pi}C_{0,\pi}}) = 194.0362°$. However, we note that if the vertex angle $\angle P_2$ would have defined by the collection of the angle $\angle B_{\alpha,\theta}$ of the triangles $\triangle A_{\alpha,\theta}B_{\alpha,\theta}C_{\alpha,\theta}$

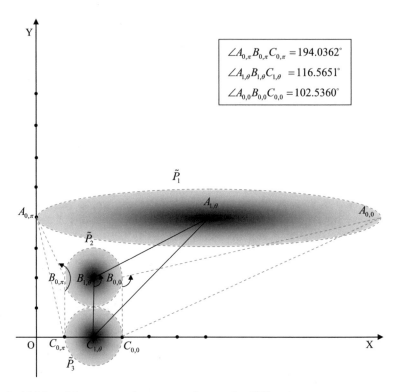

$$\angle A_{0,\pi}B_{0,\pi}C_{0,\pi} = 194.0362°$$
$$\angle A_{1,\theta}B_{1,\theta}C_{1,\theta} = 116.5651°$$
$$\angle A_{0,0}B_{0,0}C_{0,0} = 102.5360°$$

Fig. 2 Addition of the vertex angles may contain more that 180° on its support

for all possible θ and α, the measurement of $\widetilde{\angle}P_2$ cannot have more than 180° on its support.

In one another way, we may define vertex angle of a fuzzy triangle as follows. Let us consider a fuzzy triangle $\widetilde{\Delta}P_1P_2P_3$ having vertices as $\widetilde{P}_1$, $\widetilde{P}_2$, and $\widetilde{P}_3$. The vertex angle $\widetilde{\angle}P_2$ may be defined as $\widetilde{\angle}P_2 = \bigvee\{\angle ABC$ where $\triangle ABC$ is a triangle, $A \in \widetilde{P}_1$, $B \in \widetilde{P}_2$, $C \in \widetilde{P}_3$ are three same points$\}$. Similarly, $\widetilde{\angle}\widetilde{P}_1$, $\widetilde{\angle}\widetilde{P}_3$ can be defined. By this definition of fuzzy vertex angle, for the triangle considered in the Example 3, we obtain that $\widetilde{\angle}P_2(0) = [102.5306°, 180°]$. However, by this definition also addition of three vertex angles may contain more than 180° angle. Thus, in either definition $\widetilde{\angle}P_1 + \widetilde{\angle}P_2 + \widetilde{\angle}P_3 \neq \bigvee\{\angle A + \angle B + \angle C$: where $\triangle ABC$ is a triangle having vertices as same points of $\widetilde{P}_1$, $\widetilde{P}_2$, and $\widetilde{P}_3\}$, since right-hand side is the crisp number 180° and left-hand side is a fuzzy number.

Now, let us try to investigate perimeter and area of a fuzzy triangle.

Definition 10 (*Perimeter of a fuzzy triangle*) Let us consider a fuzzy triangle $\widetilde{\Delta}P_1P_2P_3$. Fuzzy perimeter of the considered fuzzy triangle may be defined by the following ways.

(**Method 1**) Let us denote the fuzzy perimeter as $\widetilde{\delta}_1$. It may be defined by the membership function: $\mu(\delta|\widetilde{\delta}_1) = \sup\{\alpha : \delta$ is the perimeter of the triangle formed by three same points $u \in \widetilde{P}_1(0), v \in \widetilde{P}_2(0),$ and $w \in \widetilde{P}_3(0)$ as its vertices with $\mu(u|\widetilde{P}_1) = \mu(v|\widetilde{P}_2) = \mu(w|\widetilde{P}_3) = \alpha\}$.

(**Method 2**) In this method, let us denote the fuzzy perimeter as $\widetilde{\delta}_2$. It may be defined by: $\widetilde{\delta}_2 = \widetilde{p}_1 + \widetilde{p}_2 + \widetilde{p}_3$.

Remark 3 Here, addition $\widetilde{a} + \widetilde{b}$ of two fuzzy numbers $\widetilde{a}$ and $\widetilde{b}$ will be performed by applying the concept of same point as defined in [6]. The definition in [6] for addition of two fuzzy numbers says that $\widetilde{a} + \widetilde{b} = \bigvee\{x + y : x, y$ are same points in $\widetilde{a}, \widetilde{b}\}$. In fact, this addition and extended addition give the same result as shown in [6].

Note 3 Fuzzy perimeters obtained by above two methods are not equal, i.e., $\widetilde{\delta}_1 \neq \widetilde{\delta}_2$. It is easily followed from the formation of $\widetilde{\Delta}P_1P_2P_3$ and the evaluation of distance between two fuzzy vertices (points). $\widetilde{\Delta}P_1P_2P_3$ is formed by taking union of all crisp triangles whose vertices are same points of the fuzzy vertices of $\widetilde{\Delta}P_1P_2P_3$, whereas distance between two fuzzy vertices is evaluated by combining distances between inverse points. Thus, addition of the side lengths, i.e., $\widetilde{p}_1 + \widetilde{p}_2 + \widetilde{p}_3$, cannot be equal to the union of all the perimeter of the crisp triangles on the support of $\widetilde{\Delta}P_1P_2P_3$.

The following example explores the fact that $\widetilde{\delta}_1 \neq \widetilde{\delta}_2$.

Example 4 Let $\widetilde{\Delta}P_1P_2P_3$ be a fuzzy triangle whose vertices $\widetilde{P}_1(1, 0), \widetilde{P}_2(2, 0),$ and $\widetilde{P}_3(1.5, 2)$ are as follows. All of them has membership function as right circular cone with support sets are: $\widetilde{P}_1(0) = \{(x, y) : (x - 1)^2 + y^2 \leq \frac{1}{4}\}, \widetilde{P}_2(0) = \{(x, y) : (x - 2)^2 + y^2 \leq \frac{1}{4}\}, \widetilde{P}_3(0) = \{(x, y) : (x - 1.5)^2 + (y - 2)^2 \leq 1\}$.

The same points with membership value $\alpha \in [0, 1]$ on $\widetilde{P}_1(1, 0), \widetilde{P}_2(2, 0),$ and $\widetilde{P}_3(1.5, 2)$ are:

$A_{\alpha,\theta} : \left(1 + \frac{(1-\alpha)}{2}\cos\theta, \frac{(1-\alpha)}{2}\sin\theta\right),$

$B_{\alpha,\theta} : (2 + \frac{(1-\alpha)}{2}\cos\theta, \frac{(1-\alpha)}{2}\sin\theta)$ and

$C_{\alpha,\theta} : (1.5 + (1 - \alpha)\cos\theta, 2 + (1 - \alpha)\sin\theta)$, respectively, where $\theta \in [0, 2\pi]$.

Thus from definition of $\widetilde{\delta}_1$ and $\widetilde{\delta}_2$, we get $\widetilde{\delta}_1(0) = \bigvee_{\alpha\in[0,1],\theta\in[0,2\pi]}|d(A_{\alpha,\theta}, B_{\alpha,\theta}) + d(B_{\alpha,\theta}, C_{\alpha,\theta}) + d(C_{\alpha,\theta}, A_{\alpha,\theta})| = [4.1623, 6.0990]$ and $\widetilde{\delta}_2(0) = \widetilde{p}_1(0) + \widetilde{p}_2(0) + \widetilde{p}_3(0) = [0.5616, 3.5616] + [0.5616, 3.5616] + [0, 2] = [1.1232, 9.1232] \neq \widetilde{\delta}_1(0)$.

The results of the following two theorems give information to get α-cuts, and hence membership functions of $\widetilde{\delta}_1$ and $\widetilde{\delta}_2$

Theorem 2 $\widetilde{\delta}_1$ *is a fuzzy number and* $\widetilde{\delta}_1(\alpha) = \{\delta : \delta$ *is the perimeter of the triangle constructed by the same points* $u \in \widetilde{P}_1(\alpha), v \in \widetilde{P}_2(\alpha)$ *and* $w \in \widetilde{P}_3(\alpha)$ *as its vertices*}.

Proof Similar to Theorem 5.1 of [6].

Theorem 3 $\widetilde{\delta}_2$ *is a fuzzy number and* $\widetilde{\delta}_2(\alpha) = \widetilde{p}_1(\alpha) + \widetilde{p}_2(\alpha) + \widetilde{p}_3(\alpha)\forall\alpha \in [0, 1]$.

Proof Theorem directly follows from the addition of fuzzy numbers using same points.

Now, let us define area of a fuzzy triangle.

Definition 11 (*Area of a fuzzy triangle*) Fuzzy area ($\widetilde{\Delta}$) of a fuzzy triangle $\widetilde{\Delta}P_1P_2P_3$ may be defined by its membership function as $\mu(\Delta|\widetilde{\Delta}) = \sup \{\alpha : \Delta$ is the area of the triangle constructed by the same points $u \in \widetilde{P}_1(0)$, $v \in \widetilde{P}_2(0)$ and $w \in \widetilde{P}_3(0)$ as its vertices with $\mu(u|\widetilde{P}_1) = \mu(v|\widetilde{P}_2) = \mu(w|\widetilde{P}_3) = \alpha\}$.

Example 5 Let us calculate the area ($\widetilde{\Delta}$) of the fuzzy triangle in Example 4. Area of the triangle $\Delta A_{\alpha,\theta} B_{\alpha,\theta} C_{\alpha,\theta}$ for a particular value of θ and α is $\frac{1}{2}|2 + \frac{1}{2}(1 - \alpha)\sin\theta|$. Thus, support of $\widetilde{\Delta}$ is

$$\widetilde{\Delta}(0) = \bigcup_{\theta \in [0,2\pi]} \bigcup_{\alpha \in [0,1]} \frac{1}{2}|2 + \frac{1}{2}(1 - \alpha)\sin\theta| = [\frac{3}{4}, \frac{5}{4}]$$

and core of $\widetilde{\Delta}$ is 1.

The results of the following theorem give information to get α-cuts, and hence membership function, of $\widetilde{\Delta}$.

Theorem 4 $\widetilde{\Delta}$ *is a fuzzy number and* $\widetilde{\Delta}(\alpha) = \{\Delta : \Delta$ *is the area of the triangle constructed by the same points* $u \in \widetilde{P}_1(\alpha)$, $v \in \widetilde{P}_2(\alpha)$ *and* $w \in \widetilde{P}_3(\alpha)$ *as its vertices*}.

Proof Similar to Theorem 5.1 of [6]. $\blacksquare$

In the next section, we will introduce basic fuzzy trigonometric functions using the proposed fuzzy triangle. It has been shown that several well-known trigonometric identities do not hold with proper equality for fuzzy angle.

4 Fuzzy Trigonometry

To define fuzzy trigonometric functions, let us suppose a fuzzy triangle $\widetilde{\Delta}P_1P_2P_3$ is given and we have to find $\sin \widetilde{\angle}P_2$, $\cos \widetilde{\angle}P_2$, $\tan \widetilde{\angle}P_2$, etc. Here, a definition of $\sin \widetilde{\angle}P_2$ is studied and other functions can be derived in a similar way.

Let a, b, c are three same points taken from $\widetilde{P}_1(0)$, $\widetilde{P}_2(0)$, $\widetilde{P}_3(0)$, respectively ($a, b, c \in \mathbb{R}^2$). Now let us consider the triangle Δabc in $\widetilde{\Delta}P_1P_2P_3$.

Let θ be the angle between $\overline{ab}$, $\overline{bc}$; and n be the foot of perpendicular from a to the line $\overline{bc}$.

Obviously, $\sin\theta = \frac{d(a,n)}{d(a,b)}$, where $d(,)$ is the usual Euclidean distance.

If $\mu(a|\widetilde{P}_1) = \alpha$, then $\mu(b|\widetilde{P}_2) = \alpha$ and $\mu(c|\widetilde{P}_3) = \alpha$.

Since membership values of a, b, and n in the fuzzy triangle $\widetilde{\Delta}P_1P_2P_3$ are α, α, and greater or equals to α, respectively, membership value of $\sin\theta$ in $\sin \widetilde{\angle}P_2$ may be assigned as minimum of membership value of a, b, and n in $\widetilde{\Delta}P_1P_2P_3$. Thus, $\mu(\sin\theta| \sin \widetilde{\angle}P_2) = \alpha$.

Now, $\sin \widetilde{\angle}P_2$ can be defined as union of the above $\sin\theta$s. Therefore, $\sin \widetilde{\angle}P_2$ is defined as follows.

Definition 12 (*Fuzzy sine function*) Let for a fuzzy triangle $\widetilde{\Delta}P_1P_2P_3$, $\angle P_2 = \widetilde{\Theta}$. Then, $\sin \widetilde{\Theta}$ may be defined by the membership function: $\mu(s|\sin \widetilde{\Theta}) = \sup\{\alpha : s = \sin \theta = \frac{d(a,n)}{d(a,b)}$ where $a \in \widetilde{P}_1(0), b \in \widetilde{P}_2(0), c \in \widetilde{P}_3(0)$ are same points with membership value α and n is the foot of perpendicular from a to the line joining b and $c\}$.

In the next section, it is proved that above-defined $\sin \widetilde{\Theta}$ is a fuzzy number. In the proof, for $0 < \alpha \le 1$, the notation $\mathbb{A}(\alpha)$ is used to represent the set $\{\frac{d(a,n)}{d(a,b)} : a \in \widetilde{P}_1(\alpha), b \in \widetilde{P}_2(\alpha), c \in \widetilde{P}_3(\alpha)$ are same points, and n is the foot of perpendicular from a to the line joining b and $c\}$. Similar to Theorem 5.1 of [6], we can prove that $\mathbb{A}(\alpha) = \sin \widetilde{\Theta}(\alpha)$. Before proving the theorem, we will observe one surprising fact that for a fuzzy triangle $\widetilde{\Delta}P_1P_2P_3$, $\sin \angle P_2$ may have singleton support.

Example 6 Let us consider a fuzzy triangle $\widetilde{\Delta}P_1P_2P_3$ with vertices $\widetilde{P}(2, 3)$, $\widetilde{P}(4, 5)$, and $\widetilde{P}(6, 7)$. All of these three fuzzy points have right circular cone as membership functions having bases $(x - 2)^2 + (y - 3)^2 \le \frac{1}{4}$, $(x - 4)^2 + (y - 5)^2 \le \frac{1}{4}$, and $(x - 6)^2 + (y - 7)^2 \le \frac{1}{4}$; and vertices at $(2, 3)$, $(4, 5)$, and $(6, 7)$, respectively.

The same points with respect to the above fuzzy points can be represented by $a = (2 + \frac{(1-\alpha)}{2}\cos\theta, 3 + \frac{(1-\alpha)}{2}\sin\theta)$, $b = (4 + \frac{(1-\alpha)}{2}\cos\theta, 5 + \frac{(1-\alpha)}{2}\sin\theta)$, and $c = (6 + \frac{(1-\alpha)}{2}\cos\theta, 7 + \frac{(1-\alpha)}{2}\sin\theta)$, respectively, with $0 \le \theta \le 2\pi, 0 \le \alpha \le 1$.

For any a, b, and c: $\angle(\overline{ab}, \overline{bc}) = \frac{\pi}{4}$. Apparently, $\widetilde{\angle}P_2 = \frac{\pi}{4}$ and $\sin \widetilde{\angle}P_2$ is the crisp number $\frac{1}{\sqrt{2}}$. Hence, support of $\sin \angle P_2$ is the singleton set $\{\frac{1}{\sqrt{2}}\}$.

Note 4 It can be easily perceived that if membership functions and supports of $\widetilde{P}_1$, $\widetilde{P}_2$, $\widetilde{P}_3$ are identical up to a translation, then all of $\sin \angle P_1$, $\sin \angle P_2$ and $\sin \angle P_3$ must have singleton support, since in this situation the angles $\angle P_1$, $\angle P_2$, and $\angle P_3$ are crisp angles.

Theorem 5 $\sin \widetilde{\Theta}$ *evaluated by the Definition 12 is a fuzzy number.*

Proof Let us take three different fuzzy points $\widetilde{P}_1$, $\widetilde{P}_2$, and $\widetilde{P}_3$ and consider a fuzzy triangle using them as vertices. Let $\widetilde{\Theta}$ be the fuzzy angle between $\widetilde{L}_{P_1P_2}$ and $\widetilde{L}_{P_2P_3}$.

As $\widetilde{P}_1$, $\widetilde{P}_2$, $\widetilde{P}_3$ are fuzzy points, their α-cuts $\widetilde{P}_1(\alpha)$, $\widetilde{P}_2(\alpha)$, $\widetilde{P}_3(\alpha)$ are non-empty compact subset of $\mathbb{R}^2$. Hence, supremum and infimum of $\mathbb{A}(\alpha)$ are attainable at $\mathbb{A}(\alpha)$. That is, if those elements are $u(\alpha)$, $l(\alpha)$, respectively, then $l(\alpha) \in \mathbb{A}(\alpha)$ and $u(\alpha) \in \mathbb{A}(\alpha)$. Therefore, $\mathbb{A}(\alpha) \subseteq [l(\alpha), u(\alpha)]$.

We prove that $\mathbb{A}(\alpha) = [l(\alpha), u(\alpha)]$ for $0 < \alpha \le 1$. To prove this, it is sufficient to prove that $\mathbb{A}(\alpha)$ is convex, closed, and bounded set.

Boundedness of $\mathbb{A}(\alpha)$ is trivially true, because it is assumed that the sets $\widetilde{P}_1(0)$, $\widetilde{P}_2(0)$ and $\widetilde{P}_3(0)$ have empty intersection.

Now as $l(\alpha) \in \mathbb{A}(\alpha)$ and $u(\alpha) \in \mathbb{A}(\alpha)$, obviously convexity of $\mathbb{A}(\alpha)$ will imply its closedness. We will prove that $\mathbb{A}(\alpha)$ is convex.

It is easy to notice that core of $\sin \widetilde{\Theta}$ is the singleton $\{s_0\}$ where $s_0 = \sin \angle(P_1P_2, P_2P_3)$. We argue that $\mathbb{A}(\alpha)$ contains all the points of $[l(\alpha), s_0]$ and also of $[s_0, u(\alpha)]$. If $s_0 = l(\alpha) = u(\alpha)$, then result is trivially true. If not, then let $\lambda \in (0, 1)$ and t_1, $t_2 \in \mathbb{A}(\alpha)$ with $t_1 < t_2 < s_0$. Obviously, $l(\alpha) < t_1 < \lambda t_1 + (1 - \lambda)t_2 < t_2 < s_0$. Let

$\theta_1 = \sin^{-1}(t_1)$, $\theta_2 = \sin^{-1}(t_2)$, $\theta_\alpha = \sin^{-1}(l(\alpha))$ and $\theta_\lambda = \sin^{-1}(\lambda t_1 + (1-\lambda)t_2)$, where $\sin^{-1}$ not necessarily represents principle value. We took θ_1 such a manner that $0 \leq \theta_1 < \theta_0 \leq \pi$. The similar restriction also followed for θ_2, θ_λ, and θ_α. It can be easily observed that $0 \leq \theta_\alpha < \theta_1 < \theta_\lambda < \theta_2 < \theta_0 \leq \pi$. As membership function of $\widetilde{\Theta}$ is continuous, it follows that $\theta_\lambda \in \widetilde{\Theta}(\alpha)$. Hence $\sin(\theta_\lambda) \in \mathbb{A}(\alpha)$, i.e., $\lambda t_1 + (1-\lambda)t_2 \in \mathbb{A}(\alpha)$. So $[l(\alpha), s_0] \subseteq \mathbb{A}(\alpha)$. Similarly, we can prove that $[s_0, u(\alpha)] \subseteq \mathbb{A}(\alpha)$.

Hence, $\mathbb{A}(\alpha) = [l(\alpha), u(\alpha)]$, a closed bounded interval. Therefore, membership function of $\sin \widetilde{\Theta}$ is upper semi-continuous.

Let $0 < \alpha \leq \beta \leq 1$. Apparently, $\widetilde{P}_i(\beta) \subseteq \widetilde{P}_i(\alpha)$ for $i = 1, 2, 3$. Therefore, $\mathbb{A}(\beta) \subseteq \mathbb{A}(\alpha)$, i.e., $[l(\beta), u(\beta)] \subseteq [l(\alpha), u(\alpha)]$. This implies, l is an increasing function and u is a decreasing function. On the other hand, apparently, $\mathbb{A}(0) = [l(0), u(0)]$ and $\mathbb{A}(1) = \{\theta_0\}$. Obviously, membership value of $\sin \theta_0$ in the fuzzy set $\sin(\widetilde{\Theta})$ is one. Hence the result is proved.

Here, a natural question may arise whether there exists any relation between the value of $\sin \widetilde{\Theta}$ evaluated by extension principle and by the Definition 12? Result of the following theorem finds this relation.

Theorem 6 *Let us consider a fuzzy triangle constructed by three different fuzzy points $\widetilde{P}_1$, $\widetilde{P}_2$, $\widetilde{P}_3$. Let $\widetilde{\Theta}$ be the fuzzy angle between $\widetilde{L}_{P_1 P_2}$ and $\widetilde{L}_{P_2 P_3}$ and $\mathbb{S}(\alpha) = \sin(\widetilde{\Theta})(\alpha)$, where $\sin(\widetilde{\Theta})$ is evaluated by the extension principle. Then, $\mathbb{S}(\alpha)$ is identical to $\mathbb{A}(\alpha)$ for $0 \leq \alpha \leq 1$.*

Proof The theorem is followed from the fact that $\sin \widetilde{\Theta} = \bigvee_{\alpha \in [0,1]} \mathbb{A}(\alpha) = \bigvee_{\alpha \in [0,1]} \{s_\alpha : s_\alpha = \sin \widetilde{\Theta}_\alpha\}$.

Therefore, the trigonometric sine functions evaluated by extension principle and by the Definition 12 are identical.

In a similar way of defining $\sin \widetilde{\Theta}$, the other fuzzy trigonometric functions, like cosine, tangent, etc., can also be defined for fuzzy angles.

Here, it is surprising to note that $\sin \widetilde{\Theta}$ may have discontinuous membership function even if membership function of $\widetilde{\Theta}$ is continuous. Following example is an counterexample of this fact.

Example 7 Let $\widetilde{\Theta} = (0/\frac{\pi}{4}/\pi)$. Then, $\sin \widetilde{\Theta}$ has following membership function which is discontinuous.

$$\mu(s|\sin \widetilde{\Theta}) = \begin{cases} \frac{4\sin^{-1}(s)}{\pi} & \text{if } 0 \leq s < \frac{1}{\sqrt{2}} \\ \frac{4(\pi - \sin^{-1}(s))}{3\pi} & \text{if } \frac{1}{\sqrt{2}} \leq s \leq 1 \\ 0 & \text{elsewhere.} \end{cases}$$

Figures 3 and 4 depict membership functions of $\widetilde{\Theta}$ and $\sin \widetilde{\Theta}$, respectively.

With the above fact, several well-known trigonometric identities do not hold with proper equality in the fuzzy environment. Let us make a point wise analysis on those identities. All of the analysis are supported by numerical illustration.

Fig. 3 Membership function of $\widetilde{\Theta}$ is continuous (in Example 7)

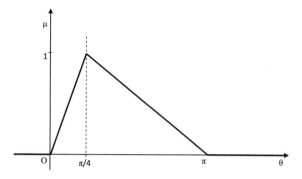

Fig. 4 Membership function of $\sin(\widetilde{\Theta})$ is discontinuous (in Example 7)

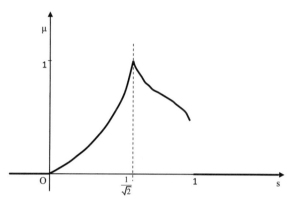

1. **Pythagorean law for a right-angled fuzzy triangle does not hold**. For instance, let us consider the fuzzy triangle $\widetilde{\Delta}P_1P_2P_3$ whose vertices have membership function as right circular cone and with support sets $\widetilde{P}_1(0,0)(0) = \{(x,y) : x^2 + y^2 \leq \frac{1}{4}\}$, $\widetilde{P}_2(1,0)(0) = \{(x,y) : (x-1)^2 + y^2 \leq \frac{1}{4}\}$ and $\widetilde{P}_3(1,1)(0) = \{(x,y) : (x-1)^2 + (y-1)^2 \leq \frac{1}{4}\}$. Here, support of the fuzzy hypotenuse is $\widetilde{p}_2(0) = [\sqrt{2} - 1, \sqrt{2} + 1]$; support of fuzzy perpendicular and fuzzy base sides are $\widetilde{p}_1(0) = \widetilde{p}_3(0) = [0, 2]$. Thus, $(\widetilde{p}_1)^2(0) + (\widetilde{p}_3)^2(0) = [0, 8] \neq [3 - 2\sqrt{2}, 3 + 2\sqrt{2}] = (\widetilde{p}_2)^2(0)$.

2. **For a fuzzy angle $\widetilde{\Theta}$, $\tan \widetilde{\Theta}$ cannot be written as ratio of** $\sin \widetilde{\Theta}$ **and** $\cos \widetilde{\Theta}$. For example, let us take the same fuzzy angle $\widetilde{\Theta}$ of Example 7. $\sin \widetilde{\Theta}$ is given in the Example 7. Membership function of $\cos \widetilde{\Theta}$ and $\tan \widetilde{\Theta}$ has the following membership functions:

$$\mu(c|\cos \widetilde{\Theta}) = \begin{cases} \frac{4(\pi - \cos^{-1}(c))}{3\pi} & \text{if } -1 \leq c < \frac{1}{\sqrt{2}} \\ \frac{4\cos^{-1}(c)}{\pi} & \text{if } \frac{1}{\sqrt{2}} \leq c \leq 1 \\ 0 & \text{elsewhere} \end{cases}$$

Fig. 5 Membership function of $\cos(\widetilde{\Theta})$

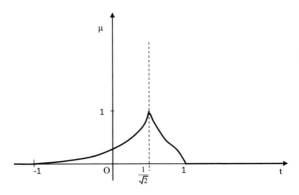

Fig. 6 Membership function of $\tan(\widetilde{\Theta})$

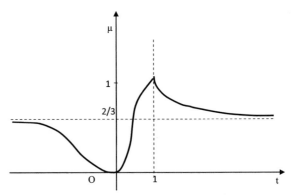

and

$$\mu(t|\tan\widetilde{\Theta}) = \begin{cases} \frac{4\tan^{-1}(t)}{\pi} & \text{if } 0 \leq t \leq 1 \\ \frac{4(\pi-\tan^{-1}(t))}{3\pi} & \text{elsewhere.} \end{cases}$$

Figures 5 and 6 depict the membership function of $\cos(\widetilde{\Theta})$ and $\tan(\widetilde{\Theta})$, respectively.

So, for each $\alpha \in [0, 1]$, $\frac{\sin\widetilde{\Theta}}{\cos\widetilde{\Theta}}(\alpha) = [-\sec(\frac{3\pi}{4}\alpha)\sin(\frac{\pi}{4}\alpha), \sec(\frac{\pi}{4}\alpha)\sin(\frac{3\pi}{4}\alpha)]$. This α-cut is not equal to α-cut of $\tan\widetilde{\Theta}$. For instance, if $\alpha = \frac{2}{3}$, then $\frac{\sin\widetilde{\Theta}}{\cos\widetilde{\Theta}}(\alpha) = (-\infty, \frac{2}{\sqrt{3}}]$, whereas $\tan\widetilde{\Theta} = [\frac{1}{\sqrt{3}}, \infty)$.

3. **For a fuzzy angle** $\widetilde{\Theta}$ **which is a vertex angle of a right-angled fuzzy triangle,** $\sin\widetilde{\Theta}$ **cannot be equal to ratio of fuzzy perpendicular side and fuzzy hypotenuse side length.** For a simple example, let us consider the fuzzy right-angled triangle taken in the Point 1 to show Pythagorean law does not hold. In that triangle $\angle\widetilde{P_2} = 45° = \widetilde{\Theta}$ (say), support of the fuzzy hypotenuse is $\widetilde{p}_2(0) = [\sqrt{2}-1, \sqrt{2}+1]$ and support of fuzzy perpendicular side is $\widetilde{p}_1(0) = [0, 2]$. Thus, $\sin\widetilde{\Theta}(0) = \frac{1}{\sqrt{2}} \neq [0, 4.8284] = \frac{\widetilde{p}_1}{\widetilde{p}_2}(0)$.

4. **For a fuzzy angle** $\widetilde{\Theta}$, $\sin^2\widetilde{\Theta} + \cos^2\widetilde{\Theta}$ **may not be equal to 1**. The result trivially follows from the observation that $\sin^2\widetilde{\Theta} + \cos^2\widetilde{\Theta}$ is a fuzzy number which cannot always be a crisp number, viz. '1'. For a numerical example, let us take the angle $\widetilde{\Theta} = (0/\frac{\pi}{4}/\pi)$. We observe that $(\sin^2\widetilde{\Theta})(0) + (\cos^2\widetilde{\Theta})(0) = [0, 1] + [0, 1] = [0, 2] \neq 1$. Similarly, it can be easily be noted that the identities $\sec^2\widetilde{\Theta} = 1 + \tan^2\widetilde{\Theta}$ and $\csc^2\widetilde{\Theta} = 1 + \cot^2\widetilde{\Theta}$ also may not hold.

Remark 4 Though $\sin^2\widetilde{\Theta} + \cos^2\widetilde{\Theta} \neq 1$, we note that if length of the support of $\widetilde{\Theta}$ is greater than or equal to π, then $(\sin^2\widetilde{\Theta} + \cos^2\widetilde{\Theta})(0) = [0, 2]$. However, core of the fuzzy number $(\sin^2\widetilde{\Theta} + \cos^2\widetilde{\Theta})$ is always 1.

5. **For a fuzzy angle** $\widetilde{\Theta}$, **the identity** $\sin^{-1}(\sin\widetilde{\Theta}) = \widetilde{\Theta}$ **of the inverse circular function holds true.** To prove the result, let θ_α be the angle whose membership value on $\widetilde{\Theta}$ is α. Now, we observe that

$$\sin^{-1}(\sin\widetilde{\Theta}) = \bigvee_{\alpha\in[0,1]} \sin^{-1}(\sin\theta_\alpha) = \bigvee_{\alpha\in[0,1]} \theta_\alpha = \widetilde{\Theta}.$$

The similar reasoning gives that $\cos^{-1}(\cos\widetilde{\Theta}) = \widetilde{\Theta}$, $\tan^{-1}(\tan\widetilde{\Theta}) = \widetilde{\Theta}$, etc.
6. Following the same way as in the Point 5, the properties $\sin(-\widetilde{\Theta}) = -\sin\widetilde{\Theta}$, $\cos(-\widetilde{\Theta}) = \cos\widetilde{\Theta}$, $\sin^{-1}(\sin\widetilde{\Theta}) = \widetilde{\Theta}$, etc., can be proved to be hold for a fuzzy angle $\widetilde{\Theta}$.
7. **Periodic properties of trigonometric functions hold for fuzzy angles**, e.g., $\sin(2\pi + \widetilde{\Theta}) = \sin\widetilde{\Theta}$, $\sin(\frac{\pi}{2} + \widetilde{\Theta}) = \cos\widetilde{\Theta}$, etc. The proof of this properties also will be the same as in the Point 5.
8. **Area of a fuzzy triangle may not be determined by the rule** $\frac{1}{2}\widetilde{bc}\sin\widetilde{A}$. Let us consider a fuzzy triangle constructed by three fuzzy points $\widetilde{A}$, $\widetilde{B}$, and $\widetilde{C}$. Lengths of the sides of fuzzy triangle are $\widetilde{a} = \widetilde{D}(\widetilde{B}, \widetilde{C})$, $\widetilde{b} = \widetilde{D}(\widetilde{A}, \widetilde{C})$, $\widetilde{c} = \widetilde{D}(\widetilde{A}, \widetilde{B})$ and while computing $\widetilde{D}(\widetilde{A}, \widetilde{B})$, $\widetilde{D}(\widetilde{A}, \widetilde{C})$, $\widetilde{D}(\widetilde{A}, \widetilde{B})$ the combinations of distances of inverse points of $\widetilde{A}$ and $\widetilde{B}$ are being taken into account. But vertex angles $\angle A$, $\angle B$, etc., of the fuzzy triangle are evaluated by considering vertex angles of the crisp triangles having vertices are same points with respect to fuzzy points $\widetilde{A}$, $\widetilde{B}$, and $\widetilde{C}$. Apparently, in general area of the fuzzy triangle $\widetilde{\Delta}$ cannot be equal to $\frac{1}{2}\widetilde{bc}\sin\widetilde{A}$. For example, let us consider the fuzzy triangle $\widetilde{\Delta}ABC$ whose vertices are three fuzzy points having right circular cone as their membership function with support sets are $\widetilde{A}(0) = \{(x, y) : (x - 2)^2 + y^2 \leq \frac{1}{4}\}$, $\widetilde{B}(0) = \{(x, y) : (x - 2)^2 + (y - 2)^2 \leq \frac{1}{4}\}$ and $\widetilde{C}(0) = \{(x, y) : x^2 + y^2 \leq 1\}$. If area of $\widetilde{\Delta}ABC$ is $\widetilde{\Delta}$, then $\widetilde{\Delta}(0) = \bigvee_{\theta\in[0,2\pi],\alpha\in[0,1]} \frac{1}{2}|4 - (1 - \alpha)\cos\theta| = [1.5, 2.5]$. However $(\frac{1}{2}\widetilde{bc}\sin\widetilde{A})(0) = \frac{1}{2}[0.5, 3.5][1, 3]\sin[56.3103°, 123.6901°] = \frac{1}{2}[0.5, 3.5][1, 3]$ $[0.8321, 1] = [0.2080, 5.2500] \neq [1.5, 2.5] = \widetilde{\Delta}(0)$.
9. **Sine law of fuzzy triangle may not hold for a fuzzy triangle**. Sine law for fuzzy triangle may not hold with proper equality, i.e., $\widetilde{a}\sin\widetilde{B} \neq \widetilde{b}\sin\widetilde{A}$. For instance, let us consider the fuzzy triangle considered in the Point 8 just above. Here, $(\widetilde{a}\sin\widetilde{B})(0) = [0.1716, 5.8284]\sin[39.8034°, 140.1970°] = [0.1716, 5.$ $8284][0.6402, 1] = [0.1099, 5.8284]$, whereas $(\widetilde{b}\sin\widetilde{A})(0) = [0.5, 3.5]\sin[56.$

$3103°, \ 123.6901°] = [0.5, 3.5][0.8321, 1] = [0.4160, 3.5] \neq [0.1099, 5.8284]$
$= (\tilde{a}\sin\tilde{B})(0).$

5 Conclusion

This paper discussed a few basic concepts of fuzzy triangle and fuzzy triangular properties. The sup-min composition of fuzzy sets and the concepts of same and inverse points are used in all the discussion. We have studied here basic ideas on formation of fuzzy triangle, its perimeter, and area and fuzzy trigonometric functions. Two different methods are proposed to find perimeter of a fuzzy triangle; the lesser imprecise value may be preferred as value of fuzzy perimeter.

In the formation of fuzzy triangle $\tilde{\Delta}P_1P_2P_3$, we note that if we consider a line $l(x, y)$ perpendicular to $\tilde{\bar{L}}_{P_1P_2}(1)$ at $(x, y) \in \tilde{\bar{L}}_{P_1P_2}(1)$, then along the line $l(x, y)$ there must exist one fuzzy number on $(x, y) \in \tilde{\bar{L}}_{P_1P_2}(0)$ given by $l(x, y) \cap \tilde{\bar{L}}_{P_1P_2}$ [6]. We denote this fuzzy number by $\tilde{l}_3(x, y)$. Thus, corresponding to each (x, y), the function $\tilde{l}_3(x, y)$ always gives one fuzzy number. Similarly, we will get two functions $\tilde{l}_1(x, y)$ and $\tilde{l}_2(x, y)$ corresponding to $\tilde{\bar{L}}_{P_2P_3}$ and $\tilde{\bar{L}}_{P_1P_3}$. Now taking the crisp triangle $\Delta P_1P_2P_3$ as prototype, fuzzy triangle can be obtained by f-transformation $(x, y) \mapsto \tilde{l}_1(x, y), \tilde{l}_2(x, y)$ or $\tilde{l}_3(x, y)$. Thus, the defined concept of fuzzy triangle is similar to Zadeh [15]. According to the methodologies and definitions proposed, measurement of the fuzzy area and fuzzy perimeter yields fuzzy numbers. All the proposed study of fuzzy triangle has been made in a coordinate reference frame of $\mathbb{R}^2$ to account the present imprecision in the fuzzy triangle very easily. Future research can focus to study fuzzy triangles in more generalized spaces.

Here, it is worthy to mention that the proposed definition of fuzzy triangle and its properties can be easily generalized to obtain and analyze fuzzy polygon. Fuzzy polygon has its application in fuzzy optimization.

In defining fuzzy trigonometric functions in fuzzy environment, proposed value of sine of a fuzzy angle is exactly same as the result obtained by direct use of extension principle. In fuzzy trigonometric properties, it is noted in the Sect. 4 that almost all the trigonometric identities/rules which hold with proper equality in the case of classical trigonometry are not holding with proper equality in the notion of fuzzy trigonometry. This happened because of that we have considered classical equality instead of fuzzy equality. To mend it, we may use a fuzzy equality. From the literature of fuzzy equality relation, we are overwhelmed by several definitions of fuzzy equality relations. Which definition will be appropriate here is not known properly. After getting an appropriate definition of fuzzy equality relation, we may also be able to investigate further. Future research work can focus on this topic.

References

1. Bogomolny, A.: On the perimeter and area of fuzzy sets. Fuzzy Sets Syst. **23**, 257–269 (1987)
2. Buckley, J.J., Eslami, E.: Fuzzy plane geometry II: circles and polygons. Fuzzy Sets Syst. **87**, 79–185 (1997)
3. Buckley, J.J., Eslami, E.: An Introduction to Fuzzy Logic and Fuzzy Systems. Physica-Verlag, Heidelberg (2002)
4. Chakraborty, D., Ghosh, D.: Analytical fuzzy plane geometry II. Fuzzy Sets Syst. **243**, 84–109 (2014)
5. Chaudhuri, B.B.: Some shape definitions in of space fuzzy geometry. Pattern Recogn. Lett. **12**, 531–535 (1991)
6. Ghosh, D., Chakraborty, D.: Analytical fuzzy plane geometry I. Fuzzy Sets Syst. **209**, 66–83 (2012)
7. Imran, B.M., Beg, M.M.S.: Estimation of f-similarity in f-triangles using fis. In: Meghanathan, N., Chaki, N., Nagamalai, D. (eds.) CCSIT 2012, Part III LNICST, vol. 86, pp. 290–299. Springer, Heidelberg (2012)
8. Imran, B.M., Beg, M.M.S.: Elements of sketching with words. In: Hu, X. (ed.) IEEE International Conference on Granular Computing, pp. 241–246. IEEE Computer Society, San Jose, California, USA (2010)
9. Li, Q., Guo, S.: Fuzzy geometric object modelling. Fuzzy Inf. Eng. (ICFIE) ASC **40**, 551–563 (2007)
10. Liu, H., Coghill, G.M.: Fuzzy qualitative trigonometry. Proceedings of the IEEE International Conference on Systems, Man and Cybernetics, Hawaii, USA vol. 2, pp. 1291–1296 (2005)
11. Pham, B.: Representation of fuzzy shapes. In: Arcelli C., et al. (eds.) IWVF4, LNCS, Vol. 2059, pp. 239–248. Springer, Heidelberg (2001)
12. Rosenfeld, A.: Fuzzy plane geometry: triangles. Pattern Recogn. Lett. **15**, 1261–1264 (1994)
13. Rosenfeld, A.: Fuzzy geometry: an updated overview. Inf. Sci. **110**, 127–133 (1998)
14. Rosenfeld, A., Haber, S.: The perimeter of a fuzzy set. Pattern Recogn. **18**, 125–130 (1985)
15. Zadeh, L.A.: Toward extended fuzzy logic-a first step. Fuzzy Sets Syst. **160**, 3175–3181 (2009)

Chapter 28
An Extension Asymptotically λ-Statistical Equivalent Sequences via Ideals

Ekrem Savas and Rabia Savas

Abstract In (Savaş in Indian J Math 56(2):1–10 (2014) [27]), we examine the asymptotically $\mathcal{I}^\lambda$-statistical equivalent of order α which is a natural combination of the definition for asymptotically equivalent of order α, where $0 < \alpha \leq 1$, $\mathcal{I}$-statistically limit, and λ-statistical convergence. In this paper, we continue to study by proving some more results.

Keywords Asymptotically statistical equivalent · λ-asymptotically statistical equivalent · Asymptotically equivalent of order α · Statistical limit points

1 Introduction and Background

Let w be the set of all sequences of real or complex numbers and ℓ_∞, c, and c_0 be, respectively, the Banach spaces of bounded, convergent, and null sequences $x = (x_j)$ with the usual norm $\|x\| = \sup |x_j|$, where $j \in \mathbb{N} = \{1, 2, \ldots\}$, the set of positive integers.

The (relatively more general) concept of $\mathcal{I}$-convergence was introduced by Kostyrko et al. [10] in a metric space as a generalized form of the concept of statistical convergence, and it is based upon the notion of an ideal of the subset of the set $\mathbb{N}$ of positive integers. This concept has been studied by many authors; see, for instance, [18–20, 22–26].

The notion of the convergence of a real sequence has been extended to statistical convergence by Fast [7] (see also [29]) as follows: Let E be a subset of $\mathbb{N}$. Then, the asymptotic density of E denoted by $\delta(E) := \lim_{n \to \infty} \frac{1}{n} |\{j \leq n : j \in E\}|$, where the vertical bars denote the cardinality of the enclosed set. A number sequence $x = (x_j)$ is said to be statistically convergent to ξ if for every $\varepsilon > 0$,

E. Savas (✉)
Department of Mathematics, Usak University, Usak, Turkey
e-mail: ekremsavas@yahoo.com

R. Savas
Department of Mathematics, Sakarya University, Sakarya, Turkey
e-mail: rabiasavass@hotmail.com

© Springer Nature Singapore Pte Ltd. 2018
D. Ghosh et al. (eds.), *Mathematics and Computing*, Springer Proceedings
in Mathematics & Statistics 253, https://doi.org/10.1007/978-981-13-2095-8_28

$\delta\left(\left\{j \in \mathbb{N}: |x_j - \xi| \geq \varepsilon\right\}\right) = 0$. If (x_j) is statistically convergent to ξ, we write st-$\lim x_j = \xi$. Statistical convergence turned out to be one of the most active areas of research in summability theory after the works of Fridy [8], Nuray and Ruckle [15], and Šalát [17].

Let $\lambda = \{\lambda_p\}_{p \in \mathbb{N}}$ be a non-decreasing sequence of positive numbers tending to ∞ such that

$$\lambda_{p+1} \leq \lambda_p + 1, \lambda_1 = 1.$$

The collection of such sequences λ will be denoted by Δ. However, the idea of λ-statistical convergence was introduced and studied by Mursaleen [14]. Mursaleen defined λ-statistical convergence as follows: A sequence (x_j) of real numbers is said to be λ-statistically convergent to ξ (or, S_λ-convergent to ξ) if for any $\epsilon > 0$,

$$\lim_{p \to \infty} \frac{1}{\lambda_p} |\{j \in I_p : |x_j - \xi| \geq \epsilon\}| = 0,$$

where $|A|$ denotes the cardinality of $A \subset \mathbb{N}$.

λ-statistical convergence is a special case of A-statistical convergence which is studied by Kolk in [9].

Later, Colak [2] introduced the notion of statistical convergence of order α, $0 < \alpha \leq 1$ by replacing n by n^α in the denominator in the definition of statistical convergence. One can also see [1, 3–5] for related works.

Marouf [13] has presented the definition of asymptotically equivalent sequences and asymptotic regular matrices. Further, in 1997, asymptotic equivalence of sequences and summability was studied by Li [12]. Also, Patterson [16], enlarged these concepts by using an asymptotically statistical equivalent and natural regularity conditions for nonnegative summability matrices. Recently, asymptotically $\mathcal{I}^\lambda$-statistical equivalent sequences was studied by Gümüs and Savaş [6] (see also, Kumar and Sharma [11]). $\mathcal{I}$-asymptotically lacunary statistical equivalent sequences and $\mathcal{I}$-asymptotically lacunary statistical equivalent of order α were studied by Savaş [20, 28], and also, Savaş [21] studied $\mathcal{I}_\lambda$-statistically convergent sequences in topological groups. Recently, Savaş [27] defined asymptotically $\mathcal{I}$-statistical equivalent sequences of order.

In the present paper, we continue to study the concept asymptotically $\mathcal{I}^\lambda$-statistical equivalent of order α. In addition, we study some more natural inclusion theorems.

2   Definitions and Preliminaries

The following definitions and notions will be needed in the sequel.

Definition 1 ([13]) Two nonnegative sequences $x = (x_j)$ and $y = (y_j)$ are said to be **asymptotically equivalent** if

$$\lim_{j} \frac{x_j}{y_j} = 1$$

(denoted by $x \sim y$).

Definition 2 ([8]) The sequence $x = (x_j)$ has **statistic limit** ξ, denoted by $st - \lim x = \xi$ provided that for every $\epsilon > 0$,

$$\lim_{n} \frac{1}{n} \{\text{the number of } j \leq n : |x_j - \xi| \geq \epsilon\} = 0.$$

The next definition is natural combination of Definitions 1 and 2.

Definition 3 ([16]) Two nonnegative sequence $x = (x_j)$ and $y = (y_j)$ are said to be asymptotically statistical equivalent of multiple ξ provided that for every $\epsilon > 0$,

$$\lim_{n} \frac{1}{n} \{\text{the number of } j < n : |\frac{x_j}{y_j} - \xi| \geq \epsilon\} = 0,$$

(denoted by $x \overset{S_\xi}{\sim} y$), and simply asymptotically statistical equivalent if $\xi = 1$.

Definition 4 ([10]) A family $\mathcal{I} \subset 2^{\mathbb{N}}$ is said to be an ideal of $\mathbb{N}$ if the following conditions hold:
(a) $P, Q \in \mathcal{I}$ implies $P \cup Q \in \mathcal{I}$,
(b) $P \in \mathcal{I}$, $Q \subset P$ implies $Q \in \mathcal{I}$,

Definition 5 ([10]) A non-empty family $F \subset 2^{\mathbb{N}}$ is said to be an filter of $\mathbb{N}$ if the following conditions hold:
(a) $\phi \notin F$,
(b) $P, Q \in F$ implies $P \cap Q \in F$,
(c) $P \in F$, $P \subset Q$ implies $Q \in F$,

Definition 6 ([10]) A proper ideal $\mathcal{I}$ is said to be admissible if $\{n\} \in \mathcal{I}$ for each $n \in \mathbb{N}$.

Definition 7 (*see* [10]) Let $\mathcal{I} \subset 2^{\mathbb{N}}$ be a proper admissible ideal in $\mathbb{N}$. Then, the sequence (x_j) of elements of $\mathbb{R}$ is said to be $\mathcal{I}$-convergent to $\xi \in \mathbb{R}$ if for each $\epsilon > 0$, the set $K(\epsilon) = \{n \in \mathbb{N} : |x_j - \xi| \geq \epsilon\} \in \mathcal{I}$.

Definition 8 ([27]) The two nonnegative sequences $x = (x_j)$ and $y = (y_j)$ are said to be asymptotically $\mathcal{I}$-statistical equivalent of order α to multiple ξ, $(0 < \alpha \leq 1)$, provided that for each $\epsilon > 0$ and $\gamma > 0$

$$\{n \in \mathbb{N} : \frac{1}{n^\alpha} |\{j \leq n : |\frac{x_j}{y_j} - \xi| \geq \epsilon\}| \geq \gamma\} \in \mathcal{I},$$

(denoted by $x \overset{S^\xi(\mathcal{I})^\alpha}{\sim} y$) and simply asymptotically $\mathcal{I}$-statistical equivalent of order α if $\xi = 1$. Furthermore, let $S^\xi(\mathcal{I})^\alpha$ denote the set of x and y such that $x \overset{S^\xi(\mathcal{I})^\alpha}{\sim} y$.

Remark 1 If $\mathcal{I} = \mathcal{I}_{fin} = \{B \subseteq \mathbf{N} : B \text{ is a finite subset}\}$, asymptotically $\mathcal{I}$-statistical equivalent of order α to multiple ξ reduces to asymptotically statistical equivalent of order α to multiple ξ. For an arbitrary ideal $\mathcal{I}$ and for $\alpha = 1$, it reduces to asymptotically $\mathcal{I}$-statistical equivalent of multiple ξ (see [6]). When $\mathcal{I} = \mathcal{I}_{fin}$ and $\alpha = 1$, it becomes only asymptotically statistical equivalent of multiple ξ, [16].

The following definition is given in [27].

Definition 9 Let $\lambda = (\lambda_p) \in \Delta$. The two nonnegative sequences $x = (x_j)$ and $y = (y_j)$ are said to be asymptotically $\mathcal{I}^\lambda$-statistical equivalent of order α, $(0 < \alpha \le 1)$, to multiple ξ provided that for any $\epsilon > 0$ and $\gamma > 0$

$$\{p \in \mathbb{N} : \frac{1}{\lambda_p^\alpha} |\{j \in I_p : |\frac{x_j}{y_j} - \xi| \ge \epsilon\}| \ge \gamma\} \in \mathcal{I},$$

(denoted by $x \overset{S^\xi(\mathcal{I})^\alpha}{\sim} y$) and simply asymptotically $\mathcal{I}^\lambda$-statistical equivalent of order α if $\xi = 1$. Furthermore, let $S^\xi_\lambda(\mathcal{I})^\alpha$ denote the set of x and y such that $x \overset{S^\xi_\lambda(\mathcal{I})^\alpha}{\sim} y$.

Remark 2 If we take $\alpha = 1$, the above definition reduces to asymptotically $\mathcal{I}^\lambda$-statistical equivalent of multiple ξ (see [6]). For $\mathcal{I} = \mathcal{I}_{fin}$, asymptotically λ-statistical equivalent of order α to multiple ξ is a special case of asymptotically $\mathcal{I}^\lambda$-statistical equivalent of order α to multiple L.

Definition 10 Let $\lambda = (\lambda_p) \in \Delta$, $\alpha \in (0, 1]$ be any real number and r be a positive real number. Two nonnegative sequences $x = (x_j)$ and $y = (y_j)$ are strong r-asymptotically $\mathcal{I}^\lambda$-equivalent of order α to multiple ξ provided that for any $\epsilon > 0$

$$\{p \in \mathbb{N} : \frac{1}{\lambda_p^\alpha} \sum_{j \in I_p} |\frac{x_j}{y_j} - \xi|^r \ge \epsilon\} \in \mathcal{I},$$

(denoted by $x \overset{V^\xi_\lambda(\mathcal{I})^\alpha_r}{\sim} y$) and simply strong r-asymptotically $\mathcal{I}^\lambda$-equivalent of order α if $\xi = 1$. Further, let $[V^\xi_\lambda](\mathcal{I})^\alpha_r$ denote the set of x and y such that $x \overset{V^\xi_\lambda(\mathcal{I})^\alpha_r}{\sim} y$.

3 Main Results

In this section, we present the main theorems of this paper.

Theorem 1 *Let* $\lambda = (\lambda_p) \in \Delta$ *and* α, β *be fixed real numbers such that* $0 < \alpha \leq \beta \leq 1$ *and let r be a positive real number, then* $[V_\lambda^\xi](\mathcal{I})_r^\alpha \subset S_\lambda^\xi(\mathcal{I})^\beta$ *and the inclusion is strict.*

Proof The inclusion part of proof is easy. Taking $\lambda_p = p$ for all p, we prove the strictness of the inclusion $[V_\lambda^\xi](\mathcal{I})_r^\alpha \subset S_\lambda^\xi(\mathcal{I})^\beta$. For this, consider the sequence $x = (x_j)$ defined by

$$x_j = \begin{cases} 1, & \text{if } j = n^2 \\ 0, & \text{if } j \neq n^2 \end{cases} \quad n = 1, 2, \ldots. \tag{1}$$

and $y_j = 1$ for all j. Then, for every $\varepsilon > 0$ and $\alpha \in \left(\dfrac{1}{2}, 1\right]$, we have

$$\frac{1}{\lambda_p^\alpha} \left| \left\{ j \in I_p : \left| \frac{x_j}{y_j} - 0 \right| \geq \varepsilon \right\} \right| \leq \frac{\sqrt{p}}{p^\alpha} = \frac{1}{p^{\alpha - \frac{1}{2}}}$$

and for any $\gamma > 0$, we get

$$\left\{ p \in \mathbb{N} : \frac{1}{\lambda_p^\alpha} \left| \left\{ j \in I_p : \left| \frac{x_j}{y_j} - 0 \right| \geq \varepsilon \right\} \right| \geq \delta \right\} \subseteq \left\{ p \in \mathbb{N} : \frac{[\sqrt{p}]}{p^\alpha} \geq \gamma \right\}.$$

Since the set on the right-hand side is a finite set and so belongs to $\mathcal{I}$, it follows that $x_j \to 0 \left(S_\lambda^\xi(\mathcal{I})^\alpha \right)$ for $\alpha \in (\dfrac{1}{2}, 1]$. On the other hand for $\alpha \in (0, \dfrac{1}{2}]$, we have

$$\frac{\sqrt{p} - 1}{p^\alpha} \leq \frac{1}{p^\alpha} \sum_{j \in I_p} \left| \frac{x_j}{y_j} \right|^r = \frac{1}{\lambda_p^\alpha} \sum_{j \in I_p} \left| \frac{x_j}{y_j} - 0 \right|^r,$$

and so we have

$$\left\{ p \in \mathbb{N} : \frac{\sqrt{p} - 1}{p^\alpha} \geq 1 \right\} \subseteq \left\{ p \in \mathbb{N} : \frac{1}{\lambda_p^\alpha} \sum_{j \in I_p} \left| \frac{x_j}{y_j} - 0 \right|^r \geq 1 \right\}$$

which belongs to $F(\mathcal{I})$, since $\mathcal{I}$ is admissible. So $x_j \nrightarrow 0[V_\lambda^\xi](\mathcal{I})_r^\alpha$.

Corollary 1 *If two nonnegative sequences* $x = (x_j)$ *and* $y = (y_j)$ *are strong r-asymptotically* $\mathcal{I}^\lambda$-*equivalent of order* α *to multiple* ξ, *then they are asymptotically* $\mathcal{I}^\lambda$-*statistical equivalent of order* α *to multiple* ξ.

Even if $x = (x_j)$ and $y = (y_j)$ are bounded sequences, the converse of Theorem 3.1 does not hold, in general. To show this, we must find two sequences that bounded (that is, $x, y \in \ell_\infty$) and $x \overset{S_\lambda^\xi(\mathcal{I})^\alpha}{\sim} y$, but need not to be $x \overset{V_\lambda^\xi(\mathcal{I})^\alpha}{\sim} y$, for some α and β real numbers such that $0 < \alpha \leq \beta \leq 1$. For this, consider a sequence $x = (x_j)$ defined by (1) and $y_j = 1$ for all j. It can be shown that $x, y \in \ell_\infty$ and asymptotically $\mathcal{I}^\lambda$-statistical equivalent of order α to multiple ξ for $\alpha \in (\frac{1}{3}, 1]$ and $x, y \notin [V_\lambda^\xi](\mathcal{I})_r^\alpha$ for $\alpha \in (0, \frac{1}{2})$. Therefore, $x, y \in S_\lambda^\xi(\mathcal{I})^\beta \setminus [V_\lambda^\xi](\mathcal{I})_r^\alpha$ for $\alpha \in \left(\frac{1}{3}, \frac{1}{2}\right)$.

Theorem 2 *Let α and β be fixed real numbers such that $0 < \alpha \le \beta \le 1$ and r be a positive real number, then $[V_\lambda^\xi](\mathcal{I})_r^\alpha \subseteq [V_\lambda^\xi](\mathcal{I})_r^\beta$ and the inclusion is strict.*

Proof The inclusion part of proof is given in [27]. Taking $\lambda_p = p$ for all p, we demonstrate the strictness of the inclusion $[V_\lambda^\xi](\mathcal{I})_r^\alpha \subseteq [V_\lambda^\xi](\mathcal{I})_r^\beta$ for a special case. Write a sequence such as in (1). Then,

$$\frac{1}{\lambda_p^\beta} \sum_{j \in I_p} \left| \frac{x_j}{y_j} - 0 \right| \le \frac{\sqrt{p}}{p^\beta} = \frac{1}{p^{\beta - 1/2}} \to 0, \ (p \to \infty) \text{ for } \beta \in \left(\frac{1}{2}, 1\right),$$

but

$$\frac{1}{\lambda_p^\alpha} \sum_{j \in I_p} \left| \frac{x_j}{y_j} - 0 \right| = \frac{1}{p^\alpha} \sum_{j \in I_p} \left| \frac{x_j}{y_j} - 0 \right| \ge \frac{\sqrt{p} - 1}{p^\alpha} \to \infty, \ (p \to \infty) \text{ for } \alpha \in \left(0, \frac{1}{2}\right)$$

So $x \in [V_\lambda^\xi](\mathcal{I})_r^\beta$ for $\frac{1}{2} < \beta < 1$ but $x \notin [V_\lambda^\xi](\mathcal{I})_r^\alpha$ for $0 < \alpha < \frac{1}{2}$.

The following result is a consequence of Theorem 2.

Corollary 2 *Let $0 < \alpha \le 1$ be a positive real number and $\lambda \in \Delta$. Then, $[V_\lambda^\xi](\mathcal{I})_r^\alpha \subseteq [V_\lambda^\xi](\mathcal{I})_r$ for each $\alpha \in (0, 1]$.*

Now, we shall prove some more inclusion relations.

Theorem 3 *Let $\lambda = (\lambda_p)$ and $\mu = (\mu_p)$ be two sequences in Δ such that $\lambda_p \le \mu_p$ for all $p \in \mathbb{N}$, and let α and β be fixed real numbers such that $0 < \alpha \le \beta \le 1$,*
  (i) If

$$\liminf_{p \to \infty} \frac{\lambda_p^\alpha}{\mu_p^\beta} > 0 \tag{2}$$

then $S_\mu^\xi(\mathcal{I})^\beta \subseteq S_\lambda^\xi(\mathcal{I})^\alpha$,
  (ii) If

$$\lim_{p \to \infty} \frac{\mu_p}{\lambda_p^\beta} = 1 \tag{3}$$

then $S_\lambda^\xi(\mathcal{I})^\alpha \subseteq S_\mu^\xi(\mathcal{I})^\beta$.

Proof (i) Suppose that $\lambda_p \le \mu_p$ for all $p \in \mathbb{N}$, and let (2) be satisfied. For given $\varepsilon > 0$, we have

$$\left\{ j \in J_p : \left| \frac{x_j}{y_j} - \xi \right| \geq \varepsilon \right\} \supseteq \left\{ j \in I_p : \left| \frac{x_j}{y_j} - L \right| \geq \varepsilon \right\}$$

where $I_p = [p - \lambda_p + 1, p]$ and $J_p = [p - \mu_p + 1, p]$. Therefore, we can write

$$\frac{1}{\mu_p^\beta} \left| \left\{ j \in J_p : \left| \frac{x_j}{y_j} - \xi \right| \geq \varepsilon \right\} \right| \geq \frac{\lambda_p^\alpha}{\mu_p^\beta} \frac{1}{\lambda_p^\alpha} \left| \left\{ j \in I_p : \left| \frac{x_j}{y_j} - \xi \right| \geq \varepsilon \right\} \right|$$

and so for all $p \in \mathbb{N}$, we have, for $\gamma > 0$,

$$\left\{ p \in \mathbb{N} : \frac{1}{\lambda_p^\alpha} \left| \left\{ j \in I_p : \left| \frac{x_j}{y_j} - \xi \right| \geq \varepsilon \right\} \right| \geq \gamma \right\} \subseteq$$

$$\left\{ p \in \mathbb{N} : \frac{1}{\mu_p^\beta} \left| \left\{ j \in J_p : \left| \frac{x_j}{y_j} - \xi \right| \geq \varepsilon \right\} \right| \geq \gamma \frac{\lambda_p^\alpha}{\mu_p^\beta} \right\} \in \mathcal{I}.$$

Hence, $S_\mu^\xi(\mathcal{I})^\beta \subseteq S_\lambda^\xi(\mathcal{I})^\alpha$.

(ii) Let $x = (x_j)$ and $y = (y_j) \in S_\lambda^\xi(\mathcal{I})^\alpha$ and (3) be satisfied. Since $I_p \subset J_p$, for $\varepsilon > 0$, we may write

$$\frac{1}{\mu_p^\beta} \left| \left\{ j \in J_p : \left| \frac{x_j}{y_j} - \xi \right| \geq \varepsilon \right\} \right| = \frac{1}{\mu_p^\beta} \left| \left\{ p - \mu_p + 1 < j \leq p - \lambda_p : \left| \frac{x_j}{y_j} - \xi \right| \geq \varepsilon \right\} \right|$$

$$+ \frac{1}{\mu_p^\beta} \left| \left\{ j \in I_p : \left| \frac{x_j}{y_j} - \xi \right| \geq \varepsilon \right\} \right|$$

$$\leq \frac{\mu_p - \lambda_p}{\mu_p^\beta} + \frac{1}{\lambda_p^\beta} \left| \left\{ j \in I_p : \left| \frac{x_j}{y_j} - \xi \right| \geq \varepsilon \right\} \right|$$

$$\leq \left(\frac{\mu_p - \lambda_p^\beta}{\lambda_p^\beta} \right) + \frac{1}{\lambda_p^\alpha} \left| \left\{ j \in I_p : \left| \frac{x_j}{y_j} - \xi \right| \geq \varepsilon \right\} \right|$$

$$\leq \left(\frac{\mu_p}{\lambda_p^\beta} - 1 \right) + \frac{1}{\lambda_p^\alpha} \left| \left\{ j \in I_p : \left| \frac{x_j}{y_j} - \xi \right| \geq \varepsilon \right\} \right|$$

for all $p \in \mathbb{N}$. Hence, we have

$$\left\{ p \in \mathbb{N} : \frac{1}{\mu_p^\beta} \left| \left\{ j \in J_p : \left| \frac{x_j}{y_j} - \xi \right| \geq \varepsilon \right\} \right| \geq \gamma \right\} \subseteq$$

$$\left\{ p \in \mathbb{N} : \frac{1}{\lambda_p^\alpha} \left| \left\{ j \in I_p : \left| \frac{x_j}{y_j} - \xi \right| \geq \varepsilon \right\} \right| \geq \gamma \right\} \in \mathcal{I}.$$

This implies that $S_\lambda^\xi(\mathcal{I})^\alpha \subseteq S_\mu^\xi(\mathcal{I})^\beta$.

From Theorem 3, we have the following.

Corollary 3 *Let* $\lambda = (\lambda_p)$ *and* $\mu = (\mu_p)$ *be two sequences in* Δ *such that* $\lambda_p \le \mu_p$ *for all* $p \in \mathbb{N}$. *If* (2) *holds, then*

(i) $S_\mu^\xi(\mathcal{I})^\alpha \subseteq S_\lambda^\xi(\mathcal{I})^\alpha$ *for each* $\alpha \in (0, 1]$,

(ii) $S_\mu^\xi(\mathcal{I}) \subseteq S_\lambda^\xi(\mathcal{I})^\alpha$ *for each* $\alpha \in (0, 1]$,

(iii) $S_\mu^\xi(\mathcal{I}) \subseteq S_\lambda^\xi(\mathcal{I})$.

Corollary 4 *Let* $\lambda = (\lambda_p)$ *and* $\mu = (\mu_p)$ *be two sequences in* Δ *such that* $\lambda_p \le \mu_p$ *for all* $p \in \mathbb{N}$. *If* (3) *holds then,*

(i) $S_\lambda^\xi(\mathcal{I})^\alpha \subseteq S_\mu^\xi(\mathcal{I})^\alpha$ *for each* $\alpha \in (0, 1]$,

(ii) $S_\lambda^\xi(\mathcal{I})^\alpha \subseteq S_\mu^\xi(\mathcal{I})$ *for each* $\alpha \in (0, 1]$,

(iii) $S_\lambda^\xi(\mathcal{I}) \subseteq S_\mu^\xi(\mathcal{I})$.

Theorem 4 *Let* $\lambda = (\lambda_p)$ *and* $\mu = (\mu_p)$ *be two sequences in* Δ *such that* $\lambda_p \le \mu_p$ *for all* $p \in \mathbb{N}$, *and let* α *and* β *be fixed real numbers such that* $0 < \alpha \le \beta \le 1$,

(i) *If* (2) *holds, then* $[V_\mu^\xi](\mathcal{I})_r^\beta \subset [V_\lambda^\xi](\mathcal{I})_r^\alpha$,

(ii) *If* (3) *holds and* $x, y \in \ell_\infty$, *then* $[V_\lambda^\xi](\mathcal{I})_r^\alpha \subset [V_\mu^\xi](\mathcal{I})_r^\beta$.

Proof (i) Omitted.

(ii) Let $x, y \in [V_\lambda^\xi](\mathcal{I})_r^\alpha$, and suppose that (3) holds. Since $x = (x_j)$, $y = (y_j) \in \ell_\infty$, there exists some $M > 0$ such that $\left| \frac{x_j}{y_j} - \xi \right| \le M$ for all j. Now, since $I_p \subseteq J_p$ and $\lambda_p \le \mu_p$ for all $p \in \mathbb{N}$, we may write

$$
\frac{1}{\mu_p^\beta} \sum_{j \in J_p} \left| \frac{x_j}{y_j} - \xi \right|^r = \frac{1}{\mu_p^\beta} \sum_{j \in J_p - I_p} \left| \frac{x_j}{y_j} - \gamma \right|^r + \frac{1}{\mu_p^\beta} \sum_{j \in I_p} \left| \frac{x_j}{y_j} - \xi \right|^r
$$

$$
\le \left(\frac{\mu_p - \lambda_p}{\mu_p^\beta} \right) M^r + \frac{1}{\mu_p^\beta} \sum_{j \in I_p} \left| \frac{x_j}{y_j} - \xi \right|^r
$$

$$
\le \left(\frac{\mu_p - \lambda_p^\beta}{\lambda_p^\beta} \right) M^r + \frac{1}{\lambda_p^\beta} \sum_{j \in I_p} \left| \frac{x_j}{y_j} - \xi \right|^r
$$

$$
\le \left(\frac{\mu_p}{\lambda_p^\beta} - 1 \right) M^r + \frac{1}{\lambda_p^\alpha} \sum_{j \in I_p} \left| \frac{x_j}{y_j} - \gamma \right|^r
$$

for all $p \in \mathbb{N}$. So we have

$$
\left\{ p \in \mathbb{N} : \frac{1}{\mu_p^\beta} \sum_{j \in J_p} \left| \frac{x_j}{y_j} - \xi \right|^r \ge \gamma \right\} \subseteq \left\{ p \in \mathbb{N} : \frac{1}{\lambda_p^\alpha} \sum_{j \in I_p} \left| \frac{x_j}{y_j} - \xi \right|^r \ge \gamma \right\} \in \mathcal{I}.
$$

Therefore, $[V_\lambda^\xi](\mathcal{I})_r^\alpha \subset [V_\mu^\xi](\mathcal{I})_r^\beta$.

Corollary 5 *Let* $\lambda = (\lambda_p)$ *and* $\mu = (\mu_p)$ *be two sequences in* Δ *such that* $\lambda_p \leq \mu_p$ *for all* $p \in \mathbb{N}$. *If (2) holds, then*

(i) $[V_\mu^\xi](\mathcal{I})_r^\alpha \subset [V_\lambda^\xi](\mathcal{I})_r^\alpha$ *for each* $\alpha \in (0, 1]$,

(ii) $[V_\mu^\xi](\mathcal{I})_r \subset [V_\lambda^\xi](\mathcal{I})_r^\alpha$ *for each* $\alpha \in (0, 1]$,

(iii) $[V_\mu^\xi](\mathcal{I})_r \subset [V_\lambda^\xi](\mathcal{I})_r$.

Corollary 6 *Let* $\lambda = (\lambda_p)$ *and* $\mu = (\mu_p)$ *be two sequences in* Δ *such that* $\lambda_p \leq \mu_p$ *for all* $p \in \mathbb{N}$. *If (3) holds and* $x, y \in \ell_\infty$, *then*

(i) $[V_\lambda^\xi](\mathcal{I})_r^\alpha \subset [V_\mu^\xi](\mathcal{I})_r^\alpha$, *for each* $\alpha \in (0, 1]$,

(ii) $[V_\lambda^\xi](\mathcal{I})_r^\alpha \subset [V_\mu^\xi](\mathcal{I})_r$, *for each* $\alpha \in (0, 1]$,

(iii) $[V_\lambda^\xi](\mathcal{I})_r \subset [V_\mu^\xi](\mathcal{I})_r$.

Finally, we conclude this paper by presenting the following theorem.

Theorem 5 *Let* $\lambda = (\lambda_p)$ *and* $\mu = (\mu_p)$ *be two sequences in* γ *such that* $\lambda_p \leq \mu_p$ *for all* $p \in \mathbb{N}$, *and let* α *and* β *be fixed real numbers such that* $0 < \alpha \leq \beta \leq 1$ *and* $0 < r < \infty$. *Then, we have*

(i) *Let (2) holds, if* $x \stackrel{V_\lambda^\xi(\mathcal{I})_r^\alpha}{\sim} y$ *then* $x \stackrel{S_\lambda^\xi(\mathcal{I})^\alpha}{\sim} y$,

(ii) *Let (3) holds and* $x = (x_j)$ *and* $y = (y_j)$ *be two bounded sequences, if* $x \stackrel{S_\lambda^\xi(\mathcal{I})^\alpha}{\sim} y$ *then* $x \stackrel{V_\lambda^\xi(\mathcal{I})_r^\alpha}{\sim} y$.

Proof (i) Omitted.

(ii) Suppose that $x \stackrel{S_\lambda^\xi(\mathcal{I})^\alpha}{\sim} y$ and that $x = (x_j)$ and $y = (y_{kj})$ be bounded and $\epsilon > 0$ is given. Since $x = (x_j)$ and $y = (y_j)$ are bounded, there exists an integer M such that $|\frac{x_j}{y_j} - \xi| \leq M$ for all j; then, we may write

$$
\frac{1}{\mu_p^\beta} \sum_{j \in J_p} \left| \frac{x_j}{y_j} - \xi \right|^r = \frac{1}{\mu_p^\beta} \sum_{j \in J_p - I_p} \left| \frac{x_j}{y_j} - \xi \right|^{pr} + \frac{1}{\mu_p^\beta} \sum_{j \in I_p} \left| \frac{x_j}{y_j} - \xi \right|^r
$$

$$
\leq \left(\frac{\mu_p - \lambda_p}{\mu_p^\beta} \right) M^r + \frac{1}{\mu_p^\beta} \sum_{j \in I_p} \left| \frac{x_j}{y_j} - \xi \right|^r
$$

$$
\leq \left(\frac{\mu_p - \lambda_p^\beta}{\mu_p^\beta} \right) M^r + \frac{1}{\mu_p^\beta} \sum_{j \in I_p} \left| \frac{x_j}{y_j} - \xi \right|^r
$$

$$
= \left(\frac{\mu_p}{\lambda_p^\beta} - 1 \right) M^r + \frac{1}{\lambda_p^\beta} \sum_{\substack{j \in I_p \\ |\frac{x_j}{y_j} - \xi| \geq \varepsilon}} \left| \frac{x_j}{y_j} - \xi \right|^r + \frac{1}{\lambda_p^\beta} \sum_{\substack{j \in I_p \\ |\frac{x_j}{y_j} - \xi| < \varepsilon}} \left| \frac{x_j}{y_j} - \xi \right|^r
$$

$$
\leq \left(\frac{\mu_p}{\lambda_p^\beta} - 1 \right) M^r + \frac{M^r}{\lambda_p^\alpha} \left| \left\{ j \in I_p : \left| \frac{x_j}{y_j} - L\xi \right| \geq \varepsilon \right\} \right| + \frac{\mu_p}{\lambda_p^\beta} \varepsilon^r
$$

for all $p \in \mathbb{N}$. So we have, $\gamma > 0$,

$$\left\{ p \in \mathbb{N} : \frac{1}{\mu_p^\beta} \sum_{j \in J_p} \left| \frac{x_j}{y_j} - \xi \right|^r \geq \gamma \right\} \subseteq \left\{ p \in \mathbb{N} : \frac{1}{\lambda_p^\alpha} \left| \left\{ j \in I_p : \left| \frac{x_j}{y_j} - \xi \right| \geq \varepsilon \right\} \right| \geq \frac{\gamma}{M^r} \right\} \in \mathcal{I}.$$

Using (3), we obtain that $x = (x_j)$ strong r-asymptotically $\mathcal{I}^\lambda$-equivalent of order α to multiple ξ, whenever $x \overset{S_\lambda^\xi(\mathcal{I})^\alpha}{\sim} y$.

Corollary 7 *Let* $\lambda = (\lambda_p)$ *and* $\mu = (\mu_p)$ *be two sequences in* Δ *such that* $\lambda_p \leq \mu_p$ *for all* $p \in \mathbb{N}$. *If* (2) *holds and let* $\alpha \in (0, 1]$, *then*

(i) *If* $x \overset{V_\mu^\xi(\mathcal{I})_r^\alpha}{\sim} y$, *then* $x \overset{S_\lambda^\xi(\mathcal{I})^\alpha}{\sim} y$,

(ii) *If* $x \overset{V_\mu^\xi(\mathcal{I})_r}{\sim} y$, *then* $x \overset{S_\lambda^\xi(\mathcal{I})^\alpha}{\sim} y$,

(iii) *If* $x \overset{V_\mu^\xi(\mathcal{I})_r}{\sim} y$, *then* $x \overset{S_\lambda^\xi(\mathcal{I})}{\sim} y$.

References

1. Bhunia, S., Das, P., Pal, S.: Restricting statistical convergenge. Acta Math. Hungar **134**(1–2), 153–161 (2012)
2. Colak, R.: Statistical Convergence of Order α, Modern Methods in Analysis and its Applications, pp. 121–129. Anamaya Publishers, New Delhi (2010)
3. Colak, R., Bektas, C.A.: λ-statistical convergence of order α. Acta Math. Scientia **31B**(3), 953–959 (2011)
4. Das, P., Savaş, E.: On $\mathcal{I}$-statistical and $\mathcal{I}$-lacunary statistical convergence of order α. Bull. Iranian Soc. **40**(2), 459–472 (2014)
5. Et, M., Çınar, M., Karakaş, M.: On λ-statistical convergence of order α of sequences of function. J. Inequal. Appl. **2013**, 204 (2013)
6. Gumus, H., Savas, E.: On $S_\lambda^L(\mathcal{I})$-asymptotically statistical equivalent sequences. Numer. Analy. Appl. Math. (ICNAAM: AIP conference proceeding, vol. 1479 (2012) pp. 936–941 (2012)
7. Fast, H.: Sur la convergence statistique. Colloq. Math. **2**, 241–244 (1951)
8. Fridy, J.A.: On statistical convergence. Analysis **5**, 301–313 (1985)
9. Kolk, E.: The statistical convergence in Banach spaces. Acta Comment. Univ. Tartu **928**, 41–52 (1991)
10. Kostyrko, P., Šalát, T., Wilczynki, W.: $\mathcal{I}$-convergence. Real Anal. Exchange **26**(2) 669–685 (2000/2001)
11. Kumar, V., Sharma, A.: On asymptotically generalized statistical equivalent sequences via ideal. Tamkang J. Math. **43**(3), 469–478 (2012)
12. Li, J.: Asymptotic equivalence of sequences and summability. Int. J. Math. Math. Sci. **20**(4), 749–758 (1997)
13. Marouf, M.: Asymptotic equivalence and summability. Int. J. Math. Math. Sci. **16**(4), 755–762 (1993)
14. Mursaleen, M.: λ-statistical convergence. Math. Slovaca **50**, 111–115 (2000)
15. Nuray, F., Ruckle, W.H.: Generalized statistical convergence and convergence free spaces. J. Math. Anal. Appl. **245**(2), 513–527 (2000)
16. Patterson, R.F.: On asymptotically statistically equivalent sequences. Demonstratio Math. **36**(1), 149–153 (2003)

17. Šalát, T.: On statistically convergent sequences of real numbers. Math. Slovaca **30**, 139–150 (1980)
18. Savaş, E., Das, P.: A generalized statistical convergence via ideals. Appl. Math. Lett. **24**, 826–830 (2011)
19. Savaş, E., Das, P., Dutta, S.: A note on strong matrix summability via ideals. Appl. Math Lett. **25**(4), 733–738 (2012)
20. Savaş, E.: On $\mathcal{I}$-asymptotically lacunary statistical equivalent sequences. Adv. Differ. Equ. 2013, 2013:111 (18 April 2013)
21. Savaş, E.: On $\mathcal{I}_\lambda$-statistically convergent sequences in topological groups. Acta Comment. Univ. Tartu. Math. **18**(1), 33–38 (2014)
22. Savaş, E.: Δ^m-strongly summable sequence spaces in 2-normed spaces defined by ideal convergence and an Orlicz function. Appl. Math. Comput. **217**, 271–276 (2010)
23. Savaş, E.: A-sequence spaces in 2-normed space defined by ideal convergence and an Orlicz function. Abst. Appl. Anal. **2011** Article ID 741382 (2011)
24. Savaş, E.: On some new sequence spaces in 2-normed spaces using Ideal convergence and an Orlicz function. J. Ineq. Appl. Article Number 482392
25. Savaş, E.: On generalized double statistical convergence via ideals. In: The Fifth Saudi Science Conference 16–18 April 2012
26. Savaş, E.: On generalized A-difference strongly summable sequence spaces defined by ideal convergence on a real n-normed space. J. Ineq. Appl. **2012**, 87 (2012)
27. Savaş, E.: On asymptotically $\mathcal{I}$-statistical equivalent sequences of order. Indian J. Math., Special Volume Dedicated to Professor Billy E. Rhoades **56**(2) 1–10 (2014)
28. Savaş, E.: On asymptotically $\mathcal{I}$-lacunary statistical equivalent sequences of order α. In: The 2014 International Conference on Pure and Applied Mathematics, Venice, Italy, March 15–17 2014
29. Schoenberg, I.J.: The integrability of certain functions and related summability methods. Amer. Math. Monthly **66**, 361–375 (1959)

Chapter 29
Fuzzy Goal Programming Approach for Resource Allocation in an NGO Operation

Vinaytosh Mishra, Tanmoy Som, Cherian Samuel and S. K. Sharma

Abstract Diabetes is a major health challenge in India. The lifetime cost of treatment in the disease management is humongous. India is presently lacking at infrastructure and resources to meet the demand created by the sudden surge of the disease. This situation makes it imperative to optimally allocate the resources so that the treatment can be made available to a maximum number of patients at affordable cost. This paper uses fuzzy goal programming with exponential membership function for resource allocation. The human and financial resources are described with fuzzy conditions for determining the future strategies for unknown situations. A fuzzy goal programming model is demonstrated using the case study of an NGO working in the area of awareness and treatment of diabetes in Varanasi.

Keywords Resource allocation · Fuzzy goal programming · Diabetes · NGO
Exponential membership function

1 Introduction

The number of people living with diabetes is increasing exponentially in India [1]. The disease has become a major health challenge in the country in the last decade [2]. Such is the prevalence of the disease in the country that it is called as the diabetes capital of the world [3]. The chronic nature of the disease makes the treatment extremely costly. The cost of treatment of diabetes can be divided into two categories, namely direct cost and indirect cost [4]. The direct cost includes the expenses related to treatment, while indirect cost includes the loss of productivity. In addition to this cost, there is an intangible cost which includes reduced quality of life due to pain, anxiety, and stress [5]. The studies indicate that there has been a significant increase in the cost of diabetes management in the recent time [6, 7]. With the progression of the disease, the cost of treatment increases many folds because of comorbidities [8–10].

V. Mishra (✉) · T. Som · C. Samuel · S. K. Sharma
Indian Institute of Technology (BHU), Varanasi 221005, India
e-mail: vinaytosh@gmail.com

© Springer Nature Singapore Pte Ltd. 2018 373
D. Ghosh et al. (eds.), *Mathematics and Computing*, Springer Proceedings
in Mathematics & Statistics 253, https://doi.org/10.1007/978-981-13-2095-8_29

The studies suggest that the increasing cost of treatment results in low adherence of medication regime [11, 12]. In a recent study, Roebuck et al. concluded that improved medication adherence by people with diabetes produced substantial medical savings as a result of reductions in hospitalization and emergency department use [13]. India lags at healthcare infrastructure, and a number of doctors and beds per patients are far below the World Healthcare Organization (WHO) guidelines. The non-government organization (NGO) can play an important role in bridging this gap. The NGOs have a limited number of resources and need to optimally allocate the resources to maximize the social welfare.

1.1 Cost of Treatment

There is an acute shortage of hospital beds and doctors in India, and more than 50% of the ambulatory care is provided by the private players. The country has witnessed spiraling medical expenses in recent years. According to National Sample Survey Office (NSSO) report, consumer expenditure on healthcare in rural India increased from 6.6% in 2004–05 to 6.9% in 2011–2012, and urban Indians' expenditure on medical care increased from 5.2% in 2004–05 to 5.5% in 2011–2012. The 70% of this cost is constituted by medicine. The diabetes patient once diagnosed undergoes the treatment regime for the rest of his life after. This scenario results are high lifetime cost of treatment of diabetes. The average cost of treatment per diabetes patient per hospital admission, with and without multiple complications, is 314.15 (USD) and 29.1 (USD), respectively, out of which 255.32 (USD) falls under the direct cost of treatment of the disease [7]. Table 1 further provides the details of constituents of the direct cost of treatment of diabetes.

From the above table, we can conclude that the reducing the risk of hospitalization can significantly reduce the cost of treatment of the disease. The self-management education of the disease can help patients reducing the risk of hospitalization in diabetes [14]. Another measure suggested in the literature for reducing hospitalization risk is income tax exemption [15].

Table 1 Direct medical cost per patient per hospitalization

Component of cost		Cost
(average)	USD	% of total cost
Lab investigations	29.45	10.15
Medication for diabetes	7.00	2.42
Medication for comorbidity	69.46	23.94
Hospitalization	143.75	49.56
Doctor's consultation	40.37	13.93

1.1.1 Healthcare Finance Models

There are three main models for healthcare finance on the basis of their funding. The first one is the Beveridge model [16], which is based on taxation and has many public providers. The second is the Bismarck 'mixed' model [17], funded by a mix of government and insurance providers. Finally, the 'private insurance model' in which the cost of the treatment is borne by the health insurance provider. Health insurance helps to spread the cost of treatment over a large time period. Properly designed and administered health insurance can act as a bridge between patients and providers balancing quality care at reasonable costs [18].

India has one of the largest private health sectors in the world, with over 80% of ambulatory care being supported through out-of-pocket expenses [19, 20]. Out-of-pocket (OOP) expenditure on health care has significant implications for poverty in many developing countries [21]. In India, three-fourth of the healthcare expenses are supported by out-of-pocket spend. The government spending on health care has been paltry as a percentage of GDP when we compare it with other developing and developed countries. India spends only 5% annual gross domestic product (GDP) on health care [22].

The diabetes patient needs, affordable and quality health care, self-management education and insurance coverage to meet the cost of diabetes management. The government in India has not been able to develop adequate infrastructure and support to manage the sudden surge in a number of diabetes patients [23]. Despite recent thrust to improve the healthcare infrastructure in India, inequalities related to socioeconomic status, geography, and gender still persist. This situation is further aggravated by high out-of-pocket expenditures [24].

2 Role of an NGO in Health care

Nonprofit organizations can be registered in India as a society, under the Registrar of Societies (Society Act 1860) or as a trust, by making a trust deed, or as a Sect. 8 Company, under the Companies Act, 2013 [25]. They can work in the capacity building, policy shaping or ensure long-term results in healthcare areas. They work in partnership with communities, health institutions, donors, academicians, and governments to achieve these results. They fund their activities through international funding, government funding, local philanthropy, and income-generating activities. NGOs carry out a range of projects including emergency management and relief; healthcare research; designing and implementing alternative funding and insurance schemes; mobilization, advocacy and raising awareness, health campaigns, protection of patient's right; and balancing private players interest. NGOs can fill the gap in diabetes care by working in areas like disease awareness, free consultation, and checkup camps and providing funding for the diabetic patients not able to meet the healthcare expenses. They can also work for bringing transparency and efficiency in healthcare supply chain so that medicine reaches the patients at affordable cost.

Thus, we can say that NGOs can bridge that gap between demand and supply of diabetes care, but they need to efficiently utilize the scarce resources available at their disposal to maximize the welfare of the patients [26].

To the best of our knowledge, there is a lack of any study in the area of resource allocation in the case of cardio-diabetes management. This study attempts to fill this gap. This study uses a case study of Indian NGO, working in the area of cardio-diabetes awareness and treatment, in the eastern part of India. The study can be used as a reference for the NGOs and government bodies working in the area of diabetes management.

3 Resource Allocation in Healthcare

Healthcare resources are limited and demand exceeds supply. The allocation of resources becomes a challenge in healthcare. In the case of private healthcare providers, the allocation is resolved on the basis of the ability to pay. The allocation on basis of ability to play is against the principle of healthcare equity [27]. This section discusses the various approaches to resource allocation found in the literature.

3.1 Hippocratic Model

According to this model, the focus of medical action revolves around the physician-patient encounter. It establishes a fiduciary relationship between the physician and the patient, which means that the physician's duty toward the individual patient overrides all other considerations except insofar as these affect the physician's ability to fulfill her or his patient-related duties [28–30].

3.2 Social Service Model

This model sees the health care in much broader perspective and considers medicine as one among several social enterprises of which the overall purpose is to advance the well-being of members of society [31, 32]. In this approach, the allocation issues assume an entirely different nature. Although the physician-patient encounter still remains an element of fiduciary duty, that element is limited by the constraints pertaining to social welfare maximization [33, 34].

There is need of a coherent and consistent model for healthcare resource allocation. As for our knowledge, there is a lack of the literature on a quantitative model for the allocation of the healthcare resources. There is also a lack of a case-based approach for resource allocation for Indian NGO. This paper attempts to fill this gap.

3.3 Business Model

This model considers health care as neither a fiduciary undertaking nor a health-oriented profession that operationalizes society's duty to do the best for its members [35]. The healthcare provider ethically works for the value maximization of its shareholder [36].

4 Methodology

Goal programming (GP) is an important method for multi-objective decision-making approaches in decision making. In a standard GP formulation, goals and constraints are defined precisely [37, 38]. In healthcare, the system aims and conditions include vague and undetermined situations as every healthcare event is unique and involves uncertainties. The study uses the fuzzy membership function suggested by Turgay and Taşkın [38] and proposes a fuzzy goal programming (FGP) model for optimizing the resource allocation of an NGO working in cardio-diabetes management and education area.

The heart of the methodology of FLP lies in the construction of membership function for objection coefficients, technical coefficients, resource variable, and decision variables [39]. The reasons behind selecting exponential membership in FLP are as follows: (1) It transforms into linear membership function when dealing with nonlinear aggregate operators, and (2) it is more realistic than the linear membership function and has been successfully used for the resource allocation in health care and other industries [37, 38, 40, 41] (Figs. 1 and 2).

The exponential membership function depends on the fuzzy restriction given to a fuzzy goal of the problem in a fuzzy decision-making situation. Let t_{ln} and t_{un} be the lower- and upper-tolerance ranges considered, respectively, for the achievement of the aspired level b_n of the nth fuzzy goal. Then, the exponential membership function for the fuzzy goal $F_n(x)$ having lower tolerance limit $(b_{un} - t_{ln})$ and upper-tolerance limit $(b_{un} + t_{ln})$ can be given as follows:

Fig. 1 Exponential membership function for minimization objective

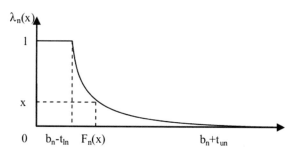

Fig. 2 Exponential membership function for maximization objective

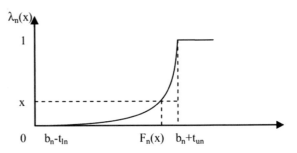

$$\mu_n(x) = \begin{cases} 1, & \text{if } F_n(x) \le b_n \\ 1 + \dfrac{e^{-\frac{\alpha i (b_n - F_n(x))}{t_n(x)}} - e^{-\alpha i}}{1 - e^{-\alpha i}}, & \text{if } b_n \le F_n(x) \le b_n + t_{un} \\ 0, & \text{if } F_n(x) \ge b_n + t_{un} \end{cases} \quad (1)$$

and

$$\mu_n(x) = \begin{cases} 1, & \text{if } F_n(x) \ge b_n \\ \dfrac{e^{-\frac{\alpha i (b_n - F_n(x))}{t_{un}(x)}} - e^{-\alpha i}}{1 - e^{-\alpha i}}, & \text{if } b_n - t_{ln} \le F_n(x) \le b_n \\ 0, & \text{if } F_n(x) \le b_n - t_{ln} \end{cases} \quad (2)$$

The exponential membership function-based fuzzy goal programming with upper and lower level conditions can be presented as follows:

Maximize λ, subject to:

$$\frac{e^{-\frac{\alpha i (b_n - F_n(x))}{t_n(x)}} - e^{-\alpha i}}{1 - e^{-\alpha i}} \le \lambda, \quad n = 1, 2, \dots, N; \quad (3)$$

$$\sum_{i=1}^{n} x_{ij} = 1, \quad j = 1, 2, \dots, N; \lambda \ge 0 \quad (4)$$

$$x_{ij} = \begin{cases} 1, & \text{if the } i\text{th resource is assigned to the } j\text{th task} \\ 0, & \text{if the } i\text{th resource is not assigned to the } j\text{th task} \end{cases} \quad (5)$$

Minimize λ, subject to:

$$\frac{e^{-\alpha_i} - e^{-\frac{\alpha_j (b_n - F_n(x))}{t_n(x)}}}{1 - e^{-\alpha_i}} \ge \lambda, n = 1, 2, \dots, N; \quad (6)$$

$$\sum_{i=1}^{n} x_{ij} = 1, j = 1, 2, \dots, N; \lambda \ge 0 \quad (7)$$

$$x_{ij} = \begin{cases} 1, & \text{if the } i\text{th resource is assigned to the } j\text{th task} \\ 0, & \text{if the } i\text{th resource is not assigned to the } j\text{th task} \end{cases} \tag{8}$$

The slack variables are minimized on the basis of the importance of achieving the aspired goal levels in the decision-making context, for the goal achievement. The fuzzy goal programming model of the problem under a preemptive priority structure can be presented as follows:

$$\text{Min } Z = \left[P_1(d^-), P_2(d^-), \ldots, P_i(d^-) \right] \tag{9}$$

$$\frac{e^{-\alpha_i} - e^{-\frac{\alpha_i(b_n - F_n(x))}{t_n(x)}}}{1 - e^{-\alpha_i}} + d_n^- - d_n^+ = 1 \tag{10}$$

$$1 - \frac{e^{-\alpha_i} - e^{-\frac{\alpha_i(b_n - F_n(x))}{t_n(x)}}}{1 - e^{-\alpha_i}} + d_n^- - d_n^+ = 1 \tag{11}$$

$$d_n^-, d_n^+ \geq 0, n = 1, 2, \ldots N \tag{12}$$

In the above formulation, Z represents the vector of i priority achievement functions and d_n^-, d_n^+ are the slack (under-deviational) and surplus (over-deviational) variables of the nth goal. $P_i(d^-)$ is a linear function of the weighted under-deviational variables, where $P_i(d^-)$ is of the form:

$$P_i(d^-) = \sum_{i=1}^{n} w_{in}^- * d_{in}^-, d_{in}^- \geq 0, (n = 1, 2, \ldots, N) \tag{13}$$

where d_{in}^- ith priority is level for d_{in}^- and w_{in}^- is the weight associated with it. Here, the numerical weight is the weight of importance of achieving the aspired level of the nth goal relative to others which are grouped together at the ith priority level [42]. The model uses the concept of preemptive priorities of the goals, and the ith priority is preferred over the higher priority irrespective of the weight associated with.

5 Model Construction

5.1 Parameter Definition

The objective of the research is to obtain a solution which minimizes the service cost as well as patient service level. Since the objective of the NGO is to include a maximum number of the patients in the social welfare program without compromising on the service quality, the first objective has a higher priority. The model includes the variable cost like salary, cost of equipment, cost of medicines, and another relevant cost (variable operating expenses). The parameters of the model are defined as below:

Table 2 Decision variable values for objective functions

Variable department	Demand per month (D_{it})	Capacity per month (U_{it})	Flexibility of the service (F_{it}) (%)	Target patients (B_{it})
Endocrinology (x_1)	90	150	15	100
Cardiology (x_2)	100	150	10	75
Internal medicine (x_3)	200	150	15	100

p_i Number of IPD patients in each service
r_i Cost of care inpatient stay
D_i Demand of each service
U_i Capacity of each service
P_i *Total budget*
F Flexibility of service quota allocation
B_i Number of patients targeted for each service
$W1_{ti}$ Number of physician in each service in t period
$W2_{ti}$ Number of nurses in each service in t period
$W3_{ti}$ Number of technicians in each service in t period
$CS1_i$ Salary of physician of department i in period t
$CS2_i$ Salary of nurses of department i in period t
$CS3_i$ Salary of technician of department i in period t
CM_i Medication cost per patient in department i
CE_i Equipment cost per patient in department i
CO_i Another relevant cost per patient in department i
a_i Arrived patient in each service.

Decision variables for the objective function and constraint are decided by taking input from the case organization and are listed in Tables 1 and 2, respectively.

5.2 Problem Statement

The first objective function minimizes the total cost to serve, while the second objective function is related to the minimization of the total patient complaints. Using the definition of the parameters in earlier section, the two objective functions of the study can be written as below:

Minimize total service cost:

$$Z_1 = \sum_{i=1} \sum_{t=1} (W1_{ti}) * (CS1_i) + (W2_{ti}) * (CS2_i) + (W3_{ti}) * (CS3_i)$$
$$+ CM_i + CE_i + CO_i \tag{14}$$

The total complaints can be written as sum product of arrived patients in each service and the complaint per patient. The objective function related to the minimization of total patient complaints is given as:

$$Z_2 = 300x_1 + 350x_2 + 400x_3 \tag{15}$$

5.3 Constraints

Constraint 1: Constraint on the demand for the healthcare services

$$\sum_{i=1}^{n} x_i = D \quad \text{(when } D \prec U \text{)} \text{ or } \sum_{i=1}^{n} x_i = U \quad \text{(when } D \succ U \text{)}$$

Constraint 2: Capacity constraint for the healthcare services

$$x_i \le U_i \quad \text{for} \quad i = 1, 2, 3$$

Constraint 3: Total budget constraint for the healthcare services

$$\sum_{i=1}^{n} r_i x_i \le P$$

Constraint 4: A constraint on the flexibility for the service quota allocation

$$\sum_{i=1}^{n} f_i x_i \ge F$$

Constraint 5: Nonnegativity constraint for all allocation quantities

$$x_i \ge 0$$

6 Objective Function

Assuming α is equal to 0.05, the maximum targeted service cost for the month is 200,000 and the current resource utilization is 0.02 for each of the services. The flexibility of the overall service is required to be more than 5% by design. The NGO also aims to serve at least 150 patients in the given time period (month). The study considers 10% of the tolerance range for all the three objectives as suggested by experts. Each of the services should be at least 10% utilized for the CSR, while the upper limit for the same is 20%. Given the above assumptions, the membership function for objective one (minimization of the service cost) can be written as:

$$\mu(z_1) = \begin{cases} 1, & \text{if } Z_1 < 180,000 \\ 1 - \dfrac{e^{-\frac{0.05(200,000-z_1)}{200,000-180,000}} - e^{-0.05}}{1-e^{-0.05}}, & \text{if } 180,000 \leq Z_1 \leq 200,000; \\ 0, \, Z_1 > 200,000 \end{cases} \quad (16)$$

The third objective function minimizing the patient complaint is given as follow:

$$(z_2) = \begin{cases} 1, & \text{if } Z_2 < 0.045 \\ 1 - \dfrac{e^{-\frac{0.05(0.5-z_2)}{0.5-0.045}} - e^{-0.05}}{1-e^{-0.05}}, & \text{if } 0.045 \leq Z_2 \leq 0.5; \\ 0, & Z_2 > 0.5 \end{cases} \quad (17)$$

Finally, the resource allocation model for the NGO can be formulated as below:

$$\text{Max} \quad f(u) = \mu_1 + \mu_2 \quad (18)$$

For small exponent, the exponential function can be transferred into a linear function as:

$$\begin{aligned} \mu_1 &: 10 - 0.00005Z_1 + d_1^- - d_1^+ = 1 \\ \mu_2 &: 10 - 0.5Z_2 + d_2^- - d_2^+ = 1 \\ x_i, d_i^-, d_i^+ &\geq 0 \end{aligned} \quad (19)$$

Other constraints:

$$90x_1 + 100x_2 + 200x_3 \leq 275 \quad (20)$$
$$150x_1 + 150x_2 + 150x_3 \leq 275 \quad (21)$$
$$400x_1 + 500x_2 + 300x_3 \leq 200 \quad (22)$$
$$0.15x_1 + 0.10x_2 + 0.15x_3 \geq 0.05 \quad (23)$$
$$\begin{aligned} 0.1 &\leq x_1 \leq 0.2; \\ 0.1 &\leq x_2 \leq 0.2; \\ 0.1 &\leq x_3 \leq 0.2; \end{aligned} \quad (24)$$

7 Results and Discussion

Using the information given in Tables 2 and 3 and problem statement, we can modify the values of Z_1 and Z_2 as below:

Table 3 Decision variable values for constraints

	$W1_{ti}$	$W2_{ti}$	$W3_{ti}$	$CS1_i$	$CS2_i$	$CS3_i$	CM_i	CE_i	CO_i	a_i
X_1	1	4	2	60	10	12	0.2	0.1	0.1	300
X_2	1	3	2	60	10	12	0.15	0.1	0.1	350
X_3	2	4	2	60	8	12	0.1	0.8	0.1	400

Note All costs are taken in thousands

Table 4 Result for the model

Variables	Results
X_1	0.2
X_2	0.12
X_3	0.2
d_1^-	0
d_2^-	0.9
Z_1 (INR)	151,000
Z_2 (Nos)	182

$$Z_1 = 274\,x_1 + 271.5\,x_2 + 316x_3 \tag{25}$$
$$Z_2 = 300\,x_1 + Z_2 = 350x_2 + 400x_3 \tag{26}$$

Solving for a preemptive solution for the problem keeping service cost as at higher priority than the patient complaints. Excel solver was used to solve the linear programing problem. The method used for solving the problem is simplex method (Table 4).

We received the following results:

The above scenario will minimize the service cost to 151 thousand. The optimal solution suggests that totally 182 patients are served during the given time period. The patient served by endocrine, cardiology, and internal medicine department is 60, 42, and 80, respectively. The answer with different objectives may give a different allocation of the resources.

8   Conclusions

The study proposes and uses a fuzzy goal programming approach as a quantitative method for the resource allocation in healthcare organization. As suggested by Turgay and Taşkın [38], the exponential membership function was used for the study. The reason behind the selection of the method is the better representation of the real-life scenarios than a linear function. Moreover, it can be easily converted into linear approximation for a small value of alpha. The case study suggests that for the given objectives, the optimal solution may be different from the most obvious solutions;

hence, a quantitative model/qualitative can help us in solving a resource allocation problem. Qualitative allocation is usually very personal to the people involved in the allocation and therefore is very subjective and quite unreliable. The quantitative models are preferred over the qualitative models because they are objective, based on data and facts, and are therefore impersonal. This model is easy to use and can be adopted in other similar organizations involved in the chronic care like diabetes, asthma, tuberculosis, and HIV.

References

1. Mishra, V., Samuel, C., Sharma, S.K.: Use of machine learning to predict the onset of diabetes. Int. J. Recent Adv. Mech. Eng. (IJMECH) **4**(2) (2015)
2. Mishra, V., Samuel, C., Sharma, S.K.: Visualization of perceived expensiveness of diabetes-fuzzy MDS approach. In: 2016 IEEE Uttar Pradesh Section International Conference on Electrical, Computer and Electronics Engineering (UPCON), pp. 67–71. IEEE, New York (2016 December)
3. Joshi, S.R., Parikh, R.M.: India; the diabetes capital of the world: now heading towards hypertension. J.-Assoc. Physicians India **55**(5), 323 (2007)
4. Jönsson, B.: Revealing the cost of type II diabetes in Europe. Diabetologia **45**(7), S5–S12 (2002)
5. Bjork, S., Kapur, A., Sylvest, C., Kumar, D., Kelkar, S., Nair, J.: The economic burden of diabetes in India: results from a national survey. Diabetes Res. Clin. Pract. **50**, 190 (2000)
6. Kapur, A.: Economic analysis of diabetes care. Indian J. Med. Res. **125**(3), 473 (2007)
7. Akari, S., Mateti, U.V., Kunduru, B.R.: Health-care cost of diabetes in South India: a cost of illness study. J. Res. Pharm. Pract. **2**(3), 114 (2013)
8. Al-Maskari, F., El-Sadig, M., Nagelkerke, N.: Assessment of the direct medical costs of diabetes mellitus and its complications in the United Arab Emirates. BMC Public Health **10**(1), 679 (2010)
9. Henriksson, F., Agardh, C.D., Berne, C., Bolinder, J., Lönnqvist, F., Stenström, P., Jönsson, B.: Direct medical costs for patients with type 2 diabetes in Sweden. J. Intern. Med. **248**(5), 387–396 (2000)
10. Hogan, P., Dall, T., Nikolov, P.: Economic costs of diabetes in the US in 2002. Diabetes Care **26**(3), 917 (2003)
11. Sokol, M.C., McGuigan, K.A., Verbrugge, R.R., Epstein, R.S.: Impact of medication adherence on hospitalization risk and healthcare cost. Med. Care **43**(6), 521–530 (2005)
12. Ho, P.M., Bryson, C.L., Rumsfeld, J.S.: Medication adherence. Circulation **119**(23), 3028–3035 (2009)
13. Roebuck, M.C., Liberman, J.N., Gemmill-Toyama, M., Brennan, T.A.: Medication adherence leads to lower health care use and costs despite increased drug spending. Health Aff. **30**(1), 91–99 (2011)
14. Norris, S.L., Lau, J., Smith, S.J., Schmid, C.H., Engelgau, M.M.: Self-management education for adults with type 2 diabetes. Diabetes Care **25**(7), 1159–1171 (2002)
15. Newhouse, J.P.: Medical care costs: how much welfare loss? J. Econ. Perspect. **6**(3), 3–21 (1992)
16. Beveridge, R.: CAEP issues. J. Emerg. Med. **16**, 507–511 (1998)
17. Vienonen, M.A., Wlodarczyk, W.C.: Health care reforms on the European scene: evolution, revolution or seesaw? world health statistics quarterly. Rapport trimestriel de statistiqu essanitaires mondiales **46**(3), 166–169 (1993)
18. Srinivasan, R.: Health insurance in India. Health Population Perspect. Issues **24**(2), 65–72 (2001)

19. Duggal, R.: Poverty & health: criticality of public financing. Indian J. Med. Res. **126**(4), 309 (2007)
20. Gangolli, L.V., Duggal, R., Shukla, A.: Review of Healthcare in India. Centre for Enquiry into Health and Allied Themes, Mumbai (2005)
21. Berman, P., Ahuja, R., Bhandari, L.: The impoverishing effect of healthcare payments in India: new methodology and findings. Econ. Political Wkly. 65–71 (2010)
22. Prinja, S., Bahuguna, P., Pinto, A.D., Sharma, A., Bharaj, G., Kumar, V., Kumar, R.: The cost of universal health care in India: a model based estimate. PLoS ONE **7**(1), e30362 (2012)
23. Patil, A.V., Somasundaram, K.V., Goyal, R.C.: Current health scenario in rural India. Aust. J. Rural Health **10**(2), 129–135 (2002)
24. Balarajan, Y., Selvaraj, S., Subramanian, S.V.: Health care and equity in India. The Lancet **377**(9764), 505–515 (2011)
25. Ganesh, S.: The myth of the non-governmental organization: governmentality and transnationalism in an Indian NGO. Int. Multicultural Organ. Commun. **7**, 193–219 (2005)
26. Delisle, H., Roberts, J.H., Munro, M., Jones, L., Gyorkos, T.W.: The role of NGOs in global health research for development. Health Res. Policy Syst. **3**(1), 3 (2005)
27. Kluge, E.H.W.: Resource allocation in healthcare: implications of models of medicine as a profession. Medscape Gen. Med. **9**(1), 57 (2007)
28. Veatch, R.M.: The Principle of Avoiding Killing. The Basics of Bioethics, pp. 88–104. Prentice Hall, Upper Saddle River, NJ (2003)
29. Beauchamp, T.L., Childress, J.F.: Principles of Biomedical Ethics. Oxford University Press, New York (2001)
30. Tauber, A.I.: Patient autonomy and the ethics of responsibility (2005)
31. Cruess, S.R.: Professionalism and medicine's social contract with society. Clin. Orthoped. Relat. Res. **449**, 170–176 (2006)
32. Bernardin, J.C.: Renewing the covenant with patients and society. Linacre Q. **63**(1), 3–10 (1996)
33. Daniels, N.: Just Health Care. Cambridge University Press, Cambridge (1985)
34. Freidson, E.: Profession of Medicine: A Study of the Sociology of Applied Knowledge. University of Chicago Press, Chicago (1988)
35. Hui, E.C.: The contractual model of the patient-physician relationship and the demise of medical professionalism. Hong Kong Med. J. (2005)
36. Carroll, C.D., Manderscheid, R.W., Daniels, A.S., Compagni, A.: Convergence of service, policy, and science toward consumer-driven mental health care. J. Mental Health Policy Econ. **9**(4), 185–192 (2006)
37. Iskander, M.G.: Exponential membership functions in fuzzy goal programming: a computational application to a production problem in the textile industry. Am. J. Comput. Appl. Math. **5**(1), 1–6 (2015)
38. Turgay, S., Taşkın, H.: Fuzzy goal programming for health-care organization. Comput. Ind. Eng. **86**, 14–21 (2015)
39. Rubin, P.A., Narasimhan, R.: Fuzzy goal programming with nested priorities. Fuzzy Sets Syst. **14**(2), 115–129 (1984)
40. Carlsson, C., Korhonen, P.: A parametric approach to fuzzy linear programming. Fuzzy Sets Syst. **20**(1), 17–30 (1986)
41. Li, R.J., Lee, E.S.: An exponential membership function for fuzzy multiple objective linear programming. Comput. Math. Appl. **22**(12), 55–60 (1991)
42. Zimmermann, H.J.: Decision making in ill-structured environments and with multiple criteria. In: Readings in Multiple Criteria Decision Aid, pp. 119–151. Springer, Berlin (1990)

Chapter 30
Stoichio Simulation of FACSP From Graph Transformations to Differential Equations

J. Philomenal Karoline, P. Helen Chandra, S. M. Saroja Theerdus Kalavathy and A. Mary Imelda Jayaseeli

Abstract In this paper, a methodology to derive ordinary differential equations (ODEs) using graph transformation technique is developed for Michaelis–Menten kinetics. This approach is based on a variant of the construction of critical pairs. It has been executed using the AGG tool and validated for FACSP.

Keywords Rate of reaction · Fuzzy artificial cell system · Parallel conflicts Sequential dependencies · Stoichiometric matrix · Place transition net

1 Introduction

Multiset processing is a simple technique, easy to be used by biologists, which contrasts with most continuous models and simulation systems. Abstract Rewriting System on Multisets (ARMS), a class of P systems based on multiset processing but with a simple membrane structure, was introduced with the aim of modelling chemical systems. It is a stochastic model where rules are applied probabilistically [1].

In particular, ARMS is based on stoichiometric chemistry, and if the number of elements in the system is large, then the behaviour of the system is similar to the behaviour of models based on differential equations [2].

J. Philomenal Karoline · P. Helen Chandra (✉) · S. M. Saroja Theerdus Kalavathy · A. Mary Imelda Jayaseeli
Jayaraj Annapackiam College for Women (Autonomous),
Periyakulam, Theni, Tamil Nadu, India
e-mail: chandrajac@yahoo.com

J. Philomenal Karoline
e-mail: philoharsh@gmail.com

S. M. Saroja Theerdus Kalavathy
e-mail: kalaoliver@gmail.com

A. Mary Imelda Jayaseeli
e-mail: imeldaxavier@gmail.com

© Springer Nature Singapore Pte Ltd. 2018
D. Ghosh et al. (eds.), *Mathematics and Computing*, Springer Proceedings in Mathematics & Statistics 253, https://doi.org/10.1007/978-981-13-2095-8_30

In [3], a new device, Fuzzy ARMS in Artificial Cell System with Proteins on Membrane (FACSP), is developed for which the structure is analysed on its parameters. In [4], a methodology has been developed to model Michaelis–Menten kinetic reactions networks in terms of DPO graph transformation.

In [5], the chemical reaction kinetics is rephrased in terms of stochastic graph transformations. The ODEs that describe the evolution of concentrations of chemical species over time are derived. It is based on stochastic graph transformation [6] which combines rules to capture the reactive behaviour of the system with a specification of rate constants governing the speed at which the reaction occur.

However, it is of great interest to study the dynamical properties of FACSP, and we have considered to apply mathematical methods developed for analysing differential equations.

In this paper, the formation of our work is designed as follows: first a background and related works are given. Then molecular representation of FACSP is deliberated, and critical pair analysis of DPO graph transformation rules using AGG tool is done. Stoichiometric matrix and the incidence matrix of the PT net are obtained.

2 Preliminaries

In [5], a stoichiometric matrix is obtained which relates each elementary reaction to each molecular species in the system by the aggregate effect the reaction has on that species population. The rate laws are extracted, and a rate law vector of length n is produced. A multiplication of this vector and the stoichiometric matrix produces a system of ordinary differential equations:

$$d[X]/dt = S \cdot R \qquad (1)$$

where $d[X]/dt$ is the differential with respect to time t, of a chemical species X in the system, S is the stoichiometric matrix, and R is the rate law vector.

In [7], the translation of Petri nets whose transitions are labelled by rate constants, to differential equations, is discussed.

2.1 The Graph Transformation System [5]

A type graph representing molecules using graph transformations is discussed in [5]. Here atoms are represented as square nodes. The round nodes are atom-specific bonding nodes. A bond between atoms is represented by an edge between two of these bonding nodes. Each bonding node is connected to only one atom node. The formal definitions of typed graph transformation system and Accountable GTS are also given in [5].

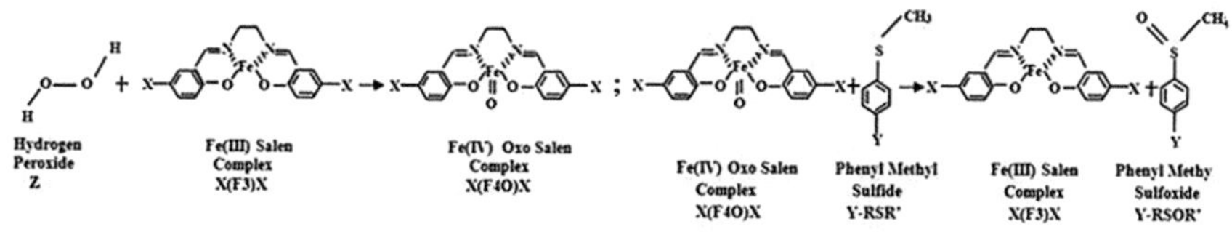

(a)

$$\begin{cases} R_{11} : [_1A_1|Z]_1 \rightarrow [_1B|\phi]_1; [_1B|S_1]_1 \rightarrow [_1A_1| \quad [_2 \quad |P_1]_2]_1 \text{ with } \omega(R_{11}) = 0.0025 \\ R_{12} : [_1A_1|Z]_1 \rightarrow [_1B|\phi]_1; [_1B|S_2]_1 \rightarrow [_1A_1| \quad [_2 \quad |P_2]_2]_1 \text{ with } \omega(R_{12}) = 0.01 \\ R_{13} : [_1A_1|Z]_1 \rightarrow [_1B|\phi]_1; [_1B|S_3]_1 \rightarrow [_1A_1| \quad [_2 \quad |P_3]_2]_1 \text{ with } \omega(R_{13}) = 0.0059 \\ R_{14} : [_1A_1|Z]_1 \rightarrow [_1B|\phi]_1; [_1B|S_4]_1 \rightarrow [_1A_1| \quad [_2 \quad |P_4]_2]_1 \text{ with } \omega(R_{14}) = 0.0016 \\ R_{15} : [_1A_1|Z]_1 \rightarrow [_1B|\phi]_1; [_1B|S_5]_1 \rightarrow [_1A_1| \quad [_2 \quad |P_5]_2]_1 \text{ with } \omega(R_{15}) = 0.0011 \\ R_{16} : [_1A_1|Z]_1 \rightarrow [_1B|\phi]_1; [_1B|S_6]_1 \rightarrow [_1A_1| \quad [_2 \quad |P_6]_2]_1 \text{ with } \omega(R_{16}) = 0.0009 \\ R_{17} : [_1A_1|Z]_1 \rightarrow [_1B|\phi]_1; [_1B|S_7]_1 \rightarrow [_1A_1| \quad [_2 \quad |P_7]_2]_1 \text{ with } \omega(R_{17}) = 0.00027 \end{cases}$$

(b)

Fig. 1 a Oxidation of sulphides and **b** evolution rules for FACSP

2.2 FACSP (Fuzzy Artificial Cell Systems With Proteins On Membranes) [3]

Oxidation of Sulphides: Oxidation of aryl methyl sulphides using iron–salen complexes as catalyst in presence of hydrogen peroxide as oxidant is followed kinetically and is described in [8]. Chemically the reaction takes place through formation of intermediate oxo compound of the catalyst and in second step the oxidation of substrate following regeneration of catalyst. The general reaction rule is presented in (a), and the structure is shown in Fig. 1a.

(a). $Z + X(F3)X \rightarrow X(F4O)X$;
$X(F4O)X + Y\text{-}RSR' \rightarrow X(F3)X + Y\text{-}RSOR'$

The structure of (a) is represented in Fig. 1a. In [3], the authors carried out catalytic reactions of aryl methyl sulphides varying the substitution at Y as H, Cl, Br, CH$_3$, OCH$_3$, F and NO$_2$ groups. In case $X = H$ and Y varying as seven substitutions, (a) consists of seven reaction rules.

Fuzzy ARMS in Artificial Cell Systems with Proteins on Membranes (FACSP) has been introduced in which the evolution rules (Fuzzy rewriting rules) are the seven reaction rules and the Fuzzy data are oxidant, catalyst and substrate (Fig. 1b).

3 Graph Transformations for FACSP

We present the molecular representation of FACSP using graph transformation system and the derivation of ordinary differential equation for the reactions through critical pair analysis using AGG tools.

3.1 Molecular Representation of FACSP Using Graphs

Let us consider the first evolution rule (R_{11}) in FACSP from Sect. 2. The structure of the corresponding reaction rule is shown in Fig. 2 in which the formation of intermediate iron (IV)–oxo salen complex of parent molecule is described. The complex acts as a catalyst for the oxidation of phenyl methyl sulphide to phenyl methyl sulphoxide. At the end of the reaction, the catalyst, iron (III)–salen complex is regenerated. The species hydrogen peroxide (Z), iron (III)–salen complex (A_1), iron (IV)–oxo salen complex (B), phenyl methyl sulphide (S_1) and phenyl methyl sulphoxide (P_1) in Fig. 2 are represented as molecules. Each molecule consists of bonds that connect two atoms. The intuitive representation of molecules consists of atoms as nodes and bonds as edges that directly connect them.

The type graph is produced in AGG (Fig. 3a) for all atoms and groups in FACSP. In this type graph, atoms and groups such as O, Fe, N, S, H, C, Cl, Br, F, CH_3, OCH_3 and NO_2 are represented as square nodes, each distinct species having its own node type. The round nodes represented are atom-specific bonding nodes. All bonding

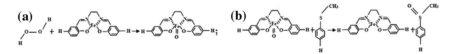

Fig. 2 Reaction rule R_{11}

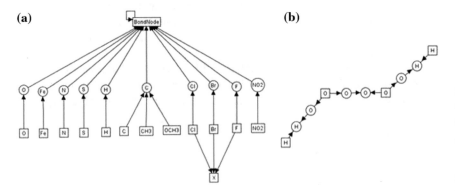

Fig. 3 **a** Type graph for FACSP and **b** type graph for the molecule hydrogen peroxide

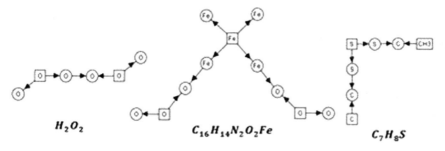

H_2O_2 $C_{16}H_{14}N_2O_2Fe$ C_7H_8S

Fig. 4 Graph depicting starting materials for FACSP, produced in AGG

nodes are subtypes of the generic bond node type. Finally, the atoms chlorine (Cl), bromine (Br) and fluorine (F) are grouped as halogens, which is denoted by X.

A bond connecting H and O is represented by an edge [arrows with filled arrowheads, (Fig. 3b)]. The bond node oxygen is connected to atom node O and that of hydrogen is connected to atom node H. Oxygen has two bonds satisfying the valency two. The atoms and groups C, CH$_3$ and OCH$_3$ have the same bonding node type (C) associated with them.

In our problem, atom C is less electronegative than atoms N and S; C and N are less electronegative atoms than O atom. H is the least electronegative atom compared with all other atoms. Thus, a bond between H and any other atom would go from H.

The type graph contains C and H node types, and so the methyl group is represented as a single CH$_3$ node type. The critical pair analysis constructs an overlap between the graphs on the left-hand side and right-hand side of the evolution rules. The single CH$_3$ node is expressed in terms of C atom nodes, H atom nodes, C bonding nodes and H bonding nodes. The node CH$_3$ and the edges between them would constitute a total of 10 nodes.

The type graphs are drawn (Fig. 4) for the starting materials (Z), (A_1) and (S_1) representing, namely, hydrogen peroxide, iron (III)–salen complex and phenyl methyl sulphide respectively taking as molecular identity rules. The LHS and RHS of this molecular identity rules are same and contain only the graph of a particular molecule.

Type graphs are obtained to all possible general rules in FACSP using the above methodology. The type graph of molecules in the reaction rule $R_{11}(a)$ and $R_{11}(b)$ is shown in Figs. 5 and 6. Each one is added as a molecular identity rule. Fig. 7 depicts the abstraction of the reaction rule R_{11} in Fig. 2.

3.2 Critical Pair Analysis for FACSP

A critical pair analysis is done between each general reaction rule and each molecular identity rule.

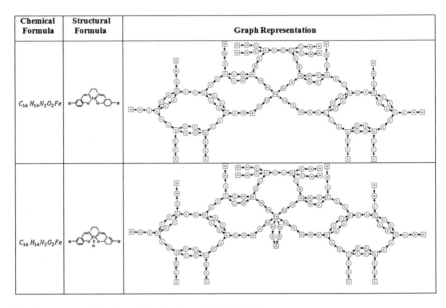

Fig. 5 Type graph for the molecule A_1 and B in FACSP

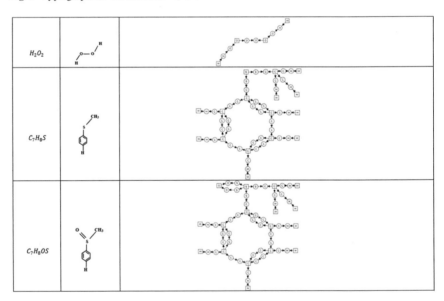

Fig. 6 Type graph for the molecule Z, S_1 and P_1 in FACSP

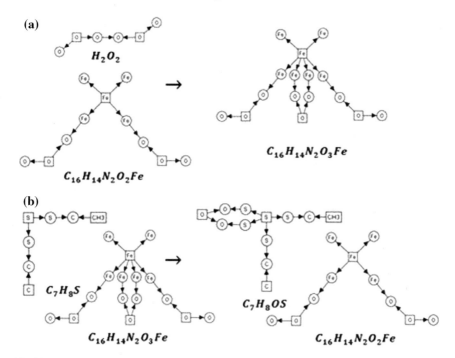

Fig. 7 $R_{11}(a)$-top, $R_{11}(b)$-bottom

Fig. 8 Critical analysis—parallel conflicts (PC) and sequential dependencies (SD)		H_2O_2 (Z)		$C_{16}H_{14}N_2O_2Fe$ (A_1)		C_7H_8S (S_1)	
		PC	SD	PC	SD	PC	SD
	$R_{11}(a)$	1	0	2	0	0	0
	$R_{11}(b)$	0	0	0	2	1	0

The parallel conflict and sequential dependencies are verified by the application of the general rule to the molecule (LHS) at the match given by the critical overlapping and the application of the general rule to the molecule (RHS) at the match given by the critical overlapping respectively. The results of the first iteration are given in Fig. 8. Each entry signifies how many of the overlappings were critical for each pair. Critical pair analysis checks all possible unions of L and M for parallel conflict analysis and R and M for sequential dependence analysis.

In a similar manner, the molecular identity rules are obtained, and hence, the critical pair analysis is done for all reaction rules in FACSP. The result of the critical pair analysis is given in Fig. 9.

For the FACSP reaction studied, the results obtained after applying the reaction rule to the overlappings at their critical matches are compared. There are two critical overlappings wherever there is a conflict with A_1 which is shown in Fig. 10. The critical nodes and edges in this overlapping (Fe) are covered by the shaded area.

	Z		A$_1$		B		S$_1$		S$_2$		S$_3$		S$_4$		S$_5$		S$_6$		S$_7$		P$_1$		P$_2$		P$_3$		P$_4$		P$_5$		P$_6$		P$_7$	
	PC	SD	PC	SD	PC	SD	PC	SD	PC	SD	PC	SD	PC	SD	PC	SD	PC	SD	PC	SD	PC	SD	PC	SD	PC	SD	PC	SD	PC	SD	PC	SD	PC	SD
$R_{11}(a)$	1	0	2	0	0	4	0	0	0	0	0	0	0	0	0	0	0	0	0	0	0	0	0	0	0	0	0	0	0	0	0	0	0	0
$R_{11}(b)$	0	0	0	2	4	0	1	0	0	0	0	0	0	0	0	0	0	0	0	0	2	0	0	0	0	0	0	0	0	0	0	0	0	0
$R_{12}(a)$	1	0	2	0	0	4	0	0	0	0	0	0	0	0	0	0	0	0	0	0	0	0	0	0	0	0	0	0	0	0	0	0	0	0
$R_{12}(b)$	0	0	0	2	4	0	0	0	1	0	0	0	0	0	0	0	0	0	0	0	0	0	2	0	0	0	0	0	0	0	0	0	0	0
$R_{13}(a)$	1	0	2	0	0	4	0	0	0	0	0	0	0	0	0	0	0	0	0	0	0	0	0	0	0	0	0	0	0	0	0	0	0	0
$R_{13}(b)$	0	0	0	2	4	0	0	0	0	0	1	0	0	0	0	0	0	0	0	0	0	0	0	0	2	0	0	0	0	0	0	0	0	0
$R_{14}(a)$	1	0	2	0	0	4	0	0	0	0	0	0	0	0	0	0	0	0	0	0	0	0	0	0	0	0	0	0	0	0	0	0	0	0
$R_{14}(b)$	0	0	0	2	4	0	0	0	0	0	0	0	1	0	0	0	0	0	0	0	0	0	0	0	0	0	2	0	0	0	0	0	0	0
$R_{15}(a)$	1	0	2	0	0	4	0	0	0	0	0	0	0	0	0	0	0	0	0	0	0	0	0	0	0	0	0	0	0	0	0	0	0	0
$R_{15}(b)$	0	0	0	2	4	0	0	0	0	0	0	0	0	0	1	0	0	0	0	0	0	0	0	0	0	0	0	0	2	0	0	0	0	0
$R_{16}(a)$	1	0	2	0	0	4	0	0	0	0	0	0	0	0	0	0	0	0	0	0	0	0	0	0	0	0	0	0	0	0	0	0	0	0
$R_{16}(b)$	0	0	0	2	4	0	0	0	0	0	0	0	0	0	0	0	1	0	0	0	0	0	0	0	0	0	0	0	0	0	2	0	0	0
$R_{17}(a)$	1	0	2	0	0	4	0	0	0	0	0	0	0	0	0	0	0	0	0	0	0	0	0	0	0	0	0	0	0	0	0	0	0	0
$R_{17}(b)$	0	0	0	2	4	0	0	0	0	0	0	0	0	0	0	0	0	0	1	0	0	0	0	0	0	0	0	0	0	0	0	0	0	2

Fig. 9 Critical pair analysis—parallel conflicts (PC) and sequential dependencies (SD)

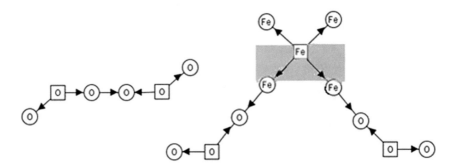

Fig. 10 Critical overlapping between R_{11} and A_1 (critical graph elements are contained within the shaded area)

The other types of nodes and edges in the critical overlappings are identical and same. Due to the symmetry around the critical Fe atom node, two overlappings arise. They have no significance to the selection of a reaction or to the outcome. So, they are equivalent and hence we have got an isomorphism between the corresponding transformations. Since the molecules involved are very small, these overlappings are reduced to 1 in all cases. We then have reduced this entry in Fig. 9 to 1 which is the obtained stoichiometric matrix (Fig. 12). It is immediate to obtain the ODEs.

4 Stochastic Graph Transformation System for FACSP

In a chemical system, the reaction speed is captured by the rate constant as a measure of the reactivity of the given components. In FACSP, reaction rules that act on the molecules are specified by rewriting rules and rate constant represents the membership value. Assigning the membership values (rate constants) to rules of a graph transformation system, we obtain a stochastic graph transformation system.

4.1   Stoichiometric Matrix for FACSP

Consider the evolution rule

$$R_{11}(a): [_1 A_1 | Z]_1 \xrightarrow[\omega_1]{} [_1 B | \phi]_1; \; R_{11}(b): [_1 B | S_1]_1 \xrightarrow[\omega_2]{} [_1 A_1 | \; [_2 \; | P_1]_2]_1 \qquad (2)$$

which comprise an example reaction mechanism for FACSP. If it is known for each reaction, how many molecules of each chemical species is created or destroyed, we can build up a matrix for the reactions in (2).

Each entry in the stoichiometric matrix corresponds to the aggregate number of molecules consumed or produced in a reaction, negative for consumption and positive for production. The first reaction in (2) with the membership value ω_1 consumed one molecule of Z, the entry for ω_1 and Z in the matrix would be -1. Similarly, the entry for ω_1 and A_1 in the matrix would be -1. Also the first reaction in (2) with the membership value ω_1 produced one molecule of B, the entry for ω_1 and B in the matrix would be 1. Proceeding like this, we build up a stoichiometric matrix for the reaction (2) which is tabulated in Fig. 11a.

In a similar way, we are able to build up the stoichiometric matrix (Fig. 12) for all the reactions in the seven evolution rules of FACSP.

The membership law for FACSP is defined such that the membership coefficient for each row is multiplied by the concentration of those species which are destroyed. For example, the membership law for the corresponding oxidation of sulphides in (2): $\omega_1[Z][A_1]$ for the reaction $R_{11}(a)$ and $\omega_2[B][S_1]$ for $R_{11}(b)$.

The membership law matrix for the reactions in (2) is shown in Fig. 11b. In a similar manner, we are able to define membership law for all reactions in the seven evolution rules of FACSP.

We multiply the membership law matrix by stoichiometric matrix, and hence, we have obtained the following ODE's.

$$d[A_1]/dt = -\omega_1[Z][A_1] + \omega_2[B][S_1];$$

$$d[Z]/dt = -\omega_1[Z][A_1];$$

$$d[B]/dt = \omega_1[Z][A_1] - \omega_2[B][S_1];$$

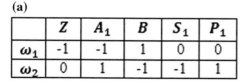

(a)

	Z	A_1	B	S_1	P_1
ω_1	-1	-1	1	0	0
ω_2	0	1	-1	-1	1

(b)

ω_1	ω_2
$\omega_1[Z][A_1]$	$\omega_2[B][S_1]$

Fig. 11 a Stoichiometric matrix for R_{11} and b membership law matrix for R_{11}

$$d[S_1]/dt = -\omega_2[B][S_1];$$

$$d[P_1]/dt = \omega_2[B][S_1].$$

4.2 Place Transition Net Representing FACSP Reaction Mechanism

In [7], it is described how a discrete Petri net can be converted into a continuous one by allowing places to have a positive real number of tokens representing the concentration of that particular chemical species in the system. The ODEs can be deduced from the incidence matrix for such a Petri net.

M M.V.	Z	A₁	B	S₁	S₂	S₃	S₄	S₅	S₆	S₇	P₁	P₂	P₃	P₄	P₅	P₆	P₇
ω_1	-1	-1	1	0	0	0	0	0	0	0	0	0	0	0	0	0	0
ω_2	0	1	-1	-1	0	0	0	0	0	0	1	0	0	0	0	0	0
ω_3	-1	-1	1	0	0	0	0	0	0	0	0	0	0	0	0	0	0
ω_4	0	1	-1	0	-1	0	0	0	0	0	0	1	0	0	0	0	0
ω_5	-1	-1	1	0	0	0	0	0	0	0	0	0	0	0	0	0	0
ω_6	0	1	-1	0	0	-1	0	0	0	0	0	0	1	0	0	0	0
ω_7	-1	-1	1	0	0	0	0	0	0	0	0	0	0	0	0	0	0
ω_8	0	1	-1	0	0	0	-1	0	0	0	0	0	0	1	0	0	0
ω_9	-1	-1	1	0	0	0	0	0	0	0	0	0	0	0	0	0	0
ω_{10}	0	1	-1	0	0	0	0	-1	0	0	0	0	0	0	1	0	0
ω_{11}	-1	-1	1	0	0	0	0	0	0	0	0	0	0	0	0	0	0
ω_{12}	0	1	-1	0	0	0	0	0	-1	0	0	0	0	0	0	1	0
ω_{13}	-1	-1	1	0	0	0	0	0	0	0	0	0	0	0	0	0	0
ω_{14}	0	1	-1	0	0	0	0	0	0	-1	0	0	0	0	0	0	1

Fig. 12 Stoichiometric matrix for FACSP, M-molecules, M.V.-membership values

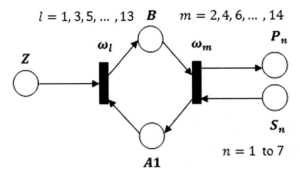

$$l = 1, 3, 5, \dots, 13 \quad B \qquad m = 2, 4, 6, \dots, 14$$

Fig. 13 PT Net for FACSP

In our problem, the incidence matrix for a place transition net with places Z, A_1, B, S_n, P_n where $n = 1$ to 7 and transitions ω_l, ω_m where $l = 1, 3, 5, \ldots, 13$ and $m = 2, 4, 6, \ldots, 14$ are shown in Fig. 13. It is obviously similar to the stoichiometric matrix (Fig. 12) and hence we are able to obtain the ODEs.

5 Conclusion

We have derived ordinary differential equations from the stoichiometric matrix by doing critical pair analysis from the graph transformation system using AGG tools. In the same way, we have obtained the stoichiometric matrix using stochastic graph transformation by assigning membership values to the evolution rules. Again we have obtained the incidence matrix of a petri net representing the FACSP mechanism which is similar to the stoichiometric matrix of the FACSP.

We have observed that once a stoichiometric matrix is established, the ODEs could be derived. Also it is understood that it is enough to encode the graph transformation system into a place transition net to find the stoichiometric matrix. This approach has been demonstrated by means of oxidation of sulphides reactions following Michaelis–Menten kinetics using the AGG tool and validated for FACSP.

Acknowledgements The author Dr. Sr. P. Helen Chandra, Principal Investigator of UGC Major Research Project (F.No. -43-412/2014(SR) dated 05 September 2015) is grateful to UGC, New Delhi, for the award of the project which enabled to execute this research work in Jayaraj Annapackiam College for Women (Autonomous), Periyakulam, Theni District, Tamil Nadu.

References

1. Suzuki, Y., Tsumoto, S., Tamaka, H.: Analysis of Cycles in Symbolic Chemical Systems Based on Abstract Rewriting Systems on Multisets, pp. 522–528. Artificial Life V, MIT Press, Cambridge, MA (1996)
2. Suzuki, Y., Fujiwara, Y., Takabayashi, J., Tanaka, H.: Artificial life applications of a class of P systems: abstract rewriting system on multisets, multiset processing. Lecture Notes in Computer Science, vol. 2235, pp. 299–346. Springer, Berlin (2001)
3. Helen Chandra, P., Saroja Theerdus Kalavathy, S.M., Mary Imelda Jayaseeli, A., Philomenal Karoline, J.: Fuzzy ACS with biological catalysts on membranes in chemical reactions. J. Netw. Innovative Comput. MIR Labs, USA **4**, 143–151 (2016)
4. Philomenal Karoline, J., Helen Chandra, P., Saroja Theerdus Kalavathy, S.M., Mary Imelda Jayaseeli, A.: Model based simulation of Michaelis-Menten kinetic reactions by DPO graph transformation. IJPAM (2017)
5. Bapordra, M., Heckel, R.: From graph transformations to differential equations. Electron. Commun. EASST **30**, 1–21 (2010)
6. Heckel, R., Lajios, G., Menge, S.: Stochastic graph transformation systems. In: International Colloquium on Theoretical Aspects of Computing 2005. Lecture Notes in Computer Science, vol. 3256, pp. 210–225 (2004)

7. Gilbert, D., Heiner, M.: From Petri nets to differential equations-an integrative approach for biochemical network analysis. In: ICATPN, Lecture Notes in Computer Science, vol. 4024, pp. 181–200 (2006)
8. Mary Imelda Jayaseeli, A., Rajagopal, S.: [Iron(III)-salen] Ion Catalyzed H_2O_2 Oxidation of organic sulfides and sulfoxides. J. Mol. Catal, A: Chem. **309**, 103–110 (2009)

Chapter 31
Fully Dynamic Group Signature Scheme with Member Registration and Verifier-Local Revocation

Maharage Nisansala Sevwandi Perera and Takeshi Koshiba

Abstract Since Bellare et al. (EUROCRYPT 2003) proposed a security model for group signature schemes, almost all the securities of group signature schemes have been discussed in their model (the BMW03 model). While the BMW03 model is for static groups, Bellare et al. in 2005 considered the case of dynamic group signature schemes and provided a solution to cope with dynamic groups. However, their scheme does not serve member revocation, serves only member registration. In this paper, we incorporate a member revocation mechanism into a group signature scheme with member registration and construct a fully dynamic group signature, which supports verifier-local revocation (VLR) to manipulate member revocation. Moreover, we achieve the security of the proposed scheme with a restricted version of full anonymity to overcome the security complications that may arise due to member revocation.

Keywords Dynamic group signature · Verifier-local revocation · Almost-full anonymity

1 Introduction

The notion of group signature was first introduced by Chaum and van Heyst [12] in 1991. Each member has a private signing key and a corresponding public key. The private signing key is used to generate signatures on messages while the public key is used as a public verification key by verifiers to authenticate the signatures. Group signatures allow group members to sign anonymously on behalf of the group (*anonymity*). Only the authorized person can reveal the identity of the member who signs (*traceability*).

M. N. S. Perera (✉)
Graduate School of Science and Engineering, Saitama University, Saitama, Japan
e-mail: perera.m.n.s.119@ms.saitama-u.ac.jp

T. Koshiba
Faculty of Education and Integrated Arts and Sciences, Waseda University, Tokyo, Japan
e-mail: tkoshiba@waseda.jp

© Springer Nature Singapore Pte Ltd. 2018
D. Ghosh et al. (eds.), *Mathematics and Computing*, Springer Proceedings
in Mathematics & Statistics 253, https://doi.org/10.1007/978-981-13-2095-8_31

Besides the naive security notions (*anonymity* and *traceability*) for group signatures, more security requirements like un-frameability, collusion resistance, and unforgeability are proposed. In 2003, Bellare et al. [2] suggested a formal security notion with *full anonymity* and *full traceability* to provide a stronger security for group signature schemes (the BMW03 model). This BMW03 model supports only for static groups, not for dynamic groups. Hence, it does not guarantee the security when group members can be flexibly reorganized.

In the setting of dynamic group signatures, neither the number of group members nor their keys should be fixed in the setup phase. Thus, a scheme should be able to register or revoke members anytime. In 2005, Bellare et al. [3] suggested a scheme by providing foundations for dynamic group signatures. The scheme in [3] helps to bridge the gap between the results in [2], and the previous works are done to deliver a dynamic group signature scheme. The dynamic groups are more complex than the static groups since they require many security concerns and deliver more issues to be focused. Schemes in [3] and [14] provide formal security definitions for dynamic group signatures to overcome those issues. Another scheme was suggested by Libert et. al. [15]. However, none of them are fully dynamic group signature schemes since they do not support member revocation. Recently, Bootle et. al. [7] suggested a security definition for fully dynamic group signature schemes and they have also provided some fixes for existing schemes. Hereafter, if a scheme supports both member registration and member revocation, we refer to it as fully dynamic, and if a scheme supports either member registration or revocation, we refer to it as dynamic.

The member revocation is an essential requirement in practice, and many researchers presented various approaches to manage member revocation in groups. One approach is replacing the group public key and the private signing keys with new keys for all existing members when a member is revoked. Since this requires to update all the existing members and the verifiers, it is not the best solution, especially not suitable for large groups. In 2001, Bresson et al. [8] provided a solution that requires signers to prove, at the time of signing, that their member certificates are not in the public revocation list. In 2002, Camenisch et al. [11] proposed a different approach, which is based on dynamic accumulators. It maps a set of values into a fixed-length string and permits efficient proofs of memberships. However, this approach requires existing members to keep track of the revoked users. Thus, it increases the workload of existing members. Moreover, schemes in [5, 10, 18] have taken some other revocation approache.

A different and simple revocation mechanism was suggested by Brickell [9], which was subsequently formalized by Boneh and Shacham [6]. This revocation mechanism is known as *Verifier-Local Revocation* (*VLR*). VLR allows the members to convince the verifiers that they are valid members, who are not revoked and eligible to sign on behalf of the group. Every member has a unique token, and when he is revoked, this token is added to a list called *Revocation List* (*RL*). Then, the group manager passes the latest *RL* to the verifiers. When a verifier needs to authenticate a signature, he checks the validity of the signer with the help of *RL*. Since the verifiers are smaller in number than the members, this mechanism is more convenient than any others, especially for large groups. Moreover, this is advantageous to the previous approaches since it does not affect on existing members.

Our Contribution

This paper presents a fully dynamic group signature scheme that allows to both add and revoke members and a new security notion to overcome some security barriers.

First, we take the scheme in [3], which includes an interactive protocol, that allows new users to join the group at any time, and we incorporate with member revocation mechanism by adapting the methods in the scheme in [3] and suggesting new methods to manage member revocation with *VLR*.

Then, we suggest a method to generate member revocation tokens in our scheme. In general, any VLR scheme consists of a token system and those tokens are generated as a part of the secret signing key. Since our intention is to apply full anonymity which requires to provide all the secret signing keys to an adversary at the anonymity game, this method is not suitable for our scheme. If we generate revocation tokens using the signing keys of the members, the adversary can obtain the tokens of the challenged indices and win the anonymity game. Thus, to present a member's token, we use his personal secret key ($usk[i]$) and his verification key (pk_i). Nevertheless, pk_i is a public attribute, revealing pk_i does not show any other information about the member. Even though $usk[i]$ is a secret key, no one can generate any secret signing key by $usk[i]$. Besides, no one can create a group member token using the secret signing key, since the token is not a part of the secret signing key. Thus, it ensures the security of the scheme.

Moreover, we present a new security notion that is somewhat weaker than the full anonymity. *VLR* relies on a weaker security notion called *selfless-anonymity*. Even our intention is to apply full anonymity for our scheme, achieving *full anonymity* suggested in the BMW03 model for *VLR* is quite difficult. In case of the anonymity game (for the definition) between a challenger and a adversary, the BMW03 model passes all the secret keys to the adversary. But, we cannot allow the adversary to reveal all the secret keys since he can corrupt the anonymity of the scheme. If we allow the adversary to reveal all the users' personal private keys (usk), which we use to create tokens he can create any token, including the challenged users' tokens. Then, he can verify the challenging signature and return the correct user index of the challenged signature. Thus, we suggest a *new* restricted version of full anonymity (*almost-full anonymity*), which will not provide all the secret keys to the adversary to ensure the security of our scheme. It will allow the adversary to reveal any member's secret signing keys not the member's personal private keys.

2 Preliminaries

In this section, we describe notations used in the paper and the primitives with which we use to construct our scheme. Construction of dynamic group signature schemes use three building blocks: public-key encryption schemes secure against chosen-ciphertext attack [13], digital signature schemes secure against chosen-message attack [1], and simulation-sound adaptive non-interactive zero-knowledge (NIZK) proofs for NP [17]. All the three primitives are based on trapdoor permutation.

2.1 Notation

We denote by λ the security parameter of the scheme and let $\mathbb{N} = \{1, 2, 3, \ldots\}$ be the set of *positive integers*. For any $k \geq 1 \in \mathbb{N}$, we denote by $[k]$ the set of integers $\{1, \ldots, k\}$. An empty string is denoted by ε. If s is a string, then $|s|$ denotes the length of the string and if $\mathcal{S}$ is a set then $|\mathcal{S}|$ denotes the size of the set. If $\mathcal{S}$ is a finite set, $b \xleftarrow{\$} \mathcal{S}$ denotes that b is chosen uniformly at random from $\mathcal{S}$. We denote experiments by **Exp**.

2.2 Digital Signature Schemes

A digital signature scheme $DS = (\mathsf{K}_s, \mathsf{Sig}, \mathsf{Vf})$ consists of three algorithms: key generation K_s, signing Sig, and verification Vf. The scheme DS should satisfy the standard notion of unforgeability under chosen-message attack.

For an adversary $\mathcal{A}$, consider an experiment $\mathbf{Exp}_{DS,\mathcal{A}}^{unforg\text{-}cma}(\lambda)$. First, a pair of a public key and the corresponding secret key for the scheme DS is obtained by executing K_s with the security parameter λ as $(\mathbf{pk}, \mathbf{sk}) \xleftarrow{\$} \mathsf{K}_s(1^\lambda)$. Then, the public key $\mathbf{pk}$ is given to the adversary, and the adversary can access the signing oracle $\mathsf{Sig}(\mathbf{sk}, \cdot)$ for any number of messages. Finally, the forging adversary $\mathcal{A}$ outputs (m, σ). He wins if σ is a valid signature on the message m and m is not queried so far. We let $\mathbf{Adv}_{DS,\mathcal{A}}^{unforg\text{-}cma}(\lambda) = \Pr[\mathbf{Exp}_{DS,\mathcal{A}}^{unforg\text{-}cma}(\lambda) = 1]$.

A digital signature scheme DS is secure against forgeries under chose message attack if $\mathbf{Adv}_{DS,\mathcal{A}}^{unforg\text{-}cma}(\lambda)$ is negligible in λ for any polynomial-time adversary $\mathcal{A}$.

2.3 Encryption Scheme

An encryption scheme $E = (\mathsf{K}_e, \mathsf{Enc}, \mathsf{Dec})$ consists of three algorithms: key generation K_e, encryption Enc, and decryption Dec. The scheme E should satisfy the standard notion of indistinguishability under adaptive chosen-ciphertext attack.

For an adversary $\mathcal{A}$, consider an experiment $\mathbf{Exp}_{E,\mathcal{A}}^{ind\text{-}cca\text{-}b}(\lambda)$. First, a pair of a public key and the corresponding secret key for the encryption scheme E is obtained by executing K_e with the security parameter λ and a randomness string r_e (where the length of r_e is bounded by some fixed polynomial $r(\lambda)$) as $(\mathbf{pk}, \mathbf{sk}) \xleftarrow{\$} \mathsf{K}_e(1^\lambda, r_e)$.

Let $\mathsf{LR}(m_0, m_1, b)$ a function which returns m_b for a bit b and messages m_0, m_1. We assume the adversary $\mathcal{A}$ never queries $\mathsf{Dec}(\mathsf{sk}, \cdot)$ on a ciphertext previously returned by $\mathsf{Enc}(\mathsf{pk}, \mathsf{LR}(\cdot, \cdot, b))$. We let $\mathbf{Adv}_{E,A}^{ind\text{-}cca}(\lambda) = |\Pr[\mathbf{Exp}_{E,\mathcal{A}}^{ind\text{-}cca\text{-}1}(\lambda) = 1] - \Pr[\mathbf{Exp}_{E,\mathcal{A}}^{ind\text{-}cca\text{-}0}(\lambda) = 1]|$.

An encryption scheme E is IND-CCA secure if $\mathbf{Adv}_{E,A}^{ind\text{-}cca}(\lambda)$ is negligible in λ for any polynomial-time adversary $\mathcal{A}$.

2.4 Simulation-Sound Non-interactive Zero-Knowledge Proof System

A two-party game between a prover and a verifier which needs to determine whether a given string belongs to a language or not is called an interactive system. The interactive system allows to exchange messages between the prover and the verifier. Besides, argument systems are like interactive proof systems, except they are required to be computationally infeasible for a prover to convince the verifier to accept inputs not in the language. Non-interactive proof systems are mono-directional [4]. The non-interactive proof systems allow a prover to convince a verifier about a truth statement while zero-knowledge ensures that the verifier learns nothing from the proof other than the truth of the statement. The non-interactive zero-knowledge proof system shows that without any interaction but using a common string computational zero-knowledge can be achieved. In a simulation-sound NIZK proof system, an adversary cannot prove any false statements even after seeing simulated proofs of arbitrary statements.

An NP-*relation over domain* Dom $\subseteq \{0, 1\}^*$ is a subset ρ of $\{0, 1\}^* \times \{0, 1\}^*$. We say that x is a *theorem* and w is a *proof* of x if $(x, w) \in \rho$. The membership of $(x, w) \in \rho$ is decidable in time polynomial in the length of the first argument for all x in Dom.

We fix an NP relation ρ over Dom and take a pair of polynomial-time algorithms (P, V), where P is randomized, and V is deterministic. Both P and V have access to a *common reference string R*. The (P, V) is a non-interactive proof system for ρ over Dom if the following two conditions are satisfied for polynomials p and l.

– *Completeness*: $\forall \lambda \in \mathbb{N}, \forall (x, w) \in \rho$ with $|x| \leq l(\lambda)$ and $x \in Dom$:

$\Pr[R \xleftarrow{\$} \{0, 1\}^{p(\lambda)}; \pi \xleftarrow{\$} P(1^\lambda, x, w, R) : V(1^\lambda, x, \pi, R) = 1] = 1$.

– *Soundness*: $\forall \lambda \in \mathbb{N}, \forall \hat{P}$ and $x \in Dom$ such that $x \notin L_\rho$:

$\Pr[R \xleftarrow{\$} \{0, 1\}^{p(\lambda)}; \pi \xleftarrow{\$} \hat{P}(1^\lambda, x, R) : V(1^\lambda, x, \pi, R) = 1] \leq 2^{-\lambda}$.

3 Our Scheme

We construct our scheme based on the scheme in [3]. In the scheme in [3], they have taken a digital signature scheme $DS = (\mathsf{K}_s, \mathsf{Sig}, \mathsf{Vf})$ and a public-key encryption scheme $E = (\mathsf{K}_e, \mathsf{Enc}, \mathsf{Dec})$ as the building blocks to construct a group signature

                    user^i Issuer (group manager)

upk[i], **usk**[i] $\leftarrow$ **UKg**(1^k)

pk$_i$, sk$_i$ $\leftarrow$ K$_s$ (1^k)

sig$_i$ $\leftarrow$ Sig(**usk**[i], pk$_i$)

$\xrightarrow{\quad \text{pk}_i, \text{sig}_i \quad}$ If pk$_i \notin reg$ and Vf(**upk**[i], pk$_i$, sig$_i$) = 1 then

                                                   $cert_i \leftarrow$ Sig(**ik**, $\langle i, \text{pk}_i \rangle$)

                                                   $reg[i] \leftarrow$ (pk$_i$, sig$_i$, 1)

                                          Else $cert_i \leftarrow \varepsilon$

$\xleftarrow{\quad cert_i \quad}$

gsk[i] $\leftarrow$ (i, pk$_i$, sk$_i$, $cert_i$)

grt[i] $\leftarrow$ (**usk**[i], pk$_i$)

Fig. 1 Group-joining protocol

scheme *GS*. Moreover, they have used NIZK proof system to convince the verifier the validity of the signature. We also use above-mentioned primitives; *DS*, *E*, and NIZK to present a new scheme *FDGS* = (GKg, UKg, Join, Issue, Revoke, Sign, Verify, Open, Judge). GKg, UKg, and Judge are same as the scheme in [3]. We provide a new algorithm Revoke to revoke members, and we modify Join, Issue, Sign, Verify, and Open to be compatible with the revocation mechanism. We use *DS* for generating the group manager's keys and *E* for generating the opener's keys. Thus, our group public key **gpk** consists of the security parameter λ, public keys of *group manager* and *opener*, and two reference strings R_1, R_2 obtained for NIZK proof.

We describe our *group-joining protocol* which executes Join and Issue in Fig. 1, and we describe other algorithms of our scheme in Fig. 2.

3.1 Coping with VLR and Making the Scheme Secure

In general, *VLR* schemes satisfy a weaker security notion called *selfless-anonymity*, which does not provide any secret keys to the adversary. Even though our scheme supports *VLR* mechanism, we make our scheme more secure by using the techniques in [3] scheme and suggesting a new security notion called *almost-full anonymity*. Making VLR scheme fully anonymous is quite difficult since the full anonymity requires to provide all the secret keys to the adversary and providing tokens to the adversary makes the scheme insecure. The adversary can execute Verify with the tokens of the challenged indices and win the game easily. Thus, we consider a new security notion called *almost-full anonymity* which will not provide tokens to the adversary, which is a restricted version of the full anonymity.

Moreover, any *VLR* scheme has an associated tracing mechanism called *implicit tracing algorithm* to trace signers. The implicit tracing algorithm requires to run Verify linear times in the number of group members. Compare to the *explicit tracing*

GKg$(1^\lambda) \to ($**gpk, ok, ik**$)$

$R_1 \xleftarrow{\$} \{0,1\}^{P1(\lambda)}$.

$R_2 \xleftarrow{\$} \{0,1\}^{P2(\lambda)}$.

$r_e \xleftarrow{\$} \{0,1\}^{r(\lambda)}$.

$(opk, osk) \leftarrow K_e(1^\lambda; r_e)$.

$(gmpk, gmsk) \xleftarrow{\$} K_s(1^\lambda)$.

gpk $\leftarrow (1^\lambda, R_1, R_2, opk, gmpk)$.

ok $\leftarrow (osk, r_e)$.

ik $\leftarrow gmsk$.

Return (**gpk,ok,ik**).

UKg$(1^\lambda) \to ($**upk, usk**$)$

$($**upk, usk**$) \xleftarrow{\$} K_s(1^\lambda)$.

Return (**upk, usk**).

Revoke$(i, $**grt**$[i], $**ik**$, RL, reg) \to (RL, reg)$

Parse **grt**$[i]$ as $(usk[i], pk[i])$

Query $reg[i] \to (i, pk_i, st)$

If $(st \neq 0$ and $pk_i = pk[i])$

then $RL \leftarrow RL \cup (usk[i], pk[i])$

update $reg[i]$ to inactive;

Return RL, reg.

Sign(**gpk, gsk**$[i], $**grt**$[i], m) \to (\sigma)$

Parse **gpk** as $(1^\lambda, R_1, R_2, opk, gmpk)$.

Parse **gsk**$[i]$ as $(i, pk_i, sk_i, cert_i)$.

$s \leftarrow \mathsf{Sig}(sk_i, m); r \xleftarrow{\$} \{0,1\}^\lambda$.

$C \leftarrow \mathsf{Enc}(opk, \langle i, pk_i, cert_i, s \rangle; r)$.

$\pi_1 \xleftarrow{\$} P_1(1^\lambda, (opk, gmpk, m, C),$

$(i, pk_i, cert_i, s, r), R_1)$.

$\sigma \leftarrow (C, \pi_1, $**grt**$[i])$.

Return σ.

Verify(**gpk**$, m, \sigma, RL) \to 1/0$

Parse **gpk** as $(1^\lambda, R_1, R_2, opk, gmpk)$.

Parse σ as $(C, \pi_1, $**grt**$[i])$.

Parse **grt**$[i]$ as $(usk[i], pk_i)$.

If $V_1(1^\lambda, (opk, gmpk, m, C), \pi_1, R_1) = 1$

and $(usk[i], pk_i) \notin RL$ then return 1

else return 0.

Open(**gpk, ok**$, reg, m, \sigma) \to (i, \tau, st)$

Parse **gpk** as $(1^\lambda, R_1, R_2, opk, gmpk)$.

Parse **ok** as (osk, r_e).

Parse σ as $(C, \pi_1, $**grt**$[i])$.

$M \leftarrow Dec(osk, C)$.

Parse M as $\langle i, pk, cert, s \rangle$.

If $reg[i] \neq \varepsilon$ then

Parse $reg[i]$ as $(pk_i, sig_i, status)$.

Else $pk_i \leftarrow \varepsilon; sig_i \leftarrow \varepsilon; st \leftarrow \varepsilon$.

$\pi_2 \leftarrow P_2(1^\lambda, (opk, C, i, pk, cert, s),$

$(osk, r_e), R_2)$.

If $V_1(1^\lambda, (opk, gmpk, m, C), \pi_1, R_1) = 0$

then return $(0, \varepsilon, 0)$.

If $pk \neq pk_i$ or $reg[i] = \varepsilon$ or $status = 0$

then return $(0, \varepsilon, 0)$.

$\tau \leftarrow (pk_i, sig_i, i, pk, cert, s, \pi_2)$.

Return (i, τ, st).

Judge(**gpk**$, i, $**upk**$[i], m, \sigma, \tau) \to 1/0$

Parse **gpk** as $(1^\lambda, R_1, R_2, opk, gmpk)$.

Parse σ as $(C, \pi_1, $**grt**$[i])$.

If $(i, \tau, st) = (0, \varepsilon, 0)$ then

return $V_1(1^\lambda, (opk, gmpk, m, C), \pi_1, R_1) =$

0.

Parse τ as $(\bar{pk}, \bar{sig}, i', pk, cert, s, \pi_2)$.

If $V_2(1^\lambda, (C, i', pk, cert, s), \pi_2, R_2) = 0$

then return 0

If all of the following are true then return

1 else return 0:

- $i = i'$

- $\mathsf{Vf}($**upk**$[i], \bar{pk}, \bar{sig})$

- $\bar{pk} = pk$.

IsActive$(i, reg) \to (0/1)$

If $reg[i] \neq \varepsilon$ then

Parse $reg[i]$ as $(pk_i, sig_i, status)$.

Return $status$.

Fig. 2 Algorithms of the new fully dynamic group signature scheme

algorithm, which is used in schemes like [16], use of the implicit tracing algorithm increases the time consumption. Hence, instead of using the implicit tracing algorithm given in *VLR*, we use algorithms provided in [3] for our scheme's tracing mechanism.

As well, *VLR* manages a token system. Thus, our scheme should consist user tokens and those tokens should be unique to the users. Furthermore, tokens should not reveal user's identity in case of disclosing to the outsiders. We generate tokens for members, which will not expose identity of the members even though tokens are opened to the outsiders. We use the combination of each group member's personal secret key and his verification key as his token, and we maintain the list *RL* with revoked members' tokens.

3.2 Description of Our Scheme

There are two authorities, *group manager* and *opener*. The trusted setup is responsible for generating the group public key and keys for the authorities. The *group manager* manages member registration and member revocation while the *opener* traces signers.

When a new user wants to join the group, he interacts with the group manager via *group-joining protocol* (Fig. 1), which allows new users to generate their public key and secret keys. We assume this interaction between the new user, and the group manager is done through a secure channel. The new user produces a signature on his verification key and sends both the signature and the key to the group manager. If the signature is acceptable, then the group manager accepts him as a new member. In the registration table *reg*, we maintain a field called *Status* for each member to identify the active status of them. Thus, the group manager stores the index i, verification key pk_i, and the signature sig_i of the new member in *reg* and makes the status of the new member as active. After that, the group manager issues member certification to the new member. Now the new member can generate signatures on messages using his secret key.

Each member has a unique token, which is the tracing key to identify the validity of signers, whether they are revoked or not. Here we use the member's personal secret key $usk[i]$ and his verification key pk_i as the token since $usk[i]$ or pk_i does not help to reveal any other information. We check the existence of the new user keys against *reg* at the joining protocol. Thus, in a situation that a revoked member wants to join again, he cannot use his previous keys, and he has to follow the process as a new user. That is to secure the scheme against adversaries who steal tokens and try to join the group. During the member revocation, the group manager adds the revoking member's token to *RL* and updates *reg* to inactive. When a member needs to sign a message, he generates the signature on a message with his secret key and passes to the verifier with his token for verification. The verifier authenticates the signature on the given message and checks the validity of the signer with the provided token against the latest *RL*. In the case of necessity to trace the signer, the opener can trace the signer using opener's key, and he can check the status of the signer in *reg*.

Our scheme is a tuple $FDGS = ($GKg, UKg, Join, Issue, Revoke, Sign, Verify, Open, Judge$)$, which consists of polynomial-time algorithms. Each algorithm is described in below. GKg, UKg, and Judge are same as in [3] and Join, Issue, Sign, Verify, and Open are different from the algorithms given in [3] since we have to generate and pass the member's token as an additional attribute in our scheme. Revoke helps to revoke the misbehaved users.

- GKg(1^λ): On input 1^λ, the trusted party obtains a group public key **gpk** and authority keys, **ik** and **ok**. Then gives secret keys, **ik** to the group manager and **ok** to the opener.
- UKg(1^λ): Every user who wants to be a member should run this algorithm before the *group-joining protocol* to obtain their personal public key and personal private key (**upk**[i], **usk**[i]). UKg takes as input 1^λ. We assume **upk** is publicly available.
- Join, Issue: The *group-joining protocol* is an *interactive protocol* between the group manager and the user who wants to be a member. Join is implemented by the user while Issue is implemented by the group manager. Join allows new users to generate keys and a signature on the keys which are needed to join the group. Issue allows the group manager to validate the keys and the signatures sent by users and generate member certifications. Each algorithm takes an incoming message as input and returns an outgoing message. Join and Issue maintains their current status for both parties. The user i generates a public / secret key pair pk_i and sk_i. Then, he produces a signature sig_i on pk_i using **usk**[i], which was obtained in UKg. Then, user sends sig_i and pk_i to the group manager to authenticate. The group manager authenticates the signature sig_i on pk_i and generates member certification by signing pk_i with his private key **ik** (**gmsk**). The group manager stores new member's informations, i, pk_i, and sig_i with the *status* as 1 (active) in *reg*. Then, he sends member certification $cert_i$ to the user who is the new member of the group. After that new user can make his secret key **gsk**[i] $= (i, pk_i, sk_i, cert_i)$, and his token **grt**[$i$] $= ($**usk**[i], $pk_i)$.
- Revoke(i, **grt**[i], **ik**, RL, reg): This algorithm takes, index i of the member, who wants to be revoked and the group manager's secret key **ik** as inputs. First, the group manager queries reg using the index i to obtain the information of the member stored. Then, he checks whether the queries are equal to the data obtained by parsing the **grt**[i]. If the data are equal and if the user is active, insert ($usk[i]$, pk_i) to RL and updates reg to 0 (inactive).
- Sign(**gpk**, **gsk**[i], **grt**[i], m): This randomized algorithm generates a signature σ on a given message m. It takes the group public key **gpk**, the group member's secret key **gsk**, and the message m as inputs. In addition, we pass the group member's token as an input to prove that the member is an active person at the time of signing.
- Verify(**gpk**, m, σ, RL): This deterministic algorithm allows anyone in possession of group public key **gpk** to verify the given signature σ on the message m and checks the validity of the signer against RL. This algorithm outputs 1 if both conditions are valid. Otherwise, it returns 0.
- Open(**gpk**, **ok**, reg, m, σ): This deterministic algorithm traces the signer by taking **gpk**, the opener's secret key **ok**, reg, the message m, and the signature σ as inputs.

It returns the index of the signer, the proof of the claim τ, and the status of the signer st at reg. If the algorithm failed to trace the signature to a particular group member, it returns $(0, \varepsilon, 0)$.

- Judge($\mathbf{gpk}$, i, $\mathbf{upk}[i]$, m, σ, τ): This deterministic algorithm outputs either 1 or 0 depending on the validity of the proof τ on σ. This takes, the group public key $\mathbf{gpk}$, the member index i, the tracing proof τ, the member verification key $\mathbf{upk}[i]$, the message m, and the signature σ as inputs. The algorithm outputs 1 if τ can proof that i produced σ. Otherwise, it returns 0.

In addition, we use the following simple polynomial-time algorithm.

- IsActive(i, reg): This algorithm determines whether the member i is active by querying the registration table and outputs either 0 or 1.

4 Security Notions of the Scheme

Even though the BMW03 model has two key requirements, full anonymity and full traceability, the scheme in [3] has three key requirements; anonymity, traceability, and non-frameability. Since *full traceability* discussed in the BMW03 model covers both traceability and non-frameability, the BMW03 model has only two requirements. In the setting of [3], traceability and non-frameability are separated since non-frameability can be achieved with lower levels of trust in the authorities than traceability as discussed below. According to the scheme in [3], the opener's secret key is provided to an adversary in traceability game but, the issuer's secret key is not provided. The scheme in [3], they assume that the opener is partially corrupted in traceability. But in non-frameability, both the opener's and the tracer's secret keys are given to the adversary. Thus, the adversary is stronger in non-frameability than in traceability. Thus, non-frameability is separated from the traceability in [3]. Moreover, anonymity allows the adversary to corrupt the issuer in [3]. Thus, we provide the issuer's secret key to the adversary but not the opener's secret key.

However, the scheme in [3] does not support member revocation but our scheme supports. Thus, we adapt the security experiments and the oracles to be compatible with *VLR*. Before we discuss the security notions, we define the set of oracles that we use. We suggest a new oracle, revoke to maintain the member revocation queried by any adversary.

For the requirement of anonymity, we suggest a restricted version of full anonymity. In the full-anonymity game, we provide all the members' secret keys to the adversary including challenged indexes' keys to the adversary. In our scheme, this may help the adversary to create the challenged indexes' tokens since he knows all the members' personal secret keys (usk) and he can execute Verify to check which index is used to generate the challenged signature. Thus, we will not provide users' personal secret keys to the adversary when he requests for user's secret keys. However, he can request for any private signing key. Hence, we suggest a new security notion *almost-full anonymity* to show the security of our scheme. Since the almost-full anonymity does not allow members' personal secret keys to the adversary, it is somewhat weaker than the full anonymity, and since, it provides members' secret signing keys including challenged indices' to the adversary, it is stronger than the selfless-anonymity.

4.1 The Oracles

All the oracles that we use are specified in Fig. 3. We maintain a set of global lists, which are manipulated by the oracles in the security experiments discussed later. **HUL** is the honest user list, which maintains the indexes of the users who are added to the group. When the adversary corrupts any user, that user's index is added to **CUL**. **SL** carries the signatures that obtained from Sign oracle. When the adversary requests a signature, the generated signature, the index, and the message are added to **SL**. When the adversary accesses **Challenge** oracle, the generated signature is added to **CL** with the message sent. We use a set S to maintain a set of revoked users.

- AddU(i): The adversary can add a user $i \in \mathbb{N}$ to the group as an honest user. The oracle adds i to **HUL** and selects keys for i. It then executes the *group-joining protocol*. If **Issue** accepts, then adds the state to *reg* and if **Join** accepts then generates **gsk**[i]. Finally, returns **upk**[i].
- CrptU(i, *upk*): The adversary can corrupt user i by setting its personal public key **upk**[i] to *upk*. The oracle adds i to **CUL** and initializes the issuer's state in *group-joining protocol*.
- SendToIssuer(i, M_{in}): The adversary acts as i and engages in *group-joining protocol* with **Issue**-executing issuer. The adversary provides i and M_{in} to the oracle. The oracle which maintains the **Issue** state returns the outgoing message and adds a record to *reg*.
- SendToUser(i, M_{in}): The adversary corrupts the issuer and engages in *group-joining protocol* with **Join**-executing user. The adversary provides i and M_{in} to the oracle. The oracle which maintains the user i state, returns the outgoing message, and sets the private signing key of i to the final state of **Join**.
- RevealU(i): The adversary can reveal secret keys of the user i. We only provide user's private signing key **gsk**[i] not his personal private key **usk**[i].
- ReadReg(i): The adversary can read the entry of i in *reg*.
- ModifyReg(i, *val*): The adversary can modify the contents of the record for i in *reg* by setting *val*.
- Sign(i, m): The adversary obtains a signature σ for a given message m and user i who is an honest user and has private signing key.
- Chal$_b$(i_0, i_1, m): This oracle is for defining anonymity and provides a group signature for the given message m under the private signing key of i_b, as long as both i_0, i_1 are active and honest users having private signing keys.
- Revoke(i): The adversary can revoke user i. The oracle updates the record for i in *reg* and adds revocation token of i to the set S.

AddU(i)
If $i \in \mathbf{HUL} \cup \mathbf{CUL}$, then return ε.
$\mathbf{HUL} \leftarrow \mathbf{HUL} \cup \{i\}$
$\mathbf{gsk}[i] \leftarrow \varepsilon$; $\mathbf{grt}[i] \leftarrow \varepsilon$
$dec_{is}^i \leftarrow$ cont;
$(\mathbf{upk}[i], \mathbf{usk}[i]) \leftarrow \mathsf{UKg}(1^\lambda)$
$St_{jn}^i \leftarrow (\mathbf{gpk}, \mathbf{upk}[i], \mathbf{usk}[i])$
$St_{is}^i \leftarrow (\mathbf{gpk}, \mathbf{ik}, i, \mathbf{upk}[i])$
$M_{jn} \leftarrow \varepsilon$.
$(St_{jn}^i, M_{is}, dec_{is}^i) \leftarrow \mathsf{Join}(St_{jn}^i, M_{jn})$.
While $(dec_{is}^i = \text{cont and } dec_{jn}^i = \text{cont})$
then do
$(St_{is}^i, M_{jn}, dec_{is}^i) \leftarrow \mathsf{Issue}(St_{is}^i, M_{is})$.
$(St_{jn}^i, M_{is}, dec_{jn}^i) \leftarrow \mathsf{Join}(St_{jn}^i, M_{jn})$.
End while.
If $dec_{is}^i = \text{accept}$ then $reg[i] \leftarrow (St_{is}^i, 1)$
If $dec_{jn}^i = \text{accept}$ then $\mathbf{gsk}[i] \leftarrow (St_{jn}^i)$ and
$\mathbf{grt}[i] = \mathbf{usk}[i]$
Return $\mathbf{upk}[i]$

CrptU(i, upk)
If $i \in \mathbf{HUL} \cup \mathbf{CUL}$ then return ε.
$\mathbf{CUL} \leftarrow \mathbf{CUL} \cup \{i\}$;
$\mathbf{upk}[i] \leftarrow upk$.
$dec_{is}^i \leftarrow$ cont.
$St_{is}^i (\mathbf{gpk}, \mathbf{ik}, i, \mathbf{upk}[i])$
Return 1.

ReadReg(i)
Return $(reg[i])$

ModifyReg(i, val)
$reg[i] \leftarrow val$

Revoke(i)
If $i \notin \mathbf{HUL}$ then return ε.
If $i \in \mathbf{CL}$ then return ε.
If $\mathsf{IsActive}(i, \text{reg}) = 0$ then return ε.
$S = S \cup \{\mathbf{grt}[i]\}$
update $reg[i]$
Return 1

SendToIssuer(i, M_{in})
If $i \notin \mathbf{CUL}$ then return ε
If $dec_{is}^i \neq cont$ then return ε
If $i \notin \mathbf{HUL}$ then return ε.
$St_{is}^i \leftarrow (\mathbf{gpk}, \mathbf{ik}, i, \mathbf{upk}[i])$
$(St_{is}^i, M_{out}, dec_{is}^i) \leftarrow \mathsf{Issue}(St_{is}^i, M_{in})$.
If $dec_{is}^i = \text{accept}$ then
$reg[i] \leftarrow (St_{is}^i, 1)$
Return (M_{out}, dec_{is}^i)

Sign(i, m)
If $i \notin \mathbf{HUL}$ then return ε.
If $\mathbf{gsk}[i] = \varepsilon$ then return ε.
If $\mathsf{IsActive}(i, \text{reg}) = 0$ then return ε
$\sigma = \mathsf{Sign}(\mathbf{gpk}, \mathbf{gsk}[i], \mathbf{grt}[i], m)$
$\mathbf{SL} = \mathbf{SL} \cup \{(i, m, \sigma)\}$
Return σ

SendToUser(i, M_{in})
If $i \in \mathbf{CUL}$ then return ε.
If $i \notin \mathbf{HUL}$ then
$\mathbf{HUL} \leftarrow \mathbf{HUL} \cup \{i\}$
$\mathbf{gsk}[i] \leftarrow \varepsilon$; $M_{in} \leftarrow \varepsilon$
$(\mathbf{upk}[i], \mathbf{usk}[i]) \leftarrow \mathsf{UKg}(1^\lambda)$
If $dec_{jn}^i \neq cont$ then return ε
$St_{jn}^i \leftarrow (\mathbf{gpk}, \mathbf{upk}[i], \mathbf{usk}[i])$
$(St_{jn}^i, M_{out}, dec_{jn}^i) \leftarrow \mathsf{Join}(St_{jn}^i, M_{in})$
If $dec_{jn}^i = \text{accept}$ then $\mathbf{gsk}[i] \leftarrow St_{jn}^i$ and
$\mathbf{grt}[i] = \mathbf{usk}[i]$
Return (M_{out}, dec_{jn}^i)

Chal$_b(i_0, i_1, m)$
If $i_0 \notin \mathbf{HUL}$ or $i_1 \notin \mathbf{HUL}$ then
return ε.
If $\mathbf{gsk}[i_0] = \varepsilon$ or $\mathbf{gsk}[i_1] = \varepsilon$ then
return ε.
If $\mathsf{IsActive}(i_0, reg) = 0$ or $\mathsf{IsActive}(i_1, reg)$
$= 0$ then
return ε
$\sigma = \mathsf{Sign}(\mathbf{gpk}, \mathbf{gsk}[i_b], \mathbf{grt}[i_b], m)$
$\mathbf{CL} = \mathbf{CL} \cup \{(m, \sigma)\}$
Return σ

RevealU(i)
If $i \notin \mathbf{HUL} \setminus (\mathbf{CUL} \cup \mathbf{CL})$ then
return ε.
Return $\mathbf{gsk}[i]$

Open(m, σ)
If $(m, \sigma) \in \mathbf{CL}$ then return $(\varepsilon, \varepsilon, \varepsilon)$
If $\mathsf{Verify}(\mathbf{gpk}, m, \sigma, S) = 0$ then return $(\varepsilon,$
$\varepsilon, \varepsilon)$
Return $\mathsf{Open}(\mathbf{gpk}, \mathbf{osk}, reg, m, \sigma)$

Fig. 3 Oracles

– Open(m, σ): The adversary can access this *opening* oracle with a message m and a signature σ to obtain the identity of the user, who generated the signature σ. If σ is queried before for Chal$_b$, oracle will abort.

4.2 Correctness

The notion of correctness requires that any signature generated by any honest and active users should be valid and Open should correctly identify the signer for a given message and a signature. Moreover, the proof returned by Open should be accepted by Judge. Hence, any scheme is correct if the advantage of the correctness game is 0, for all $\lambda \in \mathbb{N}$ and for any adversary A.

We let, $\mathbf{Adv}_{FDGS,A}^{corr}(\lambda) = \Pr[\mathbf{Exp}_{FDGS,A}^{corr}(\lambda) = 1]$.

$\underline{\mathbf{Exp}_{FDGS,A}^{corr}(\lambda)}$

(**gpk**, **ok**, **ik**) $\leftarrow$ GKg(1^λ);   **HUL** $\leftarrow$ Ø;   (i, m) $\leftarrow$ A(**gpk**; AddU, ReadReg, Revoke);

If $i \notin$ **HUL** or **gsk**[i] $= \varepsilon$ or IsActive(i, reg)$= 0$, then return 0.

$\sigma \leftarrow$ Sign(**gpk**, **gsk**[i], m);

If Verify(**gpk**, m, σ, S) $= 0$, then return 1.

(i', τ) $\leftarrow$ Open(**gpk**, **ok**, reg, m, σ);

If $i \neq i'$, then return 1.

If Judge(**gpk**, i, **upk**[i], m, σ, τ) $= 0$, then return 1 else return 0.

4.3 Anonymity

The anonymity requires the signatures do not reveal the identity of the signer. In the anonymity game, the adversary's goal is to identify the index that is used to create the signature. We allow the adversary A to corrupt any user and allow him to fully corrupt the group manager. Also, A can learn secret signing keys of any user. In full-anonymity game, adversary can access all the secret keys of any member. However, we suggest a new security notion *almost-full anonymity*, which does not allow to reveal the personal secret keys of the users since the adversary can create the tokens of the challenged ones and check with Verify. Hence, he can easily win the game. We say that *FDGS* scheme is almost-fully anonymous if the advantage of the adversary $\mathbf{Adv}_{FDGS,A}^{anon}(\lambda)$ is negligible for any polynomial-time adversary.

In the game, A selects two active group members and a message to challenge the game. He has to guess which member is used to generate the signature. He wins if he can guess the member correctly. We allow only one guess.

We let $\mathbf{Adv}_{FDGS,A}^{anon}(\lambda) = \Pr[\mathbf{Exp}_{FDGS,A}^{anon-0}(\lambda) = 1] - \Pr[\mathbf{Exp}_{FDGS,A}^{anon-1}(\lambda) = 1]$.

$$\mathbf{Exp}_{FDGS,A}^{anon\text{-}b}(\lambda)$$

$(\mathbf{gpk}, \mathbf{ok}, \mathbf{ik}) \leftarrow \mathsf{GKg}(1^\lambda); \mathbf{HUL}, \mathbf{CUL}, \mathbf{SL}, \mathbf{CL} \leftarrow \emptyset;$

$b^* \leftarrow A(\mathbf{gpk}, \mathbf{ik}; \mathsf{CrptU}, \mathsf{SendToUser}, \mathsf{RevealU}, \mathsf{Open}, \mathsf{ModifyReg}, \mathsf{Revoke}, \mathsf{Chal}_b);$

Return b^*;

4.4 Non-Frameability

The non-frameability ensures that any adversary unable to produce a signature can be attributed to an honest member, who did not produce it.

We let $\mathbf{Adv}_{FDGS,A}^{non\text{-}fram}(\lambda) = \Pr[\mathbf{Exp}_{FDGS,A}^{non\text{-}fram}(\lambda) = 1]$.

In this game, we only require that the framed member is honest. Thus, the adversary A can fully corrupt the group manager and the opener.

Formally, the *FDGS* scheme is non-frameable for all $\lambda \in \mathbb{N}$ and for any adversary A.

$$\mathbf{Exp}_{FDGS,A}^{non\text{-}fram}(\lambda)$$

$(\mathbf{gpk}, \mathbf{ok}, \mathbf{ik}) \leftarrow \mathsf{GKg}(1^\lambda); \mathbf{HUL}, \mathbf{CUL}, \mathbf{SL} \leftarrow \emptyset;$

$(m, \sigma, i, \tau) \leftarrow A(\mathbf{gpk}, \mathbf{ik}, \mathbf{ok}; \mathsf{CrptU}, \mathsf{SendToUser}, \mathsf{RevealU}, \mathsf{Sign}, \mathsf{ModifyReg});$

If $\mathsf{Verify}(\mathbf{gpk}, m, \sigma, S) = 0$, then return 0.

If $\mathsf{Judge}(\mathbf{gpk}, i, \mathbf{upk}[i], m, \sigma, \tau) = 0$, then return 0.

If $i \notin \mathbf{HUL}$ or $(i, m, \sigma, \tau) \in \mathbf{SL}$, then return 0 else 1.

4.5 Traceability

The traceability requires any adversary cannot produce a signature that unable to identify the origin of the signature. That means the adversary's challenge is to generate a signature that cannot be traced to an active member of the group. In this game, A is allowed to corrupt any user and he has the opener's key, but he is not allowed to corrupt the group manager since he can produce dummy users. He wins if he can create a signature, whose signer cannot be identified or signer is an inactive member when creating the signature, or Judge algorithm does not accept the Open algorithm's decision.

We let $\mathbf{Adv}_{FDGS,A}^{trace}(\lambda) = \Pr[\mathbf{Exp}_{FDGS,A}^{trace}(\lambda) = 1]$.

$$\mathbf{Exp}_{FDGS,A}^{trace}(\lambda)$$

$(\mathbf{gpk}, \mathbf{ok}, \mathbf{ik}) \leftarrow \mathsf{GKg}(1^\lambda); \mathbf{HUL}, \mathbf{CUL}, \mathbf{SL} \leftarrow \emptyset;$

$(m, \sigma) \leftarrow A(\mathbf{gpk}, \mathbf{ok}; \mathsf{AddU}, \mathsf{CrptU}, \mathsf{SendToIssuer}, \mathsf{RevealU}, \mathsf{Sign}, \mathsf{ModifyReg}, \mathsf{Revoke});$

If $\mathsf{Verify}(\mathbf{gpk}, m, \sigma, S) = 0$, then return 0.

$(i, \tau) \leftarrow \mathsf{Open}(\mathbf{gpk}, \mathbf{ok}, reg, m, \sigma);$

If $i = 0$ or $\mathsf{Judge}(gpk, i, upk[i], m, \sigma, \tau) = 0$, then return 1 else return 0.

5 Security Proof of Our Scheme

We can prove that our scheme is anonymous, non-frameable and traceable according to the experiments described above and which are discussed in [3] and [7]. Even though our scheme has used a token system as an additional attribute than the scheme in [3], since we are not providing the tokens to the adversary and since we have used the member's personal secret key **usk**[i] and his verification key **pk**$_i$ as his revocation token, which cannot be used to learn about the member, there is no impact on the security of the scheme from the token system. Since our scheme requires a reasonable and sufficient security notion for the problem of considering full anonymity, we use almost-full anonymity and we use security experiments provided above instead of experiments given in [3]. However, due to the page limitation, we provide only a summary of security proof and we will give a detailed proof of security in a full version of this paper.

5.1 Anonymity

On the assumption that P_1 is computational zero knowledge for ρ_1 over Dom_1 and P_2 is computational zero knowledge for ρ_2 over Dom_2, two simulations Sim_1 and Sim_2 can be fixed as $\Pi_1 = P_1, V_1, Sim_1$; $\Pi_2 = P_2, V_2, Sim_2$; Π_1 and Π_2 are the simulation-sound zero-knowledge non-interactive proof systems of them for $L_{\rho 1}$ and $L_{\rho 2}$, respectively.

For any polynomial-time adversary B, who will challenge the anonymity of our scheme and who can construct polynomial-time IND-CCA adversaries A_0, A_1 against encryption scheme E, an adversary A_s against the simulation soundness of Π and distinguishers D_1, D_2 that distinguish real proofs of Π_1 and Π_2, respectively, for all $\lambda \in \mathbb{N}$, we say

$$\mathbf{Adv}_{FDGS,B}^{anon}(k) \leq \mathbf{Adv}_{E,A_0}^{ind-cca}(k) + \mathbf{Adv}_{E,A_1}^{ind-cca}(k) + \mathbf{Adv}_{\Pi,A_s}^{ss}(k)$$
$$+ 2 \cdot (\mathbf{Adv}_{P_1,Sim_1,D_1}^{zk}(k) + \mathbf{Adv}_{P_2,Sim_2,D_2}^{zk}(k)).$$

According to the Lemma 5.1 described and proved in [3], we can say the left side function is negligible since all the functions on the right side are negligible under the assumptions on the security of building blocks described. This proves the anonymity of our scheme.

5.2 Non-Frameability

If there is a non-frameability adversary B, who creates at most $n(k)$ honest users, where n is a polynomial and who constructs two adversaries A_2, A_3 against the digital signature scheme, on the assumption that (P_1, V_1), (P_2, V_2) are sound proof systems for ρ_1, ρ_2, respectively, we say

$$\mathbf{Adv}_{FDGS,B}^{non-fram}(k) \leq 2^{-k+1} + n(k) \cdot (\mathbf{Adv}_{DS,A_2}^{unforg-cma}(k) + \mathbf{Adv}_{DS,A_3}^{unforg-cma}(k)).$$

On the assumption that the scheme DS is secure, all the functions on the right side are negligible, so the left side function. Thus, our scheme is non-frameable according to the definition of DS.

5.3 Traceability

If there is a traceability adversary B, who constructs an adversary A_1 against the scheme DS, on the assumption that (P_1, V_1) is a sound proof system for ρ_1, we say

$$\mathbf{Adv}_{FDGS,B}^{trace}(k) \leq 2^{-k+1} + \mathbf{Adv}_{DS,A_1}^{unforg-cma}(k).$$

On the assumption that DS is secure against traceability, all the functions on the right side are negligible. Because of this, the advantage of B is negligible. Thus, it proves that our scheme is traceable.

6 Conclusion

In this paper, we have presented a simple fully dynamic group signature scheme that can be used as a basic scheme to develop with different approaches. We have constructed our scheme based on the scheme in [3] and proposed *Verifier-Local revocation* mechanism, which ease member revocation and convenient for large groups. Thus, our scheme is more flexible and suitable for dynamically changing groups, even they are large. We have shown how to achieve the security with almost-fully anonymity, which is a limited version of fully anonymity.

Acknowledgements This work is supported in part by JSPS Grant-in-Aids for Scientic Research (A) JP16H01705 and for Scientic Research (B) JP17H01695.

References

1. Bellare, M., Micali, S.: How to sign given any trapdoor function. In: CRYPTO 1988, vol. 403, pp. 200–215. LNCS (1988)
2. Bellare, M., Micciancio, D., Warinschi, B.: Foundations of group signatures: formal definitions, simplified requirements, and a construction based on general assumptions. In: EUROCRYPT 2003, vol. 2656, pp. 614–629. LNCS (2003)

3. Bellare, M., Shi, H., Zhang, C.: Foundations of group signatures: the case of dynamic groups. In: CT-RSA 2005, vol. 3376, pp. 136–153. LNCS (2005)
4. Blum, M., De Santis, A., Micali, S., Persiano, G.: Noninteractive zero-knowledge. SIAM J. Comput. **20**(6), 1084–1118 (1991)
5. Boneh, D., Boyen, X., Shacham, H.: Short group signatures. In: CRYPTO 2004, vol. 3152, pp. 41–55. LNCS (2004)
6. Boneh, D., Shacham, H.: Group signatures with verifier-local revocation. In: ACM-CCS 2004, pp. 168–177. ACM (2004)
7. Bootle, J., Cerulli, A., Chaidos, P., Ghadafi, E., Groth, J.: Foundations of fully dynamic group signatures. In: ACNS 2016, pp. 117–136. LNCS (2016)
8. Bresson, E., Stern, J.: Efficient revocation in group signatures. In: PKC 2001, vol. 1992, pp. 190–206. LNCS (2001)
9. Brickell, E.: An efficient protocol for anonymously providing assurance of the container of the private key. Submitted to the Trusted Computing Group (April 2003)
10. Camenisch, J., Groth, J.: Group signatures: better efficiency and new theoretical aspects. In: SCN 2004, vol. 3352, pp. 120–133. LNCS (2004)
11. Camenisch, J., Lysyanskaya, A.: Dynamic accumulators and application to efficient revocation of anonymous credentials. In: CRYPTO 2002, vol. 2442, pp. 61–76. LNCS (2002)
12. Chaum, D., van Heyst, E.: Group signatures. In: EUROCRYPT 1991, vol. 547, pp. 257–265. LNCS (1991)
13. Dolev, D., Dwork, C., Naor, M.: Nonmalleable cryptography. SIAM Rev. **45**(4), 727–784 (2003)
14. Kiayias, A., Yung, M.: Secure scalable group signature with dynamic joins and separable authorities. Int. J. Secur. Netw. **1**(1–2), 24–45 (2006)
15. Libert, B., Ling, S., Mouhartem, F., Nguyen, K., Wang, H.: Signature schemes with efficient protocols and dynamic group signatures from lattice assumptions. In: ASIACRYPT 2016, vol. 10032, pp. 373–403. LNCS (2016)
16. Ling, S., Nguyen, K., Wang, H.: Group signatures from lattices: simpler, tighter, shorter, ring-based. In: PKC 2015, vol. 9020, pp. 427–449. LNCS (2015)
17. Sahai, A.: Non-malleable non-interactive zero knowledge and adaptive chosen-ciphertext security. In: FOCS 1999. pp. 543–553. IEEE (1999)
18. Song, D.X.: Practical forward secure group signature schemes. In: ACM-CCS 2004, pp. 225–234. ACM (2001)

Chapter 32
Fourier-Based Function Secret Sharing with General Access Structure

Takeshi Koshiba

Abstract Function secret sharing (FSS) scheme is a mechanism that calculates a function $f(x)$ for $x \in \{0, 1\}^n$ which is shared among p parties, by using distributed functions $f_i : \{0, 1\}^n \to \mathbb{G}$ ($1 \le i \le p$), where $\mathbb{G}$ is an Abelian group, while the function $f : \{0, 1\}^n \to \mathbb{G}$ is kept secret to the parties. Ohsawa et al. in 2017 observed that any function f can be described as a linear combination of the basis functions by regarding the function space as a vector space of dimension 2^n and gave new FSS schemes based on the Fourier basis. All existing FSS schemes are of (p, p)-threshold type. That is, to compute $f(x)$, we have to collect $f_i(x)$ for all the distributed functions. In this paper, as in the secret sharing schemes, we consider FSS schemes with any general access structure. To do this, we observe that Fourier-based FSS schemes by Ohsawa et al. are compatible with linear secret sharing scheme. By incorporating the techniques of linear secret sharing with any general access structure into the Fourier-based FSS schemes, we propose Fourier-based FSS schemes with any general access structure.

Keywords Function secret sharing · Distributed computation · Fourier basis
Linear secret sharing · Access structure · Monotone span program

1 Introduction

Secret sharing (SS) schemes are fundamental cryptographic primitives, which were independently invented by Blakley [4] and Shamir [21]. SS schemes involve several ordinary parties (say, p parties) and the special party called a dealer. We suppose that the dealer has a secret information s and partitions the secret information s into share information S_i ($0 \le i \le p$) which will be distributed to the ith party. In (n, p)-threshold SS scheme, the secret information S can be recovered from n shares (collected if any n parties get together), but no information on s is obtained from

T. Koshiba (✉)
Faculty of Education and Integrated Arts and Sciences,
Waseda University, 1-6-1 Nishiwaseda, Shinjuku-ku, Tokyo 169-8050, Japan
e-mail: tkoshiba@waseda.jp

© Springer Nature Singapore Pte Ltd. 2018
D. Ghosh et al. (eds.), *Mathematics and Computing*, Springer Proceedings
in Mathematics & Statistics 253, https://doi.org/10.1007/978-981-13-2095-8_32

417

at most $n - 1$ shares. This threshold property can be discussed in terms of access structures. An access structure $(\mathcal{A}, \mathcal{B})$ consists of two classes of sets of parties such that (1) if all parties in some set $A \in \mathcal{A}$ get together, then the secret information can be recovered from their shares; (2) even if all parties in any set $B \in \mathcal{B}$ get together, then any information of the secret s cannot be obtained. For example, the access structure $(\mathcal{A}, \mathcal{B})$ of the (n, p)-threshold SS scheme can be defined as $\mathcal{A} = \{A \subseteq \{1, \ldots, p\} : |A| \geq n\}$ and $\mathcal{B} = \{B \subseteq \{1, \ldots, p\} : |B| < n\}$. Besides the access structure of the threshold type, many variants have been investigated in the literature [3, 6, 7, 13, 15, 17]. As a standard technique for constructing access structures, *monotone span programs* [10, 11, 14, 18] are often used.

The idea where a secret information is secretly distributed to several parties can be applied to a function. The idea of secretly distributing a function has an application in private information retrieval (PIR) [8, 9, 16] as demonstrated in [12]. Gilboa et al. [12] consider to distribute point functions (DPFs) $f_{a,b} : \{0, 1\}^n \to \mathbb{G}$, where $f_{a,b}(x) = b$ if $x = a$ for some $a \in \{0, 1\}^n$ and $f_{a,b}(x) = 0$ otherwise. In a basic DPF scheme, the function f is partitioned into two keys f_0, f_1 and each key is distributed to the respective party of the two parties. Each party calculates the share $y_i = f_i(x)$ for common input x by using the key f_i. On the other hand, each f_i does not give any important information (e.g., the value a for $f_{a,b}$) on the original function. The functional value of the point function $f_{a,b}$ can be obtained by just summing up two shares y_0 and y_1 of the two parties. Boyle et al. [5] investigate the efficiency in the key size and extend the two-party setting into the multi-party setting. Moreover, they generalize the target functions (i.e., point functions) to other functions and propose an FSS scheme for some function family $\mathcal{F}$ in which functions $f : \{0, 1\}^n \to \mathbb{G}$ can be calculated efficiently. In the multi-key FSS scheme, we partition a function $f \in \mathcal{F}$ into p distributed functions $(f_1, \ldots, f_p)$. Likewise, an equation $f(x) = \sum_{i=1}^{p} f_i(x)$ is satisfied with respect to any x, and the information about the secret function f (except the domain and the range) does not leak out from at most $p - 1$ distributed functions. Moreover, distributed functions f_i can be described as short keys k_i, and it is required to be efficiently evaluated.

In [20], Ohsawa et al. observed that any function f from $\{0, 1\}^n$ to $\{0, 1\}$ can be described as a linear combination of the basis functions by regarding the function space as a vector space of dimension 2^n. While the point functions $f_{a,1}$ (for all $a \in \{0, 1\}^n$) constitute a (standard) basis for the vector space, any function $f : \{0, 1\}^n \to \{\pm 1\}$ can be represented as a linear combination of the Fourier basis functions $\chi_a(x) = (-1)^{\langle a, x \rangle}$, where $\langle a, x \rangle$ denotes the inner product between vectors $a = (a_1, \ldots, a_n)$ and $x = (x_1, \ldots, x_n)$. Based on the above observation, Ohsawa et al. gave new FSS schemes based on the Fourier basis. If we limit our concern to polynomial-time computable FSS schemes, functions for which the existing schemes are available would be limited. Since polynomial-time computable functions represented by combinations of point functions are quite different from ones represented by the Fourier basis functions, point function-based FSS schemes and Fourier function-based FSS schemes are complementary.

We note that properties of some functions are often discussed in the technique of the Fourier analysis. Akavia et al. [1] introduced a novel framework for proving hard-core properties in terms of Fourier analysis. Any predicates can be represented as a linear combination of Fourier basis functions. Akavia et al. show that if the number of nonzero coefficients in the Fourier representation of hard-core predicates is polynomially bounded, then the coefficients are efficiently approximable. This fact leads to the hard-core properties. Besides hard-core predicates, it is well known that low-degree polynomials are Fourier-concentrated [19].

Contribution

Since the existing FSS schemes are of (p, p)-threshold type, it is natural to consider the possibility of FSS schemes with any threshold structure of (n, p)-type and even general access structures as in the case of SS schemes.

In this paper, we affirmatively answer this question. As mentioned, Fourier-based FSS schemes in [20] are quite simpler than the previous FSS schemes. This is because Fourier basis functions have some linear structure. Shamir's threshold SS scheme can be seen as an application of the Reed–Solomon code, which is a linear code. Both the distribution phase and the reconstruction phase can be described in a linear algebraic way. From this viewpoint, we construct an (n, p)-threshold Fourier-based FSS scheme. Moreover, SS schemes with general access structure can be discussed in terms of monotone span program (MSP). The underlying structure of SS schemes by using MSP is similar to the linear algebraic view of Shamir's (n, p)-threshold SS scheme, and we can similarly construct Fourier-based FSS schemes with general access structure.

Technically speaking, Ohsawa et al. [20] consider a function from $\{0, 1\}^n$ to $\mathbb{C}$. That is, they consider Fourier transform over n-dimensional vector space of $\mathbb{F}_2$. On the other hand, we consider a function from a finite field $\mathbb{F}_q$ (of prime order q) to $\mathbb{C}$. So, in this paper, we consider the Fourier transform over $\mathbb{F}_q$ rather than $(\mathbb{F}_2)^n$. The shift of the underlying mathematical structure enables to construct FSS schemes with general access structure.

2 Preliminaries

2.1 Access Structure and Monotone Span Program

Let us assume that there are p parties in an SS (or, FSS) scheme. A *qualified* group is a set of parties who are allowed to reconstruct the secret, and a *forbidden* group is a set of parties who should not be able to get any information about the secret. The set of qualified groups is denoted by $\mathcal{A}$ and the set of forbidden groups by $\mathcal{B}$. The set $\mathcal{A}$ is said to be *monotonically increasing* if, for any set $A \in \mathcal{A}$, any set A' such that $A' \supseteq A$ is also included in $\mathcal{A}$. The set $\mathcal{B}$ is said to be *monotonically decreasing* if, for any set $B \in \mathcal{B}$, any set B' such that $B' \subseteq B$ is also included in $\mathcal{B}$. If a pair $(\mathcal{A}, \mathcal{B})$ satisfies that $\mathcal{A} \cap \mathcal{B} = \varnothing$, $\mathcal{A}$ is monotonically increasing and $\mathcal{B}$ is monotonically decreasing, then the pair is called a (monotone) *access structure*. If an access structure $(\mathcal{A}, \mathcal{B})$ satisfies that $\mathcal{A} \cup \mathcal{B}$ coincides with the power set of $\{1, \ldots, p\}$, we say that the access structure is *complete*. If we consider a complete access structure, we may simply denote the access structure by $\mathcal{A}$ instead of $(\mathcal{A}, \mathcal{B})$, since $\mathcal{B}$ is equal to the complement set of $\mathcal{A}$.

As mentioned, there are several ways to realize general access structures. Monotone span program (MSP) is a typical way to construct general access structures. Before mentioning the MSP, we prepare some basics and notations for linear algebra.

An $m \times d$ matrix M over a field $\mathbb{F}$ defines a linear map from $\mathbb{F}^d$ to $\mathbb{F}^m$. The *kernel* of M, denoted by $\ker(M)$, is the set of vectors $\boldsymbol{u} \in \mathbb{F}^d$ such that $M\boldsymbol{u} = \boldsymbol{0}$. The *image* of M, denoted by $\mathrm{im}(M)$, is the set of vectors $\boldsymbol{v} \in \mathbb{F}^m$ such that $\boldsymbol{v} = M\boldsymbol{u}$ for some $\boldsymbol{u} \in \mathbb{F}^d$.

A monotone span program (MSP) $\mathcal{M}$ is a triple $(\mathbb{F}, M, \rho)$, where $\mathbb{F}$ is a finite field, M is an $m \times d$ matrix over $\mathbb{F}$, and $\rho : \{1, \ldots, m\} \to \{1, \ldots, p\}$ is a surjective function which labels each row of M by a party. For any set $A \subseteq \{1, \ldots, p\}$, let M_A denote the submatrix obtained by restricting M to the rows labeled by parties in A. We say that $\mathcal{M}$ accepts A if $\boldsymbol{e}_1 = (1, 0, \ldots, 0)^T \in \mathrm{im}(M_A^T)$; otherwise, we say $\mathcal{M}$ rejects A. Moreover, we say that $\mathcal{M}$ accepts a (complete) access structure $\mathcal{A}$ if the following is equivalent: $\mathcal{M}$ accepts A if and only if $A \in \mathcal{A}$.

When $\mathcal{M}$ accepts a set A, there exists a *recombination* vector $\boldsymbol{\lambda}$ such that $M_A^T \boldsymbol{\lambda} = \boldsymbol{e}_1$. Also, note that $\boldsymbol{e}_1 \notin \mathrm{im}(M_B^T)$ if and only if there exists a vector $\boldsymbol{\xi}$ such that $M_B \boldsymbol{\xi} = \boldsymbol{0}$ and the first element of $\boldsymbol{\xi}$ is 1.

2.2 Function Secret Sharing

The original definition in [5] of FSS schemes are tailored for threshold schemes. We adapt the definition for general access structures. In an FSS scheme, we partition a function f into keys k_i (the succinct descriptions of f_i) which the corresponding parties P_i receive. Each party P_i calculates the share $y_i = f_i(x)$ for the common input x. The functional value $f(x)$ is recovered from shares $\boldsymbol{y}_A$ in a qualified set A of parties, which is a subvector of $\boldsymbol{y} = (y_1, y_2, \ldots, y_p)$, by using a decode function Dec. Any joint keys k_i in a forbidden set B of parties do not leak any information on function f except the domain and the range of f. We first define the decoding process from shares.

Definition 1 (*Output Decoder*) An output decoder Dec, on input a set T of parties and shares from the parties in T, outputs a value in the range R of the target function f.

Next, we define FSS schemes. We assume that $\mathcal{A}$ is a complete access structure among p parties and $T \subseteq \{1, 2, \ldots, p\}$ is a set of parties.

Definition 2 For any $p \in \mathbb{N}$, $T \subseteq \{1, 2, \ldots, p\}$, an $\mathcal{A}$-secure FSS scheme with respect to a function class $\mathcal{F}$ is a pair of PPT algorithms $(Gen, Eval)$ satisfying the following.

- The key generation algorithm $Gen(1^\lambda, f)$, on input the security parameter 1^λ and a function $f : D \to R$ in $\mathcal{F}$, outputs p keys $(k_1, \ldots, k_p)$.

– The evaluation algorithm $Eval(i, k_i, x)$, on input a party index i, a key k_i, and an element $x \in D$, outputs a value y_i, corresponding to the ith party's share of $f(x)$.

Moreover, these algorithms must satisfy the following properties:

– *Correctness*: For all $A \in \mathcal{A}, f \in \mathcal{F}$ and $x \in D$,

$$\Pr[Dec(A, \{Eval(i, k_i, x)\}_{i \in A}) = f(x) \mid (k_1, \ldots, k_p) \leftarrow Gen(1^\lambda, f)] = 1.$$

– *Security*: Consider the following indistinguishability challenge experiment for a forbidden set B of parties, where $B \notin \mathcal{A}$:

1. The adversary $\mathcal{D}$ outputs $(f_0, f_1) \leftarrow \mathcal{D}(1^\lambda)$, where $f_0, f_1 \in \mathcal{F}$.
2. The challenger chooses $b \leftarrow \{0, 1\}$ and $(k_1, \ldots, k_p) \leftarrow Gen(1^\lambda, f_b)$.
3. $\mathcal{D}$ outputs a guess $b' \leftarrow \mathcal{D}(\{k_i\}_{i \in B})$, given the keys for the parties in the forbidden set B.

The advantage of the adversary $\mathcal{D}$ is defined as $Adv(1^\lambda, \mathcal{D}) := \Pr[b = b'] - 1/2$. The scheme $(Gen, Eval)$ satisfies that there exists a negligible function ν such that for all non-uniform PPT adversaries $\mathcal{D}$ which corrupts parties in any forbidden set B, it holds that $Adv(1^\lambda, \mathcal{D}) \leq \nu(\lambda)$.

2.3 Basis Functions

The function space of functions $f : \mathbb{F}_q \to \mathbb{C}$ can be regarded as a vector space of dimension q. Therefore, the basis vectors for the function space exist, and we let $h_i(x)$ be each basis function. Any function f in the function space is described as a linear combination of the basis functions

$$f(x) = \sum_{j \in \mathbb{F}_q} \beta_j h_j(x),$$

where β_js are coefficients in $\mathbb{C}$.

The Fourier basis
Let $f : \mathbb{F}_q \to \mathbb{C}$, where q is an odd prime number. The Fourier transform of the function f is defined as

$$\hat{f}(a) = \frac{1}{q} \sum_{x \in \mathbb{F}_q} f(x) e^{-2\pi(ax/q)i}, \tag{1}$$

where i is the imaginary number. Then, $f(x)$ can be described as a linear combination of the basis functions $\chi_a(x) = e^{2\pi(ax/q)i}$, that is,

$$f(x) = \sum_{a \in \mathbb{F}_q} \hat{f}(a) \chi_a(x).$$

In the above, $\hat{f}(a)$ is called Fourier coefficient of $\chi_a(x)$. By using $\omega_q = e^{(2\pi/q)i}$, the primitive root of unity of order q, we can denote each Fourier basis function by

$$\chi_a(x) = (\omega_q)^{ax}$$

and let $\mathcal{B}_F = \{\chi_a \mid a \in \mathbb{F}_q\}$ be the sets of all the Fourier basis functions.

It is easy to see that the Fourier basis is orthonormal since

$$\frac{1}{q} \sum_{x \in \mathbb{F}_q} \chi_a(x)\chi_b(x) = \begin{cases} 1 & \text{if } a = b, \\ 0 & \text{otherwise.} \end{cases} \tag{2}$$

In this paper, we consider only Boolean-valued functions and assume that the range of the boolean function is $\{\pm 1\}$ instead of $\{0, 1\}$ without loss of generality. That is, we regard boolean functions as mappings from $\mathbb{F}_q$ to $\{\pm 1\}$. Also, we have

$$\chi_{a+b}(x) = \chi_a(x)\chi_b(x).$$

This multiplicative property plays an important role in this paper.

3 Linear Secret Sharing

3.1 Shamir's Threshold Secret Sharing

First, we give a traditional description of Shamir's (n, p)-threshold SS scheme [21], where $p \geq n \geq 2$. Let s be a secret integer which a dealer D has. First, the dealer D chooses a prime number $q > s$ and a polynomial $g(X) \in \mathbb{F}_q[X]$ of degree $n - 1$. Then, the dealer D computes $s_i = (i, g(i))$ as a share for the ith party P_i and sends s_i to each P_i. For the reconstruction, n parties get together and recover the secret s by the Lagrange interpolation from their shares.

The above procedure can be equivalently described as follows. Let M be an $n \times p$ Vandermonde matrix and m_i be the ith row in M. That is, $m_i = (1, i, i^2, \ldots, i^{n-1})$. Let $b = (b_0, b_1, \ldots, b_{n-1})^T$ be an n-dimensional vector such that $b_0 = s$ and $b_1, \ldots, b_{n-1}$ are randomly chosen elements in $\mathbb{F}_q$. Let $y = (s_1, s_2, \ldots, s_p)^T = Mb$. The share s_i for P_i is the ith element of y, that is, $s_i = \langle m_i^T, b \rangle$, where $\langle \cdot, \cdot \rangle$ denotes the inner product. Let A be a subset of $\{1, 2, \ldots, p\}$ which corresponds to a set of parties. Let M_A be a submatrix of M obtained by collecting rows m_j for all $j \in A$. We similarly define a subvector y_A by collecting elements s_j for all $j \in A$. Let $e_1 = (1, 0, 0, \ldots, 0)^T \in (\mathbb{F}_q)^n$. Then, we can uniquely determine λ such that $M_A^T \lambda = e_1$ by solving an equation system if and only if $|A| \geq n$. Then, we have

$$s = \langle b, e_1 \rangle = \langle b, M_A^T \lambda \rangle = \langle M_A b, \lambda \rangle = \langle y_A, \lambda \rangle.$$

Since y_A corresponds to all shares for P_j ($j \in A$), we can reconstruct the secret s by computing the inner product $\langle y_A, \lambda \rangle$.

3.2 Monotone Span Program and Secret Sharing

Here, we give a construction of linear secret sharing (LSS) based on monotone span program (MSP). Here, we do not mention how to construct MSP. For the construction of MSP, see the literature, e.g., [6, 10, 11, 14]. In this paper, we will use the LSS schemes. Since the LSS schemes imply MSPs [2, 22], it is sufficient to consider MSP-based SS schemes.

Let $s \in \mathbb{F}_q$ be a secret which the dealer D has and $\mathcal{M} = (\mathbb{F}_q, M, \rho)$ be an M-SP which corresponds to a complete access structure $\mathcal{A}$. The dealer D considers to partition s into several shares. In the sharing phase, the dealer D chooses a random vector $r \in (\mathbb{F}_q)^{p-1}$ and sends a share $\langle m_i^T, (s, r)^T \rangle$ to the ith party. In the reconstruction phase, using the recombination vector λ, any qualified set $A \in \mathcal{A}$ of parties can reconstruct the secret as follows:

$$\langle \lambda, M_A(s, r)^T \rangle = \langle M_A^T \lambda, (s, r)^T \rangle = \langle e_1, (s, r)^T \rangle = s.$$

Regarding the privacy, let B be a forbidden set of parties, and consider the joint information held by the parties in B. That is, $M_B b = y_B$, where $b = (s, r)^T$. Let $s' \in \mathbb{F}_q$ be an arbitrary value, and let ξ be a vector such that $M_B \xi = 0$ and the first element in ξ is equal to 1. Then, $y_B = M_B(b + \xi(s' - s))$, where the first coordinate of the vector $b + \xi(s' - s)$ is now equal to s'. This means that, from the viewpoint of the parties in B, their shares y_B are equally likely consistent with any secret $s' \in \mathbb{F}_q$.

4 Our Proposal

As mentioned, any function can be described as a linear combination of basis functions. If the function is described as a linear combination of a super-polynomial number of basis functions, then the computational cost for evaluating the function might be inefficient. We say that a function has a *succinct* description (with respect to the basis $\mathcal{B}$) if the function f is described as $f(x) = \sum_{h \in \mathcal{B}'} \beta_h h(x)$ for some $\mathcal{B}' \subset \mathcal{B}$ such that $|\mathcal{B}'|$ is polynomially bounded in the security parameter. If we can find a good basis set $\mathcal{B}$, some functions may have a succinct description with respect to $\mathcal{B}$. We consider to take the Fourier basis as such a good basis candidate.

We will provide an FSS scheme for some function class whose elements are functions with succinct description with respect to the Fourier basis $\mathcal{B}_F$. Since the Fourier basis has nice properties, our FSS scheme with general access structure can be realized.

In what follow, we assume that the underlying basis is always the Fourier basis $\mathcal{B}_F$. Moreover, we assume that $\mathcal{M} = (\mathbb{F}_q, M, \rho)$ is an MSP which corresponds to a general complete access structure $\mathcal{A}$. We will consider Fourier-based FSS schemes with this access structure.

4.1 FSS Scheme for the Fourier Basis

In this subsection, we consider to partition each Fourier basis function $\chi_a(x) = (\omega_q)^{ax}$ into several keys. That is, we give an FSS scheme with general access structure with respect to the function class $\mathcal{B}_F$.

Our FSS scheme with respect to $\mathcal{B}_F$ consists of three algorithms Gen_1^F (Algorithm 1), $Eval_1^F$ (Algorithm 2), and Dec_1^F (Algorithm 3). Gen_1^F is an algorithm that divides the secret a (for $\chi_a(x)$) into p keys $(k_1, \ldots, k_p)$ as in the SS scheme with the same access structure. Each key k_i is distributed to the ith party P_i. Note that the secret a can be recovered from the keys k_i for all i in a qualified set $A \in \mathcal{A}$.

In $Eval_1^F$, each party obtains the share by feeding x to the function distributed as the key. Dec_1^F is invoked in order to obtain the Fourier basis function $\chi_a(x)$ from the shares.

The correctness follows from

$$
\begin{aligned}
\chi_a(x) &= (\omega_q)^{ax} \\
&= (\omega_q)^{\langle y_A, \lambda \rangle x} \\
&= (\omega_q)^{(\sum k_i \lambda_i) x} \\
&= \prod \left((\omega_q)^{k_i x} \right)^{\lambda_i}.
\end{aligned}
$$

For the security, we assume that an adversary $\mathcal{D}$ chooses (f_0, f_1) where $f_0 = \chi_a$ and $f_1 = \chi_b$. Then, the challenger chooses a random bit c to select f_c and invokes $Gen_1^F(1^\lambda, a)$ if $c = 0$ and $Gen_1^F(1^\lambda, b)$ if $c = 1$. If $c = 0$, then a is divided into p keys. If $c = 1$, then b is divided into different p keys. From the argument in Sect. 3.2, the guess for the secret information a (resp., b) is a perfectly random guess. That is, the inputs to the adversary $\mathcal{D}$ are the same in the two cases. Thus, the adversary $\mathcal{D}$ cannot decide if the target function is either $\chi_a(x)$ or $\chi_b(x)$. It implies that only $\mathcal{D}$ can do for guessing the random bit c selected by the challenger is just a random guess. So, $Adv(1^\lambda, \mathcal{D}) = 0$. This concludes the security proof.

4.2 General FSS Scheme for Succinct Functions

Since we do not know how to evaluate any function efficiently, we limit ourselves to succinct functions with respect to the Fourier basis $\mathcal{B}_F$. Note that succinct functions with respect to $\mathcal{B}_F$ do not coincide with succinct functions with respect to point functions. Simple periodic functions are typical examples of succinct functions with respect to $\mathcal{B}_F$, which might not be succinct functions with respect to point functions. As mentioned, some hard-core predicates of one-way functions are succinct functions with respect to $\mathcal{B}_F$.

Let $\mathcal{F}_{\mathcal{B}_F, \ell}$ be a class of functions f which can be represented as a linear combination of ℓ basis functions (with respect to $\mathcal{B}_F$) at most, where ℓ is a polynomial in the security parameter. That is, f has the following form:

$$f(x) = \sum_{i=1}^{\ell} \beta_i \chi_{a_i}(x).$$

We construct an FSS scheme with general access structure $(Gen_{\leq \ell}^F, Eval_{\leq \ell}^F, Dec_{\leq \ell}^F)$ for a function $f \in \mathcal{F}_{\mathcal{B}_F, \ell}$ as follows. Note that the construction is a simple adaptation of the Fourier-based FSS scheme over $(\mathbb{F}_2)^n$ in [20].

Algorithm 1 $Gen_1^F(1^\lambda, a)$

Choose a random vector $r \in (\mathbb{F}_q)^{p-1}$ uniformly ;
for $i = 1$ to p **do**
  $m_i \leftarrow$ the i-th row of M ;
  $k_i \leftarrow \langle m_i, (a, r)^T \rangle$
end for
Return $(k_1, \ldots, k_p)$.

Algorithm 2 $Eval_1^F(i, k_i, x)$

$v_i \leftarrow (\omega_q)^{k_i x}$;
Return (i, v_i).

Algorithm 3 $Dec_1^F(A, \{(i, v_i)\}_{i \in A})$

Compute a recombination vector $\lambda = (\lambda_1, \ldots, \lambda_p)^T$ from A ;
Return $w = \prod_{i \in A} (v_i)^{\lambda_i}$.

- $Gen_{\leq \ell}^F(1^\lambda, f)$: On input the security parameter 1^λ and a function f, the key generation algorithm (Algorithm 4) outputs p keys $(k_1, \ldots, k_p)$.
- $Eval_{\leq \ell}^F(i, k_i, x)$: On input a party index i, a key k_i, and an input string $x \in \mathbb{F}_q$, the evaluation algorithm (Algorithm 5) outputs a value y_i, corresponding to the ith party's share of $f(x)$.

– $Dec^F_{\leq \ell}(A, \{y_i\}_{i\in A})$: On input shares $\{y_i\}_{i\in A}$ of parties in a (possibly) qualified set A, the decryption algorithm (Algorithm 6) outputs a solution $f(x)$ for x.

In the above FSS scheme $(Gen^F_{\leq \ell}, Eval^F_{\leq \ell}, Dec^F_{\leq \ell})$ for succinct functions $f \in \mathcal{F}_{\mathcal{B},\ell}$, we invoke FSS scheme $(Gen^F_1, Eval^F_1, Dec^F_1)$ for basis functions $\mathcal{B}_F$, since f can be represented as a linear combination of at most ℓ basis functions. In this construction, we distribute each basis function $\chi_{a_i}(x)$ and each coefficient β_i as follows. We invoke $(Gen^F_1, Eval^F_1, Dec^F_1)$ to distribute each basis function $\chi_{a_i}(x)$ and use any SS scheme with the same access structure to distribute each coefficient β_i.

The correctness of $(Gen^F_{\leq \ell}, Eval^F_{\leq \ell}, Dec^F_{\leq \ell})$ just comes from the correctness of each FSS scheme $(Gen^F_1, Eval^F_1, Dec^F_1)$ for the basis function $\chi_{a_i}(x)$ and the correctness of each SS scheme for the coefficients. But some care must be done. From the assumption, $f \in \mathcal{F}_{\mathcal{B}_F,\ell}$ has ℓ terms at most. If we represent f as a linear combination of exactly ℓ terms, some coefficients for basis functions must be zero. Since the zero-function $\chi_0(x) = (\omega_q)^{0\cdot x} = 1$ which maps any element $x \in \mathbb{F}_q$ to 1 can be partitioned into several functions as the ordinary basis functions can be, we can apply $(Gen^F_{\leq \ell}, Eval^F_{\leq \ell}, Dec^F_{\leq \ell})$ as well.

Algorithm 4 $Gen^F_{\leq \ell}(1^\lambda, f(\cdot) = \sum_{i=1}^{\ell} \beta_i \chi_{a_i}(\cdot))$

---

for $i = 1$ to ℓ **do**
  $(k^i_1, k^i_2, \ldots, k^i_p) \leftarrow Gen^F_1(1^\lambda, a_i)$;
  $(s^i_1, s^i_2, \ldots, s^i_p) \leftarrow$ iThe sharing phase of some SS scheme, given β_i ;
end for
for $j = 1$ to p **do**
  Set $k_j \leftarrow (k^1_j, k^2_j, \ldots, k^\ell_j)$;
  Set $s_j \leftarrow (s^1_j, s^2_j, \ldots, s^\ell_j)$;
end for
Return $((k_1, s_1), \ldots, (k_p, s_p))$.

Algorithm 5 $Eval^F_{\leq \ell}(i, (k_i, s_i), x)$

---

for $j = 1$ to ℓ **do**
  $y^i_j \leftarrow Eval^F_1(i, k^i_j, x)$;
end for
Set $y_i = (y^i_1, y^i_2, \ldots, y^i_\ell)$;
Return (i, y_i, s_i).

Algorithm 6 $Dec^F_{\leq \ell}(A, \{(i, y_i, s_i)\}_{i\in A})$

---

for $i = 1$ to ℓ **do**
  $g_i \leftarrow Dec^F_1(A, \{(j, y^j_i)\}_{j\in A})$;
  $\beta_i \leftarrow$ The reconstruction phase of the SS scheme, on input $\{s^j_i\}_{j\in A}$;
end for
Return $g = \sum_{i=1}^{\ell} \beta_i g_i$.

The security of $(Gen^F_{\leq \ell}, Eval^F_{\leq \ell}, Dec^F_{\leq \ell})$ can be discussed as follows. Without of loss of generality, we assume that all parties in a forbidden set B (where $|B| = m$) get $((k_1, s_1), \ldots, (k_m, s_m))$. For any i with $1 \leq i \leq \ell$, the m-tuples of the ith elements of $k_1, \ldots, k_m$ are identical whatever the basis function for the ith term of the target function is, because the advantage of any adversary against $(Gen^F_1, Eval^F_1, Dec^F_1)$ is 0 as discussed in Sect. 4.1. Moreover, for any i with $1 \leq i \leq \ell$, the m-tuples of the ith elements of $s_1, \ldots, s_m$ are identical whatever the coefficient for the ith term of the target function is, because of the perfect security of the underlying SS scheme with the same access structure. Furthermore, the outputs of several executions of Gen^F_1 (even for the same target basis function) are independent because each Gen^F_1 uses a fresh randomness. Thus, the information that all the parties in B can get is always the same regardless of the target function $f \in \mathcal{F}_{B_F, \ell}$. This guarantees the security of $(Gen^F \leq \ell, Eval^F_{\leq \ell}, Dec^F_{\leq \ell})$.

Remark If we do not care about the leakage of the number of terms with nonzero coefficients for f, we can omit the partitioning of zero-functions, which increases the efficiency of the scheme.

5 Conclusion

By observing that Fourier-based FSS schemes by Ohsawa et al. [20] are compatible with linear SS schemes, we have provided Fourier-based FSS schemes with general access structure, which affirmatively answers the question raised in [20].

Acknowledgements TK is supported in part by JSPS Grant-in-Aids for Scientific Research (A) JP16H01705 and for Scientific Research (B) JP17H01695.

References

1. Akavia, A., Goldwasser, S., Safra S.: Proving hard-core predicates using list decoding. In: Proceeding of the 44th Symposium on Foundations of Computer Science (FOCS 2003), pp. 146–157 (2003)
2. Beimel, A., Chor, B.: Universally ideal secret sharing schemes. IEEE Trans. Inf. Theor. **40**(3), 786–794 (1994)
3. Benaloh, J., Leichter, J.: Generalized secret sharing and monotone functions. In: Proceeding of CRYPTO '88. Lecture Notes in Computer Science, vol. 403, pp. 27–35. Springer (1990)
4. Blakley, G.R.: Safeguarding cryptographic keys. In: American Federation of Information Processing Societies: National Computer Conference, pp. 313–317 (1979)
5. Boyle, E., Gilboa, N., Ishai, Y.: Function secret sharing. In: EUROCRYPT 2015. Part II, Lecture Notes in Computer Science, vol. 9057, pp. 337–367 (2015)
6. Brickell, E.F.: Some ideal secret sharing schemes. In Proceeding of EUROCRYPT '89. Lecture Notes in Computer Science, vol. 434, pp. 468–475. Springer (1990)

7. Brickell, E.F., Davenport, D.M.: On the classification of ideal secret sharing schemes. In: Proceeding of CRYPTO '89. Lecture Notes in Computer Science, vol. 435, pp. 278–285. Springer (1990)
8. Chor, B., Gilboa, N.: Computationally private information retrieval. In: Proceeding of the 29th Annual Symposium on Theory of Computing (STOC'97), pp. 304–313 (1997)
9. Chor, B., Goldreich, O., Kushilevitz, E., Sudan, M.: Private information retrieval. J. ACM **45**(6), 965–981 (1998)
10. Fehr, S.: Span programs over rings and how to share a secret from a module. Master's thesis, ETH Zurich, Institute for Theoretical Computer Science (1998)
11. Fehr, S.: Efficient construction of the dual span program. Manuscript (1999)
12. Gilboa N., Ishai, Y.: Distributed point functions and their applications. In: Proceeding of EUROCRYPT 2014. Lecture Notes in Computer Science, vol. 8441, pp. 640–658 (2014)
13. Ito, M., Saito, A., Nishizeki, T.: Secret sharing scheme realizing general access structure. In: Proceeding of IEEE GLOBECOM '87, pp. 99–102. IEEE Communications Society (1987)
14. Karchmer, M., Wigderson, A.: On span programs. In: Proceeding of the 8th Structures in Complexity Theory Conference, pp. 102–111. IEEE Computer Society (1993)
15. Kothari, S.C.: Generalized linear threshold scheme. In: Proceeding of CRYPTO '84. Lecture Notes in Computer Science, vol. 196, pp. 231–241. Springer (1985)
16. Kushilevitz, E., Ostrovsky, R.: Replication is not needed: single database, computationally-private information retrieval. In: Proceeding of the 38th IEEE Symposium on Foundations of Computer Science (FOCS'97), pp. 364–373 (1997)
17. Nikov, V., Nikova, S., Preneel, B.: On multiplicative linear secret sharing schemes. In: Proceeding of INDOCRYPT 2003. Lecture Notes in Computer Science, vol. 2904, pp. 135–147, Springer (2003)
18. Nikov, V., Nikova, S.: New Monotone Span Programs from Old. Cryptology ePrint Archive, Report 2004/282 (2004)
19. O'Donnell, R.: Analysis of Boolean Functions. Cambridge University Press, Cambridge (2014)
20. Ohsawa, T., Kurokawa, N., Koshiba, T.: Function secret sharing using Fourier basis. In Proceeding of the 8th International Workshop on Trustworthy Computing and Security (TwCSec-2017). Lecture Notes on Data Engineering and Communications Technologies, vol. 7, pp. 865–875. Springer (2018)
21. Shamir, A.: How to share a secret. Commun. ACM **22**(11), 612–613 (1979)
22. van Dijk, M.: A linear construction of perfect secret sharing schemes. In: Proceeding of EUROCRYPT '94. Lecture Notes in Computer Science, vol. 950, pp. 23–34. Springer (1995)

Chapter 33
A Uniformly Convergent NIPG Method for a Singularly Perturbed System of Reaction–Diffusion Boundary-Value Problems

Gautam Singh and Srinivasan Natesan

Abstract In this article, we study the numerical solution of singularly perturbed system of boundary-value problems for second-order ordinary differential equations of reaction–diffusion type. The solution of these problems exhibits twin boundary layers at both the ends of the domain. To obtain the numerical solution of these problems, we apply the nonsymmetric discontinuous Galerkin FEM with interior penalties (NIPG method). Also, we proved that the method is $O(N^{-1} \ln N)^k$ accurate in energy norm, on Shishkin mesh with N number of intervals and k degree of piecewise polynomial. Numerical results are presented to support the theoretical results.

Keywords Singularly perturbed system of reaction–diffusion boundary-value problems · Shishkin mesh · Discontinuous Galerkin finite element method Uniform convergence

Subject Classification: 65L11 · 65L20 · 65L60 · 65L70

1 Introduction

The numerical solution of singularly perturbed differential equations (SPDEs) attracts many researchers in the recent years; for more details, one can see the books by Farrell et al. [1], Miller et al. [5], and Roos et al. [7]. The solution of SPDEs has a multi-scale character; it varies rapidly inside the boundary layer and varies slowly away from the boundary layers; therefore, classical numerical schemes fail to yield satisfactory numerical approximate solution on uniform meshes. Special care has to be taken to devise parameter-uniform numerical methods to these problems. There are two types of methods to solve SPDEs; one is known as fitted operator methods (FOMs) and the other one is fitted mesh methods (FMMs); see the book [5] for more details.

G. Singh (✉) · S. Natesan
Department of Mathematics, Indian Institute of Technology, Guwahati 781039, India
e-mail: gautamsingh@iitg.ernet.in

© Springer Nature Singapore Pte Ltd. 2018
D. Ghosh et al. (eds.), *Mathematics and Computing*, Springer Proceedings
in Mathematics & Statistics 253, https://doi.org/10.1007/978-981-13-2095-8_33

Classical finite element methods (FEMs) will also fail to provide parameter-uniform numerical solutions to SPDEs on uniform meshes. Either one has to use exponential basis functions as the trial functions [6], or one has to use layer-adapted nonuniform meshes for classical FEM [7]. There are several research articles which deal with the numerical solution of SPDEs by FEM; we cite a few of them [2, 10], the references ε therein.

Recently, researchers started to apply the nonsymmetric discontinuous Galerkin method with interior penalty (NIPG method) to solve SPDEs, originally designed for elliptic equations. Zarin and Roos [9] applied the NIPG method to solve singularly perturbed 2D convection–diffusion problems with parabolic layers. Zhu et. al. [12] have applied the NIPG method to solve singularly perturbed 1D convection–diffusion BVPs and showed that it converges ε—uniformly in the energy norm with optimal order. Linß and Madden [3] applied the FEM for singularly perturbed system of reaction–diffusion BVPs on S-type mesh and proved that the methods is uniformly convergent with first order.

In this article, we obtain the numerical solution of singularly perturbed system of BVPs of reaction–diffusion type, by applying the NIPG method on the layer-adapted piecewise uniform Shishkin mesh. Also, we established that the proposed method is ε—uniformly convergent of order k, where k is the degree of piecewise polynomials in finite element space. To support the theoretical findings, we carried out some numerical experiments; the results are presented in the form of tables.

This paper is organized in the following style: In Sect. 2, we describe the model problem with some basic definitions. We use the NIPG method for system of singular perturbation problems and prove its existence and uniqueness in Sect. 3. Parameter-uniform error estimate is derived in Sect. 4. Section 5 shows the numerical results obtained for a test problem.

In this paper, we use C to denote a generic positive constant that is independent of both the perturbation parameter ε and the mesh parameter N. We shall also assume that $\varepsilon \leq CN^{-1}$ as is generally the case.

2   The Model Problem and the Analytical Solution

Here, we consider the following singularly perturbed system of reaction–diffusion boundary-value problems (BVPs):

$$
\begin{cases}
-\varepsilon^2 u_1''(x) + a_{11}(x)u_1(x) + a_{12}(x)u_2(x) = f_1(x), x \in \Omega = (0,\ 1), \\
-\varepsilon^2 u_2''(x) + a_{21}(x)u_1(x) + a_{22}(x)u_2(x) = f_2(x), \\
u_1(0) = u_2(0) = u_1(1) = u_2(1) = 0,
\end{cases} \tag{1}
$$

where $0 < \varepsilon \ll 1$ is the perturbation parameter, and the coefficients a_{ij} and the source functions f_j are sufficiently smooth functions. We shall assume that reaction coefficient matrix $A = \{a_{ij}\}_{i,j=1}^2$ is an L_0—matrix with

$$\min\{a_{11} + a_{12}, a_{21} + a_{22}\} > \beta^2, \tag{2}$$

i.e., A is an M-matrix whose inverse is bounded by β^{-2} in the maximum norm. The solution $\boldsymbol{u} = (u_1, u_2)^T$ of (1) has layers at $x = 0$ and 1 of width $\mathcal{O}(\varepsilon \ln \varepsilon)$.

Lemma 1 *The solution $\boldsymbol{u} = (u_1, u_2)^T$ of (1) can be decomposed as $\boldsymbol{u} = S + E$, where S and E are smooth and layer parts, respectively. Then, the bound on the smooth and layer components are*

$$|S_i^{(l)}(x)| \leq C \tag{3}$$
$$|E_i^{(l)}(x)| \leq C\varepsilon^{-l}D_\varepsilon(x), \quad \text{for, } i = 1, 2 \text{ and } 0 \leq l \leq p \tag{4}$$

where $D_\varepsilon(x) = \exp((-\beta x)/\varepsilon) + \exp((-\beta(1 - x))/\varepsilon)$. Here, $p > 0$ depends on the smoothness of the data.

Proof The proof of this lemma can be found in [4].

The space of square integrable functions on an interval $K \subset R$ will be denoted by $L^2(K)$, with the associated inner product

$$(\boldsymbol{u}, \boldsymbol{v})_K = \int_K u_1(x)v_1(x)dx + \int_K u_2(x)v_2(x)dx.$$

We will also use the usual Sobolev space $H^k(K)$ to denote the space of functions on K whose generalized derivatives are in $L^2(K)$, for 0, 1, 2, ..., k, and it is equipped with norm and seminorm $\|.\|_{k,K}$ and $|.|_{k,K}$, respectively. For any vector-valued functions $\boldsymbol{u} = (u_1(x), u_2(x))^T$, we will write

$$\|\boldsymbol{u}\|_{k,K}^2 = \|u_1\|_{k,K}^2 + \|u_2\|_{k,K}^2.$$

Let $T_N = \{K_j = (x_{j-1}, x_j) : j = 1, \ldots, N\}$, be a partition of the domain Ω. To each element $K_j \in T_N$, denote the discrete Sobolev space of order s with

$$H^s(\Omega, T_N) = \{u \in L^2(\Omega) : u|_{K_j} \in H^s(K_j), \forall K_j \in T_N\}.$$

Discrete Sobolev norm and seminorm are given by

$$\|\boldsymbol{u}\|_{s,T_N}^2 = \sum_{j=1}^{N} \|u_1\|_{s,K_j}^2 + \sum_{j=1}^{N} \|u_2\|_{s,K_j}^2, \quad |\boldsymbol{u}|_{s,T_N}^2 = \sum_{j=1}^{N} |u_1|_{s,K_j}^2 + \sum_{j=1}^{N} |u_1|_{s,K_j}^2,$$

where $\|.\|_{0,K_j}$ and $|.|_{1,K_j}$ are the usual Sobolev norm and seminorm defined over the domain K_j, respectively.

To discretize the domain $\Omega = (0, 1)$, we use the layer-adapted piecewise uniform Shishkin mesh, which is described in the following. We divide the domain $\Omega = (0, 1)$ into three subdomains as $\Omega = \Omega_1 \cup \Omega_2 \cup \Omega_3$, where Ω_1, Ω_2 and Ω_3 are $[0, \tau_\varepsilon]$, $[\tau_\varepsilon, 1 - \tau_\varepsilon]$ and $[1 - \tau_\varepsilon, 1]$, respectively. Here the transition point τ_ε is defined by

$$\tau_\varepsilon = \min\left(\frac{1}{4}, \frac{\alpha\varepsilon}{\beta} \ln N\right).$$

In this article, we will take $\tau_\varepsilon = (\alpha\varepsilon/\beta) \ln N$.

The step size in each of the subdomain is given by

$$h_i = \begin{cases} 4\tau_\varepsilon/N, & \text{for } \Omega_1 \text{ and } \Omega_3, \\ 2(1 - 2\tau_\varepsilon)/N, & \text{for } \Omega_2. \end{cases}$$

Let us define the finite element space $V_N^k(\Omega)$ associated with the family T_N of Shishkin mesh by

$$V_N^k(\Omega) = \{u \in L^2(\Omega) : u|_{K_j} \in P^k(K_j), \forall K_j \in T_N\},$$

where $P^k(K_j)$ denotes the polynomials space of degree k on K_j. The function in $V_N^k(\Omega)$ are discontinuous at each mesh point.

3 The NIPG Method

The finite element problem corresponds to (1) by the NIPG method reads:

$$\begin{cases} \text{find } \boldsymbol{u_h} \in V_N^k(\Omega)^2, \text{ such that} \\ B(\boldsymbol{u_h}, \boldsymbol{v_h}) = L(\boldsymbol{v_h}), \forall \boldsymbol{v_h} \in V_N^k(\Omega)^2, \end{cases} \tag{5}$$

with $B(\boldsymbol{u}, \boldsymbol{v}) = B_1(\boldsymbol{u}, \boldsymbol{v}) + B_2(\boldsymbol{u}, \boldsymbol{v})$, where

$$B_1(\boldsymbol{u}, \boldsymbol{v}) = \varepsilon^2 \sum_{j=1}^{N} \int_{K_j} u_1'(x)v_1'(x)dx + \varepsilon^2 \sum_{j=0}^{N} (\{u_1'(x_j)\}[v_1(x_j)] - \{v_1'(x_j)\}[u_1(x_j)])$$

$$+ \sum_{j=0}^{N} \sigma_j [u_1(x_j)][v_1(x_j)] + \sum_{j=1}^{N} \int_{K_j} (a_{11}(x)u_1(x) + a_{12}(x)u_2(x))v_1(x)dx,$$

$$B_2(\boldsymbol{u}, \boldsymbol{v}) = \varepsilon^2 \sum_{j=1}^{N} \int_{K_j} u_2'(x)v_2'(x)dx + \varepsilon^2 \sum_{j=0}^{N} (\{u_2'(x_j)\}[v_2(x_j)] - \{v_2'(x_j)\}[u_2(x_j)])$$

$$+ \sum_{j=0}^{N} \sigma_j [u_2(x_j)][v_2(x_j)] + \sum_{j=1}^{N} \int_{K_j} (a_{21}(x)u_1(x) + a_{22}(x)u_2(x))v_2(x)dx,$$

and

$$L(v) = \sum_{j=1}^{N} \int_{K_j} (f_1 v_1 + f_2 v_2) dx,$$

here $\sigma_j \geq 0$ $(j = 0, 1, \ldots, N)$ are penalty parameter with the node x_j.

Lemma 2 *Let u be the exact solution of the problem (1), then the bilinear form $B(., .)$ defined in (5) satisfies the Galerkin orthogonality property*

$$B(u - u_h, v) = 0, \quad \forall v \in V_N^k(\Omega)^2.$$

Proof Since u is the exact solution of (1), we have $[u_1(x_j)] = [u_2(x_j)] = 0$, $0 \leq j \leq N$ and $[u_1'(x_j)] = [u_2'(x_j)] = 0$, $1 \leq j \leq N - 1$. Then, for all $v \in V_N^k(\Omega)^2$, we easily get

$$B_1(u, v) = \varepsilon^2 \sum_{j=1}^{N} \int_{K_j} u_1'(x) v_1'(x) dx + \varepsilon^2 \sum_{j=0}^{N} (\{u_1'(x_j)\}[v_1(x_j)])$$

$$+ \sum_{j=1}^{N} \int_{K_j} a_{11}(x) u_1(x) v_1(x) dx + \sum_{j=1}^{N} \int_{K_j} a_{12}(x) u_2(x) v_1(x) dx,$$

Similarly, we can write for $B_2(u, v)$. Using integration by parts and the definition of jump and average, one can show that

$$\varepsilon^2 \sum_{j=1}^{N} \int_{K_j} u_1'(x) v_1'(x) dx = -\varepsilon^2 \sum_{j=1}^{N} \int_{K_j} u_1''(x) v_1(x) dx - \varepsilon^2 \sum_{j=0}^{N} (\{u_1'(x_j)\}[v_1(x_j)]),$$

and

$$\varepsilon^2 \sum_{j=1}^{N} \int_{K_j} u_2'(x) v_2'(x) dx = -\varepsilon^2 \sum_{j=1}^{N} \int_{K_j} u_2''(x) v_2(x) dx - \varepsilon^2 \sum_{j=0}^{N} (\{u_2'(x_j)\}[v_2(x_j)]).$$

Using the above estimate and recalling our model problem, we obtain

$$B(u - u_h, v) = 0, \quad \forall v \in V_N^k(\Omega)^2.$$

∎

A natural norm associated with the bilinear form $B(.,.)$ is the energy norm

$$\||\boldsymbol{v}|\|^2 = \varepsilon^2 \sum_{j=1}^{N} (\|v_1'\|_{L^2(K_j)}^2 + \|v_2'\|_{L^2(K_j)}^2) + \beta^2 \sum_{j=1}^{N} (\|v_1\|_{L^2(K_j)}^2 + \|v_2\|_{L^2(K_j)}^2)$$

$$+ \sum_{j=0}^{N} \sigma_j [v_1(x_j)]^2 + \sum_{j=0}^{N} \sigma_j [v_2(x_j)]^2, \tag{6}$$

where σ_j are the penalty parameters and β is such that it satisfies $\min\{a_{11} + a_{12}, a_{21} + a_{22}\} > \beta^2$. It is easy to show that the bilinear form given in (5) satisfies the coercivity condition, and hence, it can be shown that

$$\||\boldsymbol{u_h}|\| \leq \|f\|_{L^2(\Omega)},$$

using the above expression one can show the uniqueness of the solution of (5), and also using rank nullity theorem, we can show the existence of the solution.

4 Error Analysis

In this section, we perform the error analysis for the NIPG method (5). We will show that the NIPG method possesses optimal order of convergence. From [8], we introduce a special interpolant on each element K_j; for any $w \in C(K_j)$, we define $k + 1$ nodal functional N_l by

$$N_0(w) = w(x_{j-1}), \quad N_k(w) = w(x_j),$$
$$N_l(w) = \frac{1}{(x_j - x_{j-1})^l} \int_{x_{j-1}}^{x_j} (x - x_{j-1})^{l-1} w(x) dx, \quad l = 1, \ldots, k - 1.$$

Now a local interpolation $w_I|_k \in P^K(K_j)$ is defined by $N_l(w_I - w) = 0, l = 0, \ldots, k.$

Lemma 3 *[8] Interpolation error has the following bounds:*

$$|u - u_I|_{l,K_j} \leq Ch_j^{k+1-l}|u|_{k+1,K_j}, l = 0, 1, \ldots, k + 1, \forall u \in H^{k+1}(K_j),$$

$$\|u - u_I\|_{L^\infty(K_j)} \leq Ch_j^{k+1}|u|_{k+1,\infty,K_j}, \forall u \in W^{k+1,\infty}(K_j),$$

here K_j are elements of T_N and its length are h_j.

Lemma 4 *Let S_I and E_I are the interpolants of S and E, respectively. Then, we can write $u_I = S_I + E_I$ and the estimates*

$$\|u_m - u_{m,I}\|_{L^\infty(\Omega_i)} \le C(N^{-1} \ln N)^{k+1}, for \ i = 1, 3, \tag{7}$$

$$\|(S_m - S_{m,I})^l\|_{L^2(\Omega)} \le CN^{l-(k+1)}, \ for \ l = 0, 1, 2, \ldots, k, \tag{8}$$

$$\|E_m\|_{L^\infty(\Omega_2)} + \varepsilon^{-1/2}\|E_m\|_{L^2(\Omega_2)} \le CN^{-\alpha}, \tag{9}$$

$$\|E_m^l\|_{L^2(\Omega_2)} \le C\varepsilon^{1/2-l}N^{-\alpha}, \tag{10}$$

$$N^{-1}\|E_{m,I}'\|_{L^2(\Omega_2)} + \|E_{m,I}\|_{L^2(\Omega_2)} \le C(\varepsilon^{1/2}N^{-\alpha} + N^{-(1/2+\alpha)}), \tag{11}$$

$$\|u_m - u_{m,I}\|_{L^\infty(\Omega_2)} \le C(N^{-(k+1)} + N^{-\alpha}), \tag{12}$$

for $m = 1, 2$ hold true.

Proof Using the linearity, we can write $u_I = (S + E)_I = S_I + E_I$. Using Lemma 3, we can have

$$\|u_m - u_{m,I}\|_{L^\infty(\Omega_i)} \le Ch_j^{(k+1)}|u_m^{(k+1)}|_{L^\infty(\Omega_i)}, \tag{13}$$

using the solution decomposition and the bounds on smooth and layer parts, *i.e.*, (3), we can obtain that

$$|u_m^{k+1}|_{L^\infty(\Omega_i)} \le |S_m^{k+1}|_{L^\infty(\Omega_i)} + |E_m^{k+1}|_{L^\infty(\Omega_i)} \le C\varepsilon^{-(k+1)}D_\varepsilon(x) \tag{14}$$

We know that the interval length on the fine part of the mesh is given by $h_i = (2\alpha\varepsilon/\beta)N^{-1} \ln N$. From Eqs. (13) and (14), we obtain

$$\|u_m - u_{m,I}\|_{L^\infty(\Omega_i)} \le C(N^{-1} \ln N)^{k+1} for \ i = 1, 3.$$

Next, by using Lemma 3 for the smooth part of the solution and the bound on the smooth part, *i.e.*, (3), we get

$$\|(S_m - S_{m,I})^l\|_{L^2(\Omega)} \le CN^{l-(k+1)}|S_m|_{k+1,K_j} \le CN^{l-(k+1)}.$$

To prove (9), we need to show that

$$\|E_m\|_{L^\infty(\Omega_2)} \le C \max_{[\tau_\varepsilon, 1-\tau_\varepsilon]} (\exp(-\beta x/\varepsilon) + \exp(-\beta(1-x)/\varepsilon))$$

$$\le CN^{-\alpha}, \text{(using } \tau_\varepsilon = (\alpha\varepsilon/\beta) \ln N),$$

by using the definition of L^2 norm and value of τ_ε, we obtain

$$\|E_m\|_{L^2(\Omega_2)}^2 \le C \int_{\tau_\varepsilon}^{1-\tau_\varepsilon} (D_\varepsilon(x))^2 dx$$

$$\le C \int_{\tau_\varepsilon}^{1-\tau_\varepsilon} (\exp(-2\beta x/\varepsilon) + \exp(-2\beta(1-x)/\varepsilon))dx$$

$$\le C\varepsilon N^{-2\alpha}.$$

By adding above two inequality, we get

$$\|E_m\|_{L^\infty(\Omega_2)} + \varepsilon^{-1/2}\|E_m\|_{L^2(\Omega_2)} \le CN^{-\alpha}.$$

In a similar way, we can prove (10). The proofs of the estimates (11)–(12) can be established following the idea used in (Lemma 12 of [8]). ∎

To obtain the error estimate of the NIPG method, we decompose the error in two parts, as

$$\|\|u - u_h\|\| \le \|\|u - u_I\|\| + \|\|u_I - u_h\|\|.$$

Let $\eta = u - u_I$ and $\xi = u_I - u_h$. For proving the interpolation error on the mesh point, we will use the following lemma.

Lemma 5 [11] *For any $w \in H^1(K_j)$, we have the following bound*

$$|w(x_s)|^2 \le 2(h_j^{-1}\|w\|_{L^2(K_j)}^2 + \|w\|_{L^2(K_j)}\|w'\|_{L^2(K_j)}), \quad s \in \{j-1, j\}.$$

Lemma 6 *Assuming $\alpha = k + 1$, we can show the following estimates for $\{\eta_m'\}$*

$$\{\eta_m'(x_j)\}^2 \le \begin{cases} C\varepsilon^{-2}(N^{-1}\ln N)^{2k}, & \text{for } x_j \in \Omega_1 \cup \Omega_3, \\ C\varepsilon^{-2}N^{-(2k+1)}, & \text{for } x_j \in \Omega_2, \end{cases}$$

where $\eta_m = u_m - u_{m,I}$.

Proof Using Lemma 5, we can write

$$\begin{aligned}
\{\eta_m'(x_j)\}^2 &= \frac{1}{4}(\eta_m'(x_j^-) + \eta_m'(x_j^+))^2 \le \frac{1}{2}(\eta_m'(x_j^-)^2 + \eta_m'(x_j^+)^2) \\
&\le h_j^{-1}\|\eta_m'\|_{L^2(K_j)}^2 + \|\eta_m'\|_{L^2(K_j)}\|\eta_m''\|_{L^2(K_j)} + h_{j+1}^{-1}\|\eta_m'\|_{L^2(K_{j+1})}^2 \\
&\quad + \|\eta_m'\|_{L^2(K_{j+1})}\|\eta_m''\|_{L^2(K_{j+1})}.
\end{aligned}$$

Now our job is to estimate $\|\eta_m'\|_{L^2(K_j)}$ and $\|\eta_m''\|_{L^2(K_j)}$ separately. Using (8), we obtain

$$\|(S_m - S_{m,I})'\|_{L^2(K_j)} \le CN^{-k},$$
$$\|(S_m - S_{m,I})''\|_{L^2(K_j)} \le CN^{-(k+1)}, \quad \text{for all } K_j, \ j = 1, \ldots, N.$$

In order to obtain the bounds for $\|\eta_m'\|_{L^2(K_j)}$ and $\|\eta_m''\|_{L^2(K_j)}$, it remains to estimate $\|(E_m - E_{m,I})'\|_{L^2(K_j)}$ and $\|(E_m - E_{m,I})''\|_{L^2(K_j)}$ inside and outside regions.

First, we will estimate $\|(E_m - E_{m,I})'\|_{L^2(K_j)}$ and $\|(E_m - E_{m,I})''\|_{L^2(K_j)}$ outside layer that is $K_j \in \Omega_2$. By using (10) and (11) and the fact that $\varepsilon \le N^{-1}$, $\alpha = k + 1$, we obtain

$$\|(E_m - E_{m,I})'\|_{L^2(K_j)} \le \|E_m'\|_{L^2(K_j)} + \|E_{m,I}'\|_{L^2(K_j)} \le C\varepsilon^{-1/2}N^{-(k+1)}.$$

Similarly, using the inverse inequality and the fact that $\varepsilon \leq N^{-1}$, $\alpha = k + 1$, we obtain

$$\|(E_m - E_{m,I})''\|_{L^2(K_j)} \leq \|E_m''\|_{L^2(K_j)} + \|E_{m,I}''\|_{L^2(K_j)}$$
$$\leq \|E_m''\|_{L^2(K_j)} + Ch_j^{-1}\|E_{m,I}'\|_{L^2(K_j)}$$
$$\leq C\varepsilon^{-3/2}N^{-(k+1)}.$$

Now, we will estimate $\|(E_m - E_{m,I})'\|_{L^2(K_j)}$ and $\|(E_m - E_{m,I})''\|_{L^2(K_j)}$ inside the boundary layer regions, that is, $K_j \in \Omega_1 \cup \Omega_3$. By using Lemmas 1 and 3, we have for $l = 1, 2$

$$\|(E_m - E_{m,I})^{(l)}\|_{L^2(K_j)} \leq Ch_j^{k+1-l}\|E_m\|_{L^2(K_j)}$$
$$\leq Ch_j^{k+1-l}\left(\int_{K_j} \varepsilon^{-2(k+1)}D_\varepsilon^2(x)\right)^{1/2}$$
$$\leq C\varepsilon^{1/2-l}(N^{-1}\ln N)^{k+3/2-l}.$$

From the triangle inequality and the above estimate, we can obtain

$$\|\eta_m'\|_{L^2(K_j)} \leq \begin{cases} C\varepsilon^{-1/2}(N^{-1}\ln N)^{k+1/2}, & \text{for } K_j \in \Omega_1 \cup \Omega_3, \\ C\varepsilon^{-1/2}N^{-k}(\varepsilon^{1/2} + N^{-1}), & \text{for } K_j \in \Omega_2, \end{cases}$$

and

$$\|\eta_m''\|_{L^2(K_j)} \leq \begin{cases} C\varepsilon^{-3/2}(N^{-1}\ln N)^{k+1/2}, & \text{for } K_j \in \Omega_1 \cup \Omega_3, \\ C\varepsilon^{-3/2}N^{1-k}(\varepsilon^{3/2} + N^{-2}), & \text{for } K_j \in \Omega_2, \end{cases}$$

and by using above estimate, we get our desired result.   ∎

Theorem 1 *By taking $\alpha = k + 1$, we have following interpolation error bound:*

$$\||\eta\|| \leq C(N^{-1}\ln N)^k,$$

where $\eta = u - u_I$.

Proof Because $u - u_I$ is continuous in Ω, hence $[\eta_1]_j = [\eta_2]_j = 0$, $j = 0, \ldots, N$. Then,

$$\||\eta\||^2 = \sum_{j=1}^{N} \varepsilon^2 \|\eta'\|_{L^2(K_j)}^2 + \beta^2 \sum_{j=1}^{N} \|\eta\|_{L^2(K_j)}^2.$$

By Lemma 4, we can easily conclude that

$$
\begin{aligned}
\|u - u_I\|_{L^2(\Omega)} &\le |\Omega_1|^{1/2}\|u - u_I\|_{L^\infty(\Omega_1)} + |\Omega_2|^{1/2}\|u - u_I\|_{L^\infty(\Omega_2)} \\
&\quad + |\Omega_3|^{1/2}\|u - u_I\|_{L^\infty(\Omega_3)} \\
&\le \tau_\varepsilon^{1/2}\|u - u_I\|_{L^\infty(\Omega_1)} + \|u - u_I\|_{L^\infty(\Omega_2)} + \tau_\varepsilon^{1/2}\|u - u_I\|_{L^\infty(\Omega_3)} \\
&\le C(N^{-1}\ln N)^k.
\end{aligned}
$$

Similarly, we can show that $\varepsilon|u - u_I|_{1,\Omega} \le C(N^{-1}\ln N)^k$. Hence, we have

$$
\||\boldsymbol{\eta}\|| \le C(N^{-1}\ln N)^k.
$$

∎

Theorem 2 *Let $\boldsymbol{\xi} = u_h - u_I$. Then, $\boldsymbol{\xi}$ satisfies the following error bound:*

$$
\||\boldsymbol{\xi}\|| \le C(N^{-1}\ln N)^k.
$$

Proof As we know the bilinear form given in (5) that for the first term,

$$
\sum_{j=1}^N \int_{K_j} \varepsilon^2 \eta_1'(x)\xi_1'(x)dx \le \left(\sum_{j=1}^N \int_{K_j} \varepsilon^2(\eta_1')^2(x)dx\right)^{1/2}\left(\sum_{j=1}^N \int_{K_j} \varepsilon^2(\xi_1')^2(x)dx\right)^{1/2}
$$
$$
\le C\|\eta'\|_{L^2(\Omega)}\||\boldsymbol{\xi}\|| \le C(N^{-1}\ln N)^k\||\boldsymbol{\xi}\||,
$$

and for the second term

$$
\sum_{j=0}^N \varepsilon^2(\{\eta_1'(x_j)\}[\xi_1(x_j)]) \le \left(\sum_{j=0}^N \frac{\varepsilon^2}{\sigma_j}\{\eta_1'\}_j^2\right)^{1/2}\left(\sum_{j=0}^N \sigma_j[\xi_1]^2(x)dx\right)^{1/2}
$$

$$
\le C(N^{-1}\ln N)^k\||\boldsymbol{\xi}\||.
$$

Similarly, we can show that the third and fourth terms satisfy

$$
\sum_{j=1}^N \int_{K_j} a_{11}(x)\eta_1(x)\xi_1(x)dx \le C(N^{-1}\ln N)^k\||\boldsymbol{\xi}\||
$$

$$
\sum_{j=1}^N \int_{K_j} a_{12}(x)\eta_2(x)\xi_1(x)dx \le C(N^{-1}\ln N)^k\||\boldsymbol{\xi}\||
$$

Hence, we have $\||\boldsymbol{\xi}\|| \le C(N^{-1}\ln N)^k$.

∎

From Theorems 1 and 2, we can obtain the parameter-uniform error estimate for the NIPG method, which is given in the following theorem.

Theorem 3 *Let u and u_h be the solution of the continuous and discrete problem, respectively. Then,*

$$\|\|u - u_h\|\| \leq C(N^{-1} \ln N)^k.$$

5 Numerical Result

In this section, we verify experimentally our convergence result by considering the numerical solution of a constant coefficient problem.

Example 1 Consider the following singularly perturbed system of BVP:

$$\begin{cases} -\varepsilon^2 u_1''(x) + 2u_1(x) - u_2(x) = f_1(x), x \in \Omega = (0, 1), \\ -\varepsilon^2 u_2''(x) - u_1(x) + 2u_2(x) = f_2(x), \\ u_1(0) = u_2(0) = u_1(1) = u_2(1) = 0, \end{cases} \tag{15}$$

where $f_1(x), f_2(x)$ are chosen such that

$$u_1(x) = \frac{2}{(1 + \exp(-1/\varepsilon))}(\exp(-x/\varepsilon) + \exp(-(1-x)/\varepsilon)) - 2,$$

$$u_2(x) = \frac{1}{(1 + \exp(-1/\varepsilon))}(\exp(-x/\varepsilon) + \exp(-(1-x)/\varepsilon)) - 1,$$

are the exact solution of the (15).

   We calculate the error in the energy norm as defined in (6). The numerical convergence rate is computed by using the formula $r = \ln(e_N/e_{2N})/\ln 2$, where e_N is the computation error with N number of interval. Tables 1 and 2 provide the numerical result with the finite element polynomials of order $k = 1, 2$.

Table 1 Energy norm error for Example 1 for $k = 1$

ε	Number of mesh intervals N				
	16	32	64	128	256
10^{-2}	4.1102e−01	2.1731e−01	1.0850e−01	5.2935e−02	2.5923e−02
	0.9194	1.0021	1.0353	1.0300	
10^{-3}	4.0432e−01	2.1116e−01	9.9547e−02	4.3265e−02	1.7935e−02
	0.9371	1.0849	1.2022	1.2705	
10^{-4}	4.0519e−01	2.1328e−01	1.0250e−01	4.5540e−02	1.8899e−02
	0.9258	1.0571	1.1704	1.2689	
$\vdots$	$\vdots$	$\vdots$	$\vdots$	$\vdots$	$\vdots$
10^{-10}	4.0530e−01	2.1356e−01	1.0307e−01	4.6427e−02	1.9855e−02
	0.92434	1.0511	1.1505	1.2255	

Table 2 Energy norm error for Example 1 for $k = 2$

ε	Number of mesh intervals N				
	16	32	64	128	256
10^{-1}	1.0326e−02	2.3568e−03	5.4424e−04	1.2900e−04	3.1266e−05
	2.1314	2.1145	2.0769	2.0447	
10^{-2}	2.6762e−02	9.7779e−03	2.9795e−03	8.0849e−04	2.0313e−04
	1.4526	1.7144	1.8818	1.9928	
10^{-3}	2.5394e−02	9.2121e−03	2.8263e−03	7.6634e−04	1.9307e−04
	1.4629	1.7046	1.8828	1.9889	
$\vdots$	$\vdots$	$\vdots$	$\vdots$	$\vdots$	$\vdots$
10^{-9}	2.5333e−02	9.1889e−03	2.8190e−03	7.6956e−04	1.8028e−04
	1.4631	1.7047	1.8731	2.0938	
10^{-10}	2.5309e−02	9.1968e−03	2.7990e−03	7.6424e−04	1.4193e−04
	1.4604	1.7162	1.8728	2.4289	

References

1. Farrell, P.A., Hegarty, A.F., Miller, J.J.H., O'Riordan, E., Shishkin, G.I.: Robust Computational Techniques for Boundary Layers. Chapman & Hall/CRC Press, Boca Raton (2000)
2. Lin, R., Stynes, M.: A balanced finite element method for singularly perturbed reaction-diffusion problems. SIAM J. Numer. Anal. **50**(5), 2729–2743 (2012)
3. Linß, T., Madden, N.: A finite element analysis of a coupled system of singularly perturbed reaction–diffusion equations. Appl. Math. Comput. **148**(3), 869–880 (2004)
4. Madden, N., Stynes, M.: A uniformly convergent numerical method for a coupled system of two singularly perturbed linear reaction-diffusion problems. IMA J. Numer. Anal. **23**(4), 627–644 (2003)
5. Miller, J.J.H., O'Riordan, E., Shishkin, G.I.: Fitted Numerical Methods for Singular Perturbation Problems. World Scientific, Singapore (1996)
6. O'Riordan, E., Stynes, M.: An analysis of some exponentially fitted finite element methods for singularly perturbed elliptic problems. In: Computational Methods for Boundary and Interior Layers in Several Dimensions, volume 1 of Adv. Comput. Methods Bound. Inter. Layers, pp. 138–153. Boole, Dublin (1991)
7. Roos, H.-G., Stynes, M., Tobiska, L.: Robust Numerical Methods for Singularly Perturbed Differential Equations, vol. 24, 2nd edn. Springer Series in Computational Mathematics, Berlin (2008)
8. Tobiska, L.: Analysis of a new stabilized higher order finite element method for advection-diffusion equations. Comput. Methods Appl. Mech. Engrg. **196**(1–3), 538–550 (2006)
9. Zarin, H., Roos, H.-G.: Interior penalty discontinuous approximations of convection-diffusion problems with parabolic layers. Numer. Math. **100**(4), 735–759 (2005)
10. Zhang, Z.: Finite element superconvergence approximation for one-dimensional singularly perturbed problems. Numer. Methods Partial Differ. Equ. **18**(3), 374–395 (2002)
11. Zhu, P., Xie, Z., Zhou, S.: A coupled continuous-discontinuous FEM approach for convection diffusion equations. Acta Math. Sci. Ser. B Engl. Ed. **31**(2):601–612 (2011)
12. Zhu, P., Yang, Y., Yin, Y.: Higher order uniformly convergent NIPG methods for 1-d singularly perturbed problems of convection–diffusion type. Appl. Math. Model. **39**(22), 6806–6816 (2015)

Chapter 34
On Solving Bimatrix Games with Triangular Fuzzy Payoffs

Subrato Chakravorty and Debdas Ghosh

Abstract The aim of this paper is to introduce the concept of bimatrix fuzzy games. The fuzzy games are defined by payoff matrices constructed using triangular fuzzy numbers. The bimatrix fuzzy game discussed in this paper is different from the one given by Maeda and Cunlin in respect that it is not a zero-sum game and two different payoff matrices are provided for the two players. Three kinds of Nash equilibriums are introduced for fuzzy games, and their existence conditions are studied. A solution method for bimatrix fuzzy games is given using crisp parametric bimatrix games. Finally, a numerical example is discussed to support the model described in the paper.

Keywords Bimatrix games · Nash equilibrium · Fuzzy set theory · Fuzzy games
Non-cooperative games

1 Introduction

Game theory is the study of mathematical models for conflict resolution among intelligent decision makers with applications in economics, finance, management, engineering, etc. It can be broadly classified into cooperative and non-cooperative games. In cooperative games, we have alliances that can be externally enforced, whereas in non-cooperative games, only self-enforcing alliances are permitted. In 1951, Nash [1] gave the solution concept of Nash equilibrium for non-cooperative games. Since then, Nash equilibrium has been one of the most fundamental concepts in game theory. An equilibrium strategy in a non-cooperative game is said to be a Nash equilibrium strategy if a player cannot improve its payoff by changing its strategy provided that all other players keep their strategy constant.

S. Chakravorty · D. Ghosh (✉)
Department of Mechanical Engineering, Indian Institute of Technology (BHU),
Varanasi 221005, Uttar Pradesh, India
e-mail: debdas.mat@iitbhu.ac.in

S. Chakravorty
e-mail: chakravorty.subrato@gmail.com

© Springer Nature Singapore Pte Ltd. 2018
D. Ghosh et al. (eds.), *Mathematics and Computing*, Springer Proceedings
in Mathematics & Statistics 253, https://doi.org/10.1007/978-981-13-2095-8_34

In traditional game theory, payoffs are assumed to be precise and well known to all the players. But in reality, due to the complexity of the problems, or due to lack of adequate information and imprecision in the knowledge of the environment, payoffs cannot be defined precisely. The lack of precision and certainty in parameters is modeled using various ways such as interval games, stochastic games, fuzzy games. In our paper, we deal with matrix games which have only two players and model our payoffs using fuzzy numbers.

Many excellent works have contributed to the field of fuzzy games. Butnariu [2, 3] modeled the beliefs of each player about other players' strategies as fuzzy sets and determined the equilibrium strategies based on the fuzzy preference relations of the investment of the players pure strategy. Campos [4], in his analysis of two-person zero-sum games with fuzzy payoffs, converted the solution of the game into a fuzzy linear programming problem using Yager's ranking index [5]. Li [6, 7] gave an efficient method to solve matrix games with triangular fuzzy numbers as payoffs. The fuzzy game value is taken as a triangular fuzzy number. Using the duality theorem of linear programming (LP), the fuzzy game value is computed by solving derived LP models. These LP models are defined using 1-cut and 0-cut of fuzzy payoffs. Sakawa and Nishizaki [8] investigated single-objective and multi-objective games with fuzzy goals and fuzzy payoffs. The models in [8] are transformed into a fractional programming problem and ultimately solved using a relaxed method. Vijay et al. [9, 10] showed that using a suitable defuzzification function, solution to zero-sum matrix games is equivalent to primal-dual pair of a fuzzy linear programming problem. Larbani [11, 12] investigated non-cooperative games with payoff functions involving fuzzy parameters. The equilibrium strategy in [11] considers the aspect of conflict as well as the aspect of decision making under uncertainty concerning the use of fuzzy parameters.

Maeda [13, 14] characterized the Nash equilibrium [1] for games with symmetric triangular fuzzy numbers as payoffs into three kinds using fuzzy max order relation. Cunlin and Qiang [15] extended the results in [13] for asymmetric triangular fuzzy numbers which was further extended by Dutta and Gupta [16] for asymmetric trapezoidal fuzzy numbers.

Above results investigate two-person zero-sum games in fuzzy domain. In crisp two-person zero-sum games, one player's gain is another player's loss. This fact is hard to accept in fuzzy games as fuzzy numbers are used to model uncertainty in payoffs and to say that one player's gain is exactly equal to another player's loss with certainty in fuzzy domain seems unrealistic. Hence, the idea of a zero-sum game using fuzzy numbers as payoffs is seldom realized. *In this paper, our goal is to extend the models of Maeda [13, 14] and Cunlin [15] for bimatrix games where we have a different payoff matrix for each player.* These payoff matrices are formed using triangular fuzzy numbers. The rest of the paper is arranged as follows. Section 2 briefly gives the definitions of some preliminaries needed to implement a bimatrix fuzzy game. Section 3 provides the definitions for three different types of equilibrium strategies, namely Nash equilibrium, non-dominated Nash equilibrium, and weak non-dominated Nash equilibrium. Conditions for existence of these strategies are given, and the relation between bimatrix fuzzy games and parametric bimatrix games is established. Further, we introduce the concept of a new parametric crisp bimatrix game whose payoffs are constructed using alpha cuts of triangular fuzzy numbers. In Sect. 4, a numerical example of a fuzzy bimatrix game is discussed. Conclusion is made in Sect. 5.

2   Preliminaries

In this section, we summarize the basic concepts of fuzzy numbers, triangular fuzzy numbers as given by Zadeh [17] and give a ranking method for the same.

Definition 1   (*Fuzzy number* [15]) A fuzzy set defined on the space of real numbers $\mathbb{R}$, is said to be a fuzzy number if its membership function $\mu_{\tilde{a}}(x)$ satisfies the following conditions:

(i) $\mu_{\tilde{a}}(x)$ is a mapping from $\mathbb{R}$ to the closed interval $[0, 1]$;
(ii)   there exists a unique real number c, called the center of $\tilde{a}$, such that

    a. $\mu_{\tilde{a}}(c) = 1$,
    b. $\mu_{\tilde{a}}(x)$ is non-decreasing on $(-\infty, c]$,
    c. $\mu_{\tilde{a}}(x)$ is non-increasing on $[c, \infty)$.

The α-cut or α-level of a fuzzy number $\tilde{a}$ for $\alpha \in [0, 1]$ is given as $\tilde{a}_\alpha = \{x | \mu_{\tilde{a}} \geq \alpha, x \in \mathbb{R}\}$. The 0-level α-cut is known as the support of $\tilde{a}$ given by $\{x | \mu_{\tilde{a}} \geq 0, x \in \mathbb{R}\}$. $a_\alpha^R = \sup \tilde{a}_\alpha$, $a_\alpha^L = \inf \tilde{a}_\alpha$, and $\tilde{a}_\alpha = [a_\alpha^L, a_\alpha^R]$.

Let $\tilde{a}$, $\tilde{b}$ be two fuzzy numbers and c be a real number, the sum of $\tilde{a}$ and $\tilde{b}$ and the scalar product of c with $\tilde{a}$ are defined as follows:

(i) $\mu_{\tilde{a}+\tilde{b}}(x) = \sup \min_{x=u+v}(\mu_{\tilde{a}}(u), \mu_{\tilde{b}}(v))$
(ii) $\mu_{c\tilde{a}}(x) = \max\{\sup_{x=cu} \mu_{\tilde{a}}(u), 0\}$, with $\sup\{\Phi\} = -\infty$.

Definition 2   (*Triangular fuzzy number*) A fuzzy number is said to be a triangular fuzzy number if its membership function is given by

$$\mu_{\tilde{a}}(x) = \begin{cases} \frac{x-l}{m-l}, & l \leq x \leq m \\ \frac{x-n}{m-n}, & m \leq x \leq n, \\ 0 & otherwise \end{cases}$$

From now onwards, we will denote a triangular fuzzy number by (l, m, n). We consider $\mathscr{F}$ to be the set of all triangular fuzzy numbers.

Lemma 1   Let $\tilde{a} = (l_1, m_1, n_1)$, $\tilde{b} = (l_2, m_2, n_2) \in \mathscr{F}, c \in \mathbb{R}^+$. Then,

(i) $\tilde{a} + \tilde{b} = (l_1 + l_2, +m_1 + m_2, n_1 + n_2)$,
(ii) $c\tilde{a} = (cl, cm, cn)$.

Definition 3 Let $x = (x_1, x_2, x_3, \ldots, x_n)$, $y = (y_1, y_2, y_3, \ldots, y_n) \in \mathbb{R}^n$. Then, we write

(i) $x \leqq y$ if $x_i \leq y_i$ $\forall i = 1, 2, \ldots, n$,
(ii) $x \leq y$ if $x_i \leq y_i$ $\forall i = 1, 2, \ldots, n$ and $x \neq y$, and
(iii) $x < y$ if $x_i < y_i$ $\forall i = 1, 2, \ldots, n$.

Definition 4 *(Maeda [13])* Let $\tilde{a}$ and $\tilde{b}$ be two fuzzy numbers. Then,

(i) $\tilde{b} \leqq \tilde{a}$ if $(b_\alpha^L, b_\alpha^R) \leqq (a_\alpha^L, a_\alpha^R)$ $\forall \alpha \in [0, 1]$;
(ii) $\tilde{b} \leq \tilde{a}$ if $(b_\alpha^L, b_\alpha^R) \leq (a_\alpha^L, a_\alpha^R)$ $\forall \alpha \in [0, 1]$;
(iii) $\tilde{b} < \tilde{a}$ if $(b_\alpha^L, b_\alpha^R) < (a_\alpha^L, a_\alpha^R)$ $\forall \alpha \in [0, 1]$;

Consequently, $\tilde{a} = \tilde{b}$ if their α-cuts are equal $\forall \alpha \in [0, 1]$.

Theorem 1 *Let $\tilde{a} = (l_1, m_1, n_1)$ and $\tilde{b} = (l_2, m_2, n_2)$ be two triangular fuzzy numbers. Then,*

(i) $\tilde{a} \leqq \tilde{b}$ if $(l_1, m_1, n_1) \leqq (l_2, m_2, n_2)$,
(ii) $\tilde{a} < \tilde{b}$ if $(l_1, m_1, n_1) < (l_2, m_2, n_2)$.

The proof is omitted in this paper as it is sufficiently intuitive.

3 Nash Equilibrium Strategies for Bimatrix Games

Let $M = \{1, 2, \ldots, m\}$ and $N = \{1, 2, \ldots, n\}$ be the sets of pure strategies of player I and player J respectively. $A = (a_{ij})_{m \times n}$ is the payoff matrix of player I, and $B = (b_{ij})_{m \times n}$ is the payoff matrix of player J such that when player I selects strategy i and player J selects strategy j, the payoff of player I is a_{ij} and the payoff of player J is b_{ij}. The mixed strategies of sets I and J are probability distributions on their sets of pure strategies given by

$$S_I = \{(x_1, x_2, \ldots, x_m) \in \mathbb{R}^m | x_i \geq 0, i = 1, 2, \ldots, m \sum_{i=1}^{m} x_i = 1\}$$

$$S_J = \{(y_1, y_2, \ldots, y_n) \in \mathbb{R}^n | y_i \geq 0, i = 1, 2, \ldots, n \sum_{i=1}^{n} y_i = 1\}$$

If the players are open to mixed strategies, we calculate the expected payoff of player I as $E(x, y) = x^T A y$ and that of player J as $E(x, y) = x^T B y$. The bimatrix game is given by

$$\Gamma \equiv (\{I, J\}, S_I, S_J, A, B)$$

A fuzzy bimatrix game is where the payoffs are triangular fuzzy numbers. The expected payoff of player I is given by $E(x, y) = x^T \tilde{A} y = \sum_{i=1}^{m} \sum_{j=1}^{n} x_i \tilde{a}_{ij} y_j$ and expected payoff of player J is given by $E(x, y) = x^T \tilde{B} y = \sum_{i=1}^{m} \sum_{j=1}^{n} x_i \tilde{b}_{ij} y_j$.

By Lemma 1, the expected payoffs will also be triangular fuzzy numbers. The bimatrix fuzzy game is denoted by

$$\widetilde{\Gamma} \equiv (\{I, J\}, S_I, S_J, \widetilde{A}, \widetilde{B}) \tag{1}$$

where $\widetilde{A} = (\widetilde{a}_{ij})_{m \times n} = ((l_{ij}, m_{ij}, n_{ij}))_{m \times n}$ and $\widetilde{B} = (\widetilde{b}_{ij})_{m \times n} = ((p_{ij}, q_{ij}, r_{ij}))_{m \times n}$. In the rest of this paper, we denote $A_0^L = (l_{ij})_{m \times n}$, $A = (m_{ij})_{m \times n}$, $A_0^R = (n_{ij})_{m \times n}$, $B_0^L = (p_{ij})_{m \times n}$, $B = (q_{ij})_{m \times n}$, and $B_0^R = (r_{ij})_{m \times n}$.

Definition 5 A point $(x^*, y^*) \in S_I \times S_J$ is said to be a Nash equilibrium strategy to game $\widetilde{\Gamma}$ if it holds that

(i) $x^T \widetilde{A} y^* \leqq x^{*T} \widetilde{A} y^* \qquad \forall x \in S_I$, and
(ii) $x^{*T} \widetilde{B} y \leqq x^{*T} \widetilde{B} y^* \qquad \forall y \in S_J$.

Definition 6 A point $(x^*, y^*) \in S_I \times S_J$ is said to be a non-dominated Nash equilibrium strategy to game $\widetilde{\Gamma}$ if it holds that

(i) there exist no $x \in S_I$ such that $x^{*T} \widetilde{A} y^* \leq x^T \widetilde{A} y^*$, and
(ii) there exist no $y \in S_J$ such that $x^{*T} \widetilde{B} y^* \leq x^{*T} \widetilde{B} y$.

Definition 7 A point $(x^*, y^*) \in S_I \times S_J$ is said to be a weak non-dominated Nash equilibrium strategy to game $\widetilde{\Gamma}$ if it holds that

(i) there exist no $x \in S_I$ such that $x^{*T} \widetilde{A} y^* < x^T \widetilde{A} y^*$, and
(ii) there exist no $y \in S_J$ such that $x^{*T} \widetilde{B} y^* < x^{*T} \widetilde{B} y$.

These three definitions of equilibrium strategies act as a natural extension of the classical Nash equilibrium in the realm of fuzzy games. We will investigate the existence of these strategies in our framework. It can be easily seen that the set of Nash equilibrium strategies is a subset of non-dominated Nash equilibrium strategies which further is a subset of weak non-dominated Nash equilibrium strategies. If the individual payoffs become crisp, then these definitions reduce to the classical one.

Theorem 2 *A pair $(x^*, y^*) \in S_I \times S_J$ is the Nash equilibrium strategy of the fuzzy game $\widetilde{\Gamma}$ if and only if the following inequalities hold true*

$$x^T A_0^L y^* \leq x^{*T} A_0^L y^*,$$

$$x^{*T} B_0^L y \leq x^{*T} B_0^L y^*,$$

$$x^T A y^* \leq x^{*T} A y^*,$$

$$x^{*T} B y \leq x^{*T} B y^*,$$

$$x^T A_0^R y^* \leq x^{*T} A_0^R y^*,$$

$$x^{*T} B_0^R y \leq x^{*T} B_0^R y^*.$$

Proof Let (x^*, y^*) be the Nash equilibrium of the game $\widetilde{\Gamma}$, then from Definition 5, we have

(i) $x^T \widetilde{A} y^* \lesssim x^{*T} \widetilde{A} y^*$ $\forall x \in S_I$, and
(ii) $x^{*T} \widetilde{B} y \lesssim x^{*T} \widetilde{B} y^*$ $y \in S_J$.

For the first condition to be true, using Theorem 1, we get

$$(x^T A_0^L y^*, x^T A y^*, x^T A_0^R y^*) \leqq (x^{*T} A_0^L y^*, x^{*T} A y^*, x^{*T} A_0^R y^*) \tag{2}$$

Analogously, for the second condition to be true, we get

$$(x^{*T} B_0^L y, x^{*T} B y, x^{*T} B_0^R y^*) \leqq (x^{*T} B_0^L y^*, x^{*T} B y^*, x^{*T} B_0^R y^*) \tag{3}$$

Clubbing Equations (2) and (3), we derive

$$x^T A_0^L y^* \leq x^{*T} A_0^L y^* \text{ and } x^{*T} B_0^L y \leq x^{*T} B_0^L y^*, \tag{4}$$

$$x^T A y^* \leq x^{*T} A y^* \text{ and } x^{*T} B y \leq x^{*T} B y^*, \tag{5}$$

$$x^T A_0^R y^* \leq x^{*T} A_0^R y^* \text{ and } x^{*T} B_0^R y \leq x^{*T} B_0^R y^*. \tag{6}$$

Therefore, we may conclude that the Nash equilibrium of the fuzzy game $\widetilde{\Gamma}$ is equivalent to the Nash equilibriums of three crisp bimatrix games given by Eqs. (4), (5), and (6). Consequently, the existence of Nash equilibrium for fuzzy bimatrix games cannot be guaranteed. This necessitates the need for non-dominated and weak non-dominated Nash equilibriums which we are going to investigate as we go further.

Theorem 3 *A pair (x^*, y^*) is said to be a non-dominated Nash equilibrium of the game $\widetilde{\Gamma}$ if and only if the following conditions hold true*

(i) *there exists no $x \in S_I$ such that*
    $x^{*T}(A_0^L, A_0^R)y^* \leq x^T(A_0^L, A_0^R)y^* \text{ and } x^{*T} A y^* \leq x^T A y^*$
(ii) *there exists no $y \in S_J$ such that*
    $x^{*T}(B_0^L, B_0^R)y^* \leq x^{*T}(B_0^L, B_0^R)y \text{ and } x^{*T} B y^* \leq x^{*T} B y.$

Proof Let the pair (x^*, y^*) be a non-dominated Nash equilibrium strategy for the game $\widetilde{\Gamma}$, we assume that there exists a $\bar{x}$ such that condition (i) holds

$$x^{*T}(A_0^L, A_0^R)y^* \leq \bar{x}^T(A_0^L, A_0^R)y^* \text{ and} \tag{7}$$

$$x^{*T} A y^* \leq \bar{x}^T A y^* \tag{8}$$

The above Inequality (7) can be broken down into following inequalities

$$x^{*T} A_0^L y^* \leq \bar{x}^T A_0^L y^*$$

$$x^{*T} A_0^R y^* \leq \bar{x}^T A_0^R y^*$$

Since the inequalities cannot occur simultaneously. Thus for any $\alpha \in [0, 1)$, we can write

$$(x^{*T}(\alpha A + (1-\alpha)A_0^L)y^*, x^{*T}(\alpha A + (1-\alpha)A_0^R)y^*)$$

$$\leq (\bar{x}^T(\alpha A + (1-\alpha)A_0^L)y^*, \bar{x}^T(\alpha A + (1-\alpha)A_0^R)y^*)$$

Hence, it follows that,

$$(x^{*T} A_\alpha^L y^*, x^{*T} A_\alpha^R y^*) \leq (\bar{x}^T A_\alpha^L y^*, \bar{x}^T A_\alpha^R y^*).$$

From Definition 4, we get, $x^{*T} \widetilde{A} y^* \leq \bar{x}^T \widetilde{A} y^*$. But, this is a contradiction.

Analogously, there exists no y such that condition (ii) holds.

For the converse, let a pair (x^*, y^*) be such that both the conditions (i) and (ii) are satisfied. Let us assume that it is not a non-dominated Nash equilibrium point. Hence, there will be a certain $\bar{x}$ such that $x^{*T} \widetilde{A} y^* \leq \bar{x}^T \widetilde{A} y^*$. From Definition 4 we get,

$$(x^{*T} A_\alpha^L y^*, x^{*T} A_\alpha^R y^*)$$

$$\leq (\bar{x}^T A_\alpha^L y^*, \bar{x}^T A_\alpha^R y^*)$$

Since, A_α^L and A_α^R are continuous with respect to α. As α tends to 1,

$$x^{*T} A y^* \leq \bar{x}^T A y^*.$$

And if $\alpha = 0$,

$$x^{*T}(A_0^L, A_0^R)y^* \leq \bar{x}^T(A_0^L, A_0^R)y^*, \text{ which is a contradiction.}$$

Analogously, we can prove that there exists no such y such that $x^{*T}(B_0^L, B_0^R)y^* \leq x^{*T}(B_0^L, B_0^R)y$ and $x^{*T} B y^* \leq x^{*T} B y$. Hence (x^*, y^*) is a non-dominated Nash equilibrium point.

Theorem 4 *A pair (x^*, y^*) is a weak non-dominated Nash equilibrium point of the game $\widetilde{\Gamma}$ if and only if the following conditions hold*

(i) *there exists no $x \in S_I$ such that*
    $x^{*T}(A_0^L, A, A_0^R)y^* < x^T(A_0^L, A, A_0^R)y^*$,
(ii) *there exists no $y \in S_J$ such that*
    $x^{*T}(B_0^L, B, B_0^R)y* < x^{*T}(B_0^L, B, B_0^R)y$.

Proof Above theorem can be proved similar to Theorem 3.

   Now, we will investigate parametric bimatrix games and transform the solutions of fuzzy games into the solutions of parametric bimatrix games and prove existence results for non-dominated and weak non-dominated Nash equilibrium. We define a parametric bimatrix game using A_0^L, A_0^R, B_0^L, and B_0^R. A parametric bimatrix game is one in which payoff matrices are given by

$$A(\lambda) = \lambda A_0^L + (1 - \lambda)A_0^R \text{ and}$$

$$B(\mu) = \mu B_0^L + (1 - \mu)B_0^R.$$

The parametric bimatrix game is denoted by $\Gamma(\lambda, \mu) = (\{I, J\}, S_I, S_J, A(\lambda), B(\mu))$.

Lemma 2 *Every parametric bimatrix game $\Gamma(\lambda, \mu)$ has at least one Nash equilibrium point. If (x^*, y^*) is a Nash equilibrium strategy then,*

$$x^T A(\lambda)y^* \leq x^{*T}A(\lambda)y^*, \ x \in S_I,$$

$$x^{*T} B(\mu)y \leq x^{*T}B(\mu)y^*, \ y \in S_J.$$

Theorem 5 *If a point $(x^*, y^*) \in S_I \times S_J$ is a Nash equilibrium of the parametric bimatrix game $\Gamma(\lambda, \mu)$ with $\lambda, \mu \in (0, 1)$, then it is also a non-dominated Nash equilibrium of the fuzzy game $\widetilde{\Gamma} = (\{I, J\}, S_I, S_J, \widetilde{A}, \widetilde{B})$.*

Proof Let a pair $(x^*, y^*) \in S_I \times S_J$ be a Nash equilibrium of the parametric bimatrix game $\Gamma(\lambda, \mu)$. Therefore,

$$x^T A(\lambda)y^* \leq x^{*T}A(\lambda)y^*, \ x \in S_I, \tag{9}$$

$$x^{*T} B(\mu)y \leq x^{*T}B(\mu)y^*, \ y \in S_J. \tag{10}$$

   Let us assume that it is not a non-dominated Nash equilibrium point for the fuzzy game $\widetilde{\Gamma}$. Therefore, we have a $\bar{x}$ such that

$$x^{*T}\widetilde{A}y^* \leq \bar{x}^T\widetilde{A}y^* \tag{11}$$

From Definition 4, it follows that

$$(x^{*T}A_\alpha^L y^*, x^{*T}A_\alpha^R y^*) \leq (\bar{x}^T A_\alpha^L y^*, \bar{x}^T A_\alpha^R y^*)$$

If $\alpha = 0$, we get

$$(x^{*T}A_0^L y^*, x^{*T}A_0^R y^*) \leq (\bar{x}^T A_0^L y^*, \bar{x}^T A_0^R y^*)$$

Since all the equalities cannot hold together, we can write using the same λ as in the parametric bimatrix game $\Gamma(\lambda, \mu)$

$$\lambda x^{*T} A_0^L y^* + (1 - \lambda) x^{*T} A_0^R y^* < \lambda \bar{x}^T A_0^L y^* + (1 - \lambda) \bar{x}^T A_0^R y^*$$

But it violates Condition (9). Hence, there exists no $\bar{x}$ such that Condition (11) holds. Analogously, there exists no $y \in S_J$ such that

$$x^{*T} \widetilde{B} y* \leq x^{*T} \widetilde{B} y \tag{12}$$

holds. Hence from Definition 6, pair (x^*, y^*) is a non-dominated Nash equilibrium of the fuzzy game $\widetilde{\Gamma}$.

Theorem 6 *If a point (x^*, y^*) is a Nash equilibrium point of the parametric bimatrix game $\Gamma(\lambda, \mu)$ with $\lambda, \mu \in [0, 1]$. Then, it is also a weak non-dominated Nash equilibrium of the fuzzy game $\widetilde{\Gamma} = (\{I, J\}, S_I, S_J, \widetilde{A}, \widetilde{B})$.*

Proof The proof to this theorem is on the lines of Theorem 5.

We can observe that non-dominated Nash equilibrium set obtained using this method is a subset of the weak non-dominated Nash equilibrium set, as the values of λ and μ need to be positive for non-dominated Nash equilibrium while for weak non-dominated Nash equilibrium, λ and μ being non-negative are sufficient. Based on Lemma 2, Theorems 5 and 6, we derive the following results.

Theorem 7 *If $\widetilde{\Gamma}$ is a fuzzy game with triangular fuzzy numbers as payoffs, then*

(i) *There exists atleast one non-dominated Nash equilibrium strategy to the game $\widetilde{\Gamma}$,*

(ii) *There exists atleast one weak non-dominated Nash equilibrium strategy to the game $\widetilde{\Gamma}$.*

4 Numerical Example

Example 1 Let us consider a bimatrix fuzzy game $\widetilde{\Gamma} \equiv (\{I, J\}, S_I, S_J, \widetilde{A}, \widetilde{B})$ given by payoff matrices $\widetilde{A}$ and $\widetilde{B}$ as follows:

$$\widetilde{A} = \begin{pmatrix} (10, 12, 17) & (5, 9, 13) \\ (7, 8, 10) & (12, 14, 15) \end{pmatrix}$$

$$\widetilde{B} = \begin{pmatrix} (2, 5, 7) & (13, 17, 18) \\ (11, 13, 15) & (7, 9, 10) \end{pmatrix}$$

We try to find whether the above game has a Nash equilibrium strategy. According to Theorem 2, Nash equilibrium of the fuzzy game $\widetilde{\Gamma}$ exists if crisp bimatrix games defined in Theorem 2 have the same equilibrium strategy. Observe that the crisp games have different Nash equilibrium strategies here; hence, the above game $\widetilde{\Gamma}$ has no Nash equilibrium strategy.

To find non-dominated and weak non-dominated Nash equilibrium strategies, we need to find the equilibrium strategies to the crisp bimatrix game

$\Gamma(\lambda, \mu) = (\{I, J\}, S_I, S_J, A(\lambda), B(\mu))$ where

$A(\lambda) = \lambda A_0^L + (1 - \lambda)A_0^R$ and

$B(\mu) = \mu B_0^L + (1 - \mu)B_0^R.$

$$A(\lambda) = \begin{pmatrix} 17 - 7\lambda & 13 - 8\lambda \\ 10 - 3\lambda & 15 - 3\lambda \end{pmatrix}$$

$$B(\mu) = \begin{pmatrix} 7 - 5\mu & 18 - 5\mu \\ 15 - 4\mu & 10 - 3\mu \end{pmatrix}$$

In order to find a mixed strategy Nash equilibrium to the above bimatrix game, we will use the property that the player who optimizes using a mixed strategy is indifferent among all pure strategies that occur in a given mixed strategy with positive probabilities. Let I chooses the row 1 strategy with probability p and consequently, row 2 strategy with probability $1 - p$ and player J chooses the column 1 strategy with probability q and consequently, column 2 strategy with probability $1 - q$. For this strategy to be a Nash equilibrium, player J must be indifferent between its strategies. Therefore,

$p(7 - 5\mu) + (1 - p)(15 - 4\mu)$

$= p(18 - 5\mu) + (1 - p)(10 - 3\mu).$

Solving the above linear equation for p, we get,

$p = \frac{5 - \mu}{16 - \mu}.$

Similarly, solving for q we get,

$q = \frac{2 + 5\lambda}{9 + \lambda}.$

Hence, the Nash equilibrium strategy for the above problem is

$$\begin{cases} x^{*T} = (\frac{5-\mu}{16-\mu}, \frac{11}{16-\mu}) \\ y^{*T} = (\frac{2+5\lambda}{9+\lambda}, \frac{7-4\lambda}{9+\lambda}). \end{cases}$$

By Theorems 5 and 6, the non-dominated Nash equilibrium and weak non-dominated Nash equilibrium strategies of the fuzzy game $\widetilde{\Gamma}$ are as follows:

$$NDN = (x^*, y^*) \, \lambda, \mu \in (0, 1)$$

$$WNDN = (x^*, y^*) \, \lambda, \mu \in [0, 1].$$

5 Conclusion

In this paper, we introduce the concept of bimatrix fuzzy games with different payoff matrices for the two players. Models by Maeda and Cunlin are extended for this game. Three different characterizations of Nash equilibrium are given for the realm of fuzzy games. These characterizations can be seen as natural extensions of Nash equilibrium for crisp games. Existence conditions are studied for the three equilibrium strategies. It was seen that the Nash equilibrium for bimatrix fuzzy games can't be guaranteed unlike the general crisp bimatrix games. Parametric bimatrix games were introduced as a solution method to find non-dominated and weak non-dominated Nash equilibrium strategies for a fuzzy bimatrix game. Further work is needed to generalize the models discussed in this paper for n players.

References

1. Nash, J.F.: Non-cooperativegames. In: Annals of Mathematics, Second Series, vol. 54, no. 2, pp. 286–295 (1951)
2. Butnariu, D.: Fuzzy games: a description of the concept. Fuzzy Sets Syst. **1**(3), 181–192 (1978). July
3. Butnariu, D.: Stability and Shapley value for an n-persons fuzzy game. Fuzzy Sets Syst. **4**(1), 63–72 (1980). July
4. Campos, L.: Fuzzylinear programming models to solve fuzzy matrix games. FuzzySets Syst. **32**, 27589 (1989)
5. Yager, R.: Ranking fuzzysubsets over the unit interval. In: Proceedings of the CDC, pp. 1435–1437 (1978)
6. Li, D.F.: A fuzzy multiobjective approach to solve fuzzy matrix games. J. Fuzzy Math. **7**, 90712 (1999)
7. Li, D.F.: A fast approach to compute fuzzy values of matrix games with payoffs of triangular fuzzy numbers. Eur. J. Oper. Res. **223**(2), 421–429 (2012)
8. Sakawa, M., Nishizaki, I.: Max-min solutions for fuzzy multiobjective matrix games. FuzzySets Syst. **61**, 26575 (1994)

9. Bector, C.R., Chandra, S.: On duality in linear programming under fuzzy environment. Fuzzy-Sets Syst. **125**, 31725 (2002)
10. Vijay, V., Chandra, S., Bector, C.R.: Matrix games with fuzzy goals and fuzzy pay offs. Omega **33**, 425429 (2005)
11. Kacher, F., Larbani, M.: Existence of equilibrium solution for a non-cooperative game with fuzzy goals and parameters. Fuzzy Sets Syst. **159**(2), 164–176 (2008). January
12. Larbani, M.: Non cooperative fuzzy games in normal form: a survey. Fuzzy Sets Syst. **160**(22), 3184–3210 (2009). November
13. Maeda, T.: On characterization of equilibrium strategy of two person zero-sum games with fuzzy pay offs. FuzzySets Syst. **139**, 28396 (2003)
14. Maeda, T.: On characterization of equilibrium strategy of bimatrix games with fuzzy pay offs. J. Math. Anal. Appl. **251**, 885896 (2000)
15. Cunlin, L., Qiang, Z.: Nash equilibrium strategy for fuzzy non-cooperative games. FuzzySets Syst. **176**, 4655 (1976)
16. Dutta, B., Gupta, S.K.: On nash equilibrium strategy of two-person zero-sum games with trapezoidal fuzzy payoffs. Fuzzy Inf. Eng. **6**, 299–314 (2014)
17. Zadeh, L.A.: Fuzzy Sets. Inf. Control **8**(3), 338–352 (1968)

Chapter 35
Comparison of Two Methods Based on Daubechies Scale Functions and Legendre Multiwavelets for Approximate Solution of Cauchy-Type Singular Integral Equation on $\mathbb{R}$

Swaraj Paul and B.N. Mandal

Abstract Two methods based on Daubechies scale functions and Legendre multiwavelet for the approximate solution of singular integral equation of the second kind with Cauchy type on the real line $\mathbb{R}$ are developed and compared. The integral equation considered here is of the form $u(x) + \lambda \int_{-\infty}^{\infty} K(x, t)u(t)dt = f(x)$, $x \in \mathbb{R}$, where $K(x, t) = \frac{1}{t-x} + h(x, t)$, $h(x, t)$ being a regular kernel. In both of the cases, two-scale relations involving the scale functions are used for the evaluation of multiscale representation of the integral operator. Then the given integral equation is converted into a system of linear algebraic equations which can be solved easily by using library function 'Solve[]' available in MATHEMATICA. The convergence of the method has been proved in L^2 spaces. Two examples are given and their approximate solutions obtained by the proposed methods have been compared with the available numerical results to assess the efficiency of the method developed here.

Keywords Singular integral equation in $\mathbb{R}$ · Cauchy type singularity · Legendre multiwavelet · Daubechies scale function

1 Introduction

Singular integral equations (SIE) with Cauchy-type kernel arises in a large class of mixed boundary value problems of mathematical physics such as contact problems in fracture mechanics, mainly crack problems in elasticity [1]. Closed form solution of this SIE is well known by complex function theoretic method [2, 3]. Unfortunately,

S. Paul (✉)
Department of Mathematics, Visva Bharati,
Santiniketan 731235, West Bengal, India
e-mail: swaraj.lie@gmail.com

B. N. Mandal
Physics and Applied Mathematics Unit,
Indian Statistical Institute, 203, B T Road, Kolkata 700108, India
e-mail: birenisical@gmail.com

© Springer Nature Singapore Pte Ltd. 2018
D. Ghosh et al. (eds.), *Mathematics and Computing*, Springer Proceedings
in Mathematics & Statistics 253, https://doi.org/10.1007/978-981-13-2095-8_35

this method makes the calculation of the corresponding singular integral (Cauchy type) required by a suitable quadrature rule quite complicated and time-consuming. So the numerical solution of SIE by avoiding quadrature rule has an important impact. Several methods have been proposed for the numerical solutions of the SIE with Cauchy-type singularity. For a thorough review of Galerkin methods, see Ioakimidis [4], Gong [5, 6], Monegato [7], etc. Collocation methods have been studied by Junghans and Kaiser [8], Scuder [9], etc. Bernstein polynomial method has been studied by Setia [10], and Taylor series expansion and Legendre polynomial method have been developed by Arzhang [11]. Method based on Daubechies scale function has been proposed by Panja and Mandal [12]. But all these methods were used in the case when the domain of the integral equation is finite. In this paper, Daubechies scale functions and Legendre multiwavelets have been used to get multiscale approximate solution of a second-kind Fredholm integral equation respectively with the Cauchy-type kernel on $\mathbb{R}$ of the form

$$u(x) + \lambda \int_{-\infty}^{\infty} \left(h(x, t) + \frac{1}{t - x} \right) u(t)dt = f(x), \quad x \in \mathbb{R}, \tag{1}$$

where $f(x)$ is a known function defined on $\mathbb{R}$, $h(x, t)$ is a regular kernel, and u is unknown. This type of integral equation arises in acoustic scattering problems [13], in the scattering of elastic waves [14, 15], in the study of unsteady water waves [16], to solve Laplace's equation in the upper half plane subject to fairly general mixed boundary data [17], etc. Bonis et al. [18] proposed a method based on interpolation process, related to zeros of Hermite polynomials to solve this equation. Sheshko and his collaborators [19, 20] investigated the exact solutions of the characteristic and dominant equation with the Cauchy kernel on the real line by transforming the domain from real line to a circle and using a complex function theoretic method.

Multiresolution analysis (MRA) of function space concerning refinable functions and wavelets became an efficient tool in several areas of mathematical physics and engineering [21, 22]. As more desirable properties of wavelet became available, this tool has been called a "mathematical microscope" in the study of smoothness of regularity of function/signal. For numerical implementation, using wavelet bases is more efficient than classical orthogonal or non-orthogonal polynomial bases due to some useful properties of wavelet, e.g., capturing the local information about the functions/signals and operators, discretization of the domain and sparseness nature of the matrix representation of the operator in wavelet bases. The pioneer workers in the development of multiwavelet involving polynomial are Alpert and his collaborators [23, 24], and they have solved the weakly singular integral equation numerically with logarithmic singularity by using these wavelets. Recently using LMW basis, Paul et al. obtained a class of Fredholm integral equation of the second kind with singular kernel, i.e., with weakly singular Abel-type kernel [25], hypersingular kernel [26], Cauchy-type kernel with constant coefficient [27], Cauchy-type kernel with variable coefficient [28]. To the authors' best knowledge, multiwavelets have not been used for getting approximate solution of Fredholm singular integral equation of the second kind on the real line $\mathbb{R}$ with the Cauchy-type singular kernel. In this work, we have successfully implemented the numerical scheme based on Daubechies scale functions and LMW for getting the solution of the singular integral equation on the real line.

In this paper, we consider the integral equation (1) and study two different methods based on Daubechies scale functions and Legendre multiwavelets by comparing the solutions. The organization of this paper is as follows: Basic properties of Daubechies scale functions and method of solution to a singular integral equation on $\mathbb{R}$ with Cauchy-type kernel by using Daubechies scale functions are discussed in Sect. 2. Transformation into a finite range of integration from infinite range under some suitable condition is discussed in Sect. 3.1. Basic definition and properties of Legendre multiwavelets are described in Sect. 3.2, while in Sect. 3.3, the multiscale approximation of a L^2 function is presented. We evaluate the double integrals involving the basis of Legendre multiwavelets and the Cauchy singular kernel in Sect. 3.4. Using the values of the double integrals obtained in Sect. 3.4, we evaluate the multiscale representation of the transformed integral operator in Sect. 3.5. In Sect. 3.6, the integral equation (1) is transformed into a system of linear algebraic equations which can be solved by a standard method. Error estimation of the proposed method is also presented. The numerical scheme has been verified through two examples in Sect. 4. Conclusion is presented in Sect. 5.

2 Solution of Singular Integral Equation on $\mathbb{R}$ with Cauchy-Type Kernel by Daubechies Scale Functions

2.1 Basic Properties of Daubechies Scale Function

The interesting property of Daubechies scale functions is refinement equation or two-scale relation

$$\phi(x) = \sqrt{2} \sum_{l=0}^{2K-1} h_l \phi(2x - l) \tag{2}$$

with the normalization $\int_{-\infty}^{\infty} \phi(x)dx = 1$. The translates of scaling function are orthogonal for a particular resolution j such that $\int_{-\infty}^{\infty} \phi_{jk_1}(x)\phi_{jk_2}(x)dx = \delta_{k_1 k_2}$, where $\phi_{jl}(x) = 2^{\frac{j}{2}}\phi(2^j x - l)$.

2.2 Method of Solution to Singular Integral Equation on $\mathbb{R}$ with Cauchy-Type Kernel

Let us consider the integral equation (1). Express $u(x)$ in terms of Daubechies scale functions $\phi_{jk}(x)$ at resolution j as

$$u(x) = \sum_{k=k_j^{\min}}^{k_j^{\max}} u_{jk} \phi_{jk}(x) \tag{3}$$

with raw image u_{jk}'s. Integrating both sides between $-\infty$ and ∞ after substituting $u(x)$ from (3) into (1) and multiplying with $\phi_{jk'}$, we get

$$u_{jk'} + \lambda \sum_{k=k_j^{\min}}^{k_j^{\max}} \left(\rho_{jkk'} + I_{jkk'} \right) u_{jk} = f_{jk'}, \tag{4}$$

where

$$\rho_{jkk'} = \int_{-\infty}^{\infty} \int_{-\infty}^{\infty} \frac{\phi_{jk'}(x) \phi_{jk}(t)}{t-x} dt dx, \tag{5}$$

$$I_{jkk'} = \int_{-\infty}^{\infty} \int_{-\infty}^{\infty} \phi_{jk'}(x) \phi_{jk}(t) h(x, t) dt dx, \tag{6}$$

$$f_{jk'} = \int_{-\infty}^{\infty} \phi_{jk'}(x) f(x) dx. \tag{7}$$

The solution of a set of linear simultaneous equation (4) is the approximate numerical solution $u(x)$ of the integral equation (1).

2.3 Evaluation of Matrix Element

The technique for calculating the integrals $f_{jk'}$ and $I_{jkk'}$ can be obtained by using a one-point quadrature rule [29] as

$$\int_{-\infty}^{\infty} \phi_{jk'}(x) f(x) dx = \frac{1}{2^{\frac{j_1}{2}}} \sum_{l_{j_1}=0}^{2K-1} \sum_{l_{j_1-1}}^{2K-1} \cdots \sum_{l_1}^{2K-1} h_{l_{j_1}} h_{l_{j_1-1}} \cdots h_{l_1} f\left(x_{j,j_1,k'}\right) \tag{8}$$

and

$$\int_{-\infty}^{\infty}\int_{-\infty}^{\infty}\phi_{jk'}(x)\phi_{jk}(y)h(x,y)dydx$$

$$= \frac{1}{2^{j_1}}\sum_{l_{j_1}=0}^{2K-1}\sum_{l_{j_1-1}=0}^{2K-1}\cdots\sum_{l_1=0}^{2K-1}\sum_{m_{j_1}=0}^{2K-1}\sum_{m_{j_1-1}=0}^{2K-1}\cdots\sum_{m_1=0}^{2K-1}h_{l_{j_1}}h_{l_{j_1-1}}\cdots h_{l_1}h_{m_{j_1}}h_{m_{j_1-1}}\cdots h_{m_1}f\left(x_{j,j_1,k'},y_{j,j_1,k}\right),$$

(9)

where

$$x_{j,j_1,k'} = \frac{\frac{2^{j_1-1}l_1+\cdots+l_{j_1}+m}{2^{j_1}}+k'}{2^j},$$

$$y_{j,j_1,k} = \frac{\frac{2^{j_1-1}m_1+\cdots+m_{j_1}+m}{2^{j_1}}+k}{2^j},$$

and $m = \int_{-\infty}^{\infty}x\phi(x)dx$. The main task for numerical solution of Eq.(1) is the numerical evaluation of the double integral $\rho_{jkk'}$ of (5). Clearly, $\rho_{jkk'} = \rho_{k-k'}$ and $\rho_{k-k'} = -\rho_{k'-k}$ where

$$\rho_n = \int_{-\infty}^{\infty}\int_{-\infty}^{\infty}\frac{\phi_n(x)\phi(t)}{x-t}dtdx.$$

(10)

For evaluation of ρ_n, we follow the technique given by [12] through the use of their asymptotic values

$$\rho_n \approx \frac{1}{n}, \quad \text{when } |n| > 20$$

(11)

in their two-scale relation

$$\rho_n = \sum_{ll'}h_lh_{l'}\rho_{2n+l-l'}.$$

(12)

For $5 < |n| < 20$, the value of the nonsingular integrals ρ_n can be evaluated by the using the formulae (11) and (12) simultaneously. The value of the singular integral ρ_n for $|n| \le 5$ can be evaluated by solving the linear equations (12).

3 Solution of Singular Integral Equation on $\mathbb{R}$ with Cauchy-Type Kernel by Legendre Multiwavelets

3.1 Transformation to the Finite Range of Integration

We consider integral equation of the form,

$$u(x) + \lambda\int_{-\infty}^{\infty}K(x,t)u(t)dt = f(x), \quad x \in \mathbb{R},$$

(13)

where $K(x, t) = h(x, t) + \frac{1}{t-x}$, and the operator

$$(Hu)(x) = \int_{-\infty}^{\infty} h(x, t)u(t)dt, \quad x \in \mathbb{R} \tag{14}$$

is assumed to be compact, $u(x)$ and $f(x)$ be bounded continuous functions on $\mathbb{R}$. We abbreviate (13) by the operator form as

$$u + \lambda Ku = f, \tag{15}$$

where K is the integral operator defined by

$$(Ku)(x) = \int_{-\infty}^{\infty} K(x, t)u(t)dt, \quad x \in \mathbb{R}. \tag{16}$$

Define the finite section approximation

$$u_\beta(x) + \lambda \int_{|t|<\beta} K(x, t)u_\beta(t)dt = f(x), \quad |x| < \beta, \tag{17}$$

where $u_\beta(x)$ converges to $u(x)$ as $\beta \to \infty$, $u_\beta(x)$ is a finite section approximation of $u(x)$, and here the required condition of the force term $f(x)$ is such that $f(x) \to 0$ as $|x| \geq \beta$. We abbreviate (17) in operator form as

$$u_\beta + \lambda K_\beta u_\beta = f, \tag{18}$$

where K_β is defined by

$$K_\beta(u(x)) = \int_{-\beta}^{\beta} K(x, t)u(t)dt. \tag{19}$$

It has been shown in [30] and [31] that, under quite general conditions on the kernel K, the convergence of u_β to u is uniform on finite intervals of $\mathbb{R}$. Condition for the existence and uniform boundedness of $(I + K_\beta)^{-1}$ have been obtained in [31] for the special case when $K = W + H$, where W is a Wiener–Hopf integral operator and H is a compact operator. Chandler-Wilde [32] proved that this finite section approximation method is stable for a perturbed equation in which the kernel K is replaced by $K + h$. Also, Chandler-Wilde [33] showed that under some condition on $K(x, t)$ and if $f(x) \to 0$ as $x \to \infty$, then $u(x) \to 0$ as $x \to \infty$. Now we are interested to solve (17) instead of (13). Under some change of variable (17) can be transformed to

$$v(x) + \int_0^1 G(x, t)v(t)dt = F(x), \quad 0 < x < 1 \tag{20}$$

where

$$v(x) = u_\beta \, (2\beta x - \beta) \, ,$$

$$G(x, t) = 2\lambda\beta \, K \, (2\beta x - \beta, 2\beta t - \beta)$$
$$= 2\lambda\beta \, h \, (2\beta x - \beta, 2\beta t - \beta) + \frac{\lambda}{x - t}. \tag{21}$$

3.2 Legendre Multiwavelets

The scaling functions in Legendre multiwavelet (LMW) basis consist of K component vectors

$$\phi^i(x) := (2i + 1)^{\frac{1}{2}} P_i(2x - 1), \quad i = 0, 1, \ldots, K - 1; \ 0 \le x < 1 \tag{22}$$

where $P_i(x)$ is the Legendre polynomial of degree i $(i = 0, 1, \ldots, K - 1)$. Their expressions at resolution j are given by

$$\phi^i_{j,k}(x) := 2^{\frac{j}{2}} \phi^i(2^j x - k), \quad j \in \{0\} \cup \mathbb{N}, \ k = 0, 1, ., 2^j - 1, \tag{23}$$

The refinement equations or the two-scale relations among the scale functions $\phi^i_{j,k}(x)$ are

$$\phi^i_{j,k}(x) = \frac{1}{\sqrt{2}} \sum_{r=0}^{K-1} \left(h^{(0)}_{i,r} \phi^r_{j+1,2k}(x) + h^{(1)}_{i,r} \phi^r_{j+1,2k+1}(x) \right) = \frac{1}{\sqrt{2}} \sum_{r=0}^{K-1} \sum_{s=0}^{1} h^{(s)}_{i,r} \phi^r_{j+1,2k+s}(x). \tag{24}$$

The elements $\psi^i_{j,k}(x) \left(:= 2^{\frac{j}{2}} \psi^i(2^j x - k) \right)$ of Legendre multiwavelets $\psi_{j,k}$ having K components for each resolution j and admissible shift k $(0 \le k \le 2^j - 1)$ are given by

$$\psi^i_{j,k}(x) = \frac{1}{\sqrt{2}} \sum_{r=0}^{K-1} \left(g^{(0)}_{i,r} \phi^r_{j+1,2k}(x) + g^{(1)}_{i,r} \phi^r_{j+1,2k+1}(x) \right) = \frac{1}{\sqrt{2}} \sum_{r=0}^{K-1} \sum_{s=0}^{1} g^{(s)}_{i,r} \phi^r_{j+1,2k+s}(x). \tag{25}$$

The elements $h^{(s)}_{i,r}$ and $g^{(s)}_{i,r}$, $(s = 0, 1)$ of the low-pass filter $\mathbf{H} = \frac{1}{\sqrt{2}} \left(\mathbf{h}^{(0)} \vdots \mathbf{h}^{(1)} \right)$

and high-pass filter $\mathbf{G} = \frac{1}{\sqrt{2}} \left(\mathbf{g}^{(0)} \vdots \mathbf{g}^{(1)} \right)$ are obtained by using (22) into (24) and the following relations at resolution 0 [23] respectively :

$$\int_0^1 \psi_{0,0}^i(x)x^m dx = 0 \quad \text{for } i = 0, 1, \ldots, K-1; \; m = 0, 1, \ldots, K-1+i, \quad (26)$$

$$\int_0^1 \psi_{0,0}^{i_1}(x)\psi_{0,0}^{i_2}(x)dx = \delta_{i_1,i_2} \quad \text{for } i_1, i_2 = 0, 1, \ldots, K-1. \quad (27)$$

Explicit values of the elements of $\mathbf{h}^{(0)}, \mathbf{g}^{(0)}$ for $K = 4, \; 5$ can be found in [25].

3.3 Multiscale Approximation of a Function

Here we represent multiscale approximation of a function $f \in L^2[0, 1]$.

$$f(x) \approx \begin{cases} (P_{V_J^K} f)(x) \equiv \sum\limits_{k=0}^{2^J-1} \sum\limits_{i=0}^{K-1} c_{J,k}^i \, \phi_{J,k}^i(x), & \text{in Legendre piecewise polynomial (LPP) basis} \\[2ex] \left(P_{V_0^K \oplus \bigoplus\limits_{j=0}^{J-1} W_j^K} f\right)(x) \equiv \sum\limits_{i=0}^{K-1} \{c_{0,0}^i \, \phi^i(x) + \sum\limits_{j=0}^{J-1}\sum\limits_{k=0}^{2^j-1} d_{j,k}^i \, \psi_{j,k}^i(x)\}, & \text{in LMW basis} \end{cases} \quad (28)$$

For the clarity of presentation, we use the following notations [25]. For a given j and $k = 0, 1, \ldots, 2^j - 1$

$$\boldsymbol{\Phi}_{J,k}(x) = \left(\phi_{J,k}^0(x), \phi_{J,k}^1(x), \ldots \phi_{J,k}^{K-1}(x)\right), \quad (29)$$

$$\boldsymbol{\Psi}_{j,k}(x) = \left(\psi_{j,k}^0(x), \psi_{j,k}^1(x), \ldots \psi_{j,k}^{K-1}(x)\right). \quad (30)$$

The bases for V_J^K, W_j^K, and $\bigoplus\limits_{j=0}^{J} W_j^K$ are then denoted by

$$\boldsymbol{\Phi}_J := \left(\boldsymbol{\Phi}_{J,0}(x), \boldsymbol{\Phi}_{J,1}(x), \ldots \boldsymbol{\Phi}_{J,2^J-1}(x)\right)_{1 \times 2^J K}, \quad (31)$$

$$\boldsymbol{\Psi}_j := \left(\boldsymbol{\Psi}_{j,0}(x), \boldsymbol{\Psi}_{j,1}(x), \ldots \boldsymbol{\Psi}_{j,2^j-1}(x)\right)_{1 \times 2^j K}, \quad (32)$$

and

$$_J\boldsymbol{\Psi} := (\boldsymbol{\Psi}_0, \boldsymbol{\Psi}_1, \ldots \boldsymbol{\Psi}_J)_{1 \times (2^{J+1}-1)K}. \quad (33)$$

The components of $\mathbf{c}_{j,k}$ and $\mathbf{d}_{j,k}$, the coefficients in multiscale approximation of $f(x)$ are the inner product of $f(x)$ with $\phi_{J,k}^i(x)$ and $\psi_{j,k}^i(x)$ respectively. Finally, we use the symbols

$$\mathbf{c}_J := \left(\mathbf{c}_{J,0}, \mathbf{c}_{J,1}, \ldots \mathbf{c}_{J,2^J-1}\right)_{1 \times 2^J K}, \quad (34)$$

$$\mathbf{d}_j := \left(\mathbf{d}_{j,0}, \mathbf{d}_{j,1}, \ldots \mathbf{d}_{j,2^j-1}\right)_{1 \times 2^j K}, \quad (35)$$

and

$$_J\mathbf{d} := (\mathbf{d}_0, \mathbf{d}_1, \ldots \mathbf{d}_J)_{1\times(2^{J+1}-1)K} . \tag{36}$$

Then (28) can be expressed as

$$f(x) \approx \begin{cases} (P_{V_J^K} f)(x) \equiv \Phi_J \, \mathbf{c}_J^T, & \text{in LPP basis} \\[2mm] \left(P_{V_0^K \oplus \bigoplus\limits_{j=0}^{J-1} W_j^K} f\right)(x) \equiv (\Phi_0, \, _{(J-1)}\Psi) \begin{pmatrix} \mathbf{c}_0^T \\ _{(J-1)}\mathbf{d}^T \end{pmatrix}, & \text{in LMW basis} \end{cases} \tag{37}$$

where the superscript T denotes the transpose. These notations and symbols have been discussed in somewhat details because of their relevance in the subsequent discussions.

3.4 Evaluation of Integrals

Our main task is to solve the given integral equation (20) which is of the form

$$v(x) + (Hv)(x) + (\mathcal{L}v)(x) = h(x), \quad 0 < x < 1$$
$$\text{where} \quad (\mathcal{L}v)(x) = \int_0^1 \frac{v(t)}{x-t} dt. \tag{38}$$

Now for evaluation of the multiscale representation of the integral operator $\mathcal{L}$, we have to evaluate the CPV integrals involving the product of elements of basis and their images under $\mathcal{L}$. The method to evaluate such integrals has been discussed in some detail by Paul et al. [27]. For the sake of completeness, the results are stated in Sect. 3.4.

Integrals involving scale functions We use the notation

$$\rho(n; l_1, l_2) = \int_0^1 \int_0^1 \frac{\phi^{l_1}(x) \, \phi^{l_2}(x)}{n+x-t} dt \, dx, \quad n \in \mathbb{Z} \tag{39}$$

Theorem 1 $\rho(n; l_1, l_2)$ satisfy the following relations

$$\rho(n; l_1, l_2) = \sum_{k_1=0}^{K-1} \sum_{k_2=0}^{K-1} \left\{ h_{l_1,k_1}^{(0)} h_{l_2,k_2}^{(1)} \rho(2n-1; l_1, l_2) + \left(h_{l_1,k_1}^{(0)} h_{l_2,k_2}^{(0)} \right. \right.$$
$$\left. \left. + h_{l_1,k_1}^{(1)} h_{l_2,k_2}^{(1)} \right) \rho(2n; l_1, l_2) + h_{l_1,k_1}^{(1)} h_{l_2,k_2}^{(0)} \rho(2n+1; l_1, l_2) \right\}. \tag{40}$$

Moreover, this relation forms an system of linear equation of $\rho(0; l_1, l_2)$, and it has unique solution.

Integrals involving product of scale functions and wavelets

Theorem 2 *We denote by*

$$\alpha(n; l_1, l_2, j, k) = \int_0^1 \int_0^1 \frac{\phi^{l_1}(x)\, \psi_{j,k}^{l_2}(x)}{n + x - t}\, dt dx, \ n \in \mathbb{Z}, \ j \geq 0, \ k = 0, ., 2^j - 1, \tag{41}$$

then

$$\alpha(n; l_1, l_2, 0, 0) = \sum_{k_1=0}^{K-1} \sum_{k_2=0}^{K-1} \left(h_{l_1,k_1}^{(0)} g_{l_2,k_2}^{(1)}\, \rho(2n - 1; k_1, k_2) \right.$$

$$\left. \left\{ h_{l_1,k_1}^{(0)} g_{l_2,k_2}^{(0)} + h_{l_1,k_1}^{(1)} g_{l_2,k_2}^{(1)} \right\} \rho(2n; k_1, k_2) + h_{l_1,k_1}^{(1)} g_{l_2,k_2}^{(0)}\, \rho(2n + 1; k_1, k_2) \right). \tag{42}$$

Moreover, the formula for the evaluation of $\alpha(n; l_1, l_2, j, k)$ *for* $j > 0$ *and* $k = 0, 1, \ldots, 2^j - 1$ *is given by*

$$\alpha(n; l_1, l_2, j, k) = \begin{cases} \sum_{k_1=0}^{K-1} \left\{ h_{l_1 k_1}^{(0)} \alpha(2n; k_1, l_2, j - 1, k) \right. \\ \left. + h_{l_1 k_1}^{(1)} \alpha(2n + 1; k_1, l_2, j - 1, k) \right\}, \\ \qquad\qquad\qquad\qquad for \ k = 0, 1, \ldots 2^{j-1} - 1, \\ \sum_{k_1=0}^{K-1} \left\{ h_{l_1 k_1}^{(0)} \alpha(2n - 1; k_1, l_2, j - 1, k - 2^{j-1}) + \right. \\ \left. h_{l_1 k_1}^{(1)} \alpha(2n; k_1, l_2, j - 1, k - 2^{j-1}) \right\}, \\ \qquad\qquad for \ k = 2^{j-1}, 2^{j-1} + 1, \ldots 2^j - 1. \end{cases} \tag{43}$$

Theorem 3 *We denote by*

$$\beta(n; l_1, l_2, j, k) = \int_0^1 \int_0^1 \frac{\psi_{j,k}^{l_1}(x) \phi^{l_2}(t)}{n + x - t}\, dt dx, \ n \in \mathbb{Z}, \ j \geq 0, \ k = 0, ., 2^j - 1, \tag{44}$$

then

$$\beta(n; l_1, j, k; l_2) = -\alpha(-n; l_2; l_1, j, k). \tag{45}$$

Integrals involving product of wavelets

Theorem 4 *We denote by*

$$\gamma(n; l_1, j_1, k_1; l_2, j_2, k_2) = \int_0^1 \int_0^1 \frac{\psi_{j_1,k_1}^{l_1}(x) \psi_{j_2,k_2}^{l_2}(t)}{n + x - t}\, dt dx, \ j_1, j_2 \in \mathbb{N}$$

$$k_1 = 0, 1, \ldots, 2^{j_1} - 1, \ k_2 = 0, 1, \ldots, 2^{j_2} - 1, \tag{46}$$

then

$$\gamma(n; l_1, 0, 0; l_2, 0, 0) = \sum_{k_1=0}^{K-1} \sum_{k_2=0}^{K-1} \left(g_{l_1,k_1}^{(0)} g_{l_2,k_2}^{(1)} \rho(2n - 1; k_1, k_2) + \right.$$

$$\left. \left\{ g_{l_1,k_1}^{(0)} g_{l_2,k_2}^{(0)} + g_{l_1,k_1}^{(1)} g_{l_2,k_2}^{(1)} \right\} \rho(2n; k_1, k_2) + g_{l_1,k_1}^{(1)} g_{l_2,k_2}^{(0)} \rho(2n + 1; k_1, k_2) \right), \tag{47}$$

$$\gamma(n; l_1, 0, 0; l_2, j, k) = \begin{cases} \sum_{k_1=0}^{K-1} \left\{ g_{l_1 k_1}^{(0)} \alpha(2n; k_1, l_2, j-1, k) \right. \\ \left. + g_{l_1 k_1}^{(1)} \alpha(2n+1; k_1; l_2, j-1, k) \right\}, \\ \qquad \text{for } k = 0, 1, \ldots 2^{j-1} - 1, \\ \sum_{k_1=0}^{K-1} \left\{ g_{l_1 k_1}^{(0)} \alpha(2n-1; k_1; l_2, j-1, k - 2^{j-1}) + \right. \\ \left. g_{l_1 k_1}^{(1)} \alpha(2n; k_1; l_2, j-1, k - 2^{j-1}) \right\}, \\ \qquad \text{for } k = 2^{j-1}, 2^{j-1} + 1, \ldots 2^j - 1 \end{cases} \tag{48}$$

$$\gamma(0; l_1, j_1, k_1; l_2, j_2, k_2) = 2^{j_1} \gamma\left(k_1 - r; l_1, 0, 0; l_2, j_2 - j_1, k_2 - 2^{j_2 - j_1} r\right). \tag{49}$$

Here r takes the value $\{0, 1, \ldots, 2^j - 1\}$ so that for given r, $k_2 \in \{r2^{j_2-j_1}, r2^{j_2-j_1} + 1, \ldots, (r+1)2^{j_2-j_1} - 1\}$.

now denote the matrices as

$$\begin{aligned} \rho &:= [\rho(0; l_1, l_2)]_{K \times K}, \\ \alpha(j) &:= [\alpha(0; l_1; l_2, j, k)]_{K \times (2^j K)}, \\ \beta(j) &:= [\beta(0; l_1, j, k; l_2)]_{(2^j K) \times K}, \\ \gamma(j_1, j_2) &:= [\gamma(0; l_1, j_1, k_1; l_2, j_2, k_2)]_{(2^{j_1} K) \times (2^{j_2} K)}. \end{aligned} \tag{50}$$

3.5 Multiscale Representation of the Operator $\mathcal{L}$

The multiscale representation

$$\langle (\Phi_0, {}_{(J-1)}\Psi), \mathcal{L}(\Phi_0, {}_{(J-1)}\Psi) \rangle$$

of $\mathcal{L}$ in the basis $(\Phi_0, {}_{(J-1)}\Psi)$ can be written in the form

$$\mathcal{L}_J^{MS} = \begin{pmatrix} \rho & \alpha(0) & \alpha(1) & \ldots & \alpha(J-1) \\ \beta(0) & \gamma(0,0) & \gamma(0,1) & \ldots & \gamma(0, J-1) \\ \beta(1) & \gamma(1,0) & \gamma(1,1) & \ldots & \gamma(1, J-1) \\ \vdots & \vdots & \vdots & \ldots & \vdots \\ \beta(J-1) & \gamma(J-1,0) & \gamma(J-1,1) & \ldots & \gamma(J-1, J-1) \end{pmatrix}_{(2^J K) \times (2^J K)} \tag{51}$$

where the sub-matrices ρ, α, β, γ are given in (50).
Also the multiscale representation

$$\langle (\Phi_0, {}_{(J-1)}\Psi), H(\Phi_0, {}_{(J-1)}\Psi) \rangle$$

of H in the basis $(\boldsymbol{\Phi}_0, {}_{(J-1)}\boldsymbol{\Psi})$ can be written in the form

$$
H_J^{MS} = \begin{pmatrix}
\rho_H & \alpha_H\,(0) & \alpha_H\,(1) & \cdots\cdots & \alpha_H\,(J-1) \\
\beta_H\,(0) & \gamma_H\,(0,0) & \gamma_H\,(0,1) & \cdots\cdots & \gamma_H\,(0,J-1) \\
\beta_H\,(1) & \gamma_H\,(1,0) & \gamma_H\,(1,1) & \cdots\cdots & \gamma_H\,(1,J-1) \\
\vdots & \vdots & \vdots & \cdots\cdots & \vdots \\
\beta_H\,(J-1) & \gamma_H\,(J-1,0) & \gamma_H\,(J-1,1) & \cdots\cdots & \gamma_H\,(J-1,J-1)
\end{pmatrix}_{(2^J K)\times(2^J K)}
\tag{52}
$$

where the sub-matrices ρ_H, α_H, β_H, γ_H be easily be calculated by any standard Gauss–Legendre quadrature rule as the kernel H is a regular kernel.

3.6 Application to Singular Integral Equation on $\mathbb{R}$ with Cauchy-Type Kernel

For the integral equation (20), we assume that $f \in L^2[0,1]$. The solution $v(x)$ of the equation (20) is also $L^2[0,1]$ so that it has multiscale expansion similar to (37). Using the multiscale representation (51) for $\mathcal{L}$ and (52) for $\mathcal{H}$, discussed in Sect. 3.5, the integral equation (20) can be recast into a system of linear algebraic equations given by

$$
(\mathbb{I} + \lambda\, H_J^{MS} + \lambda\, \mathcal{L}_{\mathcal{J}}{}^{MS}) \begin{pmatrix} \mathbf{a}_0^T \\ {}_{(J-1)}\mathbf{b}^T \end{pmatrix} = \begin{pmatrix} \mathbf{c}_0^T \\ {}_{(J-1)}\mathbf{d}^T \end{pmatrix}.
\tag{53}
$$

The matrix $\mathbb{I}$ in (53) is an identity matrix of order $(2^J K) \times (2^J K)$. The matrix $(\mathbb{I} + \lambda\, H_J^{MS} + \lambda\, \mathcal{L}_{c\,\mathcal{J}}{}^{MS})$ is well conditioned so unknown coefficient of $\mathbf{a}_0$, ${}_{(J-1)}\mathbf{b}$ can be found from

$$
\begin{pmatrix} \mathbf{a}_0^T \\ {}_{(J-1)}\mathbf{b}^T \end{pmatrix} = (\mathbb{I} + \lambda\, H_J^{MS} + \lambda\, \mathcal{L}_{\mathcal{J}}{}^{MS})^{-1} \begin{pmatrix} \mathbf{c}_0^T \\ {}_{(J-1)}\mathbf{d}^T \end{pmatrix}.
\tag{54}
$$

Error estimation The L^2- error $\epsilon_J^{L^2} = ||v - v_J^{MS}||_{L^2}$ in the approximate solution v_J^{MS} given by (53) can be derived as

$$
\epsilon_J^{L^2} = \left[\sum_{l=0}^{K-1} \sum_{j=J}^{\infty} \sum_{k=0}^{2^j-1} |b_{j,k}^l|^2 \right]^{\frac{1}{2}}.
\tag{55}
$$

4 Illustrative Examples

To test the efficiency of the numerical method developed here, we consider here two examples [18].

Example 1 We consider the Eq. (1) with

$$\lambda = -3, \quad h(x, t) = \frac{e^{-t^2}}{(1 + |t|^{\frac{7}{2}} + x^2)^3} \tag{56}$$

and

$$f(x) = \frac{x^2}{(1 + x^2)^4}. \tag{57}$$

For LMW basis, we choose $\beta = 20$, then $f(x)$ satisfies the condition for finite range transformation and $f(x) < 10^{-8}$ if $|x| > 20$. Also we have computed the L^2 error of the solution without knowing the exact solution. It is found to be .0007 from the wavelet coefficient for $J = 3$ which is presented in Table 1. The plot of approximate solution $\sqrt{e^{-x^2}} v_3(x)$ by using LMW basis is shown in Fig. 1. The values at some points in LMW basis are also displayed in Table 2 for comparison with the results of Bonis et al. [18] obtained by using interpolation process, related to zeros of Hermite polynomials. The approximate solution by using Daubechies scale functions is shown in Fig. 2 for $j = 5$, $k_5^{min} = -128$, $k_5^{max} = 123$, and $j_1 = 3$. The values of the approximate solution $\sqrt{e^{-x^2}} u(x)$ by using Daubechies scale functions at some points are also displayed in Table 3 to compare our results with these of Bonis et al. [18].

Example 2 We consider another equation (1) with

$$\lambda = -1, \quad h(x, t) = \frac{t^2 e^{-t^2}}{(1 + t^4 + x^4)^3} \tag{58}$$

and

$$f(x) = \frac{\tan^{-1}(1 + x)}{(1 + x^2)^3}. \tag{59}$$

Table 1 The coefficients $\mathbf{b}_{j,k}$ ($j = 0, 1, 2$, $k = 0, 1, 2, \ldots, 2^j - 1$) obtained by using (54) by using LMW in Example 1

j	$b_{j,k}^l$	$l = 0$	$l = 1$	$l = 2$	$l = 3$
0	$k = 0$	2×10^{-4}	-8×10^{-4}	-5×10^{-5}	9×10^{-4}
1	$k = 0$	-1×10^{-5}	-4×10^{-4}	-9×10^{-5}	-2×10^{-4}
	$k = 1$	2×10^{-4}	-3×10^{-4}	1×10^{-4}	-9×10^{-5}
2	$k = 0$	1×10^{-4}	-4×10^{-4}	6×10^{-5}	-4×10^{-4}
	$k = 1$	-3×10^{-4}	-5×10^{-4}	-3×10^{-5}	-2×10^{-4}
	$k = 2$	4×10^{-4}	-3×10^{-4}	6×10^{-5}	-6×10^{-5}
	$k = 3$	6×10^{-6}	-4×10^{-5}	6×10^{-6}	-5×10^{-5}

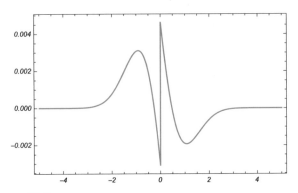

Fig. 1 The plot of $\sqrt{e^{-x^2}}\,v_3(x)$ for Example 1

Table 2 The value of the $\sqrt{e^{-x^2}}\,v_3(x)$ by using LMW in Example 1 with comparison with [18]

	$x = -1.5$	$x = 0.2$	$x = 0.5$
Present method for $J = 3$	2.2×10^{-3}	5.88×10^{-3}	2.9×10^{-4}
Method in [18] for $n = 256$	1.97×10^{-3}	5.14×10^{-3}	1.99×10^{-3}

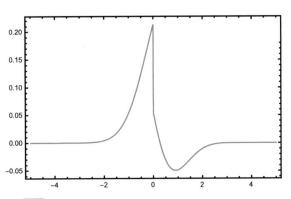

Fig. 2 The plot of $\sqrt{e^{-x^2}}\,v_3(x)$ for Example 2

Table 3 The value of the $\sqrt{e^{-x^2}}\,u(x)$ in Example 1 by using Daubechies scale functions with comparison with [18]

	$x = -1.5$	$x = 0.5$
Present method for $J = 4$ by using Daubechies scale functions	2.519×10^{-3}	3.119×10^{-3}
Present method for $J = 5$ by using Daubechies scale functions	2.723×10^{-3}	3.129×10^{-3}
Method in [18] for $n = 256$	1.97×10^{-3}	1.99×10^{-3}

For LMW basis, we also choose $\beta = 20$ for this example, $f(x)$ satisfies the condition for finite range transformation and $f(x) < 10^{-8}$ if $|x| > 20$. Also we have computed the L^2 error of the solution without knowing the exact solution. It is found to be 0.01 from the wavelet coefficient for $J = 3$ which is presented in Table 4. The plot of $\sqrt{e^{-x^2}} v_3(x)$ by using LMW is shown in Fig. 3. The values at some points are also displayed in Table 5 to compare our results with these of Bonis et al. [18]. The approximate solution by using Daubechies scale functions is shown in Fig. 4 for $j = 5$, $k_5^{\min} = -128$, $k_5^{\max} = 123$, and $j_1 = 3$. The values of the approximate solution $\sqrt{e^{-x^2}} u(x)$ by using Daubechies scale functions at some points are also displayed in Table 6 to compare our results with these of Bonis et al. [18]. From the figures, it is noted that the result (by using Daubechies scale function) agrees with these of Bonis et al. [18] compared to the result by using LMW-based method. So it is obvious that the Daubechies scale function-based method is more accurate than LMW-based method for dealing with unbounded domain.

Table 4 The coefficients $\mathbf{b}_{j,k}$ ($j = 0, 1, 2$, $k = 0, 1, 2, \ldots, 2^j - 1$) obtained by using (54) by using LMW in Example 2

j	$b_{j,k}^l$	$l = 0$	$l = 1$	$l = 2$	$l = 3$
0	$k = 0$	0.0026	−0.0068	−0.0084	0.02479
1	$k = 0$	0.0086	0.0042	0.0026	0.0002
	$k = 1$	0.0013	−0.0034	0.002	−0.0026
2	$k = 0$	0.0007	−0.0013	0.0004	−0.0012
	$k = 1$	0.0048	0.0016	0.0003	−0.0004
	$k = 2$	0.005	−0.0091	0.0032	−0.0074
	$k = 3$	0.0006	−0.0016	0.0005	−0.0015

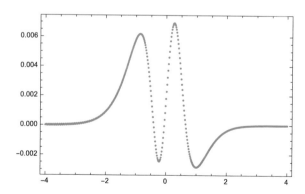

Fig. 3 The plot of $\sqrt{e^{-x^2}} u(x)$ by using Daubechies scale functions with $j = 5$ for Example 1

Table 5 The value of the $\sqrt{e^{-x^2}}v_3(x)$ in Example 2 by using LMW with comparison with [18]

	$x = -1$	$x = -0.5$	$x = 0.4$
Present method for $J = 3$	5.43×10^{-2}	1.25×10^{-1}	-1.97×10^{-2}
Method in [18] for $n = 256$	5.17×10^{-2}	1.49×10^{-1}	-4.12×10^{-2}

Fig. 4 The plot of $\sqrt{e^{-x^2}}u(x)$ by using Daubechies scale functions with $j = 5$ for Example 2

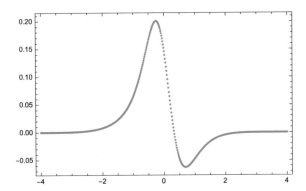

Table 6 The value of the $\sqrt{e^{-x^2}}u(x)$ in Example 2 by using Daubechies scale functions with comparison with [18]

	$x = -1$	$x = -0.5$
Present method for $J = 4$ by using Daubechies scale functions	6.448×10^{-2}	1.671×10^{-1}
Present method for $J = 5$ by using Daubechies scale functions	6.429×10^{-2}	1.668×10^{-1}
Method in [18] for $n = 256$	5.17×10^{-2}	1.49×10^{-1}

5 Conclusion

In this paper, we have presented a comparative study of two methods based on Daubechies scale functions and Legendre multiwavelets. These methods are used to solve the Fredholm integral equation of second kind with Cauchy-type kernel in a unbounded domain. The method of solution by using Daubechies scale functions is discussed. In the process of our development, the recurrence relations among elements or formulae involving the elements of the multiscale representation of the integral operator (CPV) have been derived. The efficiency and comparison of the two methods have been tested for two examples and compute the L^2-error of the solution by wavelet coefficients. It is found to be 10^{-3} order for taking 32 basis elements. For solving Fredholm integral equation of second kind with Cauchy-type kernel in a unbounded domain, it is found that Daubechies scale function-based method is more appropriate than LMW-based method.

This study motivates us to extend the scheme based on Daubechies scale function and Legendre multiwavelets to get multiscale approximation and local behavior of the solution of integro-differential equation, integro-differential-difference equations with constant or variable coefficients and regular or non-smooth input function involving weakly singular, Cauchy singular or hypersingular kernels in finite as well as infinite domain. Works in these directions are in progress and will be reported in due course.

Acknowledgements S. Paul is thankful to Dr. M. M. Panja for his idea and valuable suggestions during the preparation of this paper. This work is supported by a research grant from SERB(DST), No. SR/S4/MS:821/13.

References

1. Helsing, J., Peters, G.: Integral equation methods and numerical solution of crack and inclusion problems in plannar elastrostatics. SIAM. J. Appl. Math. **59**(3), 965–982 (1999)
2. Muskhelishvilli, N. I.: Singular Integral Equations: Boundary Problems of Function Theory and Their Application to Mathematical Physics (1953)
3. Gakhov, F.D.: Boundary Value Problems. Pergamon Press, New York (1966)
4. Ioakimidis, N.I.: On the weighted Galerkin method of numerical solution of Cauchy-type singular integral equations. SIAM. J. Numer. Anal. **18**(6), 1120–1127 (1981)
5. Gong, Y.: Galerkin solution of a singular integral equation with constant coefficients. J. Comput. Appl. Math. **230**, 393–399 (2009)
6. Gong, Y.: New errors for Galerkin method to an airfoil equation. J. Comp. Appl. Math. **206**, 278–287 (2007)
7. Monegato, G., Sloan, I.H.: Numerical Solution of the generalized airfoil equation for an airfoil with a flap. SIAM J. Numer. Anal. **34**, 2288–2305 (1997)
8. Junghans, P., Kaiser, R.: Collocation for a Cauchy singular integral equation. Lin. Alg. Appl. **439**(1), 729–770 (2013)
9. Scuderi, L.: A collocation method for the generalized airfoil equation for an airfoil with a flap. SIAM J. Numer. Anal. **35**, 1725–1739 (1998)
10. Setia, A.: Numerical solution of various cases of Cauchy type singular integral equation. Appl. Math. Comp. **230**(3), 200–207 (2014)
11. Arzhang, A.: Numerical solution of weakly singular integral equations by using Taylor series and Legendre polynomial. Math. Sci. **4**(2), 187–203 (2010)
12. Panja, M.M., Mandal, B.N.: Solution of second kind integral equation with Cauchy type kernel using Daubechies scale function. J. Comp. Appl. Math. **241**, 130–142 (2013)
13. Chandler-Wilde, S.N., Ross, C.R., Zhang, B.: Scattering by infinite one-dimensional rough surfaces. Proc. R. Lond. A. **455**, 3767–3787 (1999)
14. Arens, T.: Uniqueness of elastic wave scattering by rough surfaces. SIAM. J. Math. Anal. **33**, 461–476 (2001)
15. Arens, T.: Existence of solution in elastic wave scattering by unbounded rough surfaces. Math. Meth. Appl. Sci. **25**, 507–526 (2002)
16. Preston, M.D., Chamberlain, P. G., Chandler-Wilde, S. N.: An integral equation method for a boundary value problem arising in unsteady water wave problems. In: Advances in Boundary Integral Methods, Proceeding of the 5th UK conference on Boundary Integral methods, University of Liverpool, pp. 126–133 (2005)
17. Fariborz, S.J.: Singular integral equations with Cauchy kernel on the half line. Int. J. Engng. Sci. **25**, 123–126 (1987)

18. Bonis, M.C., Frammartino, C., Mastroianni, G.: Numerical methods for some special Fredholm integral equations on the real line. J. Comp. Appl. Math. **164–165**, 225–243 (2004)
19. Sheshko, M.A., Sheshko, S.M.: Singular integral equation with Cauchy kernel on the real axis. Diff. Eqn. **46**, 568–585 (2010)
20. Pylak, D., Karczmarek, P., Sheshko, M.A.: Cauchy-type singular integral equation with constant coefficients on the real line. Appl. Math. Comput. **217**, 2977–2988 (2010)
21. Daubechies, I.: Ten lectures on wavelets. In: CBMS Lecture notes, SIAM publication, Philadelphia (1992)
22. Meyer, Y.: Wavelets and Operators. Cambridge University Press, Cambridge (1992)
23. Alpert, B.K.: A class of bases in L^2 for the sparse representation of integral operators. SIAM. J. Math. Anal. **24**(1), 246–262 (1993)
24. Alpert, B.K., Beylikn, G., Coifman, R., Rokhlin, V.: Wavelet-like bases for the fast solution of second-kind integral equations. SIAM J. Sci. Comput. **14**, 159–184 (1993)
25. Paul, S., Panja, M.M., Mandal, B.N.: Multiscale approximation of the solution of weakly singular second kind Fredholm integral equation in Legendre multiwavelet basis. J. Comput. Appl. Math. **300**, 275–289 (2016)
26. Paul, S., Panja, M.M., Mandal, B.N.: Wavelet based numerical solution of second kind hypersingular integral equation. Appl. Math. Sci. **10**(54), 2687–2707 (2016)
27. Paul, S., Panja, M.M., Mandal, B.N.: Use of Legendre multiwavelets in solving second kind singular integral equations with Cauchy type kernel. Invest. Math. Sci. **5**, 2687–2707 (2016)
28. Paul, S., Panja, M.M., Mandal, B.N.: Use of Legendre multiwavelets to solve Carleman type singular integral equations. Appl. Math. Model. **55**, 522–535 (2018)
29. Panja, M.M., Mandal, B.N.: A note on one-point quadrature formula for Daubechies scale function with partial support. Appl. Math. Comput. **218**, 4147–4151 (2011)
30. Atkinson, K.E.: The numerical Solution of integral equations on the half line. SIAM. J. Numer. Anal. **6**, 375–397 (1969)
31. Anselone, P.M., Sloan, I.H.: Integral equations on the half line. J. Integr. Eqn. **9**, 3–23 (1985)
32. Chandler-Wilde, S.N.: On asymptotic behavior at infinity and the finite section method for integral equations on the half line. J. Integr. Eqn. **6**, 1–38 (1994)
33. Chandler-Wilde, S.N.: On the behavior at infinity of solutions of integral equations on the real line. J. Integr. Eqn. **4**(2), 1–25 (1992)

Printed in the United States
By Bookmasters